Flexible Regression and Smoothing

Using GAMLSS in R

Chapman & Hall/CRC
The R Series

Series Editors

John M. Chambers
Department of Statistics
Stanford University
Stanford, California, USA

Torsten Hothorn
Division of Biostatistics
University of Zurich
Switzerland

Duncan Temple Lang
Department of Statistics
University of California, Davis
Davis, California, USA

Hadley Wickham
RStudio
Boston, Massachusetts, USA

Aims and Scope

This book series reflects the recent rapid growth in the development and application of R, the programming language and software environment for statistical computing and graphics. R is now widely used in academic research, education, and industry. It is constantly growing, with new versions of the core software released regularly and more than 10,000 packages available. It is difficult for the documentation to keep pace with the expansion of the software, and this vital book series provides a forum for the publication of books covering many aspects of the development and application of R.

The scope of the series is wide, covering three main threads:

- Applications of R to specific disciplines such as biology, epidemiology, genetics, engineering, finance, and the social sciences.
- Using R for the study of topics of statistical methodology, such as linear and mixed modeling, time series, Bayesian methods, and missing data.
- The development of R, including programming, building packages, and graphics.

The books will appeal to programmers and developers of R software, as well as applied statisticians and data analysts in many fields. The books will feature detailed worked examples and R code fully integrated into the text, ensuring their usefulness to researchers, practitioners and students.

Published Titles

Stated Preference Methods Using R, *Hideo Aizaki, Tomoaki Nakatani, and Kazuo Sato*

Using R for Numerical Analysis in Science and Engineering, *Victor A. Bloomfield*

Event History Analysis with R, *Göran Broström*

Extending R, *John M. Chambers*

Computational Actuarial Science with R, *Arthur Charpentier*

Testing R Code, *Richard Cotton*

The R Primer, Second Edition, *Claus Thorn Ekstrøm*

Statistical Computing in C++ and R, *Randall L. Eubank and Ana Kupresanin*

Basics of Matrix Algebra for Statistics with R, *Nick Fieller*

Reproducible Research with R and RStudio, Second Edition, *Christopher Gandrud*

R and MATLAB® *David E. Hiebeler*

Statistics in Toxicology Using R *Ludwig A. Hothorn*

Nonparametric Statistical Methods Using R, *John Kloke and Joseph McKean*

Displaying Time Series, Spatial, and Space-Time Data with R,
Oscar Perpiñán Lamigueiro

Programming Graphical User Interfaces with R, *Michael F. Lawrence and John Verzani*

Analyzing Sensory Data with R, *Sébastien Lê and Theirry Worch*

Parallel Computing for Data Science: With Examples in R, C++ and CUDA,
Norman Matloff

Analyzing Baseball Data with R, *Max Marchi and Jim Albert*

Growth Curve Analysis and Visualization Using R, *Daniel Mirman*

R Graphics, Second Edition, *Paul Murrell*

Introductory Fisheries Analyses with R, *Derek H. Ogle*

Data Science in R: A Case Studies Approach to Computational Reasoning and Problem Solving, *Deborah Nolan and Duncan Temple Lang*

Multiple Factor Analysis by Example Using R, *Jérôme Pagès*

Customer and Business Analytics: Applied Data Mining for Business Decision Making Using R, *Daniel S. Putler and Robert E. Krider*

Flexible Regression and Smoothing: Using GAMLSS in R, *Mikis D. Stasinopoulos, Robert A. Rigby, Gillian Z. Heller, Vlasios Voudouris, and Fernanda De Bastiani*

Implementing Reproducible Research, *Victoria Stodden, Friedrich Leisch, and Roger D. Peng*

Flexible Regression and Smoothing

Using GAMLSS in R

Mikis D. Stasinopoulos
Robert A. Rigby
Gillian Z. Heller
Vlasios Voudouris
Fernanda De Bastiani

CRC Press
Taylor & Francis Group
Boca Raton London New York

CRC Press is an imprint of the
Taylor & Francis Group, an **informa** business

A CHAPMAN & HALL BOOK

CRC Press
Taylor & Francis Group
6000 Broken Sound Parkway NW, Suite 300
Boca Raton, FL 33487-2742

First issued in paperback 2020

© 2017 by Taylor & Francis Group, LLC
CRC Press is an imprint of Taylor & Francis Group, an Informa business

No claim to original U.S. Government works

ISBN 13: 978-0-367-65806-9 (pbk)
ISBN 13: 978-1-138-19790-9 (hbk)

Visit the Taylor & Francis Web site at
http://www.taylorandfrancis.com

and the CRC Press Web site at
http://www.crcpress.com

To
Jill, Harry and Melissa,
Jake,
Steven, Ilana and Monique,
Sylvain,
Nelson, Angélica and Felipe.

Contents

Preface

This book is about statistical modelling and learning using the Generalized Additive Models for Location, Scale and Shape (GAMLSS) framework, and the associated software in **R**. GAMLSS is a modern *distribution*-based approach to (semiparametric) regression analysis, in which a parametric distribution is assumed for the response (target) variable, and the parameters of this distribution can vary according to explanatory variables (e.g. as nonparametric smooth functions of explanatory variables). In data science language, the book is about supervised machine learning. It is the first in a series of three texts on GAMLSS, the second (forthcoming) being on the GAMLSS family of distributions and the third (forthcoming) on the different inferential procedures available for GAMLSS models, i.e. Bayesian, boosting and classical approaches.

A guiding principle of the book is how to *learn from data* generated in many fields. The standard linear regression analysis is one of the oldest and also most popular and powerful statistical techniques for learning from data by exploring the linear relationship between a response variable and explanatory/predictor variables (or covariates). The standard regression analysis has served us well for a good part of the 20th century.

The 21st century brings new challenges. With the advent of big data and the information age, practitioners who use the standard regression model find that many of its assumptions seldom hold. Learning from data requires statistical learning frameworks that challenge the usual assumptions that the response variable has a normal distribution, with its mean expressed as the sum of linear functions of the explanatory variables, and a constant variance.

To learn from data (i.e. extract important relationships), especially with the large data sets generated today, we need flexible statistical frameworks to capture:

- The heavy-tailed or light-tailed characteristics of the distribution of the data. This means that the probability of rare events (e.g. an outlier value) occurs with higher or lower probability compared with the normal distribution. Furthermore, the probability of occurrence of an outlier value might change as a function of the explanatory variables;

- The skewness of the response variable, which might change as a function of the explanatory variables;

- The nonlinear or smooth relationship between the response variable and the explanatory variables.

GAMLSS, and its associated **R** software, is a flexible statistical framework for learning from data and enhancing data analytics. In particular, the GAMLSS statistical framework enables flexible regression and smoothing models to be fitted to the data. The GAMLSS model assumes that the response variable has any parametric distribution which might be heavy- or light-tailed, and positively or negatively skewed. In addition, all the parameters of the distribution (location (e.g. mean), scale (e.g. standard deviation) and shape (skewness and kurtosis)) can be modelled as linear, nonlinear or smooth functions of explanatory variables.

The book includes a large number of practical examples which reflect the range of problems addressed by GAMLSS models. This also means that the examples provide a practical illustration of the process of using flexible GAMLSS models for statistical learning.

This book is written for:

- practitioners and researchers who wish to understand and use the GAMLSS models to learn from data,

- students who wish to learn GAMLSS through practical examples, and

- for us the authors who often forget what we have done in the past, and require documentation to remember it.

We assume that readers are familiar with the basic concepts of regression and have a working knowledge of **R**. Generally **R** commands are given within the text, and data sets used are available in the software. The reader is encouraged to learn by repeating the examples given within the book.

Structure of this book

This book is not designed to be read necessarily from beginning to end. What we are hoping to achieve is an easy and comprehensive introduction to GAMLSS models, describing the different functionalities of the GAMLSS **R** packages. With this in mind we have divided the book into six parts dealing with different aspects of the statistical 'regression type' modelling:

Part I Introduction to models and packages: This part provides an explanation of why GAMLSS models are needed and information about the GAMLSS **R** packages, using two practical examples.

Part II Algorithms, functions and inference: This part is designed to help users to familiarize themselves with the GAMLSS algorithms, basic functions of the **gamlss** package and the inferential tools.

Part III Distributions: This part describes the different available distributions for the response variable. They are the distributions available in the **gamlss.dist**

package, and also distributions which are generated by transforming, truncating and finite mixing. They comprise continuous, discrete and mixed (i.e. continuous-discrete) distributions, which can be highly skewed (positively or negatively) and/or highly platykurtotic or leptokurtotic (i.e. light or heavy tails).

Part IV Additive terms: This part shows the different ways in which additive terms can be used to model a distribution parameter within a GAMLSS model. In particular it explains linear parametric terms, nonlinear smoothing terms and random effects.

Part V Model selection and diagnostics: Model selection is crucial in statistical modelling. This part explains the different methods and tools within the GAMLSS packages for model selection and diagnostics.

Part VI Applications: Centile estimation and some further interesting applications of the GAMLSS models are covered in this part.

Readers interested in practical applications are advised to read parts I and VI first, followed by the technical parts III, IV, V and finally II.

Software information

All software and datasets used in this book are contained within the GAMLSS packages. There is an exception of this rule: Exercises 4, 5, and 6 of Chapter 13, which need data that can be downloaded from the "Global Lung Function Initiative" website.

More about latest developments, further examples and exercises using GAMLSS can be found on `www.gamlss.org`. The GAMLSS software is available from:

- The Comprehensive R Archive Network: `https://cran.r-project.org`.

- GitHub: `https://github.com/gamlss` (source code of **gamlss** in R and Java)

Notation used in this book

In this book we distinguish between statistical models, **R** packages and **R** functions. We use capital letters for models, bold characters for packages and code type characters (with extra brackets) for functions. For example

- GAMLSS refers to the statistical model,

- **gamlss** refers to the **R** package, and

- `gamlss()` to the **R** function.

Vectors in general will be represented in lower-case bold letters, e.g. $\mathbf{x} = (x_1, x_2, \ldots, x_n)$ and matrices in an upper-case bold letter, for example $\mathbf{X}$. Scalar random variables are represented by upper case, for example Y. The observed value of a random variable is represented by lower case, for example y.

The following table shows the notation that is used throughout this book.

	Systematic part
Y	a univariate response variable
$\mathbf{y}$	the vector of observed values of the response variable, i.e. $(y_1, y_2, \ldots, y_n)^\top$
n	total number of observations
K	the total number of parameters in the distribution of Y
k	the parameter number $(k = 1, \ldots, K)$
p_k	the number of columns in the design matrix $\mathbf{X}_k$ for the kth parameter vector, $\boldsymbol{\theta}_k$
J_k	the total number of smoothers for the kth distribution parameter $\boldsymbol{\theta}_k$
q_{kj}	the dimension of the random effect vector $\boldsymbol{\gamma}_{kj}$
$\mathbf{x}_{kj}$	the jth explanatory variable vector for the kth parameter, $\boldsymbol{\theta}_k$
$\mathbf{X}_k$	an $n \times J_k'$ fixed effects design matrix for the kth parameter, $\boldsymbol{\theta}_k$
$\boldsymbol{\beta}_k$	a vector of fixed effect parameters for the kth parameter, $\boldsymbol{\theta}_k$ of length J_k', i.e. $(\beta_{k1}, \beta_{k2}, \ldots, \beta_{kJ_k'})^\top$
$\boldsymbol{\beta}$	a vector of all fixed effects parameters, i.e. $(\boldsymbol{\beta}_1^\top, \boldsymbol{\beta}_2^\top, \ldots, \boldsymbol{\beta}_K^\top)^\top$
$\boldsymbol{\gamma}_{kj}$	the jth random effect parameter vector for the kth parameter, $\boldsymbol{\theta}_k$, of length q_{kj}
$\boldsymbol{\gamma}$	a vector of all random effects parameters, i.e. $(\boldsymbol{\gamma}_{11}^\top, \ldots, \boldsymbol{\gamma}_{KJ_K}^\top)^\top$
$\mathbf{Z}_{kj}$	an $n \times q_{kj}$ random effect design matrix for the jth smoother of the kth parameter, $\boldsymbol{\theta}_k$
$\mathbf{G}_{kj}$	an $q_{kj} \times q_{kj}$ penalty matrix for $\boldsymbol{\gamma}_{kj}$
$\boldsymbol{\eta}_k$	the predictor for the kth distribution parameter, i.e. $\boldsymbol{\eta}_k = g_k(\boldsymbol{\theta}_k)$
$\mathbf{H}_k$	the hat matrix for the kth parameter
$\mathbf{z}_k$	the adjusted dependent variable for the kth distribution parameter $\boldsymbol{\theta}_k$
$g_k()$	link function applied to model the kth distribution parameter $\boldsymbol{\theta}_k$
$s_{kj}()$	the jth nonparametric smooth function (in the predictor $\boldsymbol{\eta}_k$)
$\mathbf{W}$	a $n \times n$ diagonal matrix of weights
$\mathbf{w}$	a n dimensional vector of weights (the diagonal elements of $\mathbf{W}$)
$\mathbf{S}_{kj}$	the jth smoothing matrix for the kth distribution parameter $\boldsymbol{\theta}_k$
	Distributions
$f(y)$	theoretical probability (density) function of the random variable Y [1] (d function)

---

[1] Occasionally, for clarity, the subscript Y is used, i.e. $f_Y(\cdot)$.

$F(y)$ cumulative distribution function of the random variable Y (p function)

$\mathcal{D}(\cdot)$ a generic distribution

$\mathcal{E}(\cdot)$ exponential family of distributions

$Y \overset{\text{ind}}{\sim} \mathcal{D}(\boldsymbol{\mu}, \boldsymbol{\sigma}, \boldsymbol{\nu}, \boldsymbol{\tau})$ $Y_1, \ldots, Y_n$ are independently distributed with distributions $\mathcal{D}(\mu_i, \sigma_i, \nu_i, \tau_i)$, for $i = 1, \ldots, n$

$Q(.)$ inverse cumulative distribution function of the random variable Y (q function), i.e. $F^{-1}(\cdot)$

$E(Y)$ Expectation of random variable Y

$\text{Var}(Y)$ Variance of random variable Y

$\text{SD}(Y)$ Standard deviation of random variable Y

$f_{Y|X}(\cdot)$ conditional probability (density) function of the random variable Y given X

$\mathcal{N}(\mu, \sigma^2)$ normal distribution with mean μ and variance σ^2

$\phi(\cdot)$ probability density function of a standard normal distribution $\mathcal{N}(0, 1)$

$\Phi(\cdot)$ cumulative distribution function of a standard normal distribution

π_k the kth prior probability in a finite mixture

$\boldsymbol{\pi}$ vector of prior (or mixing) probabilities in a finite mixture $\boldsymbol{\pi} = (\pi_1, \pi_2 \ldots, \pi_{\mathcal{K}})^{\top}$

Distribution parameters

θ_k the kth distribution parameter, where $\theta_1 = \mu$, $\theta_2 = \sigma$, $\theta_3 = \nu$ and $\theta_4 = \tau$

$\boldsymbol{\theta}_k$ a vector of length n of the kth distribution parameter, e.g. $\boldsymbol{\theta}_1 = \boldsymbol{\mu}$, $\boldsymbol{\theta}_2 = \boldsymbol{\sigma}$

μ the first parameter of the distribution (usually location)

σ the second parameter of the distribution (usually scale)

ν the third parameter of the distribution (usually shape, e.g. skewness)

τ the fourth parameter of the distribution (usually shape, e.g. kurtosis)

λ a hyper-parameter

$\boldsymbol{\lambda}$ the vector of all hyper-parameters in the model

σ_b standard deviation of a normal random effect term for a parameter

u standard normal (Gaussian) quadrature mass point

Likelihood and information criteria

L likelihood function

ℓ log-likelihood function

Λ generalized likelihood ratio test statistic

$\boldsymbol{i}$ Fisher's expected information matrix

I observed information matrix

GDEV   global deviance, i.e. minus twice the fitted log-likelihood

GAIC generalized Akaike information criterion, i.e. GDEV $+ (\kappa \times df)$

df total (effective) degrees of freedom used in the model

κ penalty for each degree of freedom in the model

Residuals

$\mathbf{u}$ vector of (randomized) quantile residuals

$\mathbf{r}$ vector of normalized (randomized) quantile residuals

$\boldsymbol{\varepsilon}$ vector of (partial) residuals

Q Q statistic calculated from the residuals

Z Z-statistic calculated from the residuals

GAMLSS model components

$\mathcal{M}$ a GAMLSS model containing $\{\mathcal{D}, \mathcal{G}, \mathcal{T}, \mathcal{L}\}$

$\mathcal{D}$ the specification of the distribution of the response variable

$\mathcal{G}$ the different link functions, e.g. $g_k()$ where $g_k(\boldsymbol{\theta}_k) = \boldsymbol{\eta}_k$

$\mathcal{T}$ the explanatory variable terms in the model

$\mathcal{L}$ the specification of the smoothing parameters

Vector operators

$\circ$ the Hadamard element by element product, i.e. let $\mathbf{y} = (y_1, y_2, y_3)^\top$
and $\mathbf{x} = (x_1, x_2, x_3)^\top$ then $(\mathbf{y} \circ \mathbf{x}) = (y_1 x_1, y_2 x_2, y_3 x_3)^\top$

Acknowledgements

The authors would like to thank: Paul Eilers for his valuable contribution to smoothing functions in GAMLSS and his encouragement and advice for the book; Bob Gilchrist for his helpful comments and for providing the right research environment for a long time; Tim Cole, Elaine Borghie, Stefan van Buuren and Huiqi Pan for helping with the development of the centiles functions; Popi Akanziliotou, Marco Enea, Daniil Kiose, Luiz Nakamura, Majid Djennad, Raydonal Ospina, Konstantinos Pateras and Nicoleta Mortan, for their past or current contributions to the GAMLSS software; Christian Kiffne, Albert Wong, Willem Vervoort, Steve Ellison, Michael Hohle and Larisa Kosidou for suggesting changes in the **R** functions or for small contributions to the packages; Gareth Amber, John Chambers, Trevor Hastie, Jochen Einbeck, Ross Darnell, John Hinde, Jim Lindsey, Philippe Lambert, Brian Ripley, Simon Wood and Brian Marx, for providing **R** functions which we have adapted for our own purpose; and Thomas Kneib, Nadja Klein, Stefan Lang and Andreas Mayr for promoting GAMLSS in the Bayesian and 'boosting' communities. Finally we would like to thank Harry Stasinopoulos for providing us with the cover figure for this book.

Part I

Introduction to models and packages

1

Why GAMLSS?

CONTENTS

1.1 Introduction

This chapter shows the evolution of statistical modelling from the linear model (LM) through the generalized linear model (GLM) and the generalized additive model (GAM) to the generalized additive model for location, scale and shape (GAMLSS). It provides

1. a discussion on the historical evolution of GAMLSS through a simple example,

2. an introduction to the GAMLSS models in **R**, and

3. the definition of a GAMLSS model.

This chapter is the starting point for using GAMLSS in **R**.

This chapter serves as an introduction to generalized additive models for location, scale and shape (GAMLSS). It builds up the GAMLSS model using ideas from its predecessors, in particular from linear regression models, generalized linear models and generalized additive models. It uses a relatively simple example, the Munich **rent** data, to demonstrate why GAMLSS is needed.

1.2   The 1980s Munich rent data

The **rent** data come from a survey conducted in April 1993 by Infratest Sozial-
forschung, where a random sample of accommodation with new tenancy agreements
or increases of rents within the last four years in Munich was selected, including
single rooms, small apartments, flats and two-family houses. The accommodation
is subsequently referred to as a flat. The data were analysed by Stasinopoulos et al.
[2000] and they are in the package **gamlss.data** (which is automatically loaded when
gamlss is loaded by library(gamlss)). There are 1,969 observations on nine variables in the data set but, for the purpose of demonstrating GAMLSS, we will use
only the following five variables:

R data file: rent in package **gamlss.data** of dimensions 1969×9

var R : the response variable which is the monthly net rent in Deutsche Marks
       (DM), i.e. the monthly rent minus calculated or estimated utility cost

  Fl : the floor space in square metres

  A : the year of construction

  H : a two-level factor indicating whether there is central heating (0=yes,
       1=no)

  loc : a factor indicating whether the location is below average (1), average
        (2), or above average (3)

purpose: to demonstrate the need for using GAMLSS models.

```
PPP <- par(mfrow=c(2,2))
plot(R~Fl, data=rent, col=gray(0.7), pch=15, cex=0.5)
plot(R~A, data=rent, col=gray(0.7), pch=15, cex=0.5)
plot(R~H, data=rent, col=gray(0.7), pch=15, cex=0.5)
plot(R~loc, data=rent, col=gray(0.7), pch=15, cex=0.5)
par(PPP)
```

Fig.
1.1

Figure 1.1 shows plots of the rent, R, against each of the explanatory variables.
Although these are bivariate exploratory plots and take no account of the interplay
between the explanatory variables, they give an indication of the complexity of these
data. The first two explanatory variables, Fl and A, are continuous. The plot of rent,
R, against floor space, Fl, suggests a positive relationship, with increased variation
for larger floor spaces, with the result that an assumption of homogeneity of variance
would be violated here. There is also some indication of positive skewness in the
distribution of rent, R. The peculiarity of the plot of R against year of construction,
A, is due to the method of data collection. Many of the observations of A were

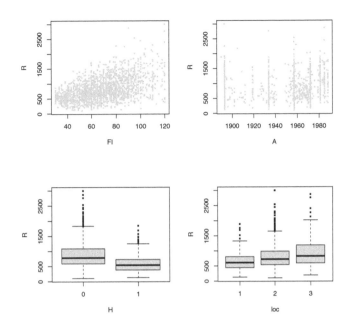

R
). 4

FIGURE 1.1: Plot of the rent R against explanatory variables F1, A, H and loc.

collected on an interval scale and assigned the value of the interval midpoint, while
for the rest the actual year of construction was recorded. The plot suggests that
for flats up to 1960 the median rent is roughly constant but, for those constructed
after that year, there is an increasing trend in the median rent. The two boxplots
display how the rent varies according to the explanatory factors. The median rent
increases if the flat has central heating, and increases as the location changes from
below average to average and then to above average. There are no surprises in the
plots here but again the problem of skewness is prominent, with asymmetrical boxes
about the median and longer upper than lower whiskers.

In summary, any statistical model used for the analysis of the rent data should be
able to deal with the following statistical problems:

Complexity of the relationship between rent and the explanatory variables. The dependence of the median of the response variable rent on floor space
and age of construction is nonlinear, and nonparametric smoothing functions
may be needed. Median rent may also depend on linear or nonlinear interactions
between the explanatory variables.

Non-homogeneity of variance of rent. There is clear indication of non-
homogeneity of the variance of rent. The variance of rent may depend on its
mean and/or explanatory variables. A statistical model in which this dependence can be modelled explicitly, is needed.

Skewness in the distribution of rent. There is clear indication of positive skewness in the distribution of rent which may depend on explanatory variables and this has to be accounted for within the statistical model.

1.3 The linear regression model (LM)

Linear regression is a simple but effective model, which served the statistical community well for most of the last century. With response variable Y, r covariates $x_1, \ldots, x_r$ and sample size n, it is defined as

$$Y_i = \beta_0 + \beta_1 x_{i1} + \ldots + \beta_r x_{ir} + \epsilon_i$$

$$\text{where} \qquad \epsilon_i \stackrel{\text{ind}}{\sim} \mathcal{N}(0, \sigma^2) , \qquad \text{for } i = 1, 2, \ldots, n$$

i.e. ϵ_i for $i = 1, 2, \ldots, n$ are independently distributed each with a normal distribution with mean zero and variance σ^2. This specification is equivalent to

$$Y_i \stackrel{\text{ind}}{\sim} \mathcal{N}(\mu_i, \sigma^2)$$

$$\text{where} \qquad \mu_i = \beta_0 + \beta_1 x_{i1} + \ldots + \beta_r x_{ir} , \qquad \text{for } i = 1, 2, \ldots, n . \qquad (1.1)$$

We rewrite model (1.1) in vector form as:

$$\mathbf{Y} \stackrel{\text{ind}}{\sim} \mathcal{N}(\boldsymbol{\mu}, \boldsymbol{\sigma}^2) \qquad (1.2)$$

$$\boldsymbol{\mu} = \mathbf{X}\boldsymbol{\beta}$$

where $\mathbf{Y} = (Y_1, \ldots, Y_n)^\top$ is the response vector, $\mathbf{X}$ is the $n \times p$ design matrix ($p = r + 1$) containing the r covariate columns, plus a column of ones (if the constant is required), $\boldsymbol{\beta} = (\beta_0, \ldots, \beta_r)^\top$ is the coefficient vector, $\boldsymbol{\mu} = (\mu_1, \ldots, \mu_n)^\top$ is the mean vector and $\boldsymbol{\sigma}^2 = (\sigma^2, \ldots, \sigma^2)^\top$ is a vector of (constant) variance. Note that in order for the model to be fitted, both $\boldsymbol{\beta}$ and σ^2 have to be estimated from the data. The usual practice is to estimate $\boldsymbol{\beta}$ using the least squares estimator, obtained by minimizing the sum of squared differences between the observations Y_i and the means $\mu_i = \beta_0 + \beta_1 x_{i1} + \ldots + \beta_r x_{ir}$, with respect to the β's . In vector form this is written as

$$\hat{\boldsymbol{\beta}} = \text{argmin}_{\boldsymbol{\beta}} \, (\mathbf{Y} - \mathbf{X}\boldsymbol{\beta})^\top (\mathbf{Y} - \mathbf{X}\boldsymbol{\beta})$$

which has solution

$$\hat{\boldsymbol{\beta}} = (\mathbf{X}^\top \mathbf{X})^{-1} \mathbf{X}^\top \mathbf{Y} . \qquad (1.3)$$

It can be shown that $\hat{\boldsymbol{\beta}}$ is also the maximum likelihood estimator (MLE) of $\boldsymbol{\beta}$. Let $\hat{\boldsymbol{\mu}} = \mathbf{X}\hat{\boldsymbol{\beta}}$ be the fitted values of the model and $\hat{\boldsymbol{\epsilon}} = \mathbf{Y} - \hat{\boldsymbol{\mu}}$ the simple residuals (i.e. fitted errors). Then the MLE for σ^2 is

$$\hat{\sigma}^2 = \frac{\hat{\boldsymbol{\epsilon}}^\top \hat{\boldsymbol{\epsilon}}}{n} , \qquad (1.4)$$

which is a biased estimator, i.e. $E\left(\hat{\sigma}^2\right) \neq \sigma^2$. An unbiased estimator of σ^2 is given by

$$s^2 = \frac{\hat{\epsilon}^{\top}\hat{\epsilon}}{n-p} \ . \tag{1.5}$$

Sometimes s^2 is referred as the REML (restricted maximum likelihood) estimator of σ^2.

A linear regression model can be fitted in **R** using the function `lm()`. Here we compare the results from `lm()` to the ones obtained by `gamlss()`. The notation

```
R ~ Fl+A+H+loc
```

is explained in Section 8.2.

```
r1 <- gamlss(R ~ Fl+A+H+loc, family=NO, data=rent, trace=FALSE)
l1 <- lm(R ~ Fl+A+H+loc,data=rent)
coef(r1)
```

```
## (Intercept)          Fl           A           H1
## -2775.038803    8.839445    1.480755  -204.759562
##          loc2        loc3
##   134.052349   209.581472
```

```
coef(l1)
```

```
## (Intercept)          Fl           A           H1
## -2775.038803    8.839445    1.480755  -204.759562
##          loc2        loc3
##   134.052349   209.581472
```

The coefficient estimates of the two fits are identical. Note that the two factors of the **rent** data, H and `loc`, are fitted as dummy variables as explained in more detail in Section 8.2.

The fitted objects `r1` and `l1` use the methods `fitted()` and `resid()` to obtain fitted values and residuals, respectively. Note that the `gamlss` object residuals are the normalized (randomized) quantile residuals as explained in Section 12.2, and not the simple residuals $\hat{\epsilon}$ that might be expected.

Important: GAMLSS uses normalized (randomized) quantile residuals.

The MLE of σ can be obtained from a `gamlss` fitted object using the command `fitted(r1, "sigma")[1]`. (Here [1] shows the first element of the fitted vector for σ). `summary()` will show the standard errors and t-tests of the estimated coefficients. The method used to calculate standard errors in the `summary()` function in a `gamlss` model is explained in Section 5.2.

```
fitted(r1, "sigma")[1]

##         1
## 308.4768

summary(r1)

## *******************************************************************
## Family:  c("NO", "Normal")
##
## Call:
## gamlss(formula = R ~ Fl + A + H + loc, family = NO,
##     data = rent, trace = FALSE)
##
## Fitting method: RS()
##
## -----------------------------------------------------------------
## Mu link function:  identity
## Mu Coefficients:
##                Estimate Std. Error t value Pr(>|t|)
## (Intercept) -2775.0388   470.1352  -5.903 4.20e-09 ***
## Fl              8.8394     0.3370  26.228  < 2e-16 ***
## A               1.4808     0.2385   6.208 6.55e-10 ***
## H1           -204.7596    18.9858 -10.785  < 2e-16 ***
## loc2          134.0523    25.1430   5.332 1.09e-07 ***
## loc3          209.5815    27.1286   7.725 1.76e-14 ***
## ---
## Signif. codes:
## 0 '***' 0.001 '**' 0.01 '*' 0.05 '.' 0.1 ' ' 1
##
## -----------------------------------------------------------------
## Sigma link function:  log
## Sigma Coefficients:
##              Estimate Std. Error t value Pr(>|t|)
## (Intercept)  5.73165    0.01594   359.7   <2e-16 ***
## ---
## Signif. codes:
## 0 '***' 0.001 '**' 0.01 '*' 0.05 '.' 0.1 ' ' 1
##
## -----------------------------------------------------------------
## No. of observations in the fit:  1969
## Degrees of Freedom for the fit:  7
##         Residual Deg. of Freedom:  1962
##                       at cycle:  2
##
## Global Deviance:      28159
```

```
##              AIC:        28173
##              SBC:        28212.1
## *****************************************************************
```

The fitted model is given by

$$Y \sim \mathcal{N}(\hat{\mu}, \hat{\sigma}^2)$$

where

$$\hat{\mu} = -2775.04 + 8.84\,\text{F1} + 1.48\,\text{A} - 204.76\,(\text{if } \texttt{H=1}) + 134.05\,(\text{if } \texttt{loc=2})$$
$$+ 209.58\,(\text{if } \texttt{loc=3})$$
$$\log(\hat{\sigma}) = 5.73165\ .$$

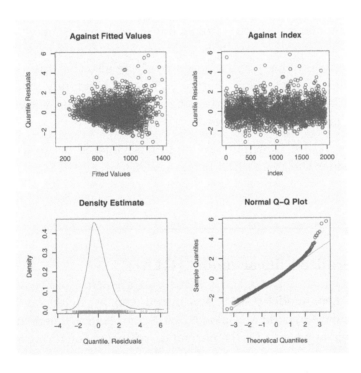

FIGURE 1.2: Residual plots of the linear model `r1`.

Note that σ is fitted on the log scale (indicated by the log link function, see Section 1.4), so its fitted value is computed from its intercept as

$$\hat{\sigma} = \exp(5.73165) = 308.48\ .$$

R^2 is obtained from the `gamlss` fitted object as

```
Rsq(r1)
```

```
## [1] 0.3372028
```

One way of checking the adequacy of a model is to examine the residuals.

```
plot(r1)
```

```
## ******************************************************************
##          Summary of the Quantile Residuals
##                         mean    =  4.959549e-13
##                     variance    =  1.000508
##              coef. of skewness  =  0.7470097
##              coef. of kurtosis  =  4.844416
## Filliben correlation coefficient  =  0.9859819
## ******************************************************************
```

Fig.
1.2

More about the interpretation of the four plots in Figure 1.2 can be found in Section 12.3. The important issue here is that the distributional assumption of normality is easily rejected by looking at the normal Q-Q plot (bottom right panel, Figure 1.2). There is a systematic departure from a linear relationship between the ordered observed (normalized quantile) residuals and their approximate expected values, indicating that the residuals are positively skewed. Note also that the plot of residuals against fitted values (top left panel, Figure 1.2) is not randomly scattered about a horizontal line at 0, but fans out, indicating variance heterogeneity, in particular that the variance increases with the mean.

Given that the normal (or Gaussian) assumption is violated because of the positive skewness, we consider the generalized linear model next.

1.4   The generalized linear model (GLM)

The generalized linear model (GLM) was introduced by Nelder and Wedderburn [1972] and further developed in McCullagh and Nelder [1989]. There are three major innovations in their approach: (i) the normal distribution for the response variable is replaced by the exponential family of distributions (denoted here as $\mathcal{E}$), (ii) a monotonic *link* function $g(\cdot)$ is used in modelling the relationship between $\mathrm{E}(Y)$ and the explanatory variables, and finally (iii) in order to find the MLE for the parameters β it uses an iteratively reweighted least squares algorithm, which can be implemented easily in any statistical package having a good weighted least squares algorithm. The GLM can be written as:

$$Y_i \stackrel{\text{ind}}{\sim} \mathcal{E}(\mu_i, \phi)$$

where $g(\mu_i) = \beta_0 + \beta_1 x_{i1} + \ldots + \beta_r x_{ir} ,$ for $i = 1, 2, \ldots, n ,$

and ϕ is the *dispersion* parameter. We rewrite this in vector form as

$$\mathbf{Y} \stackrel{\text{ind}}{\sim} \mathcal{E}(\boldsymbol{\mu}, \boldsymbol{\phi}) \qquad (1.6)$$

$$\boldsymbol{\eta} = g(\boldsymbol{\mu}) = \mathbf{X}\boldsymbol{\beta} .$$

where $\boldsymbol{\eta}$ is called the *linear predictor* and $\boldsymbol{\phi} = (\phi, \ldots, \phi)^\top$ is a vector of constant ϕ. The exponential family distribution $\mathcal{E}(\mu, \phi)$ is defined by the probability (density) function $f(y \mid \mu, \phi)$ having the form:

$$f(y \mid \mu, \phi) = \exp\left\{\frac{y\theta - b(\theta)}{\phi} + c(y, \phi)\right\} \tag{1.7}$$

where $\mathrm{E}(Y) = \mu = b'(\theta)$ and $\mathrm{Var}(Y) = \phi V(\mu)$, where $V(\mu) = b''[\theta(\mu)]$. ($V(\mu)$ is called the *variance function*.) The form of (1.7) includes many important distributions including the normal, Poisson, gamma, inverse Gaussian and Tweedie [Tweedie, 1984] distributions having variance functions $V(\mu) = 1, \mu, \mu^2, \mu^3$ and μ^p for $p < 0$ or $p > 1$, respectively, and also the binomial with variance function $V(\mu) = \frac{\mu(1-\mu)}{N}$, where N is the binomial denominator.

Within the GLM framework the normal (or Gaussian) distribution, used in the previous section to fit the **rent** data, might be replaced by the gamma distribution with probability density function (pdf) given by

$$f(y; \mu, \phi) = \frac{y^{1/\phi - 1} \exp(-\frac{y}{\phi\mu})}{(\phi\mu)^{(1/\phi)} \Gamma(1/\phi)}, \qquad y > 0, \ \mu > 0, \ \phi > 0.$$

(In **gamlss** the gamma distribution is parameterized with scale parameter σ, where $\sigma = \sqrt{\phi}$.)

Link functions were introduced by Nelder and Wedderburn [1972] for GLMs, but are appropriate for all regression models since they guarantee that parameter estimates remain within the appropriate range. For example if a parameter θ has range $0 < \theta < \infty$, the logarithmic transformation

$$\eta = \log(\theta)$$

produces $-\infty < \eta < \infty$. In parameter estimation, if the logarithmic link is used, η is estimated and transformed back to θ as

$$e^\eta = \theta \ ,$$

which is guaranteed to be in the range $(0, \infty)$. For the logarithmic link, $\log(\theta)$ is the link function and e^η is the inverse link function. In general, the link function is denoted as $\eta = g(\theta)$, and the inverse link as $g^{-1}(\eta) = \theta$. Generally, for a **gamlss.family** distribution, each model parameter has its own link function appropriate to its range. For the gamma distribution we have two model parameters μ and σ. GAMLSS uses the log link function as default for both parameters, since the range of both is $(0, \infty)$. Their link functions are denoted as

$$\eta_1 = g_1(\mu) = \log(\mu)$$
$$\eta_2 = g_2(\sigma) = \log(\sigma) \ .$$

(Note that a log link assumes that the relationship between the parameter and the predictor variables is multiplicative.) All currently available distributions within the package **gamlss.dist** together with the default link functions for their parameters are shown in Tables 6.1, 6.2 and 6.3. The glm function has as default the *canonical* link function for μ, which is different for each distribution and for the Gamma is the "inverse", i.e. $g(\mu) = 1/\mu$.

Important: The GAMLSS model as implemented in the **gamlss** package does not use canonical links as default for μ as in the glm() function, but generally uses links reflecting the range of the parameter values, i.e. "identity" for $(-\infty, \infty)$, "log" for $(0, \infty)$, "logit" for $(0, 1)$, etc.

We fit the gamma distribution, using both the gamlss() and glm() functions. For gamlss(), the gamma distribution is specified as the argument family=GA, whereas for glm() it is specified as family=Gamma. We use the log link for μ.

```
### using gamlss
r2 <-  gamlss(R ~ Fl+A+H+loc, family=GA,  data=rent)

## GAMLSS-RS iteration 1: Global Deviance = 27764.59
## GAMLSS-RS iteration 2: Global Deviance = 27764.59

coef(r2)

## (Intercept)          Fl           A          H1        loc2
##   2.86497701  0.01062319  0.00151005 -0.30007446  0.19076406
##         loc3
##   0.26408285

coef(r2, "sigma")   ### extract log(sigma)

## (Intercept)
##   -0.9821991

deviance(r2)

## [1] 27764.59

### using glm
l2 <- glm(R ~ Fl+A+H+loc, family=Gamma(link="log"),  data=rent)
coef(l2)

## (Intercept)          Fl           A          H1
##   2.864943806  0.010623194  0.001510066 -0.300074001
##         loc2         loc3
##   0.190764594  0.264083376

summary(l2)$dispersion  ### extract phi

## [1] 0.1377881
```

```
deviance(12)
```

```
## [1] 282.5747
```

The fitted coefficients for $\log(\mu)$ from the two models are essentially the same as, in this example, are the estimates of dispersion. From `gamlss` we have $\hat{\sigma} = \exp(-0.9822) = 0.374$, and hence $\hat{\phi} = \hat{\sigma}^2 = 0.14$. From `glm` we have $\hat{\phi} = 0.138$. Note however that for estimation of ϕ, `gamlss` uses maximum likelihood estimation whereas `glm` uses the method of moments.

The deviances for models `r2` and `12` are different because they are defined differently. The GLM deviance is defined as

$$
D_{\text{GLM}} = -2\log\left(\frac{\hat{L}_c}{\hat{L}_s}\right)
$$

where $\hat{L}_c$ is the fitted likelihood of the current fitted model, and $\hat{L}_s$ that of the *saturated* model (the model where in modelling μ a parameter is fitted for each observation.) The GAMLSS deviance is defined as

$$
D_{\text{GAMLSS}} = -2\log\hat{L}_c ,
$$

which we refer to as the *global deviance* or GDEV. To compare models one can use the generalized Akaike information criterion (GAIC) given by

$$
\text{GAIC}(\kappa) = -2\log\hat{L}_c + (\kappa \times \text{df}) ,
$$

where df denotes the total effective degrees of freedom (i.e. the effective number of parameters) of the model and κ is the penalty for each degree of freedom used. Hence $\text{GAIC}(\kappa = 2)$ gives the Akaike information criterion (AIC) and $\text{GAIC}(\kappa = \log(n))$ gives the Schwarz Bayesian Criterion (SBC) or Bayesian Information Criterion (BIC). The model with the lowest value of $\text{GAIC}(\kappa)$ for a chosen value of κ is selected as 'best'. To get the coefficients with their standard errors use:

```
summary(r2)
```

```
## ******************************************************************
## Family:  c("GA", "Gamma")
##
## Call:
## gamlss(formula = R ~ Fl + A + H + loc, family = GA,
##     data = rent)
##
## Fitting method: RS()
##
## --------------------------------------------------------------------
## Mu link function:  log
## Mu Coefficients:
```

```
##                   Estimate Std. Error  t value Pr(>|t|)
## (Intercept)      2.8649770  0.5688561    5.036 5.18e-07 ***
## F1               0.0106232  0.0004128   25.733  < 2e-16 ***
## A                0.0015100  0.0002890    5.226 1.92e-07 ***
## H1              -0.3000745  0.0231287  -12.974  < 2e-16 ***
## loc2             0.1907641  0.0305204    6.250 5.01e-10 ***
## loc3             0.2640828  0.0329211    8.022 1.78e-15 ***
## ---
## Signif. codes:
## 0 '***' 0.001 '**' 0.01 '*' 0.05 '.' 0.1 ' ' 1
##
## -----------------------------------------------------------------
## Sigma link function:  log
## Sigma Coefficients:
##               Estimate Std. Error t value Pr(>|t|)
## (Intercept) -0.98220    0.01558   -63.05   <2e-16 ***
## ---
## Signif. codes:
## 0 '***' 0.001 '**' 0.01 '*' 0.05 '.' 0.1 ' ' 1
##
## -----------------------------------------------------------------
## No. of observations in the fit:  1969
## Degrees of Freedom for the fit:  7
##         Residual Deg. of Freedom:  1962
##                       at cycle:  2
##
## Global Deviance:      27764.59
##            AIC:       27778.59
##            SBC:       27817.69
## ****************************************************************
```

The fitted model is given by

$$Y \sim \texttt{GA}(\hat{\mu}, \hat{\sigma})$$

where

$$\log(\hat{\mu}) = 2.865 + 0.0106\,\texttt{F1} + 0.00151\,\texttt{A} - 0.3\,(\texttt{if H=1}) + 0.191\,(\texttt{if loc=2})$$
$$+ 0.264\,(\texttt{if loc=3})$$
$$\log(\hat{\sigma}) = -0.9822\;.$$

To check which of the normal, gamma and inverse Gaussian distributions is best for the data, compare the three models using the GAIC. In this case, because all three models have the same number of parameters we could just compare the global deviance, i.e. GAIC(κ) with $\kappa = 0$.

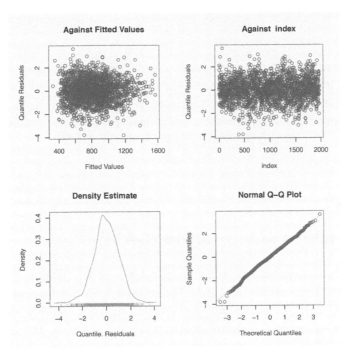

FIGURE 1.3: Residual plots of the model **r2**.

```
r22 <- gamlss(R ~ Fl+A+H+loc, family=IG, data=rent, trace=FALSE)
GAIC(r1, r2, r22, k=0) # GD

##       df      AIC
## r2     7  27764.59
## r22    7  27991.56
## r1     7  28159.00
```

The conclusion is that the gamma provides the best fit. Now we check the residuals:

```
plot(r2)

## ***********************************************************
##        Summary of the Quantile Residuals
##                             mean  =  0.0004795675
##                         variance  =  1.000657
##            coef. of skewness  =  -0.1079453
##            coef. of kurtosis  =  3.255464
## Filliben correlation coefficient  =  0.9990857
## ***********************************************************
```

The residuals at this stage look a lot better than the residuals of Figure 1.2 in that at least some of the heterogeneity in the residuals, in the plot of residuals against

R
. 15

Fig.
1.3

fitted values, has disappeared. Also the curvature in the normal Q-Q plot has been substantially reduced.

Even though presence of heterogeneity and skewness in the residuals has been (partially) addressed, we next introduce the generalized additive model (GAM), which allows more flexible modelling of the relationship between the distribution parameter μ and the continuous explanatory variables. This extra flexibility might be needed to improve fit and the resulting residuals of the GLM.

1.5   The generalized additive model (GAM)

Smoothing techniques became popular in the late 1980s. Hastie and Tibshirani [1990] introduced them within the GLM framework, coining the term generalized additive models (GAM). Wood [2006a] has contributed extensively to GAM theory and popularity by allowing, in his implementation of GAM in **R** (package **mgcv**), the automatic selection of the smoothing parameters in the model. (In the original implementation of GAM in S-plus and **R** the smoothing parameters λ or equivalently the effective degrees of freedom had to be fixed.) The GAM can be written as:

$$\mathbf{Y} \stackrel{\text{ind}}{\sim} \mathcal{E}(\boldsymbol{\mu}, \boldsymbol{\phi})$$
$$\boldsymbol{\eta} = g(\boldsymbol{\mu}) = \mathbf{X}\boldsymbol{\beta} + s_1(\mathbf{x}_1) + \ldots + s_J(\mathbf{x}_J) \qquad (1.8)$$

where s_j is a nonparametric smoothing function applied to covariate $\mathbf{x}_j$, for $j = 1, \ldots, J$. The idea is to let the data determine the relationship between the *predictor*[1] $\eta = g(\mu)$ and the explanatory variables, rather than enforcing a linear (or polynomial) relationship. In Chapter 2 some examples of different smoothers are given. More detail about smoothers within the **gamlss** package can be found in Chapter 9. Here we will use the smoothing function **pb()**, which is an implementation of a P-splines smoother in GAMLSS [Eilers and Marx, 1996].

We now model the **rent** parameter μ using smooth functions for **Fl** and **A**, and compare this model with the simple GLM fitted using **gamlss** (i.e. model **r2**) in the previous section.

```
r3 <-  gamlss(R ~ pb(Fl)+pb(A)+H+loc, family=GA,  data=rent,
              trace=FALSE)
AIC(r2,r3)

##              df      AIC
## r3 11.21547 27705.65
## r2  7.00000 27778.59
```

---

[1]Note that we refer to η as the predictor rather than the linear predictor, since the smoothing terms introduce nonlinearities in the model.

According to the AIC, the GAM model with smoothers is better than the simple GLM with linear terms for `Fl` and `A`. The summary of the fit is shown below:

```
summary(r3)

## *******************************************************************
## Family:  c("GA", "Gamma")
##
## Call:
## gamlss(formula = R ~ pb(Fl) + pb(A) + H + loc, family = GA,
##     data = rent, trace = FALSE)
##
## Fitting method: RS()
##
## -------------------------------------------------------------------
## Mu link function:  log
## Mu Coefficients:
##              Estimate Std. Error t value Pr(>|t|)
## (Intercept)  3.0851197  0.5696362    5.416 6.85e-08 ***
## pb(Fl)       0.0103084  0.0004031   25.573  < 2e-16 ***
## pb(A)        0.0014062  0.0002895    4.858 1.28e-06 ***
## H1          -0.3008111  0.0225803  -13.322  < 2e-16 ***
## loc2         0.1886692  0.0299295    6.304 3.58e-10 ***
## loc3         0.2719856  0.0322863    8.424  < 2e-16 ***
## ---
## Signif. codes:
## 0 '***' 0.001 '**' 0.01 '*' 0.05 '.' 0.1 ' ' 1
##
## -------------------------------------------------------------------
## Sigma link function:  log
## Sigma Coefficients:
##              Estimate Std. Error t value Pr(>|t|)
## (Intercept)  -1.00196     0.01559   -64.27    <2e-16 ***
## ---
## Signif. codes:
## 0 '***' 0.001 '**' 0.01 '*' 0.05 '.' 0.1 ' ' 1
##
## -------------------------------------------------------------------
## NOTE: Additive smoothing terms exist in the formulas:
##  i) Std. Error for smoothers are for the linear effect only.
## ii) Std. Error for the linear terms maybe are not accurate.
## -------------------------------------------------------------------
## No. of observations in the fit:  1969
## Degrees of Freedom for the fit:  11.21547
##       Residual Deg. of Freedom:  1957.785
```

```
##                              at cycle:   3
##
## Global Deviance:       27683.22
##              AIC:       27705.65
##              SBC:       27768.29
## ********************************************************************
```

There is a "NOTE" on the output warning users that because smoothers are included in the model, the standard errors given should be treated with care. There are two issues associated with the output given by the `summary()` function. The first is that the resulting coefficients of each smoother and its standard error refer only to the linear part of the smoother and not to the smoother's contribution as a whole, which is decomposed into a linear plus a nonlinear smoothing part. To test the contribution of the smoother as a whole (including the linear term), use the function `drop1()` as shown below. The second issue has to do with the standard errors of the linear part of the model, i.e. of the terms H and loc. Those standard errors are estimated assuming that the smoother terms are fixed at their fitted values and therefore do not take into the account the uncertainty introduced by estimating the smoothing terms. Some suggestions for correcting this are given in Section 5.2.3.

> **Important**: When smoothers are fitted all standard errors shown should be treated with caution.

Now we use `drop1()` to check for the approximate significance of the contribution of the smoothers (including the linear terms).

```
drop1(r3)

## Single term deletions for
## mu
##
## Model:
## R ~ pb(Fl) + pb(A) + H + loc
##            Df    AIC    LRT   Pr(Chi)
## <none>           27706
## pb(Fl) 1.4680  28261 558.59  < 2.2e-16 ***
## pb(A)  4.3149  27798 101.14  < 2.2e-16 ***
## H      1.8445  27862 160.39  < 2.2e-16 ***
## loc    2.0346  27770  68.02  1.825e-15 ***
## ---
## Signif. codes:
## 0 '***' 0.001 '**' 0.01 '*' 0.05 '.' 0.1 ' ' 1
```

All terms contribute significantly to modelling the predictor $\log(\mu)$. Note that `drop1()` can be very slow for large data sets and for models with many smoother terms. Note also that the degrees of freedom reduction (1.8445) resulting from dropping H from the model **r3** is different from 1 because dropping H results in a slight changes in the automatically chosen degrees of freedom used in the smoothing terms **pb(Fl)** and **pb(A)**.

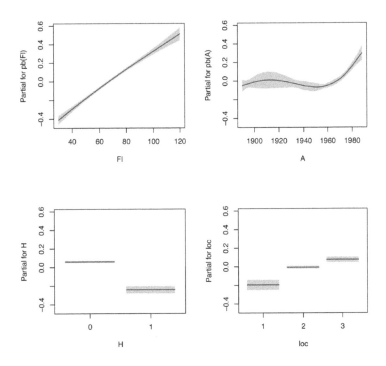

FIGURE 1.4: A plot of the fitted terms for model **r3**.

One of the properties of the fitted nonparametric smooth functions is that they cannot be described simply in a mathematical form. However they can be displayed. Here we plot them using the the function **term.plot()**:

Fig.
1.4

```
term.plot(r3, pages=1, ask=FALSE)
```

The plot shows that the predictor $\eta = \log(\mu)$ for the mean **rent** rises almost linearly with floor space **Fl**, but nonlinearly with age **A**, remaining stable if the flat was built before the 1960s and rising after that. The contribution of the two factors **H** and **loc** are what we would expect, i.e. lower rent if the flat does not have central heating (**H=1**) and increasing rent as the location of the flat changes from below average to average and then to above average (**loc**=1, 2 and 3, respectively). The shaded areas are the pointwise 95% confidence bands for the smoothing curves and factor levels. The GAMs in general allow for a flexible specification of the dependence of the parameter predictors on different explanatory terms. To check the adequacy of the fitted GAM we use a *worm plot*, which is a de-trended QQ-plot of the residuals [van Buuren and Fredriks, 2001]:

Fig.
1.5

```
wp(r3, ylim.all=.6)
```

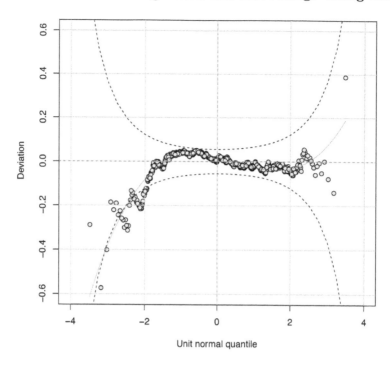

R

p. 20

FIGURE 1.5: Worm plot for model r3.

Chapter 12 explains in detail how to interpret a worm plot. Here it is sufficient to say that for an adequate fitted model we would expect the dots (which look like a worm) to be close to the middle horizontal line and 95% of them to lie between the upper and lower dotted curves, which act as 95% pointwise confidence intervals, with no systematic departure. This does not appear to be the case for the fitted GAM, where the worm is below the lower curve on the left of the figure. Multiple worm plots allow investigation of the adequacy of the model within ranges of the explanatory variables, and are often used in centile estimation (see Chapter 13). To improve the residuals of the GAM, we shall next model the parameter σ of the gamma distribution as a function of the explanatory variables.

1.6   Modelling the scale parameter

The gamma distribution has two parameters: μ, the mean of the distribution and σ, a *scale* parameter which is related to the variance by $\mathrm{Var}(Y) = \sigma^2 \mu^2$. Hence σ is the coefficient of variation of the gamma distribution. Up to now we have modelled only μ as a function of explanatory variables, but there are occasions (as for the Munich rent data) in which the assumption of a constant scale parameter is not appropriate. On these occasions modelling σ as a function of explanatory variables could solve the problem. Modelling σ started in the 1970s: Harvey [1976] and Aitkin [1987] were the first to model the variance of the normal distribution. As a solution to the problem of heteroscedasticity, Engle [1982, 1995] proposed a time series model for σ (volatility) for financial data. Engle's ARCH (autoregressive conditional heteroscedastic) model has created a whole industry of related models in finance. Modelling the dispersion parameter within the GLM was done by Nelder and Pregibon [1987], Smyth [1989] and Verbyla [1993]. Rigby and Stasinopoulos [1996a,b] introduced smooth functions for modelling both μ and σ, which they called the mean and dispersion additive model (MADAM). In the original MADAM formulation the response distribution had to be in the exponential family, but the method of fitting was quasi-likelihood rather than full maximum likelihood, which is used in GAMLSS.

We now consider the following submodel of the GAMLSS model:

$$\mathbf{Y} \overset{\text{ind}}{\sim} \mathcal{D}(\boldsymbol{\mu}, \boldsymbol{\sigma})$$
$$\eta_1 = g_1(\boldsymbol{\mu}) = \mathbf{X}_1 \boldsymbol{\beta}_1 + s_{11}(\mathbf{x}_{11}) + \ldots + s_{1J_1}(\mathbf{x}_{1J_1})$$
$$\eta_2 = g_2(\boldsymbol{\sigma}) = \mathbf{X}_2 \boldsymbol{\beta}_2 + s_{21}(\mathbf{x}_{21}) + \ldots + s_{2J_2}(\mathbf{x}_{2J_2}) \tag{1.9}$$

where $\mathcal{D}(\boldsymbol{\mu}, \boldsymbol{\sigma})$ denotes any two-parameter distribution and both μ and σ are linear and/or smooth functions of the explanatory variables. We model the Munich rent data using the gamma (`GA`) and inverse Gaussian (`IG`) distributions in model (1.9). The following models for μ and σ are specified:

$$\log(\mu) = \beta_{10} + s_{11}(\text{Fl}) + s_{12}(\text{A}) + \beta_{11}(\text{if } \text{H=1}) + \beta_{12}(\text{if } \text{loc=2}) +$$
$$\beta_{13}(\text{if } \text{loc=3})$$
$$\log(\sigma) = \beta_{20} + s_{21}(\text{Fl}) + s_{22}(\text{A}) + \beta_{21}(\text{if } \text{H=1}) + \beta_{22}(\text{if } \text{loc=2}) +$$
$$\beta_{23}(\text{if } \text{loc=3}) .$$

In the **R** code, the model for μ is specified after the response variable R ~, while the model for σ is specified after `sigma.fo=~` .

```
r4 <-  gamlss(R ~ pb(Fl)+pb(A)+H+loc, sigma.fo=~pb(Fl)+pb(A)+H+loc,
              family=GA, data=rent, trace=FALSE)
r5 <-  gamlss(R ~ pb(Fl)+pb(A)+H+loc, sigma.fo=~pb(Fl)+pb(A)+H+loc,
              family=IG, data=rent, trace=FALSE)
AIC(r3, r4, r5)

##            df       AIC
## r4 22.25035 27614.78
## r3 11.21547 27705.65
## r5 21.82318 27716.66
```

The smallest value of AIC for **r4** indicates that the gamma distribution fits better than the inverse Gaussian (model **r5**). To plot the fitted terms for σ use:

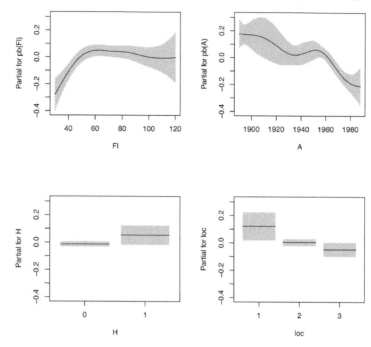

FIGURE 1.6: A plot of the fitted terms for σ for model r4.

R
p. 2?

```
term.plot(r4, pages=1, what="sigma", ask=FALSE)
```
Fig.
1.6

The approximate significance of the terms for σ can be tested using the drop1()
function:

```
drop1(r4, what="sigma")

## Single term deletions for
## sigma
##
## Model:
## ~pb(Fl) + pb(A) + H + loc
##              Df     AIC    LRT    Pr(Chi)
## <none>            27615
## pb(Fl) 4.02694 27631 24.683 5.997e-05 ***
## pb(A)  3.87807 27659 52.167 1.067e-10 ***
## H      0.88335 27615  1.866   0.14788
## loc    2.03694 27619  8.036   0.01872 *
## ---
## Signif. codes:
## 0 '***' 0.001 '**' 0.01 '*' 0.05 '.' 0.1 ' ' 1
```

Every term apart from H seems to contribute significantly to explaining the behaviour of the σ parameter. To check the adequacy of the distribution use the worm plot:

Fig.
1.7

```
wp(r4, ylim.all=.6)
```

R
. 23

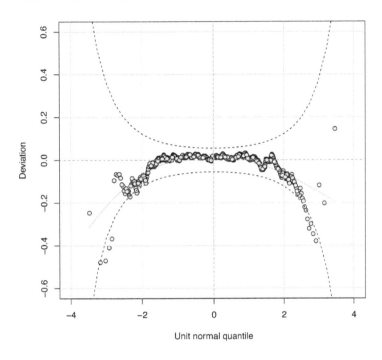

FIGURE 1.7: Worm plot for model **r4**.

There are a few points of the worm plot falling outside the 95% pointwise confidence intervals, indicating that the distribution may be inadequate. Furthermore, the inverted U-shape of the residuals indicates negative skewness in the residuals and suggests that the gamma distribution might not be flexible enough to capture the skewness in the data. Thus, we will fit a more general GAMLSS model because it allows for greater flexibility compared with the GLM and GAM models.

1.7 The generalized additive model for location, scale and shape (GAMLSS)

One of the problems of a two-parameter distribution is the fact that the skewness and kurtosis of the distribution are fixed for fixed μ and σ. With a relatively large set of data we would like to have the option of a model with more flexible skewness and/or kurtosis. Model (1.9) can be extended as follows:

$$\mathbf{Y} \overset{\text{ind}}{\sim} \mathcal{D}(\boldsymbol{\mu}, \boldsymbol{\sigma}, \boldsymbol{\nu}, \boldsymbol{\tau})$$

$$\boldsymbol{\eta}_1 = g_1(\boldsymbol{\mu}) = \mathbf{X}_1\boldsymbol{\beta}_1 + s_{11}(\mathbf{x}_{11}) + \ldots + s_{1J_1}(\mathbf{x}_{1J_1})$$

$$\boldsymbol{\eta}_2 = g_2(\boldsymbol{\sigma}) = \mathbf{X}_2\boldsymbol{\beta}_2 + s_{21}(\mathbf{x}_{21}) + \ldots + s_{2J_2}(\mathbf{x}_{2J_2}) \qquad (1.10)$$

$$\boldsymbol{\eta}_3 = g_3(\boldsymbol{\nu}) = \mathbf{X}_3\boldsymbol{\beta}_3 + s_{31}(\mathbf{x}_{31}) + \ldots + s_{3J_3}(\mathbf{x}_{3J_3})$$

$$\boldsymbol{\eta}_4 = g_4(\boldsymbol{\tau}) = \mathbf{X}_4\boldsymbol{\beta}_4 + s_{41}(\mathbf{x}_{41}) + \ldots + s_{4J_4}(\mathbf{x}_{4J_4})$$

where $\mathcal{D}(\boldsymbol{\mu}, \boldsymbol{\sigma}, \boldsymbol{\nu}, \boldsymbol{\tau})$ is a four-parameter distribution and where $\boldsymbol{\nu}$ and $\boldsymbol{\tau}$ are shape parameters which are often related to the skewness and kurtosis of the distribution. Model (1.10) defines the generalized additive model for location, scale and shape (GAMLSS) first introduced by Rigby and Stasinopoulos [2005]. GAMLSS, and in particular its implementation in **R**, is the main subject of this book. The following comments related to model (1.10) are appropriate here:

Distributions The form of the distribution $\mathcal{D}(\boldsymbol{\mu}, \boldsymbol{\sigma}, \boldsymbol{\nu}, \boldsymbol{\tau})$ is general and only implies that the distribution should be in parametric form. In the current implementation there are around 100 *discrete, continuous* and *mixed* distributions implemented in `gamlss.family`, including some highly skew and kurtotic distributions. Chapter 6 describes the available distributions, while more detail about individual distributions can be found in Rigby et al. [2017]. In addition:

- Creating a *new* distribution is relatively easy (Section 6.4);

- Any distribution in `gamlss.family` can be left-, right- or both sides- *truncated* (Section 6.2.2);

- A *censored* version of any `gamlss.family` distribution can be created, allowing modelling of censored and interval response variables (Section 6.2.2);

- Any distributions in `gamlss.family` can be mixed to create a new finite mixture distribution as described in Chapter 7;

- *Discretized* continuous distributions can be created for modelling discrete response variables, for example Section 14.2;

- Any continuous `gamlss.family` distribution on $(-\infty, \infty)$ can be transformed to a distribution in $(0, \infty)$ or $(0, 1)$ using the arguments `type` with options `log` or `logit`, respectively, of the function `gen.Family()` (Section 6.2.2).

Additive terms Explanatory variables can affect the parameters of the specified distribution in different ways. GAMLSS models allow linear or nonlinear parametric functions, or nonparametric smoothing functions of explanatory variables. The **gamlss** package allows the following additive terms: (i) P-splines (penalized B-splines), (ii) monotone P-splines, (iii) cycle P-splines, (iv) varying coefficient P-splines, (v) cubic smoothing splines, (vi) loess curve fitting, (vii) fractional polynomials, (viii) random effects, (ix) ridge regression and (x) nonlinear parametric fits. In addition, through appropriate interfaces installed automatically with package **gamlss.util**, the software allows fitting of (i) neural networks, via package **nnet**, (ii) decision trees, via **rpar**, (iii) random effects, via **nlme** and (iv) multidimensional smoothers, via **mgcv**.

Fitting methods and algorithms A parametric GAMLSS model (i.e. (1.10) without smoothing functions) is fitted by maximum likelihood estimation. Note that $\mathbf{Y} \overset{\text{ind}}{\sim} \mathcal{D}(\boldsymbol{\mu}, \boldsymbol{\sigma}, \boldsymbol{\nu}, \boldsymbol{\tau})$ implies that the *likelihood function* which is defined as the likelihood of observing the sample, is $L(\boldsymbol{\mu}, \boldsymbol{\sigma}, \boldsymbol{\nu}, \boldsymbol{\tau}) = \prod_{i=1}^{n} f(y_i \mid \mu_i, \sigma_i, \nu_i, \tau_i)$. Therefore the log-likelihood function is $\ell = \sum_{i=1}^{n} \log f(y_i \mid \mu_i, \sigma_i, \nu_i, \tau_i)$. The more general model (1.10) is generally fitted by maximum penalized likelihood estimation. Chapter 9 shows that most of the smoothers used within GAMLSS can be written as $\mathbf{s}(\mathbf{x}) = \mathbf{Z}\boldsymbol{\gamma}$, where $\mathbf{Z}$ is a basis matrix depending on values of $\mathbf{x}$, and $\boldsymbol{\gamma}$ is a set of coefficients subject to the quadratic penalty $\boldsymbol{\gamma}^{\top} \mathbf{G}(\boldsymbol{\lambda}) \boldsymbol{\gamma}$ where $\boldsymbol{\lambda}$ is a vector or scalar of hyperparameter(s). Rigby and Stasinopoulos [2005] have shown that the algorithm used for fitting the GAMLSS model for fixed values of the hyperparameters $\boldsymbol{\lambda}$ maximizes a penalized likelihood function ℓ_p given by

$$\ell_p = \ell - \frac{1}{2} \sum_{k=1}^{4} \sum_{j=1}^{J_k} \boldsymbol{\gamma}_{kj}^{\top} \mathbf{G}_{kj}(\boldsymbol{\lambda}_{kj}) \boldsymbol{\gamma}_{kj} \tag{1.11}$$

where $\ell = \sum_{i=1}^{n} \log f(y_i \mid \mu_i, \sigma_i, \nu_i, \tau_i)$ is the log-likelihood function. Rigby and Stasinopoulos [2005] suggested two basic algorithms for fitting GAMLSS model (1.10). The first, the CG algorithm, is a generalization of the Cole and Green [1992] algorithm and uses the first derivatives and the (exact or approximate) expected values of the second and cross derivatives of the likelihood function with respect to $\boldsymbol{\theta} = (\mu, \sigma, \nu, \tau)$. However for many probability (density) functions $f(y_i \mid \mu_i, \sigma_i, \nu_i, \tau_i)$, the parameters are information orthogonal (where the expected values of the cross derivatives of the likelihood function are zero), for example location and scale models and dispersion family models. In this case the second, the RS algorithm, which is a generalization of the algorithm used by Rigby and Stasinopoulos [1996a,b] for fitting the MADAM models, is more suited as it does not use the expected values of the cross derivatives.

We now return to the Munich data to see if we can improve the model by fitting a three-parameter distribution. We use the Box–Cox Cole and Green (BCCGo) distribution, which is based on Cole and Green [1992] who were the first to fit a single smoothing term to each of the three parameters of the distribution. They called their model the "LMS method" and it is widely used for centile estimation; see Chapter 13. The first model (**r6**) fits a constant ν while the second (**r7**) fits the same model for ν as was fitted for μ and σ.

```
r6 <-   gamlss(R ~ pb(Fl)+pb(A)+H+loc, sigma.fo=~pb(Fl)+pb(A)+H+loc,
              nu.fo=~1, family=BCCGo,  data=rent, trace=FALSE)


r7 <-   gamlss(R ~ pb(Fl)+pb(A)+H+loc,sigma.fo=~pb(Fl)+pb(A)+H+loc,
```

```
                 nu.fo=~pb(Fl)+pb(A)+H+loc, family=BCCGo,  data=rent,
                 trace=FALSE)

AIC(r4, r6, r7)

##             df       AIC
## r7 28.41391 27608.15
## r6 22.48092 27611.02
## r4 22.25035 27614.78
```

Based on the AIC values, the BCCGo distribution provides a superior fit compared
to the gamma, and modelling the ν parameter as a function of the explanatory
variables improves slightly the fit. To check the adequacy of the fitted distribution
we use worm plots:

```
wp(r6, ylim.all=.6) ;  title("r6: BCCG(mu, sigma)")
wp(r7, ylim.all=.6) ;  title("r7:  BCCG(mu, sigma, nu)")
```

Fig.
1.8

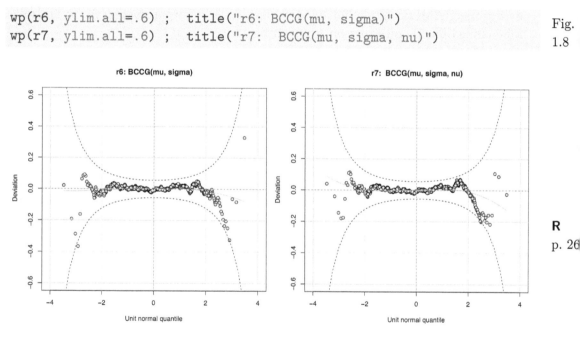

R
p. 26

FIGURE 1.8: Worm plots for models **r6** and **r7**.

Both worm plots show adequate fits, so we finish our demonstration here.

We have used the Munich rent data to demonstrate how GAMLSS can be used, and
we arrived at a more sophisticated model than using only GLM or GAM. In particu-
lar, modelling both μ and σ of a gamma ($GA(\mu, \sigma)$) distribution in model **r4** provides
a substantially improved fit to the rent response variable as compared to the GAM
model **r3**. Also a three-parameter distribution model using the $BCCGo(\mu, \sigma, \nu)$ distribution marginally improves the fit. GAMLSS provides greater flexibility in regression modelling, but with this flexibility comes more responsibility for the statistician.

This is not a bad thing. The philosophy of GAMLSS is to allow the practitioner to have a wide choice of regression models.

We conclude this chapter with a summary of some of the basic properties of GAMLSS:

- GAMLSS is a very flexible unifying framework for univariate regression models.

- It allows *any* distribution for the response variable. *All* the parameters of the distribution can be modelled as functions of explanatory variables.

- It allows a variety of additive terms in the models for the distribution parameters.

- The fitted algorithm is modular, where different components can be added easily.

- It extends basic statistical models allowing flexible modelling of overdispersion, excess of zeros, skewness and kurtosis in the data.

1.8 Bibliographic notes

Regression analysis is attributed to Francis Galton in the 1870s but *least squares* goes back to the early 1800s and the German mathematician Karl Gauss, who used the technique to predict astronomical phenomena. Lehmann [2008] presents the history and use of some standard regression statistical models. Many statistics texts discuss the linear regression model, for example, Weisberg [1980], Wood [2006b] and Faraway [2009]. Muller and Stewart [2006] present linear model theory for univariate, multivariate and mixed model cases. Generalized linear models were introduced by Nelder and Wedderburn [1972] and popularized by McCullagh and Nelder [1989] and Dobson and Barnett [2008]. Details about generalized additive models (GAM) can be found in Hastie and Tibshirani [1990] and Wood [2006b]. Generalized additive models for location, scale and shape (GAMLSS) were introduced by Rigby and Stasinopoulos [2005]. These models have been applied in different areas, for example Gilchrist et al. [2009] present an insurance type model for the health cost of cold housing; Hudson et al. [2010] analyzed the climatic influences on the flowering phenology of four eucalypts; de Castro et al. [2010] fitted long-term survival models under the GAMLSS framework; Barajas et al. [2015] applied GAMLSS models in the treatment of agro-industrial waste; Hu et al. [2015] used GAMLSS models for the analysis of speech intelligibility tests. Computational details about GAMLSS are given in Stasinopoulos and Rigby [2007] and Rigby and Stasinopoulos [2013]. The GAM and GAMLSS models allow penalized smoothers, based on splines; for details see Wahba [1978, 1985], Parker and Rice [1985], Hastie and Tibshirani [1990], Wahba [1990] and Marx and Eilers [1998]. Eilers and Marx [1996] present flexible smoothing with penalties using P-splines. Eilers and Marx [2010] present an overview of

splines, knots and penalties. Eilers et al. [2015] give an overview of many of the
central developments during the first two decades of P-splines.

1.9   Exercises

1. **The rent data example:** Familiarize yourself with **R** and the **gamlss** functions
   by repeating the **R** commands given in this chapter using the **rent** data example.
   First download the **gamlss** package and the **rent** data.

2. **The air quality data:** The data set `airquality` is one of the data frames
   available in **R** within the standard package **datasets**. It has the daily air quality
   measurements in New York, from May to September 1973.

> **R data file:** `airquality` in package **datasets** of dimensions 154×6
>
> **variables**
>
> > `Ozone` : in ppb
> >
> > `Solar.R` : in lang
> >
> > `Wind` : in mph
> >
> > `Temp` : in F
> >
> > `Month` : Month (1–12)
> >
> > `Day` : Day of month (1–31)
>
> **purpose:** to demonstrate the need for smooth functions.

(a) Here we will use `Ozone` as the response variable and `Solar.R`, `Wind` and `Temp`
    as explanatory variables. (We will not consider `Month` and `Day`.) The data
    can be plotted using:

```
data(airquality)
plot(airquality[,-c(5,6)])
```

Comment on the plot.

(b) To fit a standard regression model (i.e. with a normal distribution and con-
    stant variance) use the function `lm()`:

```
# Fit the standard linear model
air.lm <- lm(Ozone~Temp+Wind+Solar.R,data=airquality)
summary(air.lm)
```

The `summary()` provides information about the coefficients and their standard errors. To plot the fitted model terms use `termplot()`:

```
op<-par(mfrow=c(1,3))
termplot(air.lm,partial.resid=TRUE,se=T)
par(op)
```

Comment on the term plot.

(c) Check the residuals using `plot()`:

```
op<-par(mfrow=c(1,2))
plot(air.lm,which=1:2)
par(op)
```

(d) Fit the same model using the `gamlss()` function, but note that the data set `airquality` has some missing observations (i.e. NA values). The `gamlss()` function does not work with NA's, so before fitting the model the cases with missing values have to be removed:

```
library(gamlss)
da <- na.omit(airquality)  # clear the data of NA's
mno<-gamlss(Ozone~Temp+Wind+Solar.R, data=da)  # fit the model
summary(mno)
```

Summarize the fitted `gamlss` model using `summary()`. Plot the fitted terms using the corresponding function for `gamlss` called `term.plot()`:

```
term.plot(mno, pages=1, partial=T)  # plot the fitted terms
```

(e) Check the residuals using the `plot()` and `wp()` functions:

```
plot(mno)
wp(mno)
```

Comment on the worm plot. Note the warning message that some points are missed out of the worm plot. Increase the limits in the vertical axis by using the argument `ylim.all=2` in `wp()`.

(f) Since the fitted normal distribution seems not to be correct, try to fit different distributions (e.g. gamma (`GA`), inverse Gaussian (`IG`) and Box Cox Cole and Green (`BCCGo`)) to the data. Compare them with the normal distribution using GAIC with penalty $\kappa = 2$ (i.e. AIC).

```
# fit different distributions
mga <- gamlss(Ozone~Temp+Wind+Solar.R, data=da, family=GA)
mig <- gamlss(Ozone~Temp+Wind+Solar.R, data=da, family=IG)
mbccg <- gamlss(Ozone~Temp+Wind+Solar.R, data=da, family=BCCGo)
GAIC(mno, mga, mig, mbccg)
```

(g) For the selected distribution, fit smoothing terms, i.e. pb(), for Solar.R, Wind and Temp.

```
# fit smoothers
mga1=gamlss(Ozone~pb(Temp)+pb(Wind)+pb(Solar.R),data=da,
            family=GA)
term.plot(mga1, pages=1)
plot(mga1)
wp(mga1)
```

Is the model improved according to the AIC? Use term.plot() output to see the fitted smooth functions for the predictor of μ for your chosen distribution. Use plot() and wp() output to check the residuals.

2

Introduction to the **gamlss** *packages*

CONTENTS

2.1 Introduction

This chapter provides:

1. an introduction to the **gamlss** packages; and

2. an introduction to some of the facilities of the **gamlss** packages using a simple regression model.

This chapter uses a simple example of a continuous response variable and a single continuous explanatory variable to demonstrate some of the facilities that the **gamlss** packages provide. Section 2.2 describes the different **gamlss** packages in **R**. Section 2.3 demonstrates the `gamlss()` function and other functionalities in the **gamlss** packages.

2.2   The gamlss packages

The **gamlss** packages comprise the original **gamlss** and other add-on packages:

1. The original **gamlss** package for fitting a GAMLSS model contains the main function `gamlss()` for fitting a GAMLSS model, and methods for dealing with fitted `gamlss` objects. Chapter 3 describes the algorithms used by `gamlss()`. Chapter 4 describes the `gamlss()` function, its arguments and other methods and functions which can be used for a `gamlss` fitted object. Chapter 5 describes the functions which are related to inference and prediction from `gamlss` fitted objects.

2. The **gamlss.add** package for fitting extra additive terms provides extra additive terms for fitting a parameter of the distribution of the response variable. This is mainly achieved by providing interfaces with other **R** packages. For example, neural networks, decision trees and multidimensional smoothers can be fitted within `gamlss()` by using the packages **nnet**, **rpart** and **mgcv**, respectively, which are loaded automatically with the **gamlss.add** package. The use of those terms is explained in Chapter 9.

3. The **gamlss.cens** package for fitting left-, right- or interval-censored response variables generates `gamlss.family` distributions suitable for fitting censored data within a GAMLSS model.

4. The **gamlss.data** package contains the data used in this book, and is automatically loaded when **gamlss** is loaded.

5. The **gamlss.demo** package for teaching demos: the purpose of this package is twofold. Firstly, it provides a visual presentation of the `gamlss.family` distributions. That is, the user can visualise how the shape of the distribution changes when any of the parameters of the distribution changes. Secondly, it provides a visual presentation of some of the smoothing and P-splines ideas.

6. The **gamlss.dist** package for `gamlss.family` distributions. This package contains all the distributions available in GAMLSS models and is automatically loaded with **gamlss**. More information about the distributions available can be found in Chapter 6 and in the book by Rigby et al. [2017].

7. The **gamlss.mx** package for fitting finite mixture distributions and nonparametric random effects. Chapter 7 provides examples of fitting finite mixtures to data, while Chapter 10 provides information for fitting nonparametric random effects.

8. The **gamlss.nl** package for fitting nonlinear parametric models within the GAMLSS framework. This package is not discussed in this book but the reader can find more information on the GAMLSS website `www.gamlss.org`.

9. The **gamlss.spatial** package for spatial models. This package provides facilities

for Gaussian Markov random fields (GMRF) terms to be fitted within GAMLSS models. GMRF are appropriate when a factor provides geographical information, for example areas in a region, and we wish to take the neighbourhood information into account. Section 9.4.9 describes the function for fitting GMRF terms in a model.

10. The **gamlss.tr** package for fitting truncated distributions. This package can take any `gamlss.family` distribution and truncate it (left, right or both).

Note that more add-on packages will be added in the future, including time series modelling and more flexible inflated and other distributions. Please check the GAMLSS website `www.gamlss.org` for more details.

The **gamlss** packages can be downloaded and installed from the **R** library CRAN at `www.r-project.org`. Help files are provided for all functions in **gamlss** in the usual way. For example using

```
help(package="gamlss")
?gamlss
```

gives the user information about the package **gamlss** and the function `gamlss()`, respectively.

2.3   A simple example using the gamlss packages

The `gamlss()` function allows modelling of up to four parameters in a distribution family, which are conventionally called μ, σ, ν and τ. Here we give a simple demonstration using the `film90` data set.

R data file: `film90` in package **gamlss.data** of dimension 4015×4.
variables
   lnosc : the log of the number of screens in which the film was played
   lboopen : the log of box office opening week revenues
   lborev1 : the log of box office revenues after the first week (the response variable which has been randomized)
   dist : a factor indicating whether the distributor of the film was an "Independent" or a "Major" distributor
purpose: to demonstrate the fitting of a simple regression model in the **gamlss** package.

The original data were analysed in Voudouris et al. [2012], where more information about the data and the purpose of the original study can be found. Here for demonstrating some of the features of **gamlss** we analysed only two variables: `lborev1` as the response variable, and `lboopen` as an explanatory variable.

We start by plotting the data in Figure 2.1. Two key features are suggested: (i) the relationship between the response and the explanatory variable is nonlinear, and (ii) the shape of the response variable distribution changes for different levels of the explanatory variable. As we will see in Section 2.3.7, a GAMLSS model has the flexibility to model these features.

```
library(gamlss)
data(film90)
plot(lborev1~lboopen, data=film90, col="lightblue",
     xlab="log opening revenue", ylab="log extra revenue")
```

Fig.
2.1

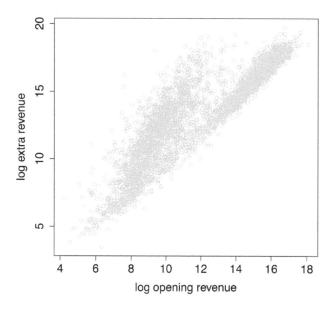

R
p. 34

FIGURE 2.1: Scatterplot of the `film90` revenues.

2.3.1   Fitting a parametric model

Below we fit a simple linear regression model with normal errors. It is clear from Figure 2.2 that the model does not fit well, especially for low values of `lboopen`.

```
m <- gamlss(lborev1~lboopen, data=film90, family=NO)

## GAMLSS-RS iteration 1: Global Deviance = 15079.74
## GAMLSS-RS iteration 2: Global Deviance = 15079.74
```

R

. 35

```
plot(lborev1~lboopen, data=film90, col = "lightblue")
lines(fitted(m)~film90$lboopen)
```
Fig.
2.2

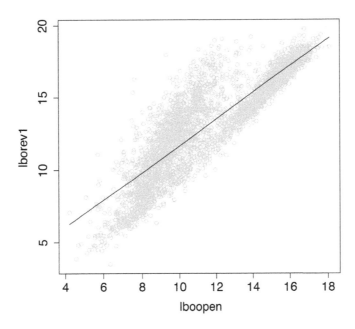

FIGURE 2.2: Scatterplot of the `film90` data with the fitted linear model for the mean.

The problem seems to be the linear term in **lboopen**, so next we fit a cubic polynomial. One method of fitting polynomial curves in **R** is by using the function `I()`. A different method is by using the function `poly()` which fits orthogonal polynomials (see later).

```
m00 <- gamlss(lborev1~lboopen+I(lboopen^2)+I(lboopen^3), data=film90,
                        family=NO)

## GAMLSS-RS iteration 1: Global Deviance = 14518.26
## GAMLSS-RS iteration 2: Global Deviance = 14518.26

summary(m00)

## *******************************************************************
## Family:  c("NO", "Normal")
##
## Call:
## gamlss(formula = lborev1 ~ lboopen + I(lboopen^2) +
##     I(lboopen^3), family = NO, data = film90)
##
## Fitting method: RS()
##
```

```
## ------------------------------------------------------------------
## Mu link function:  identity
## Mu Coefficients:
##                 Estimate Std. Error t value Pr(>|t|)
## (Intercept)  -2.232e+01  1.271e+00  -17.57   <2e-16 ***
## lboopen       7.147e+00  3.516e-01   20.32   <2e-16 ***
## I(lboopen^2) -4.966e-01  3.153e-02  -15.75   <2e-16 ***
## I(lboopen^3)  1.270e-02  9.142e-04   13.89   <2e-16 ***
## ---
## Signif. codes:
## 0 '***' 0.001 '**' 0.01 '*' 0.05 '.' 0.1 ' ' 1
##
## ------------------------------------------------------------------
## Sigma link function:  log
## Sigma Coefficients:
##              Estimate Std. Error t value Pr(>|t|)
## (Intercept)   0.38189    0.01114   34.29   <2e-16 ***
## ---
## Signif. codes:
## 0 '***' 0.001 '**' 0.01 '*' 0.05 '.' 0.1 ' ' 1
##
## ------------------------------------------------------------------
## No. of observations in the fit:  4031
## Degrees of Freedom for the fit:  5
##        Residual Deg. of Freedom:  4026
##                       at cycle:  2
##
## Global Deviance:     14518.26
##            AIC:      14528.26
##            SBC:      14559.77
## ******************************************************************
```

Note that for large data sets it could be more efficient (and may be essential) to calculate the polynomial terms in advance prior to using the **gamlss()** function, e.g.

```
x2<-x^2; x3<-x^3
```

and then use them within the **gamlss()** function, since the evaluation is then done only once:

```
film90 <- transform(film90, lb2=lboopen^2, lb3=lboopen^3)
m002 <- gamlss(lborev1~lboopen + lb2 + lb3, data=film90, family=NO)
```

The fitted model is displayed in Figure 2.3. Although the new model is an improvement, the polynomial line does not fit well for smaller values of **lboopen**. This behaviour, i.e. erratic fitting in the lower or upper end of the covariate, is very common in fitting parametric polynomial curves.

```
plot(lborev1~lboopen, col="lightblue", data=film90)
lines(fitted(m002)[order(film90$lboopen)]~
                    film90$lboopen[order(film90$lboopen)])
```

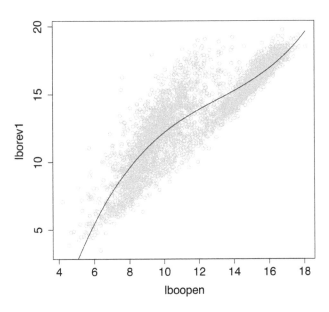

FIGURE 2.3: Scatterplot of the `film90` data with the fitted cubic model for the mean.

Using the notation $y = $ `lborev1` and $x = $ `lboopen`, the fitted model `m00` is given by

$$Y \sim \mathcal{N}(\hat{\mu}, \hat{\sigma}^2)$$

where

$$\hat{\mu} = \hat{\beta}_{10} + \hat{\beta}_{11}x + \hat{\beta}_{12}x^2 + \hat{\beta}_{13}x^3$$
$$= -22.320 + 7.147x - 0.497x^2 + 0.013x^3$$
$$\log(\hat{\sigma}) = 0.3819 \ ,$$

giving $\hat{\sigma} = \exp(0.3819) = 1.465$.

The `summary()` function is useful for providing standard errors for the fitted coefficient parameters. The `summary()` function has two ways of producing standard errors: (i) `type="vcov"` (the default) and (ii) `type="qr"`. The way the standard errors are produced using the `vcov` method is described in detail in Section 5.2. It starts by defining the likelihood function at the maximum (using `gen.likelihood()`) and then obtaining the full (numerical) Hessian matrix of second and cross derivatives of the likelihood function with respect to the beta coefficient parameters in the model. Standard errors are obtained from the inverse of the observed information matrix (i.e. the inverse of the negative of the Hessian matrix). The standard errors obtained this way are more reliable than those produced by the `qr` method, since they take into account the information about the interrelationship between the distribution parameters, i.e. μ and σ in the above example. On occasions when the above procedure fails, the standard errors are obtained from `type= "qr"`, which uses the individual fits of the distribution parameters and therefore should be used with caution. The `summary()` output gives a warning when this happens, as the standard errors produced this way do not take into account the correlation between the estimates of the distribution parameters μ, σ, ν and τ. (In the example above the estimates of μ and σ of the normal distribution are asymptotically uncorrelated.)

Robust ("*sandwich*" or "*Huber sandwich*") standard errors can be obtained using the argument `robust=TRUE` of the `summary()` function. Robust standard errors were introduced by Huber [1967] and White [1980] and are, in general, more reliable than the usual standard errors when the variance model is suspected not to be correct (assuming the mean model is correct). The sandwich standard errors are usually (but not always) larger than the usual ones.

Next we demonstrate how `vcov()` can be used to obtain the variance-covariance matrix, the correlation matrix and the (usual and robust) standard errors of the estimated parameters:

```
# the variance-covariance matrix of the parameters
print(vcov(m00), digit=3)
```

```
##               (Intercept)    lboopen I(lboopen^2)
## (Intercept)      1.61e+00 -4.43e-01     3.90e-02
## lboopen         -4.43e-01  1.24e-01    -1.10e-02
## I(lboopen^2)     3.90e-02 -1.10e-02     9.94e-04
## I(lboopen^3)    -1.10e-03  3.15e-04    -2.87e-05
## (Intercept)      2.24e-11 -6.15e-12     5.40e-13
##              I(lboopen^3) (Intercept)
## (Intercept)     -1.10e-03    2.24e-11
## lboopen          3.15e-04   -6.15e-12
## I(lboopen^2)    -2.87e-05    5.40e-13
## I(lboopen^3)     8.36e-07   -1.53e-14
## (Intercept)     -1.53e-14    1.24e-04
```

```
# the correlation matrix
print(vcov(m00, type="cor"), digit=3)
```

```
##               (Intercept)    lboopen I(lboopen^2)
## (Intercept)      1.00e+00 -9.93e-01     9.74e-01
## lboopen         -9.93e-01  1.00e+00    -9.94e-01
## I(lboopen^2)     9.74e-01 -9.94e-01     1.00e+00
## I(lboopen^3)    -9.49e-01  9.79e-01    -9.95e-01
## (Intercept)      1.58e-09 -1.57e-09     1.54e-09
##              I(lboopen^3) (Intercept)
```

```
## (Intercept)      -9.49e-01      1.58e-09
## lboopen          9.79e-01      -1.57e-09
## I(lboopen^2)     -9.95e-01      1.54e-09
## I(lboopen^3)     1.00e+00      -1.50e-09
## (Intercept)      -1.50e-09      1.00e+00
```

```
# standard errors
print(vcov(m00, type="se"), digits=2)
```

```
##   (Intercept)        lboopen I(lboopen^2) I(lboopen^3)
##       1.27058        0.35164      0.03153      0.00091
##   (Intercept)
##       0.01114
```

```
print(vcov(m00, type="se", robust=TRUE), digits=2)
```

```
##   (Intercept)        lboopen I(lboopen^2) I(lboopen^3)
##        1.9702         0.5217       0.0446       0.0012
##   (Intercept)
##        0.0135
```

Note that in the final row and/or column of the above output, `Intercept` refers to the intercept of the predictor model for σ ($\hat{\beta}_{20}$), while the first row and/or column `Intercept` refers to the intercept of the predictor for μ ($\hat{\beta}_{10}$).

Now we fit the same model as in `m00`, but using orthogonal polynomials (see Section 8.3) using function `poly()`, i.e. `poly(x,3)`:

```
m0 <- gamlss(lborev1~poly(lboopen,3), data=film90, family=NO)
```

```
## GAMLSS-RS iteration 1: Global Deviance = 14518.26
## GAMLSS-RS iteration 2: Global Deviance = 14518.26
```

It is of some interest to compare the correlations between the parameter estimates for the two fitted models `m00` and `m0`. Visual representation of the correlation coefficients can be obtained using the package **corrplot**.

```
library(corrplot)
col1 <- colorRampPalette(c("black","grey"))
corrplot(vcov(m00, type="cor"), col=col1(2), outline=TRUE,
         tl.col = "black", addCoef.col = "white")
corrplot(vcov(m0, type="cor"),  col=col1(2), outline=TRUE,
         tl.col = "black", addCoef.col = "white")
```

Fig. 2.4

Figure 2.4 shows the resulting graphical displays. Because, μ and σ in the normal distribution are information independent (i.e. asymptotically uncorrelated), the first four estimated parameters (μ model) are effectively not correlated with the fifth, the constant in the model for $\log(\sigma)$, in both models `m0` and `m00`. In addition all

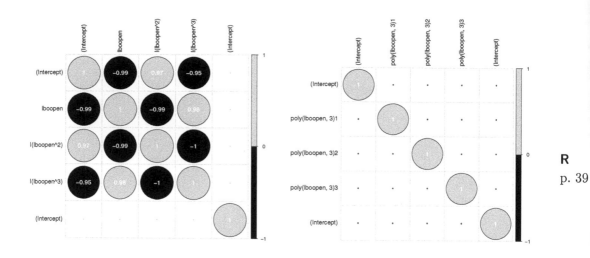

R
p. 39

FIGURE 2.4: Graphical displays of the correlation coefficient matrices for models
m00 (left) and m0 (right).

the parameters of the μ model for m0 are uncorrelated because we used orthogonal
polynomials, but for m00 they are highly correlated.

2.3.2   Fitting a nonparametric smoothing model

In this section, we outline a few of the nonparametric smoothing functions implemented in GAMLSS. In particular, we discuss the pb() (P-splines), cs() (cubic
splines), lo() (locally weighted regression) and nn() (neural networks) functions.
For a comprehensive discussion (and list of smoothing functions within GAMLSS),
see Chapter 9.

2.3.2.1   P-splines

Model m0 is a linear parametric GAMLSS model, which we have seen does not fit
particularly well. Another approach is to fit a smooth term to the covariate lboopen.
Eilers and Marx [1996] introduced nonparametric penalized smoothing splines (P-
splines), which are described in Section 9.4.2. In order to fit the mean of lborev1
with a P-spline for lboopen, use:

```
m1<-gamlss(lborev1~pb(lboopen), data=film90, family=NO)

## GAMLSS-RS iteration 1: Global Deviance = 14109.58
## GAMLSS-RS iteration 2: Global Deviance = 14109.58

summary(m1)

## ****************************************************************
```

```
## Family:  c("NO", "Normal")
##
## Call:
## gamlss(formula = lborev1 ~ pb(lboopen), family = NO,
##     data = film90)
##
## Fitting method: RS()
##
## ------------------------------------------------------------------
## Mu link function:  identity
## Mu Coefficients:
##             Estimate Std. Error t value Pr(>|t|)
## (Intercept) 2.347147   0.087053   26.96   <2e-16 ***
## pb(lboopen) 0.928889   0.007149  129.93   <2e-16 ***
## ---
## Signif. codes:
## 0 '***' 0.001 '**' 0.01 '*' 0.05 '.' 0.1 ' ' 1
##
## ------------------------------------------------------------------
## Sigma link function:  log
## Sigma Coefficients:
##             Estimate Std. Error t value Pr(>|t|)
## (Intercept)  0.33120    0.01114   29.74   <2e-16 ***
## ---
## Signif. codes:
## 0 '***' 0.001 '**' 0.01 '*' 0.05 '.' 0.1 ' ' 1
##
## ------------------------------------------------------------------
## NOTE: Additive smoothing terms exist in the formulas:
##  i) Std. Error for smoothers are for the linear effect only.
## ii) Std. Error for the linear terms maybe are not accurate.
## ------------------------------------------------------------------
## No. of observations in the fit:  4031
## Degrees of Freedom for the fit:  12.73672
##       Residual Deg. of Freedom:  4018.263
##                      at cycle:  2
##
## Global Deviance:     14109.58
##             AIC:     14135.05
##             SBC:     14215.32
## ******************************************************************
```

In the smoothing function pb() the smoothing parameter (and therefore the effective degrees of freedom) are estimated automatically using the default local maximum likelihood method described in Section 3.4.2.1 and Rigby and Stasinopoulos [2013].

Within the `pb()` function there are also alternative ways of estimating the smoothing parameter, such as the local generalized AIC (GAIC), and the local generalized cross validation (GCV). See Section 3.4 for details.

The fitted model is displayed in Figure 2.5.

```
plot(lborev1~lboopen, col="lightblue", data=film90)
lines(fitted(m1)[order(film90$lboopen)]~
            film90$lboopen[order(film90$lboopen)])
```

Fig. 2.5

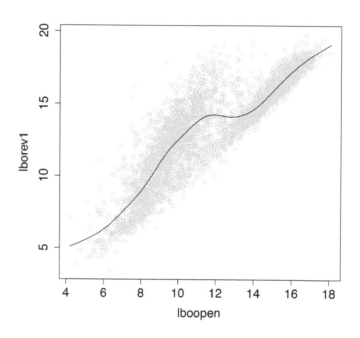

R
p. 42

FIGURE 2.5: P-splines fit: the `film90` data with the fitted smooth mean function fitted using `pb()`.

The effective degrees of freedom (i.e. the effective number of parameters) used in the fitting smooth function given by `pb()` can be obtained using `edf()`:

```
edf(m1, "mu")
```

```
## Effective df for mu model
## pb(lboopen)
##     11.73672
```

One of the important things to remember when fitting a smooth nonparametric term in `gamlss()` is that the displayed coefficient of the smoothing term and its standard error (s.e.) refer only to the linear component of the term. For example the coefficient 0.9289 and its s.e. 0.0071 in the above output should be interpreted with care. They are an artefact of the way the fitting algorithm works with the `pb()` function. This is because the linear part of the smoothing is fitted together with all other linear terms (in the above case only the intercept). One should try to

interpret the whole smoothing function, which can be obtained using `term.plot()`.
The effect that the smoothing function has on the specific parameters can also
be checked using the function `getPEF()`, which calculates the partial effect of a
continuous variable given the rest of the explanatory variables are fixed at specified
values. The same function can be used to obtain the first and second derivatives for
the partial effects. Approximate significance of smoothing terms is obtained using
the function `drop1()`, but this may be slow for a large data set with many fitted
smoothing terms.

> **Important**: Do not try to interpret the linear coefficients or the standard errors
> of the smoothing terms.

Note also that when smoothing additive terms are involved in the fitting, both
methods (default and robust) used in **summary** to obtain standard errors are ques-
tionable. The reason is that the way `vcov()` is implemented effectively assumes that
the estimated smoothing terms were fixed at their estimated values. The functions
`prof.dev()` and `prof.term()` can be used for obtaining more reliable individual
parameter confidence intervals, by fixing the smoothing degrees of freedom at their
previously selected values.

2.3.2.2 Cubic splines

Other smoothers are also available. For details on cubic smoothing splines see Section 9.4.6. In order to fit a nonparametric smoothing cubic spline with 10 effective
degrees of freedom in addition to the constant and linear terms, use

```
m2<-gamlss(lborev1~cs(lboopen,df=10), data=film90, family=NO)

## GAMLSS-RS iteration 1: Global Deviance = 14107.72
## . . .
## GAMLSS-RS iteration 2: Global Deviance = 14107.72
```

The effective degrees of freedom used in the fitting of μ in the above model are
12 (one for the constant, one for the linear and 10 for smoothing). Note that the
`gamlss()` notation is different from the `gam()` notation in S-PLUS where the equivalent model is fitted using `s(x,11)`.

The total degrees of freedom used for model m2 is 13, i.e. 12 for μ and 1 for σ. The
fitted values of μ for models m1 and m2 are displayed in Figure 2.6.

Fig.
2.6
```
plot(lborev1~lboopen, col="lightblue", data=film90)
lines(fitted(m1)[order(film90$lboopen)]~
                film90$lboopen[order(film90$lboopen)])
lines(fitted(m2)[order(film90$lboopen)]~
                film90$lboopen[order(film90$lboopen)],
col="red", lty=2,  lwd=2)
```

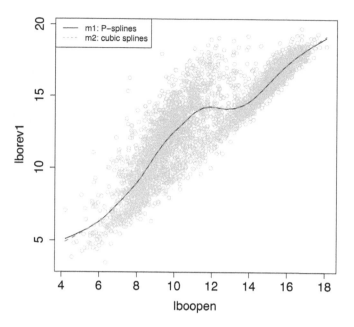

R

p. 43

FIGURE 2.6: P-splines and cubic splines fits: plot of the `film90` data together with the fitted smooth mean functions of model `m1` fitted by `pb()` (continuous line) and model `m2` fitted by `cs()` (dashed line).

```
legend("topleft",legend=c("m1: P-splines","m2: cubic splines"),
       lty=1:2,col=c("black","red"),cex=1)
```

2.3.2.3　Loess

Locally weighted scatterplot smoothing [Cleveland and Devlin, 1988], or loess, is described in Section 9.6.3. Loess curves are implemented as

```
m4 <- gamlss(lborev1~lo(~lboopen,span=.4), data=film90, family=NO)
```

2.3.2.4　Neural networks

Neural networks can be considered as another type of smoother. For details see Section 9.6.1. Here a neural network smoother is fitted using an interface of **gamlss** with the **nnet** package [Venables and Ripley, 2002]. The additive function to be used with **gamlss()** is **nn()**, which is part of the package **gamlss.add**. The following example illustrates its use.

```
library(gamlss.add)
mnt <- gamlss(lborev1~nn(~lboopen,size=20,decay=0.1), data=film90,
              family=NO)
```

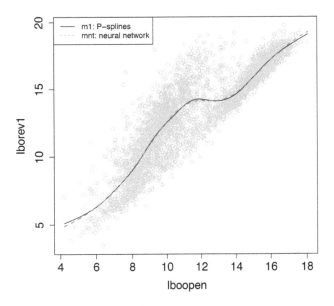

FIGURE 2.7: Neural network fit: a plot of the `film90` data together with the fitted smooth mean functions of model `m1` fitted by `pb()` (continuous line) and the neural network model `mnt` fitted by `nn()` (dashed line).

```
## GAMLSS-RS iteration 1: Global Deviance = 14187.07
## . . .
## GAMLSS-RS iteration 3: Global Deviance = 14124.96
```

This fits a neural network model with one covariate and 20 hidden variables. The `decay` argument is used for penalizing the fitted coefficients. The fitted values of models `mnt` and `m1` are displayed in Figure 2.7.

```
plot(lborev1~lboopen, col="lightblue", data=film90)
lines(fitted(m1)[order(film90$lboopen)]~
                film90$lboopen[order(film90$lboopen)])
lines(fitted(mnt)[order(film90$lboopen)]~
                film90$lboopen[order(film90$lboopen)],
      col="red", lty=2,  lwd=2)
legend("topleft",legend=c("m1: P-splines","mnt: neural network"),
      lty=1:2,col=c("black","red"),cex=1)
```

The function `getSmo()` is used to get more information about the fitted neural network model. This function retrieves the last fitted object within the backfitting GAMLSS algorithm (in this case a `"nnet"` object). Reserved methods such as `print()`, `summary()` or `coef()` can be used to get information for the objects. Here we retrieve its 61 coefficients. (There are 40 parameters from the relationship between the 20 hidden variables and the explanatory variable (constant and slope parameters), together with 21 parameters from the relationship between the response variable and the 20 hidden variables (constant and 20 slope parameters).)

```
coef(getSmo(mnt))
```

```
##          b->h1            i1->h1             b->h2             i1->h2
##   6.7845145255 -0.7598112866   1.2732354424 -0.1585005113
##  . . .
```

2.3.3 Extracting fitted values

Fitted values of the distribution parameters of a GAMLSS model (for all cases) can
be obtained using the `fitted()` function. For example

```
plot(lboopen, fitted(m1,"mu"))
```

will plot the fitted values of μ distribution parameter against x (`lboopen`). The constant estimated scale parameter (the standard deviation of the normal distribution
in this case) can be obtained:

```
fitted(m1,"sigma")[1]
```

```
##          1
## 1.392632
```

where [1] indicates the first element of the vector. The same value can be obtained
using the more general function `predict()`:

```
predict(m1,what="sigma", type="response")[1]
```

```
##          1
## 1.392632
```

The function `predict()` can also be used to predict the response variable distribution parameters for both old and new data values of the explanatory variables.
This is explained in Section 5.4.

One of the flexibilities offered by GAMLSS is the modelling of all the distribution
parameters (rather than just μ). This means that the scale and shape of the distribution can vary as a (linear or smooth) function of explanatory variables. Below,
we show how to model both μ and σ of a normal response distribution. Figure 2.1
suggests that this flexibility of a GAMLSS model might be required.

2.3.4 Modelling both μ and σ

To model the predictors of both the mean μ and the scale parameter σ as nonparametric smoothing P-spline functions of `lboopen` (with a normal response distribution) use:

```
m3 <- gamlss(lborev1~pb(lboopen),sigma.formula=~pb(lboopen),
             data=film90, family=NO)
edfAll(m3)
```

```
## GAMLSS-RS iteration 1: Global Deviance = 12263.21
## . . .
## GAMLSS-RS iteration 4: Global Deviance = 12263.54
## $mu
## pb(lboopen)
##      12.1442
##
## $sigma
## pb(lboopen)
##      10.67769
```

R
. 47

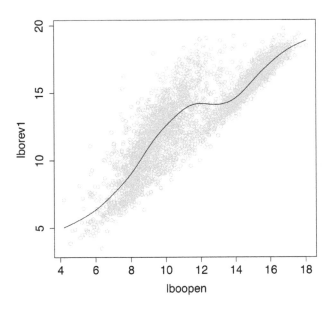

FIGURE 2.8: The `film90` data with the fitted smooth mean function of model `m3`, in which both the mean and variance models are fitted using `pb(lboopen)`.

The function `edfAll()` is used to obtain the effective degrees of freedom for all the distribution parameters. These are 12.14 and 10.68 for μ and σ, respectively. The fitted model for μ is displayed in Figure 2.8.

Fig.
2.8

```
plot(lborev1~lboopen, col="lightblue", data=film90)
lines(fitted(m3)[order(film90$lboopen)]~
            film90$lboopen[order(film90$lboopen)])
```

2.3.5 Diagnostic plots

Once a GAMLSS model is fitted, it is important to assess the adequacy of the fitted model by examining the model residuals. See Chapter 12 for more details. The function `resid()` (or `residuals()`) can be used to obtain the fitted (normalized randomized quantile) residuals of a model, referred to as residuals throughout this book. See Dunn and Smyth [1996] and Chapter 12 for more details. Residual plots are graphed using `plot()`:

```
plot(m3)
```

Fig. 2.9

```
## ***************************************************************
##          Summary of the Quantile Residuals
##                         mean    =  0.0006979142
##                     variance    =  1.000248
##           coef. of skewness     =  0.5907226
##           coef. of kurtosis     =  3.940587
## Filliben correlation coefficient =  0.9909749
```

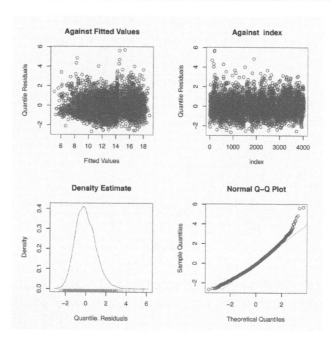

R

p. 48

FIGURE 2.9: Residual plots from the fitted normal model `m3`, using `pb(lboopen)` for both μ and $\log(\sigma)$.

Figure 2.9 shows plots of the residuals: (top left) against the fitted values of μ; (top right) against an index (i.e. case number); (bottom left) a nonparametric kernel density estimate; (bottom right) a normal Q-Q plot. Note that the `plot()` function does not produce additive term plots (as it does, for example, in the `gam()` function of **mgcv**). The function which does this in the **gamlss** package is `term.plot()`.

The worm plot (see Section 12.4) is a de-trended normal Q-Q plot of the residuals. Model inadequacy is indicated when many points plotted lie outside the (dotted) point-wise 95% confidence bands or when the points follow a systematic shape. The worm plot is obtained using `wp()`:

```
wp(m3)

## Warning in wp(m3): Some points are missed out
## increase the y limits using ylim.all

title("(a)")
```

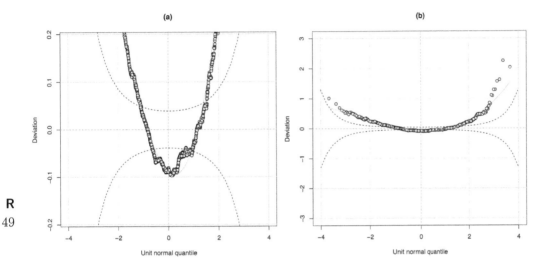

FIGURE 2.10: Worm plots from model **m3**.

To include all points in the worm plot, change the "Deviation" axis range by increasing the value of `ylim.all` until all points are included in the plot (avoiding a warning message):

```
wp(m3, ylim.all=3)
title("(b)")
```

Since there is no warning message, all points have been included in the worm plot. Model inadequacy is indicated by the fact that many points lie outside the 95% confidence bands and the points have a systematic quadratic type shape.

2.3.6 Fitting different distributions

One of the most important modelling decisions for a GAMLSS model is the choice of the distribution for the response variable. See Chapter 6 for a discussion of available distributions in GAMLSS. To use a distribution other than the normal (the default), use the `family` option of `gamlss()`. For example, to fit the Box–Cox Cole and Green (BCCG), a three-parameter continuous distribution, use:

```
m5 <-gamlss(lborev1~pb(lboopen), sigma.formula=~pb(lboopen),
            nu.formula=~pb(lboopen), data=film90, family=BCCG)
```

```
## GAMLSS-RS iteration 1: Global Deviance = 11888.56
## . . .
## GAMLSS-RS iteration 5: Global Deviance = 11809.64
```

To fit the Box–Cox power exponential (BCPE) distribution, a four-parameter continuous distribution:

```
m6 <-gamlss(lborev1~pb(lboopen), sigma.formula=~pb(lboopen),
            nu.formula=~pb(lboopen), tau.formula=~pb(lboopen),
            data=film90, start.from=m5, family=BCPE)
```

```
## GAMLSS-RS iteration 1: Global Deviance = 11738.54
## . . .
## GAMLSS-RS iteration 20: Global Deviance = 11733.63
```

Note that we have used the argument `start.from=m5` to start the iterations from
the previous fitted `m5` model. The details of all the distributions currently available
in `gamlss()` are given in Rigby et al. [2017].

2.3.7   Selection between models

Once different models in GAMLSS have been fitted (either by using different distributions and/or smoothing terms), models may be selected by using, for example, an
information criterion. See Chapter 11 for model selection techniques in GAMLSS.

For example, different models can be compared by a test based on their global
deviances: $\text{GDEV} = -2\hat{\ell}$ (if they are nested), or by selecting the model with lowest
generalized Akaike information criterion: $\text{GAIC} = -2\hat{\ell} + \kappa \cdot \text{df}$, where $\hat{\ell}$ is the
fitted log-likelihood function, df is the effective degrees of freedom used in the fitted
model and κ is a required penalty, e.g. $\kappa = 2$ for the AIC, $\kappa = \log n$ for the SBC,
or $\kappa = 3.84$ (corresponding to a Chi-squared test with one degree of freedom for
a single parameter). The function `deviance()` provides the global deviance of the
model.

Note that the `gamlss()` global deviance is different from the deviance provided
by `glm()` and `gam()`; see Section 1.4. The global deviance is *exactly* minus twice
the fitted log-likelihood function, *including* all constant terms in the log-likelihood.
The `glm()` deviance is calculated as a deviation from the saturated model. It does
not include 'constant' terms (which do not depend on the mean of distribution
but do depend on the scale parameter) in the fitted log-likelihood, and so cannot
be used to compare different distributions. The functions `AIC()` or `GAIC()` (which
are identical) are used to obtain the generalized Akaike information criterion. For
example to compare the models `m0` to `m6`:

```
GAIC(m0,m1,m2,m3,m4,m5,m6)
```

```
##          df      AIC
## m6 44.97879 11823.59
## m5 36.06436 11881.77
## m3 22.82189 12309.19
## m2 12.99817 14133.72
## m1 12.73672 14135.05
## m4 10.08556 14139.34
## m0  5.00000 14528.26
```

GAIC() uses default penalty $\kappa = 2$, resulting in the AIC. Hence according to the AIC model m6 is selected as best (smallest value of AIC). To change the penalty in GAIC() use the argument k:

```
GAIC(m0,m1,m2,m3,m4,m5,m6, k=log(4031))
```

```
##          df      AIC
## m6 44.97879 12107.03
## m5 36.06436 12109.04
## m3 22.82189 12453.00
## m4 10.08556 14202.89
## m1 12.73672 14215.32
## m2 12.99817 14215.63
## m0  5.00000 14559.77
```

In this case with GAIC ($\kappa = \log n$) we have the SBC. Models selected using SBC are generally simpler than those selected using AIC.

Other model selection criteria based on training, validation and test samples are discussed on Chapter 11.

Chosen Model

Using the AIC, model m6 is selected with $Y = \text{lborev} \sim \text{BCPE}(\mu, \sigma, \nu, \tau)$ where each of μ, σ, ν and τ are modelled as smooth functions of $x = \text{lboopen}$. The fitted smooth functions for μ, σ, ν and τ for models m5 and m6 are shown in Figure 2.11.

Fig. `fittedPlot(m5, m6, x=film90$lboopen, line.type = TRUE)`
2.11

Since, in this example, only one explanatory variable is used in the fit, centile estimates for the fitted distribution can be shown using the functions centiles() or centiles.fan().

Fig. `centiles.fan(m6,  xvar=film90$lboopen,  cent=c(3,10,25,50,75,90,97),`
2.12 `colors="terrain",ylab="lborev1", xlab="lboopen")`

Figure 2.12 shows centile curves for lborev1 against lboopen from the fitted model

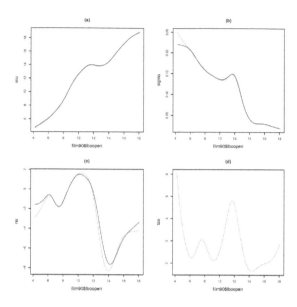

R

p. 51

FIGURE 2.11: A plot of the smooth fitted values for all the parameters (a) μ, (b) σ, (c) ν and (d) τ from models m5 (dashed line) and m6 (continuous line). The distribution for model m5, BCCG, has only three parameters so does not appear in panel (d).

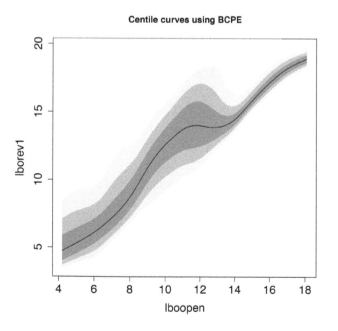

R

p. 51

FIGURE 2.12: Centile fan plot for the m6 model showing the 3%, 10%, 25%, 50%, 75%, 90% and 97% centiles for the fitted BCPE distribution.

m6. For example the lowest curve is the fitted 3% centile curve, defined by 3% of the values of `lborev1` lying below the curve for each value of `lboopen`, for the fitted model `m6` if it was the correct model. For more details on centile curves see Chapter 13. Figure 2.13 also shows how the fitted conditional distribution for the response variable `lborev1` changes according to variable `lboopen`. The function `plotSimpleGamlss()` from the package **gamlss.util** is used here.

```
library(gamlss.util)
library(colorspace)
plotSimpleGamlss(lborev1,lboopen, model=m6,    data=film90,
               x.val=seq(6,16,2), val=5, N=1000, ylim=c(0,25),
               cols=heat_hcl(100))
```

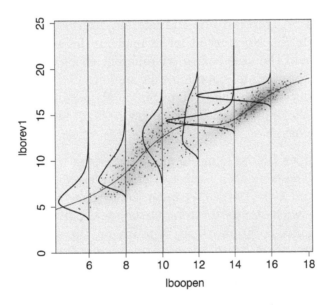

FIGURE 2.13: Fitted conditional distribution of the response variable `lborev1`, showing how it changes for different values of the covariate `lboopen`.

Figure 2.13 highlights how the fitted conditional distribution of `lborev1` changes with `lboopen`. This is the essence of GAMLSS modelling.

Important: Within GAMLSS, the shape of the conditional distribution of the response variable can vary according to the values of the explanatory variables.

2.4   Bibliographic notes

The **R** software was created by Ross Ihaka and Robert Gentleman at the University of Auckland, New Zealand, and is currently developed by the R Development Core Team [R Core Team, 2016]. **R** is an implementation of the **S** language (combined with lexical scoping semantics) which was created by John Chambers while he was working at Bell Labs in the 1980s [Becker and Chambers, 1984]. The source code for **R** is written primarily in C, Fortran and **R**. The beauty of **R** as a language is it allows the creation of libraries (or packages as they are known within **R**), for adding new methods and techniques. The pdf file within the **R** help system 'Writing **R** Extensions' provides information on the creation of packages. The **gam** package was created for `Splus` by Trevor Hastie and later transferred to **R** [Chambers and Hastie, 1992, Hastie, 2015]. No method for estimating the smoothing parameters was available in **gam** and the user had to try different effective degrees of freedom in order to find the appropriate smoothing parameter. Estimation of smoothing parameters was introduced with the work of Simon Wood and the creation of the **mgcv** package [Wood, 2006a]. In addition the **mgcv** package also includes smoothers of more than one dimension.

Fitting polynomial curves can be found in many regression textbooks, for example Draper et al. [1966]. The 'sandwich' standard errors ideas can be found in Huber [1967], White [1980] and Cox [1961]. Freedman [2006] gives a more comprehensive description of the sandwich standard errors. Orthogonal polynomials and their estimated diagonal variance-covariance matrix touch on the subject of 'parameter orthogonality'. Cox and Reid [1987] is a good reference for the subject. In general parameter orthogonality helps estimation and makes inference easier. However sometimes the interpretation of the results are more difficult, if the orthogonal parameters are less interpretable than the original parameters.

P-splines were introduced by Eilers and Marx [1996] and their popularity has grown steadily since then because of their simplicity and ease of implementation. For a review of the P-splines methodology and comparison with other methods, see Eilers et al. [2015]. More detail on cubic smoothing splines can be found in, inter alia, Hastie and Tibshirani [1990], Green and Silverman [1994], and Wood [2006a]. The loess curves were introduced by Cleveland [1979], and they are still one of the most reliable ways of smoothing based on local polynomial regression. For more discussion on loess, see Chapter 9. Neural networks are parameterized nonlinear regression models. Their overparameterization makes them very flexible and capable of approximating any smooth function. More details on neural networks can be found in Ripley [1993], Ripley [1996] and Bishop [1995]. See Hastie et al. [2009] for a comparison of neural networks with other similar methods and models.

2.5   Exercises

1. **The `film90` example:** Familiarize yourself with the **gamlss** functions and packages by repeating the commands given in this chapter analyzing the `film90` data.

2. **The abdom data:** Information on the abdominal data is given on page 90. Fit different response distributions and choose the 'best' model according to the GAIC criterion:

   (a) Load the **abdom** data, print the variable names and plot the data.

   (b) Fit the normal distribution model, using `pb()` to fit P-spline smoothers for the predictors for μ and σ with automatic selection of smoothing parameters:

   `mNO<- gamlss(y~pb(x), sigma.fo=~pb(x), data=abdom, family=NO)`

   (c) Try fitting alternative distributions:

        i. two-parameter distributions: GA, IG, GU, RG, LO,

        ii. three-parameter distributions: PE, TF, BCCG,

        iii. four-parameter distributions: BCT, BCPE.

   Apply `pb()` to all parameters of each distribution. Make sure to use different model names.

   (d) Compare the fitted models using GAIC with each of the penalties k=2, k=3 and k=log(length(abdom$y)), e.g.

   `GAIC(mNO,mGA,mIG,mGU,mRG,mLO,mPE,mTF,mBCCG,mBCT,mBCPE,k=2)`

   (e) Check the residuals for your chosen model, say m, by `plot(m)` and `wp(m)`.

   (f) For a chosen model, say m, look at the total effective degrees of freedom used in the fitted model `edfAll(m)`, plot the fitted parameters, `fittedPlot(m,x=abdom$x)`, and plot the data by `plot(y~x,data=abdom)`, and the fitted μ against x by `lines(fitted(m)~x, data=abdom)`.

   (g) For a chosen model, examine the centile curves using `centiles(m,abdom$x)`.

3. **The `fabric` data:** The data are 32 observations on faults in rolls of fabric.

   R data file: `fabric` in package **gamlss.data** of dimensions 32×3

   variables

        `leng` : the length of the fabric roll

y : the number of faults in the roll of the fabric

x the log of the length of the roll

purpose: to fit count data distributions and choose the 'best' model (according to GAIC criterion).

(a) Load the **fabric** data, print the variable names and plot the data.

(b) Fit the Poisson distribution model using first a linear term in **x**, and then a smooth function in **x**, i.e. **pb(x)**:

```
mPO1<-gamlss(y~x, data=fabric, family=PO)
mPO2<-gamlss(y~pb(x), data=fabric, family=PO)
```

(c) Compare the models using the AIC.

(d) Check the residuals of your chosen model.

(e) Try fitting alternative distributions instead of the Poisson,

    i. two-parameter distributions: **NBI, NBII, PIG, ZIP, ZAP**,

    ii. three-parameter distributions: **SICHEL, ZINBI, ZANBI**.

Apply **pb()** to all parameters of each distribution. Make sure to use different model names.

(f) Compare the fitted models using GAIC with each of the penalties **k=2, k=3** and **k = log(length(fabric$y))**.

(g) For a chosen model look at the total effective degrees of freedom, plot the fitted parameters and plot the data and fitted μ against **x**.

(h) Check the model using diagnostic plots.

Part II

Algorithms, functions and inference

3

The algorithms

CONTENTS

3.1 Introduction

This chapter

- defines the random effects GAMLSS model;

- describes the two algorithms for maximizing the penalized log-likelihood function with respect to the fixed effect parameters β and the random effect parameters γ; and

- describes local and global methods for estimation of the hyperparameters λ.

The material provided here will help the user to get an inside view of how the fitting algorithms of GAMLSS work.

The GAMLSS model was first introduced in Section 1.7 as

$$\mathbf{Y} \overset{\text{ind}}{\sim} \mathcal{D}(\boldsymbol{\mu},\boldsymbol{\sigma},\boldsymbol{\nu},\boldsymbol{\tau})$$
$$\eta_1 = g_1(\boldsymbol{\mu}) = \mathbf{X}_1\boldsymbol{\beta}_1 + s_{11}(\mathbf{x}_{11}) + \ldots + s_{1J_1}(\mathbf{x}_{1J_1})$$
$$\eta_2 = g_2(\boldsymbol{\sigma}) = \mathbf{X}_2\boldsymbol{\beta}_2 + s_{21}(\mathbf{x}_{21}) + \ldots + s_{2J_2}(\mathbf{x}_{2J_2})$$
$$\eta_3 = g_3(\boldsymbol{\nu}) = \mathbf{X}_3\boldsymbol{\beta}_3 + s_{31}(\mathbf{x}_{31}) + \ldots + s_{3J_3}(\mathbf{x}_{3J_3}) \quad (3.1)$$
$$\eta_4 = g_4(\boldsymbol{\tau}) = \mathbf{X}_4\boldsymbol{\beta}_4 + s_{41}(\mathbf{x}_{41}) + \ldots + s_{4J_4}(\mathbf{x}_{4J_4})$$

where $\mathcal{D}(\boldsymbol{\mu},\boldsymbol{\sigma},\boldsymbol{\nu},\boldsymbol{\tau})$ is the distribution of the response variable $\mathbf{Y}$, $\mathbf{X}_k$ are the design matrices incorporating the linear additive terms in the model (see Chapter 8), $\boldsymbol{\beta}_k$ are the linear coefficient parameters and $s_{kj}(\mathbf{x}_{kj})$ represent smoothing functions for explanatory variables $\mathbf{x}_{kj}$, for $k = 1,2,3,4$ and $j = 1,\ldots,J_k$. Note that the quantitative explanatory variables in the $\mathbf{X}$'s can be the same or different from the $\mathbf{x}_{kj}$ in the smoothers. The vectors $\boldsymbol{\eta}_1$, $\boldsymbol{\eta}_2$, $\boldsymbol{\eta}_3$ and $\boldsymbol{\eta}_4$ are called the predictors of $\boldsymbol{\mu}$, $\boldsymbol{\sigma}$, $\boldsymbol{\nu}$ and $\boldsymbol{\tau}$, respectively.

It turns out that most of the smooth functions used within GAMLSS can be written in the form $s(\mathbf{x}) = \mathbf{Z}\boldsymbol{\gamma}$ where $\mathbf{Z}$ is the basis matrix which depends on the explanatory variable $\mathbf{x}$ (see Chapter 8.2 for the definition of a basis). $\boldsymbol{\gamma}$ is a parameter vector to be estimated, subject to a quadratic penalty of the form $\lambda\boldsymbol{\gamma}^\top\mathbf{G}\boldsymbol{\gamma}$, for a known matrix $\mathbf{G} = \mathbf{D}^\top\mathbf{D}$ and where the hyperparameter λ regulates the amount of smoothing needed for the fit. We shall refer to functions in this form as *penalized smooth functions* (or *penalized smoothers*). Penalized smoothers are the subject of Chapter 9, where it is shown that different formulations for the $\mathbf{Z}$'s and the $\mathbf{D}$'s lead to different types of smoothing functions with different statistical properties.

The model (3.1) can be generalized and written as the random effects GAMLSS model:

$$\mathbf{Y}|\boldsymbol{\gamma} \overset{\text{ind}}{\sim} \mathcal{D}(\boldsymbol{\mu},\boldsymbol{\sigma},\boldsymbol{\nu},\boldsymbol{\tau})$$
$$\eta_1 = g_1(\boldsymbol{\mu}) = \mathbf{X}_1\boldsymbol{\beta}_1 + \mathbf{Z}_{11}\boldsymbol{\gamma}_{11} + \ldots + \mathbf{Z}_{1J_1}\boldsymbol{\gamma}_{1J_1}$$
$$\eta_2 = g_2(\boldsymbol{\sigma}) = \mathbf{X}_2\boldsymbol{\beta}_2 + \mathbf{Z}_{21}\boldsymbol{\gamma}_{21} + \ldots + \mathbf{Z}_{2J_2}\boldsymbol{\gamma}_{2J_2}$$
$$\eta_3 = g_3(\boldsymbol{\nu}) = \mathbf{X}_3\boldsymbol{\beta}_3 + \mathbf{Z}_{31}\boldsymbol{\gamma}_{31} + \ldots + \mathbf{Z}_{3J_3}\boldsymbol{\gamma}_{3J_3} \quad (3.2)$$
$$\eta_4 = g_4(\boldsymbol{\tau}) = \mathbf{X}_4\boldsymbol{\beta}_4 + \mathbf{Z}_{41}\boldsymbol{\gamma}_{41} + \ldots + \mathbf{Z}_{4J_4}\boldsymbol{\gamma}_{4J_4}$$

where the 'betas' are fixed effect parameters:

$$\boldsymbol{\beta} = (\boldsymbol{\beta}_1^\top,\boldsymbol{\beta}_2^\top,\boldsymbol{\beta}_3^\top,\boldsymbol{\beta}_4^\top)^\top$$

and the 'gammas' are random effect parameters:

$$\boldsymbol{\gamma} = (\boldsymbol{\gamma}_{11}^\top,\ldots,\boldsymbol{\gamma}_{1J_1}^\top,\boldsymbol{\gamma}_{21}^\top,\ldots,\boldsymbol{\gamma}_{4J_4}^\top)^\top.$$

Assume in (3.2) that the γ_{kj}'s are independent of each other, each with (prior) distribution

$$\gamma_{kj} \sim \mathcal{N}\left(\mathbf{0}, [\mathbf{G}_{kj}(\boldsymbol{\lambda}_{kj})]^{-1}\right) \qquad (3.3)$$

where $[\mathbf{G}_{kj}(\boldsymbol{\lambda}_{kj})]^{-1}$ is the (generalized) inverse of a $q_{kj} \times q_{kj}$ symmetric matrix $\mathbf{G}_{kj}(\boldsymbol{\lambda}_{kj})$ which may depend on a vector of hyperparameters $\boldsymbol{\lambda}_{kj}$. If $\mathbf{G}_{kj}(\boldsymbol{\lambda}_{kj})$ is singular, then γ_{kj} is understood to have an improper density function proportional to $\exp\left(-\frac{1}{2}\gamma^{\top}\mathbf{G}_{kj}(\boldsymbol{\lambda}_{kj})\gamma\right)$. An important special case of (3.3) is given when $\mathbf{G}_{kj}(\boldsymbol{\lambda}_{kj}) = \lambda_{kj}\mathbf{G}_{kj}$ and $\mathbf{G}_{kj}$ is a known matrix for all k, j.

If there are no random effects in the model (3.2) it simplifies to:

$$\begin{aligned}
\mathbf{Y} &\overset{\text{ind}}{\sim} \mathcal{D}(\boldsymbol{\mu}, \boldsymbol{\sigma}, \boldsymbol{\nu}, \boldsymbol{\tau}) \\
\boldsymbol{\eta}_1 &= g_1(\boldsymbol{\mu}) = \mathbf{X}_1\boldsymbol{\beta}_1 \\
\boldsymbol{\eta}_2 &= g_2(\boldsymbol{\sigma}) = \mathbf{X}_2\boldsymbol{\beta}_2 \\
\boldsymbol{\eta}_3 &= g_3(\boldsymbol{\nu}) = \mathbf{X}_3\boldsymbol{\beta}_3 \\
\boldsymbol{\eta}_4 &= g_4(\boldsymbol{\tau}) = \mathbf{X}_4\boldsymbol{\beta}_4 \,.
\end{aligned} \qquad (3.4)$$

We refer to (3.4) as the *parametric* GAMLSS model, and to (3.2) as the *random effects* GAMLSS model.

Fitting the parametric model (3.4) requires only estimates for the 'betas' $\boldsymbol{\beta}$. Fitting the random effects GAMLSS model (3.2) requires estimates for the 'betas' $\boldsymbol{\beta}$, the 'gammas' $\boldsymbol{\gamma}$, and also the 'lambdas' $\boldsymbol{\lambda}$ (see (3.6) below), where:

$$\boldsymbol{\lambda} = (\boldsymbol{\lambda}_{11}^{\top}, \dots, \boldsymbol{\lambda}_{1J_1}^{\top}, \boldsymbol{\lambda}_{21}^{\top}, \dots, \boldsymbol{\lambda}_{4J_4}^{\top})^{\top}\,.$$

Within **gamlss**, the parametric GAMLSS model is fitted by maximum likelihood estimation with respect to $\boldsymbol{\beta}$, while the more general random effects model is fitted by maximum penalized likelihood estimation (or equivalently posterior mode or maximum a posteriori (MAP), estimation, see Rigby and Stasinopoulos [2005]) with respect to $\boldsymbol{\beta}$ and $\boldsymbol{\gamma}$ for fixed $\boldsymbol{\lambda}$. The log-likelihood function for model (3.4), under the assumption that observations are independent, is given by

$$\ell = \sum_{i=1}^{n} \log f(y_i \mid \mu_i, \sigma_i, \nu_i, \tau_i)\,. \qquad (3.5)$$

The penalized log-likelihood function for model (3.2) is given by

$$\ell_p = \ell - \frac{1}{2}\sum_{k=1}^{4}\sum_{j=1}^{J_k}\gamma_{kj}^{\top}\mathbf{G}_{kj}(\boldsymbol{\lambda}_{kj})\gamma_{kj}\,. \qquad (3.6)$$

The two basic algorithms for fitting the parametric model with respect to $\boldsymbol{\beta}$, and the nonparametric model with respect to $\boldsymbol{\beta}$ and $\boldsymbol{\gamma}$ for fixed $\boldsymbol{\lambda}$, are the RS and the

CG algorithms. For the parametric model these produce maximum likelihood estimators for β, whereas for the nonparametric model they produce MAP (maximum a posteriori) estimators for β and γ for fixed λ.

Section 3.2 explains how to estimate β and γ for fixed λ; Section 3.3 covers the MAP estimators for β and γ; and Section 3.4 gives methods for estimating the smoothing parameters λ.

3.2   Estimating β and γ for fixed λ

Rigby and Stasinopoulos [2005] provided two basic algorithms for maximizing the penalized log-likelihood (3.6) with respect to β and γ for given λ:

- The RS algorithm, which is a generalization of the algorithm used by Rigby and Stasinopoulos [1996a,b] for fitting mean and dispersion additive models (MADAM). This algorithm does not use the cross derivatives of the log-likelihood.

- The CG algorithm, which is a generalization of the Cole and Green [1992] algorithm. This algorithm requires information about the first and (expected or approximated) second and cross derivatives of the log-likelihood function with respect to the distribution parameters μ, σ, ν and τ.

Rigby and Stasinopoulos [2005, Appendix C] show that both algorithms lead, for given λ, to the maximum penalized log-likelihood estimates for the betas and the gammas, i.e. $\hat{\beta}$ and $\hat{\gamma}$.

Figure 3.1 demonstrates the different ways in which the two algorithms reach the (penalized) maximum likelihood parameter estimates. The contours are equal global deviance contours (equal to minus twice the log-likelihood). The two figures are generated using a random sample from a Weibull distribution, WEI(μ, σ). The RS algorithm maximizes the (penalized) log-likelihood over each of μ, σ, ν and τ in turn, cycling until convergence. For example in Figure 3.1(a) the global deviance is minimized (and hence the likelihood is maximized) over each of μ and σ in turn, alternating until convergence. The CG algorithm has the ability, since it uses the information about the cross derivatives, to jointly update (μ, σ) as demonstrated in Figure 3.1(b). On the basis of the evidence in Figure 3.1 it seems that the CG algorithm should be preferable, but in practice this is not the case. The CG algorithm is rather unstable, especially at the beginning of the iterations, and diverges easily. The RS algorithm is generally far more stable and in most cases faster, so it is used as the default. Note though that for highly correlated fitted distribution parameters the RS algorithm can be slower, and may converge early before reaching the maximum log-likelihood.

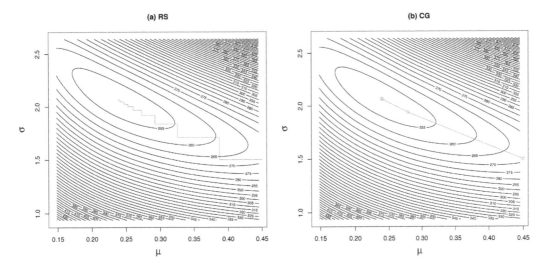

FIGURE 3.1: How the two GAMLSS algorithms (a) RS and (b) CG reach the maximum likelihood parameter estimates.

The algorithm used in `gamlss()` is specified in the `method` argument, with default `method=RS()`. The user may specify `method=CG()`, or a combination of both algorithms with `method=mixed()`; see Section 4.2. This option uses RS for the early iterations but later switches to CG, and is recommended for highly correlated fitted distribution parameters.

Next we describe the two algorithms in more detail.

3.2.1 The RS algorithm

The RS algorithm can be described using the following three nested components:

- the *outer iteration*, depicted in Figure 3.2, which calls

- the *inner iteration* (or local scoring or GLIM algorithm), depicted in Figure 3.3, which calls

- the *modified backfitting* algorithm, depicted in Figure 3.4.

The outer iteration repeatedly calls the inner iteration, which in turn repeatedly calls the modified backfitting algorithm. Convergence occurs when all three algorithms have converged.

3.2.1.1 The outer iteration (GAMLSS iteration)

Figure 3.2 describes the outer iteration diagrammatically. After initialization of the parameter vectors, say $(\hat{\boldsymbol{\mu}}, \hat{\boldsymbol{\sigma}}, \hat{\boldsymbol{\nu}}, \hat{\boldsymbol{\tau}}) = (\boldsymbol{\mu}_0, \boldsymbol{\sigma}_0, \boldsymbol{\nu}_0, \boldsymbol{\tau}_0)$, the outer iteration proceeds as follows:

1. fit a model for $\boldsymbol{\mu}$ (i.e. maximize the (penalized) log-likelihood over $\boldsymbol{\mu}$) given the latest estimates $\hat{\boldsymbol{\sigma}}$, $\hat{\boldsymbol{\nu}}$ and $\hat{\boldsymbol{\tau}}$, then

2. fit a model for $\boldsymbol{\sigma}$ given the latest estimates $\hat{\boldsymbol{\mu}}$, $\hat{\boldsymbol{\nu}}$ and $\hat{\boldsymbol{\tau}}$, then

3. fit a model for $\boldsymbol{\nu}$ given the latest estimates $\hat{\boldsymbol{\mu}}$, $\hat{\boldsymbol{\sigma}}$ and $\hat{\boldsymbol{\tau}}$, and finally

4. fit a model for $\boldsymbol{\tau}$ given the latest estimates $\hat{\boldsymbol{\mu}}$, $\hat{\boldsymbol{\sigma}}$ and $\hat{\boldsymbol{\nu}}$.

It then calculates the global deviance (equal to minus twice the current fitted log-likelihood). If the global deviance has converged then the algorithm stops, otherwise it repeats the process.

Note that the algorithm only needs initial values for the distribution parameter vectors $\boldsymbol{\mu}$, $\boldsymbol{\sigma}$, $\boldsymbol{\nu}$ and $\boldsymbol{\tau}$ rather than for the $\boldsymbol{\beta}$ parameters. The algorithm has generally been found to be stable and fast using very simple starting values (e.g. constants) for the parameter vectors. Default starting values can be changed by the user if necessary.

3.2.1.2 The inner iteration (GLM or GLIM iteration)

For convenience, the notation $\boldsymbol{\theta}_1 = \boldsymbol{\mu}$, $\boldsymbol{\theta}_2 = \boldsymbol{\sigma}$, $\boldsymbol{\theta}_3 = \boldsymbol{\nu}$, $\boldsymbol{\theta}_4 = \boldsymbol{\tau}$ will be used. For each fitting of a distribution parameter $\boldsymbol{\theta}_k$, the inner iteration is used. This is a local scoring algorithm very similar to the one used to fit generalized linear models (GLMs). This also explains the name 'GLIM algorithm'. GLIM [Francis et al., 1993] was a computer package belonging to the Royal Statistical Society, suitable for fitting GLMs. The first version of **gamlss** was written in the late 1990s in GLIM, which by now is almost an extinct species.

The idea of the local scoring algorithm is repeated weighted fits to a modified response variable, using modified weights, until convergence of the global deviance. This procedure within the GLM literature is known as iteratively reweighted least squares [McCullagh and Nelder, 1989].

The modified (iterative) response variable (sometimes called the *working variable*) for fitting the parameter θ_k is given by

$$\mathbf{z}_k = \boldsymbol{\eta}_k + \mathbf{w}_k^{-1} \circ \mathbf{u}_k \tag{3.7}$$

where

- $\mathbf{z}_k$, $\boldsymbol{\eta}_k$, $\mathbf{w}_k$ and $\mathbf{u}_k$ are all vectors of length n, e.g. weights vector $\mathbf{w}_k = (w_{k1}, w_{k2}, \ldots, w_{kn})^\top$,

- $\mathbf{w}_k^{-1} \circ \mathbf{u}_k = (w_{k1}^{-1} u_{k1}, w_{k2}^{-1} u_{k2}, \ldots, w_{kn}^{-1} u_{kn})^\top$ is the Hadamard element by element product,

- $\boldsymbol{\eta}_k = g_k(\boldsymbol{\theta}_k)$ is the predictor vector of the kth parameter vector $\boldsymbol{\theta}_k$ for $k =$

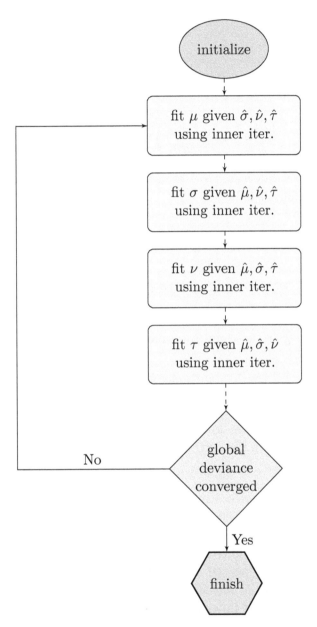

FIGURE 3.2: The outer iteration within the GAMLSS RS algorithm.

1, 2, 3, 4, corresponding to parameter vectors $\boldsymbol{\mu}$, $\boldsymbol{\sigma}$, $\boldsymbol{\nu}$ and $\boldsymbol{\tau}$, respectively, and

$$\mathbf{u}_k = \frac{\partial \ell}{\partial \boldsymbol{\eta}_k} = \left(\frac{\partial \ell}{\partial \boldsymbol{\theta}_k}\right) \circ \left(\frac{d\boldsymbol{\theta}_k}{d\boldsymbol{\eta}_k}\right) \tag{3.8}$$

is the score function (the first derivative of the log-likelihood with respect to the predictor).

Note that $\partial \ell / \partial \boldsymbol{\eta}_k$, $\partial \ell / \partial \boldsymbol{\theta}_k$ and $d\boldsymbol{\theta}_k/d\boldsymbol{\eta}_k$ are vectors of length n with elements $\partial \ell_i / \partial \eta_{ki}$, $\partial \ell_i / \partial \theta_{ki}$ and $d\theta_{ki}/d\eta_{ki}$ for $i = 1, \ldots, n$, respectively. The $\mathbf{w}_k$ are the *iterative weights* for $k = 1, 2, 3, 4$ defined as:

$$\mathbf{w}_k = -\mathbf{f}_k \circ \left(\frac{d\boldsymbol{\theta}_k}{d\boldsymbol{\eta}_k}\right) \circ \left(\frac{d\boldsymbol{\theta}_k}{d\boldsymbol{\eta}_k}\right), \tag{3.9}$$

where there are three different ways to define $\mathbf{f}_k$ depending on the information available for the specific distribution:

$$\mathbf{f}_k = \begin{cases} \mathrm{E}\left(\frac{\partial^2 \ell}{\partial \boldsymbol{\theta}_k^2}\right) & \text{if the expectation exists, leading to a Fisher's scoring algorithm} \\ \frac{\partial^2 \ell}{\partial \boldsymbol{\theta}_k^2} & \text{leading to the standard Newton–Raphson scoring algorithm} \\ -\left(\frac{\partial \ell}{\partial \boldsymbol{\theta}_k}\right) \circ \left(\frac{\partial \ell}{\partial \boldsymbol{\theta}_k}\right) & \text{leading to a quasi Newton scoring algorithm,} \end{cases}$$

where $\frac{\partial^2 \ell}{\partial \boldsymbol{\theta}_k^2}$ is a vector of length n with elements $\frac{\partial^2 \ell_i}{\partial \theta_{ki}^2}$ for $i = 1, 2, \ldots, n$.

Occasionally numerical derivatives are used to define $\mathbf{f}_k$ in the quasi Newton algorithm, but this, in general, slows down the algorithm and can make it more unstable. (Note that $\frac{\partial^2 \ell}{\partial \boldsymbol{\theta}_k^2}$ is not used in the current implementation of the `gamlss()` algorithm because it can give negative weights, which are not allowed in the backfitting.)

Figure 3.3 describes the local scoring algorithm. Given the current estimates for all the distribution parameter vectors $\hat{\boldsymbol{\mu}}$, $\hat{\boldsymbol{\sigma}}$, $\hat{\boldsymbol{\nu}}$ and $\hat{\boldsymbol{\tau}}$, the iterative weights, $\mathbf{w}_k$, and iterative working variable $\mathbf{z}_k$ for the current distribution parameter vector $\boldsymbol{\theta}_k$ are recalculated and used in a weighted fit against all the explanatory variables needed for this parameter. This is repeated until there is no change in the global deviance. (Note that other distribution parameter vectors, $\boldsymbol{\theta}_s$ for $s \neq k$, are fixed at their current values throughout the inner iteration).

There are two tuning methods within the inner iteration algorithm to avoid over-jumping (i.e. going further away from the maximum). Both of them adjust the predictor η. The first method is based on the *step* parameter $0 < \phi \leq 1$, which can be specified by the arguments `mu.step`, `sigma.step`, `nu.step` and `tau.step`. The second is automatically activated if the `autostep` argument is set to `TRUE`, which is the default value. To demonstrate how they work, let η_o, η_f and η_n be the

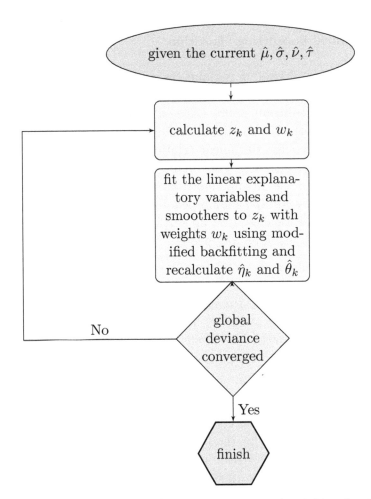

FIGURE 3.3: The inner iteration (or GLIM iteration) within the GAMLSS RS algorithm.

predictors from the previous iteration fit, the current iteration fit and the proposed new predictor, respectively. The first method uses

$$\eta_n = \phi \eta_f + (1 - \phi)\eta_o$$

for the predictor. The default value for each step parameter ϕ is 1. The second method automatically halves the step (up to 5 times) to $\eta_n = (\eta_f + \eta_o)/2$ if the deviance is still increasing.

3.2.1.3 The modified backfitting algorithm

Estimation of the beta and gamma parameters is done within the modified backfitting part of the algorithm. The backfitting algorithm is a version of the Gauss–Seidel algorithm [Hastie and Tibshirani, 1990]. (The entire RS algorithm may be said to be a Gauss–Seidel algorithm). The modification is that for most penalized smoothers, the design matrix $\mathbf{X}$ used for the linear relationships contains the linear part of the relevant explanatory variable. That helps the convergence of the algorithm. The components that the backfitting algorithm needs are (i) a good weighted least squares (WLS) algorithm, and (ii) a good penalized weighted least squares (PWLS) algorithm. (In Section 9.4 we show that all smoothers with a quadratic penalty can be fitted by least squares using an augmented data model.)

The backfitting algorithm works as follows. We wish to fit linear explanatory variables and smoothers to $\mathbf{z}_k$ with working weights $\mathbf{w}_k$ using backfitting (within the inner iteration for updating the estimate of distribution parameter $\boldsymbol{\theta}_k$). Figure 3.4 demonstrates how the process works within the RS algorithm. $\mathbf{X}_k$ represents the design matrix for the linear part of the model with coefficients $\boldsymbol{\beta}_k$, and for simplicity we assume only two smoothers with parameter sets $\boldsymbol{\gamma}_{k1}$ and $\boldsymbol{\gamma}_{k2}$ with basis matrices $\mathbf{Z}_{k1}$ and $\mathbf{Z}_{k2}$, respectively.

For given iterative weights $\mathbf{w}_k$, working response variable $\mathbf{z}_k$ and previously initialized or estimated values for the coefficients of the two smoothers $\hat{\boldsymbol{\gamma}}_{k1}$ and $\hat{\boldsymbol{\gamma}}_{k2}$, calculate the partial residuals $\boldsymbol{\varepsilon}$ for the beta parameters $\boldsymbol{\beta}_k$ (equivalently offsetting for $\hat{\boldsymbol{\gamma}}_{k1}$ and $\hat{\boldsymbol{\gamma}}_{k2}$) and fit a WLS to the residuals to obtain a new estimate for $\hat{\boldsymbol{\beta}}_k$. Now obtain the partial residual with respect to the first smoother and use PWLS to obtain a new estimate of $\hat{\boldsymbol{\gamma}}_{k1}$. Then obtain the partial residual with respect to the second smoother and use PWLS to obtain a new estimate of $\hat{\boldsymbol{\gamma}}_{k2}$. Repeat the process until the $\hat{\boldsymbol{\beta}}_k$, $\hat{\boldsymbol{\gamma}}_{k1}$ and $\hat{\boldsymbol{\gamma}}_{k2}$ converge.

The question arises here as to why backfitting was used, rather than trying to fit both linear and smoother components simultaneously. This is, for example, what `gam()` in **mgcv** does. The answer is that, while this will work with penalized smoothers (i.e. smoothers using quadratic penalties) and will probably speed up the algorithm, backfitting gives us the opportunity to use other smoothers such as loess, multidimensional smoothing splines, decision trees and neural networks; see Chapter 9.

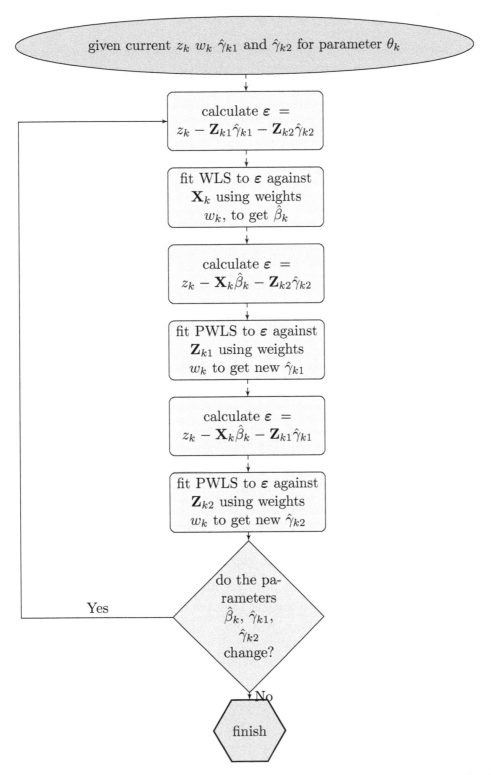

FIGURE 3.4: The modified backfitting algorithm, within the GAMLSS RS algorithm.

3.2.2 The CG algorithm

The CG algorithm is a local scoring algorithm performed within an outer iteration, an inner iteration and a modified backfitting algorithm for fitting each distribution parameter. Unlike the RS algorithm, CG needs the (expected or approximate) cross derivatives of the log-likelihood with respect to each pair of parameters of the distribution. Rigby and Stasinopoulos [2005, Appendix C] show that the CG algorithm, described below, maximizes the penalized log-likelihood (3.6) with respect to the betas and gammas for fixed $\boldsymbol{\lambda}$.

3.2.2.1 The outer iteration

In the outer iteration of the CG algorithm, the working variable and the iterative weights for the parameter vectors $\boldsymbol{\mu}$, $\boldsymbol{\sigma}$, $\boldsymbol{\nu}$ and $\boldsymbol{\tau}$ are defined (or updated) by:

$$\mathbf{z}_k = \boldsymbol{\eta}_k + \mathbf{w}_{kk}^{-1} \circ \mathbf{u}_k \ ,$$

equivalent to $\mathbf{z}_k$ defined in equation (3.7) for $k = 1, 2, 3, 4$ and the $\mathbf{w}_{ks}$ vectors contain the elements of the iterative weights, for $k = 1, 2, 3, 4$, $s = 1, 2, 3, 4$ and $s \le k$ defined by

$$\mathbf{w}_{ks} = -\mathbf{f}_{ks} \circ (\partial \boldsymbol{\theta}_k / \partial \boldsymbol{\eta}_k) \circ (\partial \boldsymbol{\theta}_s / \partial \boldsymbol{\eta}_s)$$

where $\mathbf{f}_{ks}$ is defined in one of three ways depending on the information available for the specific distribution:

$$\mathbf{f}_{ks} = \begin{cases} \mathrm{E}\left(\frac{\partial^2 \ell}{\partial \boldsymbol{\theta}_k \partial \boldsymbol{\theta}_s}\right) \\[2ex] \frac{\partial^2 \ell}{\partial \boldsymbol{\theta}_k \partial \boldsymbol{\theta}_s} \\[2ex] -\left(\frac{\partial \ell}{\partial \boldsymbol{\theta}_k}\right) \circ \left(\frac{\partial \ell}{\partial \boldsymbol{\theta}_s}\right) \end{cases} \tag{3.10}$$

and $\frac{\partial^2 \ell}{\partial \boldsymbol{\theta}_k \partial \boldsymbol{\theta}_s}$ is a vector of length n with elements $\frac{\partial^2 \ell_i}{\partial \theta_{ki} \partial \theta_{si}}$ for $i = 1, \ldots, n$. (Note that $\frac{\partial^2 \ell}{\partial \boldsymbol{\theta}_k \partial \boldsymbol{\theta}_s}$ is not used in the current implementation of the gamlss() algorithm.) Note that, for four-parameter distributions, only ten distinct vectors $\mathbf{w}_{ks}$ exist: $\mathbf{w}_{11}$, $\mathbf{w}_{12}$, $\mathbf{w}_{13}$, $\mathbf{w}_{14}$, $\mathbf{w}_{22}$, $\mathbf{w}_{23}$, $\mathbf{w}_{24}$, $\mathbf{w}_{33}$, $\mathbf{w}_{34}$, $\mathbf{w}_{44}$. The inner iteration is performed until convergence of the inner global deviance. Then the outer iteration is repeated, (updating the working variable, iterative weights and predictors) until convergence of the outer global deviance.

3.2.2.2 The inner iteration

The inner iteration process is as follows: The current working variable $\mathbf{z}_k$, current weights $\mathbf{w}_{ks}$ and current predictors denoted $\boldsymbol{\eta}_k^o$ for $k = 1, 2, 3, 4$ and $s = 1, 2, 3, 4$ are fixed. Now, for $k = 1, 2, 3, 4$, it calculates, new adjusted working variables as

$$\mathbf{z}_k' = \mathbf{z}_k + \mathbf{z}_k^a$$

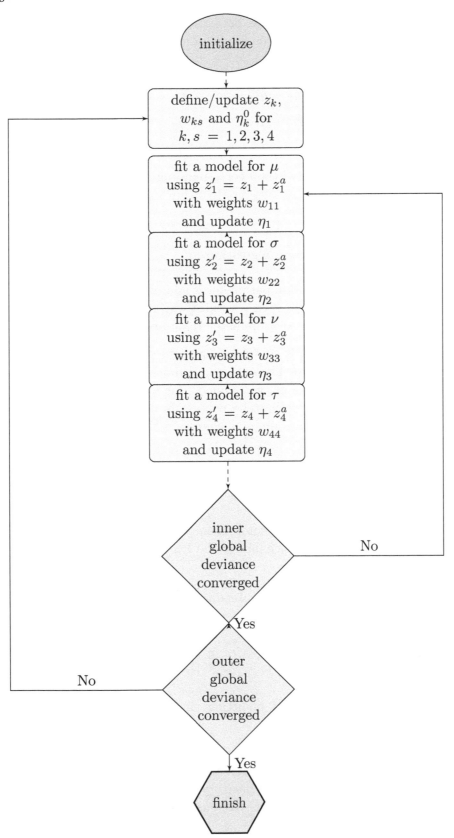

FIGURE 3.5: The outer and inner iterations within the GAMLSS CG algorithm.

where $\mathbf{z}_k^a$ is a combination of the 'cross derivatives' multiplied by the difference in the relevant predictors, defined for the four parameters as:

$$\mu : \mathbf{z}_1^a = -\mathbf{w}_{11}^{-1} \circ [\mathbf{w}_{12} \circ (\boldsymbol{\eta}_2 - \boldsymbol{\eta}_2^o) + \mathbf{w}_{13} \circ (\boldsymbol{\eta}_3 - \boldsymbol{\eta}_3^o) + \mathbf{w}_{14} \circ (\boldsymbol{\eta}_4 - \boldsymbol{\eta}_4^o)]$$

$$\sigma : \mathbf{z}_2^a = -\mathbf{w}_{22}^{-1} \circ [\mathbf{w}_{12} \circ (\boldsymbol{\eta}_1 - \boldsymbol{\eta}_1^o) + \mathbf{w}_{23} \circ (\boldsymbol{\eta}_3 - \boldsymbol{\eta}_3^o) + \mathbf{w}_{24} \circ (\boldsymbol{\eta}_4 - \boldsymbol{\eta}_4^o)]$$

$$\nu : \mathbf{z}_3^a = -\mathbf{w}_{33}^{-1} \circ [\mathbf{w}_{13} \circ (\boldsymbol{\eta}_1 - \boldsymbol{\eta}_1^o) + \mathbf{w}_{23} \circ (\boldsymbol{\eta}_2 - \boldsymbol{\eta}_2^o) + \mathbf{w}_{34} \circ (\boldsymbol{\eta}_4 - \boldsymbol{\eta}_4^o)]$$

$$\tau : \mathbf{z}_4^a = -\mathbf{w}_{44}^{-1} \circ [\mathbf{w}_{14} \circ (\boldsymbol{\eta}_1 - \boldsymbol{\eta}_1^o) + \mathbf{w}_{24} \circ (\boldsymbol{\eta}_2 - \boldsymbol{\eta}_2^o) + \mathbf{w}_{34} \circ (\boldsymbol{\eta}_3 - \boldsymbol{\eta}_3^o)].$$

Then, in a cycle, models are fitted (and corresponding predictor $\boldsymbol{\eta}_k$ updated) for μ, σ, ν and τ using the modified backfitting algorithm until convergence of the inner global deviance.

3.2.2.3 The modified backfitting algorithm

The fit of an individual distribution parameter is analogous to Section 3.2.1.3 ,using the modified backfitting algorithm with working variable $\mathbf{z}_k'$ and weights $\mathbf{w}_{kk}$ and after each fit the corresponding new predictor $\boldsymbol{\eta}_k$ is updated.

In summary the inner iteration is continued until the global deviance evaluated at this inner stage does not change. Then the algorithm returns to the outer iteration which recalculates the quantities $\mathbf{z}_k$, $\mathbf{w}_{ks}$ and $\boldsymbol{\eta}_k^o$, for $k = 1, 2, 3, 4$ and $s = 1, 2, 3, 4$, and starts the inner iteration again. The process is described in Figure 3.5. The outer iteration stops when there is no more change in the global deviance evaluated at the outer iteration stage.

3.2.3 Fish species example

The fish species data set is analyzed comprehensively in Chapter 14. Here we use it to demonstrate the differing performance of the RS and CG algorithms.

R data file: `species` in package **gamlss.data** of dimension 70×2

variables

   `fish` : the number of different species in 70 lakes in the world

   `lake` : the lake area

purpose: to demonstrate results of fitting using RS, CG and 'mixed' algorithms

The reader is referred to Section 14.2 for an analysis of the data. The response variable `fish` is a count, and the explanatory variable is `lake`. Here we use the Poisson inverse Gaussian, $\text{PIG}(\mu, \sigma)$, as response distribution, and we model both the mean μ and dispersion σ as functions of `log(lake)`. A feature of the `PIG` distribution is that μ and σ are not orthogonal parameters [Heller et al., 2016].

```
data(species)
### default estimation method=RS()
m1 <- gamlss(fish~log(lake), sigma.fo=~log(lake), family=PIG,
             data=species)

## GAMLSS-RS iteration 1: Global Deviance = 617.3683
## . . .
## GAMLSS-RS iteration 6: Global Deviance = 608.8315

m1

##
## Family:  c("PIG", "Poisson.Inverse.Gaussian")
## Fitting method: RS()
##
## Call:
## gamlss(formula = fish ~ log(lake), sigma.formula = ~log(lake),
##     family = PIG, data = species)
##
## Mu Coefficients:
## (Intercept)     log(lake)
##      2.5475        0.1444
## Sigma Coefficients:
## (Intercept)     log(lake)
##     -2.0252        0.1925
##
##   Degrees of Freedom for the fit: 4 Residual Deg. of Freedom   66
## Global Deviance:     608.831
##           AIC:       616.831
##           SBC:       625.825

### method=CG() doesn't converge in 20 iterations. More
### iterations specified in n.cyc argument.
m2 <- gamlss(fish~log(lake), sigma.fo=~log(lake), family=PIG,
             data=species, method=CG(), n.cyc=100)

## GAMLSS-CG iteration 1: Global Deviance = 1561.463
## . . .
## GAMLSS-CG iteration 34: Global Deviance = 608.8315

m2

##
## Family:  c("PIG", "Poisson.Inverse.Gaussian")
## Fitting method: CG()
```

```
##
## Call:
## gamlss(formula = fish ~ log(lake), sigma.formula = ~log(lake),
##      family = PIG, data = species, method = CG(), n.cyc = 100)
##
##
## Mu Coefficients:
## (Intercept)      log(lake)
##      2.5476          0.1444
## Sigma Coefficients:
## (Intercept)      log(lake)
##     -2.0259          0.1926
##
##  Degrees of Freedom for the fit: 4 Residual Deg. of Freedom   66
## Global Deviance:      608.831
##              AIC:      616.831
##              SBC:      625.825
```

method=mixed()
```
m3 <- gamlss(fish~log(lake), sigma.fo=~log(lake), family=PIG,
             data=species, method=mixed(1,100))
```

```
## GAMLSS-RS iteration 1: Global Deviance = 617.3683
## GAMLSS-CG iteration 1: Global Deviance = 652.0548
## . . .
## GAMLSS-CG iteration 9: Global Deviance = 608.8648
## GAMLSS-CG iteration 10: Global Deviance = 608.8339
## GAMLSS-CG iteration 11: Global Deviance = 608.8316
## GAMLSS-CG iteration 12: Global Deviance = 608.8315
```

```
m3
```

```
##
## Family:  c("PIG", "Poisson.Inverse.Gaussian")
## Fitting method: mixed(1, 100)
##
## Call:
## gamlss(formula = fish ~ log(lake), sigma.formula = ~log(lake),
##      family = PIG, data = species, method = mixed(1,
##          100))
##
## Mu Coefficients:
## (Intercept)      log(lake)
##      2.5475          0.1444
## Sigma Coefficients:
## (Intercept)      log(lake)
```

```
##        -2.0259           0.1926
##
##   Degrees of Freedom for the fit: 4 Residual Deg. of Freedom    66
## Global Deviance:        608.831
##               AIC:      616.831
##               SBC:      625.825
```

The RS, CG and mixed RS-CG algorithms arrive at the same solution, to three decimal places, all with global deviance of 608.831. RS reaches the solution in six iterations, CG in 34 iterations, and mixed RS-CG performs one RS and 12 CG iterations.

3.2.4   Remarks on the GAMLSS algorithms

The following are general comments related to the fitting algorithms RS and CG:

1. Both RS and CG algorithms can be easily implemented in any computer program which has a good (penalized) weighted linear least squares algorithm.

2. The fitting procedure is modular, making checking easy.

3. Additional distributions can be added easily, since their contribution comes through the first and exact or approximate expected second (and optionally cross) derivatives and therefore is orthogonal to the main algorithm.

4. The modified backfitting (Gauss–Seidel) algorithm can be adapted easily to fit any extra additive terms, including terms which are not necessarily based on quadratic penalties, as long as the algorithm or method used has weights.

5. Starting values are easily found, requiring initial values for $\boldsymbol{\theta} = (\boldsymbol{\mu}, \boldsymbol{\sigma}, \boldsymbol{\nu}, \boldsymbol{\tau})$ rather than for the $\boldsymbol{\beta}$ parameters. The algorithms have generally been found to be stable and fast using very simple starting values (e.g. constants) for the $\boldsymbol{\theta}$ parameters. Default values can be changed by the user if necessary.

6. The function `nlgamlss()` in the package **gamlss.nl** provides a third algorithm for fitting *parametric* linear or nonlinear GAMLSS models. However the algorithm needs starting values for all the $\boldsymbol{\beta}$ parameters, rather than $\boldsymbol{\theta}$, which can be difficult for the user to choose. This method uses the `nlm()` function for likelihood maximization, which uses numerical derivatives (if the actual derivatives are not provided).

7. For a specific data set and model, the (penalized) log-likelihood can potentially have multiple local maxima. This can be investigated using different starting values and has generally not been found to be a problem in the data sets analyzed, possibly due to the relatively large sample sizes often used.

8. Singularities in the likelihood function similar to the ones reported by Crisp and Burridge [1994] can potentially occur in specific cases within the GAMLSS

framework, especially when the sample size is small. For example occasionally the scale parameter σ can go towards zero. The problem can be alleviated by appropriate restrictions on the scale parameter. For example, the link function `logS`, a shifted log link from 0.00001, does not allow values less than 0.00001 to occur.

9. Introducing local methods for estimating the smoothing hyperparameters can sometimes make RS and CG more unstable, and occasionally the global deviance increases.

3.3 MAP estimators of β and γ for fixed λ

In this section, following Rigby and Stasinopoulos [2005, Appendix A.1], we will justify the GAMLSS algorithms of Section 3.2 as producing posterior mode or maximum a posteriori (MAP) estimates for the beta and gamma parameters for fixed lambdas for model (3.2) with (3.3). Let $f(y_i \,|\, \beta, \gamma)$ be the conditional probability (density) function of Y_i given β and γ, which can be any `gamlss.family` distribution. Assume that the observations Y_i, for $i = 1, 2, \ldots, n$, are conditionally independent given (β, γ). Assume that the γ_{kj}'s have independent (prior) normal distributions given by (3.3), for $k = 1, 2, 3, 4$ (over the distribution parameters) and $j = 1, 2, \ldots, J_k$ (over the different smoothers). Also assume a constant improper prior distribution for β and assume that λ is fixed. Then the posterior distribution for the parameters β and γ given $\mathbf{y}$ and λ is:

$$f(\beta, \gamma \,|\, \mathbf{y}, \lambda) \propto f(\mathbf{y} \,|\, \beta, \gamma) f(\gamma \,|\, \lambda)$$

$$\propto \underbrace{L(\beta, \gamma)}_{\text{likelihood}} \underbrace{\prod_k \prod_j f(\gamma_{kj} \,|\, \lambda_{kj})}_{\text{prior for } \gamma} \,, \tag{3.11}$$

where $f(\mathbf{y} \,|\, \beta, \gamma) = L(\beta, \gamma) = \prod_i f(y_i \,|\, \beta, \gamma)$ is the likelihood function and

$$f(\gamma \,|\, \lambda) = \prod_k \prod_j f(\gamma_{kj} \,|\, \lambda_{kj})$$

is the 'prior' probability density function of the (random) variables γ. That is, the posterior distribution for the parameters β and γ is proportional to the 'joint' probability density function of $\mathbf{y}$ and γ given β and λ. The sign of proportionality is there because the right-hand side of (3.11) should be divided by the marginal for $\mathbf{y}$, $f(\mathbf{y} \,|\, \lambda)$, for the posterior to be properly defined.

Now the posterior log-likelihood for $\boldsymbol{\beta}$ and $\boldsymbol{\gamma}$ is given by:

$$\log f(\boldsymbol{\beta}, \boldsymbol{\gamma} \mid \mathbf{y}, \boldsymbol{\lambda}) = \underbrace{\ell(\boldsymbol{\beta}, \boldsymbol{\gamma})}_{\text{log-likelihood}} + \sum_k \sum_j \log f(\boldsymbol{\gamma}_{kj} \mid \boldsymbol{\lambda}_{kj}) + \underbrace{c(\mathbf{y}, \boldsymbol{\lambda})}_{\text{constant}} \quad (3.12)$$

$$= \underbrace{\ell_h(\boldsymbol{\beta}, \boldsymbol{\gamma} \mid \boldsymbol{\lambda})}_{\text{hierarchical log-likelihood}} + \underbrace{c(\mathbf{y}, \boldsymbol{\lambda})}_{\text{constant}} \quad (3.13)$$

$$= \underbrace{\ell(\boldsymbol{\beta}, \boldsymbol{\gamma})}_{\text{log-likelihood}} \underbrace{- \frac{1}{2} \sum_k \sum_j \boldsymbol{\gamma}_{kj}^{\top} \mathbf{G}_{kj}(\boldsymbol{\lambda}_{kj}) \boldsymbol{\gamma}_{kj}}_{\text{log-normal exponent}} + \underbrace{c_1(\mathbf{y}, \boldsymbol{\lambda})}_{\text{constant}} (3.14)$$

$$= \underbrace{\ell_p(\boldsymbol{\beta}, \boldsymbol{\gamma} \mid \boldsymbol{\lambda})}_{\text{penalized log-likelihood}} + \underbrace{c_1(\mathbf{y}, \boldsymbol{\lambda})}_{\text{constant}} . \quad (3.15)$$

Some explanation is needed for each line of the above equation:

(3.12) This is the log-likelihood of the data given $\boldsymbol{\beta}$ and $\boldsymbol{\gamma}$, i.e. $\ell(\boldsymbol{\beta}, \boldsymbol{\gamma}) = \log L(\boldsymbol{\beta}, \boldsymbol{\gamma})$, together with the logarithm of the assumed probability density function for $\boldsymbol{\gamma}$ given $\boldsymbol{\lambda}$. The constant $c(\mathbf{y}, \boldsymbol{\lambda}) = -\log f(\mathbf{y} \mid \boldsymbol{\lambda})$ accounts for the proportionality symbol of equation (3.11).

(3.13) By combining the first two terms of (3.12), we have a hierarchical likelihood as defined by Lee et al. [2006].

(3.14) The general form of the distribution $f(\boldsymbol{\gamma}_{kj} \mid \boldsymbol{\lambda}_{kj})$ is substituted with the assumed normal distribution, resulting in the quadratic penalty $\boldsymbol{\gamma}_{kj}^{\top} \mathbf{G}_{kj}(\boldsymbol{\lambda}_{kj}) \boldsymbol{\gamma}_{kj}$. This is because the exponent of the multivariate normal probability density function is quadratic. The rest of the elements of the log of the normal pdf are absorbed by the constant, which changes from c to c_1.

(3.15) The log-posterior for the $\boldsymbol{\beta}$ and $\boldsymbol{\gamma}$ is equal to the penalized likelihood of equation (3.6), apart from a constant term which does not depend on $\boldsymbol{\beta}$ or $\boldsymbol{\gamma}$.

The conclusion we can draw from the above equations is that for fixed $\boldsymbol{\lambda}$, maximizing the penalized log-likelihood (3.6) for $\boldsymbol{\beta}$ and $\boldsymbol{\gamma}$ is equivalent to maximizing the hierarchical log-likelihood of (3.13) for $\boldsymbol{\beta}$ and $\boldsymbol{\gamma}$, which is also equivalent to maximizing the log-posterior for $\boldsymbol{\beta}$ and $\boldsymbol{\gamma}$. Therefore the estimators we obtain with such maximization are maximum a posteriori (MAP) estimators, maximum hierarchical likelihood estimators and also maximum penalized likelihood estimators.

3.4 Estimating the hyperparameters λ

What we have shown up to now is two algorithms, RS and CG, for estimating $\boldsymbol{\beta}$ and $\boldsymbol{\gamma}$ given the hyperparameters $\boldsymbol{\lambda}$. For fixed $\boldsymbol{\lambda}$ both methods lead to (penalized) maximum likelihood estimators for β and γ. More generally, it is desirable to estimate the smoothing hyperparameters λ automatically. There are various strategies for estimating λ:

Globally: when the method is applied outside the RS or CG GAMLSS algorithm.

Locally: when the method of estimation of each λ_{kj} is applied each time within the backfitting algorithm of the RS or CG GAMLSS algorithm.

In addition there are (at least) three different methodologies for estimating the (random effect or smoothing) hyperparameters:

- Likelihood-based methods (ML/REML);

- Generalized Akaike information criterion (GAIC); and

- Validation techniques such as the validation global deviance (VDEV) defined in Section 3.4.1.3, or generalized cross validation (GCV) defined in Section 3.4.2.3.

Table 3.1 shows where information about the different methods can be obtained. In our experience the local methods are much faster and often produce similar results to the global methods. The global methods can sometimes be more reliable but they are computationally intensive.

TABLE 3.1: References for the different approaches to estimating the (random effect or smoothing) hyperparameters

Global/ Local	Method	Reference
Global	ML /REML (e.g. Laplace)	Rigby and Stasinopoulos [2005]
	GAIC (e.g. AIC, SBC)	Rigby and Stasinopoulos [2004]
		Rigby and Stasinopoulos [2006]
	VDEV	Stasinopoulos and Rigby [2007]
Local	ML	Rigby and Stasinopoulos [2013]
	GAIC	Rigby and Stasinopoulos [2013]
	GCV	Wood [2006a]

The global and local and local methods are described in Section 3.4.1 and 3.4.2, respectively. The current facilities within the GAMLSS packages allow only the global GAIC through the function `find.hyper()` described in Section 11.9 and global VDEV through functions given in Section 11.8, and the local methods through different options when smoothers are used. For example `pb(x)` and `pb(x, method="GAIC")` will allow the use of a local ML (the default) and a local GAIC method, respectively, to estimate the smoothing hyperparameter when the P-splines method is used for smoothing `x`. See Section 9.4.2 for more details.

3.4.1 Global estimation

3.4.1.1 Maximum likelihood

In order to estimate the parameter(s) $\boldsymbol{\lambda}$ using maximum likelihood, we firstly have to integrate $\boldsymbol{\gamma}$ out of the joint likelihood of $\boldsymbol{\beta}$, $\boldsymbol{\gamma}$ and $\boldsymbol{\lambda}$ given in equation (3.11), i.e.

$$L(\boldsymbol{\beta}, \boldsymbol{\lambda}) = \int_{\boldsymbol{\gamma}} L(\boldsymbol{\beta}, \boldsymbol{\gamma}) \prod_{k} \prod_{j} f(\gamma_{kj} \,|\, \lambda_{kj}) \, d\boldsymbol{\gamma} \,, \tag{3.16}$$

and then maximize $L(\boldsymbol{\beta}, \boldsymbol{\lambda})$, the marginal likelihood, over $\boldsymbol{\beta}$ and $\boldsymbol{\lambda}$. This will result in maximum (marginal) likelihood estimators (MLEs) for both $\boldsymbol{\beta}$ and $\boldsymbol{\lambda}$. Note that the MLEs for $\boldsymbol{\beta}$ will be slightly different from the MAP estimators obtained by maximizing (3.11), even if $\boldsymbol{\lambda}$ is fixed at the maximum likelihood point. Now by integrating out the $\boldsymbol{\beta}$ parameters using a flat improper prior distribution we obtain the marginal likelihood for $\boldsymbol{\lambda}$:

$$L(\boldsymbol{\lambda}) = \int_{\boldsymbol{\beta}} \int_{\boldsymbol{\gamma}} L(\boldsymbol{\beta}, \boldsymbol{\gamma}) \prod_{k} \prod_{j} f(\gamma_{kj} \,|\, \lambda_{kj}) \, d\boldsymbol{\gamma} \, d\boldsymbol{\beta} \,. \tag{3.17}$$

Maximizing $L(\boldsymbol{\lambda})$ over $\boldsymbol{\lambda}$ produces the restricted maximum likelihood (REML) estimator for $\boldsymbol{\lambda}$, which will be slightly different from the MLE.

Given the generality of the GAMLSS model and the potential for different $\boldsymbol{\gamma}$'s in different parameters of the model, integrating out parameters is a very difficult task. Laplace approximation of the likelihood or Markov chain Monte Carlo (MCMC) simulations are two possible solutions to the problem. Rigby and Stasinopoulos [2005, Section 7.4] used the Laplace approximation to estimate the smoothing parameters in a rather simple random effect model. Unfortunately generalization of the method is difficult. There is a considerable literature on using MCMC to fit GAMLSS models; see for example Klein et al. [2015b] and the Bibliographic notes of this chapter.

3.4.1.2 Generalized Akaike information criterion

The Generalized Akaike information criterion (GAIC) is defined as

$$\mathrm{GAIC}(\kappa) = -2\,\ell\,(\hat{\boldsymbol{\beta}}, \hat{\boldsymbol{\gamma}}) + \kappa \cdot \mathrm{df} \tag{3.18}$$

where κ is a constant and df is the total (effective) degrees of freedom used for the fit. Global minimization of $\mathrm{GAIC}(\kappa)$ with respect to $\boldsymbol{\lambda}$ will provide estimates for $\boldsymbol{\lambda}$. The current facilities within the GAMLSS packages allow global $\mathrm{GAIC}(\kappa)$ estimation through the function `find.hyper()`; see Chapter 11. The $\mathrm{GAIC}(\kappa)$ is a conditional GAIC given the fitted fixed and random effects, $\boldsymbol{\beta}$ and $\boldsymbol{\gamma}$, which depend on $\boldsymbol{\lambda}$, i.e. $\hat{\boldsymbol{\beta}} = \hat{\boldsymbol{\beta}}(\boldsymbol{\lambda})$ and $\hat{\boldsymbol{\gamma}} = \hat{\boldsymbol{\gamma}}(\boldsymbol{\lambda})$. The $\mathrm{GAIC}(\kappa)$ is maximized globally over $\boldsymbol{\lambda}$.

3.4.1.3   Validation

To perform global validation we need either the existence of a 'new' sample, or
the re-use of the existing sample by separating it into different sub-samples (K-
fold cross validation). Section 11.1 provides more information about the different
sampling schemes.

Defining the 'different' versions of the global deviance is important as the global
deviance is used to produce estimates for the λ. The global deviance was first defined
in Chapter 1, as minus twice the fitted log-likelihood. Here let us be more specific.
For a given training data set, with response variable $\mathbf{y}$ and design matrices $(\mathbf{X}, \mathbf{Z})$,
the fitted global deviance is calculated by evaluating the predictors:

$$\hat{\eta}_k = \mathbf{X}_k \hat{\beta}_k + \mathbf{Z}_{k1} \hat{\gamma}_{k1} + \ldots + \mathbf{Z}_{kJ_k} \hat{\gamma}_{kJ_k} \ , \qquad k = 1, 2, 3, 4$$

at the fitted values $\hat{\beta}$ and $\hat{\gamma}$ and fixed $\lambda = \lambda_o$. Given the predictors, the fitted
parameter values can be calculated from $\hat{\theta}_k = g_k^{-1}(\hat{\eta}_k)$ for $k = 1, 2, 3, 4$ and therefore
the fitted global deviance (GDEV), for fixed $\lambda = \lambda_o$, can be defined as:

$$\text{GDEV} = -2\,\hat{\ell}(\mathbf{y} \,|\, \lambda_o) \ , \qquad (3.19)$$

where $\hat{\ell}(\mathbf{y} \,|\, \lambda_o) = \ell(\hat{\beta}, \hat{\gamma})$ is the log-likelihood $\ell(\beta, \gamma)$ evaluated at the fitted values
$\hat{\beta}$ and $\hat{\gamma}$, given $\lambda = \lambda_o$, $\mathbf{y}$ represents the current data and λ_o emphasizes that the
value of $\hat{\ell}$ depends on the current value for λ. Note that GDEV is a conditional
global deviance given the fitted fixed and random effects β and γ.

For a new validation or test sample (see Chapter 11 for explanation), with response
variable $\tilde{\mathbf{y}}$ and design matrices $(\tilde{\mathbf{X}}, \tilde{\mathbf{Z}})$ the predictor is

$$\tilde{\eta}_k = \tilde{\mathbf{X}}_k \hat{\beta}_k + \tilde{\mathbf{Z}}_{k1} \hat{\gamma}_{k1} + \ldots + \tilde{\mathbf{Z}}_{kJ_k}, \hat{\gamma}_{kJ_k}$$

and the predictive values for the parameters are $\tilde{\theta}_k = g_k^{-1}(\tilde{\eta}_k)$ for $k = 1, 2, 3, 4$. The
validation (or *test*) global deviance is defined as:

$$\text{VDEV} = \text{TDEV} = -2\,\tilde{\ell}(\tilde{\mathbf{y}} \,|\, \lambda_o) \qquad (3.20)$$

where $\tilde{\ell}$ emphasizes the fact that the global deviance is evaluated using predictive
values for the parameters at the new sample. Note that $\hat{\beta}$ and $\hat{\gamma}$ should be the
same fitted parameters in both the original sample and the new sample. In practice
because of the way the function `predict()` works (see Section 5.4), the values for
$\hat{\beta}$ and $\hat{\gamma}$ can be slightly different. In general a *validation* data set is used for tuning
the model, while the *test* data set is used for testing the predictive power of the
model. So the names VDEV and TDEV above reflect the purpose of use and not
the definition of the 'predictive' global deviance. If one wants to use a validation
sample, VDEV can be minimized with respect to λ_o to obtain estimates for λ. This
minimization is achieved using a numerical minimization algorithm in which for
each fixed λ_o, the β and γ are estimated using the training response variable values
$\mathbf{y}$. VDEV is then computed using the validation response variable values $\tilde{\mathbf{y}}$. Chapter
11 provides more details on the validation and alternative K-fold cross validation
methods.

3.4.2   Local estimation

In this section we assume model (3.2) with the special case of (3.3) where $\mathbf{G}_{kj}(\boldsymbol{\lambda}_{kj}) = \lambda_{kj}\mathbf{G}_{kj}$ for $j = 1, 2, \ldots, J_k$ and $k = 1, 2, \ldots, K$. All local methods assume that locally, when fitting random effect parameters $\boldsymbol{\gamma}_{kj}$ with the modified backfitting algorithm of Section 3.2.1.3, the current partial residuals $\boldsymbol{\varepsilon}_{kj}$ (defined in Rigby and Stasinopoulos [2005, Appendix B] and Figure 3.4), behave like a normally distributed random variable. Note that 'current' refers to the fact that the partial residuals are calculated within the backfitting algorithm. Below we omit the subscripts kj to simplify the presentation.

3.4.2.1   Maximum likelihood

On the predictor scale in the $\boldsymbol{\gamma}$ fitting part of the backfitting algorithm, the following (approximate) internal random effects model is assumed in order to estimate the current smoothing parameter λ:

$$\boldsymbol{\varepsilon} = \mathbf{Z}\boldsymbol{\gamma} + \mathbf{e} \ ; \qquad \mathbf{e} \sim \mathcal{N}(\mathbf{0}, \sigma_e^2\,\mathbf{W}^{-1}) \ ; \qquad \boldsymbol{\gamma} \sim \mathcal{N}(\mathbf{0}, \sigma_b^2\,\mathbf{G}^{-1}) \qquad (3.21)$$

where $\boldsymbol{\varepsilon}$ are the partial residuals (within backfitting), $\mathbf{Z}$ is the basis for smoothing the current explanatory variable, $\mathbf{W} = \mathrm{diag}(\mathbf{w})$ is a diagonal matrix with iterative weights $\mathbf{w}$, and $\mathbf{G}$ is a known matrix depending on which method for smoothing is used (see Chapter 9 for the definition of $\mathbf{G}$). This simple random effect model (3.21) has the following unknown parameters to be estimated by fitting the model: σ_e^2, σ_b^2 and $\boldsymbol{\gamma}$. The smoothing parameter λ, for smoothing the explanatory variable x, is the ratio of the two variances, i.e. $\lambda = \sigma_e^2/\sigma_b^2$. The parameters σ_e^2, σ_b^2 and $\boldsymbol{\gamma}$ can be estimated using the following simple algorithm (see for example Rigby and Stasinopoulos [2013]):

Step 1 Given the current $\hat{\lambda}$, estimate the $\boldsymbol{\gamma}$ parameters using a penalized least squares procedure:

$$\hat{\boldsymbol{\gamma}} = \left(\mathbf{Z}^\top\mathbf{W}\mathbf{Z} + \hat{\lambda}\mathbf{G}\right)^{-1}\mathbf{Z}^\top\mathbf{W}\boldsymbol{\varepsilon}$$

Step 2 Calculate:

$$\hat{\boldsymbol{\varepsilon}} = \mathbf{Z}\hat{\boldsymbol{\gamma}}$$

$$\hat{\sigma}_e^2 = (\boldsymbol{\varepsilon} - \hat{\boldsymbol{\varepsilon}})^\top (\boldsymbol{\varepsilon} - \hat{\boldsymbol{\varepsilon}}) / (n - \mathrm{tr}(\mathbf{S}))$$

$$\hat{\sigma}_b^2 = \hat{\boldsymbol{\gamma}}^\top\hat{\boldsymbol{\gamma}} / \mathrm{tr}(\mathbf{S})$$

where $\mathbf{S} = \mathbf{Z}\left(\mathbf{Z}^\top\mathbf{W}\mathbf{Z} + \hat{\lambda}\mathbf{G}\right)^{-1}\mathbf{Z}^\top\mathbf{W}$ is the smoothing matrix (so $\hat{\boldsymbol{\varepsilon}} = \mathbf{S}\boldsymbol{\varepsilon}$)

and therefore a new

$$\hat{\lambda} = \hat{\sigma}_e^2/\hat{\sigma}_b^2 \ .$$

Step 3 Stop if there is no change in $\hat{\lambda}$; otherwise go back to step 1.

3.4.2.2 Generalized Akaike information criterion

The local GAIC minimizes with respect to λ, and for given penalty κ, the quantity:

$$G_{AIC} = \left|\left|\sqrt{\mathbf{w}} \circ (\varepsilon - \mathbf{Z}\hat{\gamma})\right|\right|^2 + \kappa \operatorname{tr}(\mathbf{S}) \ .$$

Hence $\kappa = 2$ gives the local AIC and $\kappa = \log n$ gives the local SBC.

3.4.2.3 Generalized cross validation

The local GCV minimizes the following quantity with respect to λ:

$$V_g = \frac{n \left|\left|\sqrt{\mathbf{w}} \circ (\varepsilon - \mathbf{Z}\hat{\gamma})\right|\right|^2}{[n - \operatorname{tr}(\mathbf{S})]^2} \ .$$

Note that by using any of the above methods to calculate the smoothing parameters locally, the RS or CG algorithms are not necessarily optimal in the sense that they will lead to the global solution. In practice, however, the algorithms generally work well and lead to sensible results.

3.5 Bibliographic notes

The `gamlss` algorithms are maximum (penalized) likelihood-based algorithms. Maximum likelihood estimation was popularized by Ronald Fisher in the early 20th century, but was used earlier by Gauss and Laplace. For more details about the likelihood-based approach to inference and modelling, see Pawitan [2001]. Penalized likelihood estimation was proposed by Good and Gaskins [1971] while Cox and O'Sullivan [1990] studied asymptotic properties of penalized likelihood and related estimators. Chambers [1977] provides a useful summary of numerical analysis of the linear least squares problem, from a statistical viewpoint.

The backbones of the `gamlss` algorithms are (i) iteratively reweighted least squares (IRLS), and (ii) backfitting. IRLS was introduced for fitting generalized linear models (GLMs) by Nelder and Wedderburn [1972] and was implemented in the statistical software GLIM [Baker and Nelder, 1978, Francis et al., 1993]. Wedderburn [1974] explained the connection between the IRLS algorithm for maximum likelihood estimation and the Gauss–Newton method for least-squares fitting of nonlinear regressions. Wedderburn [1974] also proposed the use of IRLS algorithms for inference based on the concept of quasi-likelihood, and McCullagh [1983] extended to the multivariate case. Green [1984] describes the use of IRLS for maximum likelihood estimation, Newton–Raphson, Fisher's scoring method and some other alternatives. For the quasi-Newton method, see Gill and Murray [1974]. Osborne [1992] gives an analysis of the computational properties of Fisher's scoring method for maximizing likelihoods and solving estimating equations based on quasi-likelihoods.

The backfitting algorithm, which can be seen as a version of the Gauss–Seidel algorithm, was introduced by Breiman and Friedman [1985]. The Gauss–Seidel method, also known as successive relaxation, is an iterative method used to solve systems of linear equations but it can be applied more generally to nonlinear situations. The modified backfitting algorithm (introduced to avoid concurvity) is described in Hastie and Tibshirani [1990].

Aitkin [1987] used a successive relaxation algorithm (in GLIM) to fit a linear parametric model for both parameters μ and σ of the normal distribution. Nelder and Pregibon [1987] proposed the extended quasi-likelihood function as a way to estimate the mean μ and dispersion ϕ for the exponential family of distributions. Nelder and Lee [1992] compared the likelihood, quasi-likelihood and pseudo-likelihood approaches. One of the problems with the extended quasi-likelihood approach is the fact that the distribution of the response variable is not a 'proper' distribution but an approximation which for certain values of the dispersion parameter can be inappropriate. Smyth [1989] was the first to model the mean and dispersion for the gamma distribution using a proper likelihood rather than an extended quasi-likelihood. Rigby and Stasinopoulos [1996a] and Rigby and Stasinopoulos [1996b] extended the Aitkin [1987] and Nelder and Pregibon [1987] methodology to fit additive cubic spline smoothing terms, and implemented the algorithm in GLIM [Rigby and Stasinopoulos, 1996].

Gange et al. [1996] used the beta-binomial distribution to model appropriateness of hospital stays, and modelled both the mean and the dispersion as linear functions of the explanatory variables. It was the limitations of the extended quasi-likelihood approach to model the dispersion model in this particular case that led to the RS algorithm and the creation of GAMLSS.

Cole and Green [1992] introduced the LMS method for modelling growth curves using a modified Box–Cox transformation to deal with skewness in the data. Their algorithm forms the basis of the CG algorithm in **gamlss**. Rigby and Stasinopoulos [2004, 2005] extended their methodology by (i) introducing backfitting to cope with more than one explanatory variable, (ii) ensuring that the LMS method maximizes a proper distribution (called the Box–Cox Cole and Green (**BCCG**) distribution in **gamlss**), and (iii) introducing an extra parameter to cope with kurtosis in the data.

Selection of the smoothing parameters has been discussed in the smoothing literature since the 1980s. Cross validation and generalized cross validation are discussed by Green [1987] and Wahba [1985]. Wood [2006a] gives a comprehensive review of the methodology. The local maximum likelihood based methods GAIC and GCV are discussed in Rigby and Stasinopoulos [2004, 2005, 2006], Wood [2006a], Stasinopoulos and Rigby [2007] and Rigby and Stasinopoulos [2013]. For the mixed model approach for GAMs and GAMLSS, see for example Eilers et al. [2015], Fahrmeir et al. [2013] and Chapter 10 of this book.

There are several papers using MCMC to fit GAMLSS models. Klein et al. [2014] use Bayesian GAMLSS for modelling insurance data. Sohn et al. [2016] and Klein et al. [2015c] use Bayesian GAMLSS for modelling income distributions. For a Bayesian GAMLSS model with count data with overdispersion and/or zero inflation distributions, see Klein et al. [2015b]. For a Bayesian GAMLSS model with multivariate responses, see Klein et al. [2015a]. An extension of multivariate GAMLSS using copula specifications is given in Klein and Kneib [2016b]. For choosing the hyperpriors for the smoothing parameters in a Bayesian GAMLSS model, see Klein and Kneib [2016a].

3.6   Exercises

1. For the **species** data set, the different performances of the RS, CG and mixed algorithms were shown in Section 3.2.3. Compare the system time for the three models fitted:

```
system.time(capture.output(m1 <- gamlss(fish~log(lake),
        sigma.fo=~log(lake),family=PIG, data=species)))
system.time(capture.output(m2 <- gamlss(fish~log(lake),
        sigma.fo=~log(lake), family=PIG, data=species,
        method=CG(), n.cyc=100)))
system.time(capture.output(m3 <- gamlss(fish~log(lake),
        sigma.fo=~log(lake), family=PIG, data=species,
        method=mixed(1,100))))
```

Note that **m1** uses the default RS algorithm, i.e. **method=RS()**. Comment on the results.

2. **The oil data**: fitting GAMLSS models using the RS, CG and **mixed** algorithms. The **oil** data set contains the daily prices of front month WTI (West Texas Intermediate) oil price traded by NYMEX (New York Mercantile Exchange). The description of the variables in the data set is given on page 414 of Chapter 11. Here we use only two explanatory variables: **respLAG** (lag 1 of the response variable) and **HO1_log**, the logarithm price of front month heating oil contract traded by NYMEX.

    (a) Fit the following model using the **CG** algorithm and the **SHASHo** distribution:

```
m1.cg=gamlss(OILPRICE ~ pb(respLAG) + pb(HO1_log),
        sigma.formula = ~pb(respLAG) + pb(HO1_log),
        nu.formula = ~pb(respLAG) + pb(HO1_log),
        tau.formula = ~pb(respLAG) + pb(HO1_log),
        family = SHASHo, data = oil, method=CG(20))
```

(b) Because the model failed using the `CG` algorithm, try to fit the model with the `RS` algorithm.

```
m1.rs=gamlss(OILPRICE ~ pb(respLAG) + pb(HO1_log),
             sigma.formula = ~pb(respLAG) + pb(HO1_log),
             nu.formula = ~pb(respLAG) + pb(HO1_log),
             tau.formula = ~pb(respLAG) + pb(HO1_log),
             family = SHASHo, data = oil, method=RS(20))
```

(c) Since the `RS` algorithm works, try to fit the model using the `mixed` algorithm.

```
m1.mx=gamlss(OILPRICE ~ pb(respLAG) + pb(HO1_log),
             sigma.formula = ~pb(respLAG) + pb(HO1_log),
             nu.formula = ~pb(respLAG) + pb(HO1_log),
             tau.formula = ~pb(respLAG) + pb(HO1_log),
             family = SHASHo, data = oil, method=mixed(10,10),
             gd.tol=Inf)
```

(d) Compare the models.

```
GAIC(m1.rs,m1.mx, k=2)
GAIC(m1.rs,m1.mx, k=3)
GAIC(m1.rs,m1.mx, k=4)
GAIC(m1.rs,m1.mx, k=log(nrow(oil)))
```

Comment on the results.

4

The `gamlss()` *function*

CONTENTS

4.1   Introduction to the `gamlss()` function

This chapter:

- provides an introduction to the `gamlss()` function;

- shows how the information stored in a `gamlss` class object can be used;

- explores some of the functions associated with `gamlss` class objects; and

- provides information on how to extract information from a `gamlss` fitted object.

The `gamlss()` function, introduced in Chapters 1 and 2, is the main function of the **gamlss** package. In the following sections more explanation is given on its use. Section 4.2 explains the arguments of the function and Section 4.3 introduces the functions `refit()` and `update()`. Section 4.4 describes the components of a `gamlss` object (i.e. a fitted GAMLSS model) and Section 4.5 gives methods and functions for it.

4.2   The arguments of the gamlss() function

The usage of the function is

```
gamlss(formula = formula(data), sigma.formula = ~1,
      nu.formula = ~1, tau.formula = ~1, family = NO(),
      data = sys.parent(), weights = NULL,
      contrasts = NULL, method = RS(),  start.from = NULL,
      mu.start = NULL,  sigma.start = NULL,
      nu.start = NULL, tau.start = NULL,
      mu.fix = FALSE, sigma.fix = FALSE, nu.fix = FALSE,
      tau.fix = FALSE, control = gamlss.control(...),
      i.control = glim.control(...), ...)
```

where the arguments are defined as follows:

formula
: the standard **R** model formula for the predictor of the μ parameter, with the response on the left of a $\sim$ operator and the terms separated by + or * operators, on the right;

sigma.formula
: model formula for the predictor of the σ parameter;

nu.formula
: model formula for the predictor of the ν parameter;

tau.formula
: model formula for the predictor of the τ parameter;

family
: a **gamlss.family** object which defines the distribution of the response variable; see Chapter 6;

data
: data frame containing the variables in the formulae (see also Section 4.2.2);

weights
: a vector of prior weights. Note that this argument is not equivalent to the same argument of the **glm()** and **gam()** functions. Here **weights** can be used

1. to weight out observations (with weights equal to 1 or 0), or

2. for a weighted likelihood analysis where the contribution of the observations to the likelihood is weighted by the **weights**. Typically this is appropriate if some rows of the data are identical and the weights represent the frequencies of these rows (see also Section 4.2.2).

Any other use of **weights** is not recommended since this could have side effects. In particular, **glm()** weights do not in general translate

<table>
<tr><td></td><td>to gamlss() weights and such models should instead be fitted using offset(s) for the parameters μ and/or σ appropriately;</td></tr>
<tr><td>contrasts</td><td>list of contrasts to be used for some or all of the factors appearing as variables in the parameter(s) model formula;</td></tr>
<tr><td>method</td><td>There are three different algorithms which can be specified:</td></tr>
</table>

method=RS() The default method is the RS algorithm, which does not require accurate starting values for μ, σ, ν and τ to ensure convergence (the default starting values, often constants, are usually adequate). This method is more stable in the initial stage of the fitting and faster for larger data sets.

method=CG() The CG algorithm, which can be better for distributions with potentially highly correlated parameter estimates, but which is very unstable in the beginning of the process.

method=mixed() This is a mixture of the above two algorithms which starts with the RS algorithm and finishes with the CG.

The **RS()** and **CG()** algorithms are explained in detail in Chapter 3;

<table>
<tr><td>start.from</td><td>a fitted GAMLSS model from which to take the starting values for μ, σ, ν and τ for the current model;</td></tr>
<tr><td>mu.start</td><td>vector or scalar of starting values for the location parameter μ;</td></tr>
<tr><td>sigma.start</td><td>vector or scalar of starting values for the scale parameter σ;</td></tr>
<tr><td>nu.start</td><td>vector or scalar of starting values for the shape parameter ν;</td></tr>
<tr><td>tau.start</td><td>vector or scalar of starting values for the shape parameter τ;</td></tr>
<tr><td>mu.fix</td><td>whether the μ parameter should be kept fixed at mu.start during the fitting;</td></tr>
<tr><td>sigma.fix</td><td>whether the σ parameter should be kept fixed at sigma.start during the fitting;</td></tr>
<tr><td>nu.fix</td><td>whether the ν parameter should be kept fixed at nu.start during the fitting;</td></tr>
<tr><td>tau.fix</td><td>whether the τ parameter should be kept fixed at tau.start during the fitting;</td></tr>
<tr><td>control</td><td>control parameters of the outer iterations of the algorithm. The default setting is the gamlss.control function (see below);</td></tr>
<tr><td>i.control</td><td>control parameters of the inner iterations of the RS and CG algorithms. The default setting is the glim.control() function (see below).</td></tr>
</table>

`gamlss()` accepts all `glm()` type formulas, plus several smoothing function formulas (see Chapter 9).

> **Important**: Note that the `na.action` and `subset` arguments common to other statistical modelling functions such as `lm()` and `glm()` have been removed as arguments in the `gamlss()` function.
>
> This is because while there is only one data set in the model, there are up to four different model frames created (one for each distribution parameter) and therefore for consistency it is easier to apply subsetting and `na.action` to the whole data set and not to the individual frames.
> For subsets use `data=subset(mydata, subset=<the relevant condition>)` and for
> `na.action` use `data=na.omit(mydata)`.

Example: the abdominal data

> **R data file:** abdom in package **gamlss.data** of dimensions 610×2
>
> **variables**
>
> > y : abdominal circumference
> >
> > x : gestational age
>
> **purpose:** to demonstrate the fitting of a simple regression type model in GAMLSS

We illustrate the use of `gamlss()` with the **abdom** data, kindly provided by Dr. Eileen M. Wright. The response variable is abdominal circumference (**y**) and the explanatory variable is gestational age in weeks (**x**). The data comprises 610 observations. The following code fits smooth terms in **x**, for both μ and σ, with a normal response distribution. The default RS algorithm is used.

```
data(abdom)
h<-gamlss(y~pb(x), sigma.fo=~pb(x), family=NO, data=abdom)

## GAMLSS-RS iteration 1: Global Deviance = 4786.697
## GAMLSS-RS iteration 2: Global Deviance = 4785.695
## GAMLSS-RS iteration 3: Global Deviance = 4785.696
```

Note that the global deviance can increase slightly during the iterations. This can happen if smoothing additive terms are involved since the degrees of freedom in the different fits could change slightly. The CG algorithm is specified by:

```
h<-gamlss(y~pb(x),sigma.fo=~pb(x),family=NO,data=abdom,method=CG())
```

```
## GAMLSS-CG iteration 1: Global Deviance = 6165.522
## . . .
## GAMLSS-CG iteration 9: Global Deviance = 4785.695
```

and the mixed algorithm by:

```
h<-gamlss(y~pb(x),sigma.fo=~pb(x),family=NO,data=abdom,
          method=mixed(2,20))
```

```
## GAMLSS-RS iteration 1: Global Deviance = 4786.697
## GAMLSS-RS iteration 2: Global Deviance = 4785.695
## GAMLSS-CG iteration 1: Global Deviance = 4785.696
```

In the above example the mixed method uses two cycles of the RS algorithm, followed by one cycle of the CG algorithm. All methods end up essentially with the same fitted model, a useful check.

4.2.1   The algorithmic control functions

The `gamlss.control` function controls outer iterations of the algorithm and is defined as:

```
gamlss.control(c.crit=0.001, n.cyc=20, mu.step=1, sigma.step=1,
               nu.step=1, tau.step=1, gd.tol=Inf, iter=0,
               trace=TRUE, ...)
```

where

c.crit	is the convergence criterion for the outer iteration of the algorithm;
n.cyc	is the maximum number of cycles of the outer iteration of the algorithm;
mu.step	is the inner iteration step length for the parameter μ;
sigma.step	is the inner iteration step length for the parameter σ;
nu.step	is the inner iteration step length for the parameter ν;
tau.step	is the inner iteration step length for the parameter τ;
gd.tol	global deviance tolerance level. This allows the global deviance to temporarily increase, which is often useful when fitting complicated models with many smoothing parameters;
iter	not normally used by the user. It is used when the **refit** function is used, to count the right number of iterations;
trace	whether to print the global deviance at each outer iteration of the RS and CG algorithms. The users are advised to keep the default

value `trace=TRUE` so they can check if the algorithm is converging properly. (Sometimes in this book we specify `trace=FALSE` for reasons of space.)

The `glim.control` function controls the inner iterations of the RS and CG algorithms and is defined as:

```
glim.control(cc = 0.001, cyc = 50,   trace = FALSE, bf.cyc = 30,
             bf.tol = 0.001, bf.trace = FALSE,...)
```

where

`cc`	is the convergence criterion for the inner iteration or GLIM part of the algorithm;
`cyc`	the number of cycles of the inner iteration or GLIM part of the algorithm;
`trace`	whether to print at each inner iteration of the GLIM part of the algorithm, with default `trace=FALSE`;
`bf.cyc`	the number of cycles of the backfitting algorithm (see Sections 3.2.1.3 and 3.2.2.3);
`bf.tol`	the convergence criterion for the backfitting algorithm with default 10^{-3} (see Sections 3.2.1.3 and 3.2.2.3);
`bf.trace`	whether to print at each iteration of the backfitting (`bf.trace=TRUE`) or not (`bf.trace=FALSE`, the default).

The following is an example of how to change the convergence criterion `c.crit`. Firstly fit the model with the default convergence criterion value of 0.001:

```
h<-gamlss(y~pb(x), sigma.fo=~pb(x), family=NO, data=abdom)
```

```
## GAMLSS-RS iteration 1: Global Deviance = 4786.697
## GAMLSS-RS iteration 2: Global Deviance = 4785.695
## GAMLSS-RS iteration 3: Global Deviance = 4785.696
```

Arguments of `gamlss.control` or `glim.control` can be changed directly within the `gamlss()` function as shown below:

```
h<-gamlss(y~pb(x), sigma.fo=~pb(x), family=NO, data=abdom,
          c.crit=0.000001)
```

```
## GAMLSS-RS iteration 1: Global Deviance = 4786.697
## . . .
## GAMLSS-RS iteration 5: Global Deviance = 4785.696
```

Alternatively, we can change the convergence criterion using the `control` argument with the criterion defined within `gamlss.control()`:

```
control1<-gamlss.control(c.crit=0.000001)
h<-gamlss(y~pb(x), sigma.fo=~pb(x), family=NO, data=abdom,
        gamlss.control=control1)
```

```
## GAMLSS-RS iteration 1: Global Deviance = 4786.697
## GAMLSS-RS iteration 2: Global Deviance = 4785.695
## GAMLSS-RS iteration 3: Global Deviance = 4785.696
```

Now we change the default value of the **trace** option directly within **gamlss()**:

```
h<-gamlss(y~pb(x), sigma.fo=~pb(x), family=NO, data=abdom,
        glm.trace=TRUE)
```

```
## GLIM iteration 1 for mu: Global Deviance = 6607.265
## GLIM iteration 2 for mu: Global Deviance = 6607.265
## GLIM iteration 1 for sigma: Global Deviance = 6036.217
## . . .
## GAMLSS-RS iteration 2: Global Deviance = 4785.695
## GLIM iteration 1 for mu: Global Deviance = 4785.696
## GLIM iteration 1 for sigma: Global Deviance = 4785.696
## GAMLSS-RS iteration 3: Global Deviance = 4785.696
```

Alternatively **trace** can be changed using the **i.control** argument defined within **glim.control()**:

```
control2<-glim.control(glm.trace=TRUE)
h<-gamlss(y~pb(x), sigma.fo=~pb(x), family=NO, data=abdom,
        i.control=control2)
```

```
## GLIM iteration 1 for mu: Global Deviance = 6607.265
## GLIM iteration 2 for mu: Global Deviance = 6607.265
## GLIM iteration 1 for sigma: Global Deviance = 6036.217
## . . .
## GAMLSS-RS iteration 2: Global Deviance = 4785.695
## GLIM iteration 1 for mu: Global Deviance = 4785.696
## GLIM iteration 1 for sigma: Global Deviance = 4785.696
## GAMLSS-RS iteration 3: Global Deviance = 4785.696
```

This trick is useful when checking the convergences for the individual distribution parameters but, unless a problem is suspected, it is better to leave **trace** at the default value.

Useful advice: If a large data set is used (say more than 10,000 observations), and the user is at an exploratory stage of the analysis where many models have to be fitted relatively fast, it is advisable to change **c.crit** in **gamlss.control()** to something like 0.01 or even 0.1.

Let us now fit the t family distribution to the above data. The `family` option for the t distribution is `TF` and the degrees of freedom parameter is ν (i.e. `nu`), which is fitted as a constant by default.

```
h<-gamlss(y~pb(x),sigma.fo=~pb(x),family=TF,data=abdom,trace=FALSE)
```

The fitted value for ν is 11.42, obtained as

```
fitted(h,"nu")[1]
```

```
##           1
## 11.42469
```

or

```
exp(coef(h,"nu"))
```

```
## (Intercept)
##     11.42469
```

There are occasions where the user wants to fix the parameter(s) of a distribution at specific value(s). For example, one might want to fix the degrees of freedom of the t distribution say at 10. This can be done as follows with the `nu.start` and `nu.fix` arguments:

```
h1<-gamlss(y~pb(x), sigma.fo=~pb(x), family=TF, data=abdom,
              nu.start=10, nu.fix=TRUE, trace=FALSE)
```

Note: The t distribution may be unstable if ν is fixed close to one (usually indicating that this is an inappropriate value of ν for the particular data set).

4.2.2 Weighting out observations: the `weights` and `data=subset()` arguments

There are two ways in which the user can weight observations out of an analysis. The first relies on the `subset()` function and can be used in the `data` argument of `gamlss()`, i.e. `data=subset(mydata, condition)`, where `condition` is a relevant R code restricting the case numbers of the data.

> **Important**: It was mentioned earlier that the `subset` argument of `lm()` and `glm()` functions is not an argument in `gamlss()`. Always use `data=subset(mydata, condition)`.

The second way is through the `weights` argument. Note that the `weights` do not behave in the same way as in the `glm()` or `lm()` functions. In those functions they are prior weights used to fit only the mean of the model, while in `gamlss()` the same weights are applied for fitting all (possibly four) distribution parameters. The `weights` here can be used for a weighted likelihood analysis where the contribution of the observations to the log-likelihood is weighted according to `weights`. Typically this is appropriate in the following cases:

frequencies: if some rows of the data are identical and the weights represent the frequencies of these rows;

zero weights: A more common application of the weights is to set them equal to zero or one (`FALSE` or `TRUE`), so observations can be weighted out of the analysis;

weighted log-likelihood: This is the case where different weights in the log-likelihood for different observations is required. One example is in the fitting of finite mixtures; see Chapter 7 and package **gamlss.mx**.

Note than in general a model fitted to the original uncollapsed data frame or to the collapsed data frame using frequencies as weights should produce identical results for a fitted parametric model. An exception to this is potentially slightly different fitted models in the presence of nonparametric smooth P-spline functions, due to the knots position depending on all explanatory variable values, including those with zero weight. (Note however that the fitted models can potentially be quite different if the explanatory variable values for the zero weight cases are well outside the range of the non-zero weight cases, i.e. extrapolation). The fitted values and the residuals of the two different models do not have to have the same length, as we will demonstrate in this section.

Note that using `data=subset()` only fits the data cases in the subset, so fitted values for the distribution parameters μ, σ, ν and τ are only calculated for the subset data cases. However using the `weights` option fits all the data cases (although cases with zero weights do not contribute to the fit) and so fitted values for the distribution parameters are calculated for all data cases. This is one method for producing predicted values for the distribution parameters. (The other method uses the function `predict()`.)

Let us assume that in our abdominal circumference example we want to weight out all observations in which the x variable is less than or equal to 20. We can do this using `subset()`:

```
h2<-gamlss(y~pb(x), sigma.fo=~pb(x), family=TF,
           data=subset(abdom, x>20))

## GAMLSS-RS iteration 1: Global Deviance = 3706.584
## . . .
## GAMLSS-RS iteration 4: Global Deviance = 3706.763

c(length(fitted(h2)), length(resid(h2)), h2$noObs, h2$N)

## [1] 456 456 456 456
```

Note that `h2$N` gives the length of the response variable while `h2$noObs` is the sum of the weights. Now we use `weights`:

```
h3<-gamlss(y~pb(x), sigma.fo=~pb(x), family=TF, data=abdom,
          weights=x>20)
```

```
## GAMLSS-RS iteration 1: Global Deviance = 3706.698
## . . .
## GAMLSS-RS iteration 4: Global Deviance = 3706.827
```

```
c(length(fitted(h3)), length(resid(h3)), h3$noObs, h3$N)
```

```
## [1] 610 456 456 610
```

The response variable has 610 cases, but only 456 of them have $x > 20$ and hence weights equal to 1. Assume now that we want to weight out only a few observations, say the 200th and 400th. We can do this either using **subset** or **weights**. The advantage of using the argument **weights** is that we can get predictions for those values:

```
w <- rep(1, 610)
w[c(200,400)] <- 0
h41<-gamlss(y~pb(x),sigma.fo=~pb(x),family=TF,
          data=subset(abdom,w==1))
```

```
## GAMLSS-RS iteration 1: Global Deviance = 4766.151
## GAMLSS-RS iteration 2: Global Deviance = 4763.481
## GAMLSS-RS iteration 3: Global Deviance = 4763.506
## GAMLSS-RS iteration 4: Global Deviance = 4763.507
```

```
h42<-gamlss(y~pb(x),sigma.fo=~pb(x),family=TF,weights=w,data=abdom)
```

```
## GAMLSS-RS iteration 1: Global Deviance = 4766.151
## GAMLSS-RS iteration 2: Global Deviance = 4763.481
## GAMLSS-RS iteration 3: Global Deviance = 4763.506
## GAMLSS-RS iteration 4: Global Deviance = 4763.507
```

```
fitted(h42, "mu")[c(200,400)]
```

```
## [1] 176.8339 278.3580
```

If the variables in the reduced data frame are to be used extensively later on, it would make more sense to use the **subset** function in advance of the fitting to create a reduced data set.

The following simple artificial example demonstrates the use of the **weights** argument when frequencies are involved in the data. (The approach is particularly suited to fitting discrete distributions to frequency count data.)

```
y <- c(3,3,7,8,8,9,10,10,12,12,14,14,16,17,17,19,19,18,22,22)
x <- c(1,1,2,3,3,4, 5, 5, 6, 6, 7, 7, 8, 9, 9,10,10,11,12,12)
ex1 <- data.frame(y=y,x=x)
```

The 20×2 data frame `ex1` contains some identical rows, e.g. rows 1 and 2, and rows 7 and 8. A new data frame, containing the same information as in `ex1`, but with an extra variable called `freq` indicating the number of identical rows in `ex1` can be created as:

```
yy <- c(3, 7, 8, 9, 10, 12, 14, 16, 17, 19, 18, 22)
xx <- c(1, 2, 3, 4, 5, 6, 7, 8, 9, 10, 11, 12)
ww <- c(2, 1, 2, 1, 2, 2, 2, 1, 2, 2, 1, 2)
ex2 <- data.frame(y=yy, x=xx, freq=ww)
```

Fitting a statistical model using each of the two data frames should produce identical results. This is demonstrated below where prior `weights` are used to fit the data in `ex2`:

```
m1 <- gamlss(y~x, data=ex1, family=PO)

## GAMLSS-RS iteration 1: Global Deviance = 90.8238
## GAMLSS-RS iteration 2: Global Deviance = 90.8238

m2 <- gamlss(y~x, weights=freq, data=ex2, family=PO)

## GAMLSS-RS iteration 1: Global Deviance = 90.8238
## GAMLSS-RS iteration 2: Global Deviance = 90.8238

all.equal(deviance(m1),deviance(m2))

## [1] TRUE

c(length(fitted(m1)), length(resid(m1)), m1$noObs, m1$N)

## [1] 20 20 20 20

c(length(fitted(m2)), length(resid(m2)), m2$noObs, m2$N)

## [1] 12 20 20 12
```

Note the lengths of the fitted values and the residuals of the two models. In the case of model `m2` the residuals are expanded to represent all 20 original observations. Note that `resid(m1)` and `resid(m2)` are not going to be identical in this case since they are randomized residuals; see Chapter 12.

The user may be tempted to scale the weights but this can have undesirable consequences and should be avoided, as we demonstrate below:

```
m3<- gamlss(y~x, weights=freq/2, data=ex2, family=PO)

## GAMLSS-RS iteration 1: Global Deviance = 45.4119
## GAMLSS-RS iteration 2: Global Deviance = 45.4119
```

```
summary(m2)
```

```
## ****************************************************************
## . . .
## Mu link function:  log
## Mu Coefficients:
##              Estimate Std. Error t value Pr(>|t|)
## (Intercept)  1.62331    0.16548   9.809 1.20e-08 ***
## x            0.12904    0.01914   6.741 2.56e-06 ***
```

```
summary(m3)
```

```
## ****************************************************************
## . . .
## --------------------------------------------------------------
## Mu link function:  log
## Mu Coefficients:
##              Estimate Std. Error t value Pr(>|t|)
## (Intercept)  1.62331    0.23403   6.936 4.01e-05 ***
## x            0.12904    0.02707   4.767 0.000761 ***
```

```
deviance(m2)
```

```
## [1] 90.82379
```

```
deviance(m3)
```

```
## [1] 45.41189
```

In this example the fitted coefficients of the two models are the same. However, the deviances and more importantly the standard errors have been affected by the change in weights. Also, because the weights are not frequencies the length of the residuals remains 12. In general, using weights that are not frequencies is **not** recommended unless the user knows what he/she is doing and is aware of the problems.

4.3   The refit and update functions

4.3.1   refit()

The function refit() is used if the converged component of the gamlss fitted object is FALSE, that is, when the maximum number of iterations (n.cyc) has been reached without convergence. The default value for n.cyc is 20, which is sufficient for most problems. Here we give an artificial example in which we force the algorithm to stop in the third iteration and continue with `refit()`:

```
h<-gamlss(y~pb(x), sigma.fo=~pb(x), family=TF, data=abdom, n.cyc=3 )

## GAMLSS-RS iteration 1: Global Deviance = 4780.234
## GAMLSS-RS iteration 2: Global Deviance = 4777.493
## GAMLSS-RS iteration 3: Global Deviance = 4777.519

## Warning in RS(): Algorithm RS has not yet converged

h<-refit(h)

## GAMLSS-RS iteration 4: Global Deviance = 4777.52
```

4.3.2 update()

The function `update()` is used to update formulae or other arguments of a `gamlss` fitted object. To update formulae, `update()` uses the **R** `update.formula()` function for the specified distribution parameter.

The **gamlss** `update()` function is defined as

```
update.gamlss(object, formula., ..., what = c("mu", "sigma", "nu",
              "tau"), evaluate = TRUE)
```

where

object   a **gamlss** fitted object;

formula   the formula to update;

...   for updating argument in `gamlss()`;

what   what parameter of the distribution is required for updating in the formula: mu, sigma, nu or tau. The default is `what="mu"`;

evaluate   whether to evaluate the call or not (the default is `evaluate=TRUE`).

The aids data

We illustrate `update()` using the **aids** data, which consist of the quarterly reported AIDS cases in the U.K. from January 1983 to March 1994 obtained from the Public Health Laboratory Service, Communicable Disease Surveillance Centre, London.

R data file: aids in package **gamlss.data** of dimensions 45×3

variables

  y : the number of quarterly aids cases in England and Wales

  x : time in quarters from January 1983

qrt : a factor for the quarterly seasonal effect (1, 2, 3, 4)

purpose: to demonstrate the fitting of a simple regression model in GAMLSS

We start by using the Poisson family to model the number of reported cases (the response variable). Explanatory variables are time (a continuous variable), which we smooth with a P-spline smoother, and `qrt`, a factor representing a quarterly seasonal effect. We then (i) change the family to negative binomial type I, (ii) update to a cubic spline smoother with `df=8`, (iii) remove the quarterly seasonal effect, and (iv) finally fit a normal family model with response `log(y)`.

```
data(aids)
# fit a Poisson model
h.po <-gamlss(y~pb(x)+qrt, family=PO, data=aids)

## GAMLSS-RS iteration 1: Global Deviance = 387.1462
## GAMLSS-RS iteration 2: Global Deviance = 387.1547
## GAMLSS-RS iteration 3: Global Deviance = 387.1547

# update with a negative binomial
h.nb <-update(h.po, family=NBI)

## GAMLSS-RS iteration 1: Global Deviance = 373.1785
## . . .
## GAMLSS-RS iteration 5: Global Deviance = 366.9258

# update the smoothing using cs()
h.nb1 <-update(h.nb,~cs(x,8)+qrt)

## GAMLSS-RS iteration 1: Global Deviance = 362.9323
## . . .
## GAMLSS-RS iteration 5: Global Deviance = 359.2348

# remove qrt
h.nb2 <-update(h.nb1,~.-qrt)

## GAMLSS-RS iteration 1: Global Deviance = 379.5915
## . . .
## GAMLSS-RS iteration 4: Global Deviance = 379.5626

# put back qrt, take log of y and fit a normal distribution
h.nb3 <-update(h.nb1,log(.)~.+qrt, family=NO)

## GAMLSS-RS iteration 1: Global Deviance = -42.3446
## GAMLSS-RS iteration 2: Global Deviance = -42.3446

# verify that it is the same
h.no<-gamlss(log(y)~cs(x,8)+qrt,data=aids )

## GAMLSS-RS iteration 1: Global Deviance = -42.3446
## GAMLSS-RS iteration 2: Global Deviance = -42.3446
```

The whiteside data

Finally we give an example taken from Venables and Ripley [2002, Section 6.1], to demonstrate how **update()** can be used to fit two different lines in an analysis of covariance. Each model fits a separate regression of gas consumption on temperature for the two different levels of the factor **Insul**. The response variable is gas consumption (**Gas**) and the explanatory variables are average outside temperature in degrees Celsius (**Temp**) and a factor before or after insulation (**Insul**):

R data file: whiteside in package **MASS** of dimensions 56×3

variables

    Gas : the weekly gas consumption in 1000s of cubic feet

    Temp : the average outside temperature in degrees Celsius

    Insul : a factor, before or after insulation

purpose: to demonstrate the use of **subset()** in GAMLSS

```
library(MASS)
data(whiteside)
gasB <- gamlss(Gas~Temp, data=subset(whiteside, Insul=="Before"))

## GAMLSS-RS iteration 1: Global Deviance = 5.7566
## GAMLSS-RS iteration 2: Global Deviance = 5.7566

gasA <- update(gasB,data=subset(whiteside, Insul=="After"))

## GAMLSS-RS iteration 1: Global Deviance = 20.9026
## GAMLSS-RS iteration 2: Global Deviance = 20.9026
```

Figure 4.1 shows the gas consumption against the average outside temperature in degrees Celsius, before and after insulation.

```
with(whiteside, plot(Temp,Gas,pch=(16:17)[unclass(Insul)],
                  col=c("red","green3")[unclass(Insul)]))
with(whiteside, lines(Temp[Insul=="Before"],fitted(gasB),col="red",
                  lwd=2))
with(whiteside, lines(Temp[Insul=="After"],fitted(gasA),col="green3",
                  lty=2,lwd=2))
legend("topright",pch=16:17,lty=1:2,lwd=2,col=c("red","green3"),
          legend=c("before","after"))
```

Fig.
4.1

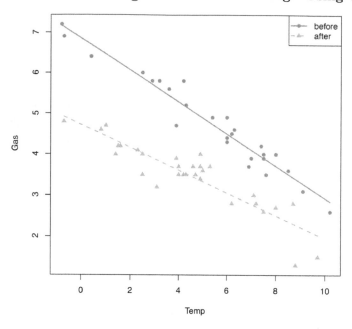

R

p. 1(

FIGURE 4.1: A linear interaction model for gas consumption against the average outside temperature in degrees Celsius for before or after insulation.

4.4   The gamlss object

The function **gamlss()** returns a **gamlss** S3 object. To demonstrate the composition of a **gamlss** object, we fit a model to the **abdom** data set:

```
h<-gamlss(y~pb(x), sigma.fo=~pb(x), family=TF, data=abdom)

## GAMLSS-RS iteration 1: Global Deviance = 4780.234
## . . .
## GAMLSS-RS iteration 4: Global Deviance = 4777.52
```

By calling the **names** function we are able to check on the components of the object h.

```
names(h)

##  [1] "family"          "parameters"
##  [3] "call"            "y"
##  [5] "control"         "weights"
##  [7] "G.deviance"      "N"
##  [9] "rqres"           "iter"
## [11] "type"            "method"
## [13] "contrasts"       "converged"
```

```
## [15]  "residuals"        "noObs"
## [17]  "mu.fv"            "mu.lp"
## [19]  "mu.wv"            "mu.wt"
## [21]  "mu.link"          "mu.terms"
## [23]  "mu.x"             "mu.qr"
## [25]  "mu.coefficients"  "mu.offset"
## [27]  "mu.xlevels"       "mu.formula"
## [29]  "mu.df"            "mu.nl.df"
## [31]  "mu.s"             "mu.var"
## [33]  "mu.coefSmo"       "mu.lambda"
## [35]  "mu.pen"           "df.fit"
## [37]  "pen"              "df.residual"
## [39]  "sigma.fv"         "sigma.lp"
## [41]  "sigma.wv"         "sigma.wt"
## [43]  "sigma.link"       "sigma.terms"
## [45]  "sigma.x"          "sigma.qr"
## [47]  "sigma.coefficients" "sigma.offset"
## [49]  "sigma.xlevels"    "sigma.formula"
## [51]  "sigma.df"         "sigma.nl.df"
## [53]  "sigma.s"          "sigma.var"
## [55]  "sigma.coefSmo"    "sigma.lambda"
## [57]  "sigma.pen"        "nu.fv"
## [59]  "nu.lp"            "nu.wv"
## [61]  "nu.wt"            "nu.link"
## [63]  "nu.terms"         "nu.x"
## [65]  "nu.qr"            "nu.coefficients"
## [67]  "nu.offset"        "nu.formula"
## [69]  "nu.df"            "nu.nl.df"
## [71]  "nu.pen"           "P.deviance"
## [73]  "aic"              "sbc"
```

More generally any **gamlss** object has the following components:

family   The distribution family of the **gamlss** object (see Chapter 6), e.g. for the object **h** we have:

```
h$family

## [1] "TF"       "t Family"
```

parameters The name of the fitted parameters as a character list:

```
h$parameters

## [1] "mu"       "sigma" "nu"
```

call   The call of the **gamlss()** function:

```
h$call

## gamlss(formula = y ~ pb(x),  sigma.formula = ~pb(x),
##       family = TF,   data = abdom)
```

y   The response variable as a vector (or matrix), accessed by h$y.

control   The gamlss() fit control settings, accessed by h$control.

weights   The vector of prior weights, accessed by h$weights.

G.deviance The value of global deviance, extracted by h$G.deviance, or by us-
            ing the generic function deviance() or deviance(gamlss.object,
            "G"):

```
deviance(h)

## [1] 4777.519
```

N   The length of the response variable (or the number of observations
            in the fit unless weights are used):

```
h$N

## [1] 610
```

noObs   The actual number of observations if (frequency count) weights are
            used (e.g. to weight out observations) in the fit, equal to the sum
            of the weights. If no weights are used this is equal to h$N:

```
h$noObs

## [1] 610
```

rqres   A function to calculate the normalized (randomized) quantile resid-
            uals of the object, accessed by h$rqres. (See Dunn and Smyth
            [1996] and Section 12.2.)

iter   The number of outer (or GAMLSS) iterations in the fitting process
            (by an outer iteration we mean the refitting of all the distribution
            parameters μ, σ, ν and τ; see Sections 3.2.1 and 3.2.2):

```
h$iter

## [1] 4
```

type   The type of the distribution of the response variable (continuous,
            discrete or mixed):

```
h$type

## [1] "Continuous"
```

method   Which algorithm is used for the fit, RS(), CG() or mixed():

```
h$method
## RS()
```

contrasts | Which contrasts were used in the fit, NULL if they have not been set in gamlss() function.

converged | Whether the model has converged:

```
h$converged
## [1] TRUE
```

residuals | The normalized (randomized) quantile residuals of the model which can be extracted by h$residuals or by using the generic function residuals(), (also abbreviated as resid()).

df.fit | The total degrees of freedom used by the model, e.g. in model h there are 3 distribution parameters μ, σ and ν. The total degrees of freedom for the fit is the total of all the degrees of freedom used to fit the individual parameters. Those degrees of freedom are stored in h$mu.df, h$sigma.df and h$nu.df, respectively.

```
h$df.fit
## [1] 8.789107
h$mu.df+h$sigma.df+h$nu.df
## [1] 8.789107
```

df.residual | The residual degrees of freedom left after the model is fitted:

```
h$df.residual
## [1] 601.2109
```

pen | The sum of the quadratic penalties for all the parameters (if appropriate additive terms are fitted):

```
h$pen
## [1] 3.839226
```

P.deviance | The penalized deviance (global deviance plus penalties), which can be extracted by h$P.deviance or by using the generic function deviance(gamlss.object,"P"):

```
h$P.deviance
## [1] 4781.359
deviance(h,"P")
```

```
## [1] 4781.359
```

aic   The Akaike information criterion, which can also be obtained using
             the functions `AIC()` or `GAIC()`:

```
h$aic

## [1] 4795.098

AIC(h)

## [1] 4795.098

GAIC(h)

## [1] 4795.098
```

sbc   The Schwarz Bayesian criterion (SBC) (or Bayesian information cri-
             terion (BIC)), which can also be extracted using `AIC()` or `GAIC()`:

```
h$sbc

## [1] 4833.888

AIC(h, k=log(length(abdom$y)))

## [1] 4833.888

GAIC(h, k=log(length(abdom$y)))

## [1] 4833.888
```

The rest of the components refer to the parameters of the model (if they exist). The
name **par** below should be replaced with the appropriate parameter, which can be
any of the `mu`, `sigma`, `nu` or `tau`).

par.fv The fitted values of the appropriate parameter, accessed by, e.g.
   `h$mu.fv`. The fitted values can also be extracted using the generic function
   `fitted()`; for example, `fitted(h,"mu")` extracts the `mu` fitted values while
   `fitted(h,"sigma")` extracts the `sigma` fitted values:

```
head(fitted(h)) #equivalent to the first values of fitted(h,"mu")

## [1] 60.81081 60.81081 60.81081 62.58732 66.13772 66.13772

tail(fitted(h,"sigma")) #equivalent to the last 6 values of

## [1] 20.46207 20.46207 20.46207 20.46207 20.58861 20.70743

                                             #fitted(h, "sigma")
```

par.lp The predictor of the appropriate parameter, accessed by, e.g.:

```
head(h$mu.lp)
```

```
## [1] 60.81081 60.81081 60.81081 62.58732 66.13772 66.13772
```

`par.wv` The working variable of the appropriate parameter.

`par.wt` The iterative weights of the appropriate parameter.

`par.link` The link function for the appropriate parameter, e.g. :

```
h$sigma.link
```

```
## [1] "log"
```

`par.terms` The terms for the appropriate parameter model.

`par.x` The design matrix for the appropriate parameter.

`par.qr` The QR decomposition of the appropriate parameter model.

`par.coefficients` The linear coefficients of the appropriate parameter model which can also be extracted using the generic function `coef()`:

```
coef(h, "mu")
```

```
## (Intercept)        pb(x)
##   -55.61858    10.34939
```

`par.formula` The formula for the appropriate parameter model, e.g. :

```
h$mu.formula
```

```
## y ~ pb(x)
```

`par.df` The appropriate parameter degrees of freedom (see above):

```
h$mu.df
```

```
## [1] 5.787446
```

`parameter.nl.df` The nonlinear (e.g. smoothing) degrees of freedom for the appropriate parameter. Note that this does not include two degrees of freedom for the fitted constant and linear part:

```
h$mu.nl.df
```

```
## [1] 3.787446
```

`par.pen` The sum of the quadratic penalties for the specific parameter (if appropriate additive terms are fitted).

`par.xlevels` (only where relevant) a record of the levels of the factors used in fitting of this parameter.

4.5   Methods and functions for `gamlss` objects

The `gamlss()` function creates an S3 class object. There are several "methods" [1] and functions which can be applied to a `gamlss` object. A method, in **R**, is a generic function which can be used to display information from a specific class object. The following is a list of functions and methods which can be applied to a `gamlss` object. Many examples of their use are given in Chapter 5 and throughout this book.

`AIC()`   Method for extracting the generalized AIC

`coef()`   Method for extracting the linear coefficients from any distribution parameter

`confint()`   Method for extracting confidence intervals

`deviance()`   Method for extracting the global deviance

`edf()`, `edfAll()`
                 Method and function, respectively, for extracting the effective degrees of freedom

`extractAIC()`   Method for extracting AIC

`fitted()`, `fv()`
                 Method and function, respectively, for extracting the fitted values

`formula()`   Method for extracting a model formula

`gen.likelihood()`
                 Function for generating the likelihood function of the fitted model (used by `vcov()`)

`get.K()`   Function for extracting the K matrix (meat) for sandwich standard errors

`getSmo()`   Function for extracting information from a fitted smoothing term

`logLik()`   Method for extracting the log-likelihood

`lp()`   Function for extracting the linear predictor for a distribution parameter (see also `lpred()`)

`lpred()`   Function for extracting the fitted values, linear predictor or specified terms (with standard errors) for a distribution parameter

[1] A method is a generic function in **R** which dispatches to different functions according to the class of the first argument of the function. For example, the `plot()` function is a method with a default function `plot.default()`. If the object m is a fitted `gamlss` object, then the command `plot(m)` will use `plot.gamlss()` rather than `plot.default()`.

`model.frame()`
> Method for extracting the model frame of a specified distribution parameter

`model.matrix()`
> Method to extract the design matrix of a specified distribution parameter

`plot()`   Method for plotting residual diagnostics

`predict()`, `predictAll()`
> Method and function, respectively, to predict from new data individual distribution parameter values (see also `lpred()` above)

`print()`   Method for printing a `gamlss` object

`residuals()`   Method to extract the normalized (randomized) quantile residuals from a fitted `gamlss` model object; see Chapter 12.

`Rsq()`   Function for getting the generalized R^2

`rvcov()`   Function for extracting the robust (sandwich) variance-covariance matrix of the beta estimates (for all distribution parameter models). This can be done also with `vcov()`.

`summary()`   Method to summarize the fit in a `gamlss` object

`terms()`   Method for extracting terms from a `gamlss` object

`vcov()`   Method to extract the variance-covariance matrix of the beta estimates (for all distribution parameter models)

The list is not exhaustive and other functions and methods will be covered in the relevant chapters.

4.6   Bibliographic notes

The function `gamlss()` is based on similar **R** functions such as `lm()`, `glm()` and `gam()`. It was first created during 2001–2002 and its main structure has remained the same since then. The RS and CG algorithms are described in detail in Chapter 3. The algorithms were first fully descibed in Rigby and Stasinopoulos [2005, Appendix B], while their theoretical justification is given in [Rigby and Stasinopoulos, 2005, Appendix C]. An important aspect of any statistical modelling function is how it handles prior weights. Section 4.2.2 describes the reason and how the prior weights are handled in `gamlss()`. Here we emphasize that prior weights in GAMLSS are different from prior weights in GLMs (see Aitkin et al. [2009] for the latter).

4.7   Exercises

1. **Reanalyzing the whiteside data.** These are the gas consumption data in the **MASS** package, analyzed on page 101. Models gasB and gasA fitted in Section 4.3.2 can be fitted simultaneously rather than individually, by fitting an interaction model for Temp and Insul.

    (a) Use gamlss() to fit the complete model:

    ```
    a1 <- gamlss(Gas~Temp*Insul, sigma.fo=~Insul, data=whiteside)
    ```

    and verify that the deviances of gasB and gasA add up to the deviance of the combined model.

    ```
    gasB <- gamlss(Gas~Temp, data=subset(whiteside,
                                     Insul=="Before"))
    gasA <- gamlss(Gas~Temp, data=subset(whiteside,
                                     Insul=="After"))
    ```

    (b) Verify that the fitted coefficients produce identical results using coef() for models a1, gasB and gasA.

    (c) Plot the fitted model:

    ```
    with(whiteside, plot(Temp,Gas,pch=21,
                      bg=c("red","green3")[unclass(Insul)]))
    posB<-which(whiteside$Insul=="Before")
    posA<-which(whiteside$Insul=="After")
    with(whiteside, lines(fitted(a1)[posB]~whiteside$Temp[posB],
                      pch=19))
    with(whiteside, lines(fitted(a1)[posA]~whiteside$Temp[posA],
                      pch=19))
    ```

    and compare it with Figure 4.1.

    (d) Compare the estimates for the σ parameter. Note that models gasB and gasA have different estimates for this parameter.

2. **EU 15 data:** The purpose is to estimate the importance of labor, capital and useful energy in explaining economic growth (quantified by the gross domestic product (GDP)) of the EU 15 from 1960 to 2009. The response variable here is GDP, while the independent variables are the UsefulEnergy, Labor and Capital. The EU 15 comprises Austria, Belgium, Denmark, Finland, France, Germany, Greece, Ireland, Italy, Luxembourg, Netherlands, Portugal, Spain, Sweden and United Kingdom (before Brexit). The data were analyzed by Voudouris et al. [2015].

> **R data file:** eu15 in package **gamlss.data** of dimensions 50×6
>
> **variables**
>
> > GDP : Sum of GDP of the EU 15 countries
> >
> > Year : Year (1960 to 2009)
> >
> > UsefulEnergy : Total amount of useful energy (energy that performs some form of 'work') for the EU 15 countries
> >
> > Labor : Total hours worked in the EU 15 countries
> >
> > Capital : Sum of net capital stock of the EU 15 countries
>
> **purpose:** to familiarize with the **gamlss** functions, and the **weight** argument.

(a) Get the data and transform it:

```
data(eu15)
eu15<-transform(eu15, lGDP=log(GDP), lCapital=log(Capital),
        lLabor=log(Labor), lUsefulEnergy=log(UsefulEnergy))
```

(b) To estimate the effects of independent variables on the response, fit a simple linear GAMLSS model using the **TF2** distribution (see Chapter 6 for the different choices of distributions).

```
mod1 <- gamlss(lGDP~lCapital+lUsefulEnergy+lLabor,data=eu15,
            family=TF2)
```

(c) Plot the model terms.

```
term.plot(mod1,pages=1)
```

(d) Plot the fitted model with data.

```
with(eu15, plot(Year,lGDP,pch=21))
lines(fitted(mod1)[order(eu15$Year)]~
    eu15$Year[order(eu15$Year)],lwd=2)
```

(e) Update **mod1** by using **weights** to emphasize recent observations more than past observations.

```
w<-1*exp(-(2/(nrow(eu15)))*seq(from=nrow(eu15)-1,to=0))
mod2<-gamlss(lGDP~lCapital+lUsefulEnergy+lLabor, data=eu15,
            family=TF2, weights=w)
```

(f) Plot the fitted models **mod1** and **mod2** with the data.

(g) Compare the generalized R^2 of the fitted models **mod1** and **mod2**.

```
Rsq(mod1)
Rsq(mod2)
```

5

Inference and prediction

CONTENTS

5.1   Introduction

This chapter:

- provides information about the inferential tools which can be used in a fitted **gamlss** object;

- shows how standard errors, confidence intervals and predictions work within the **gamlss** packages.

This chapter concerns how inference is drawn from a fitted `gamlss` object. We firstly discuss some theoretical issues and then show the relevant functions for obtaining the required inferential information.

The GAMLSS framework and its implementation in **R** follow the statistical aphorism that *"all models are wrong but some are useful"* [Box, 1979]. That is, the user is encouraged to try different models and choose the ones that she/he thinks are more suitable for the data and of course capable of answering the question at hand. The chosen model should adequately fit the data and its basic assumptions should always be checked using residual diagnostics. Awareness of these assumptions is crucial for model checking and deciding whether we should accept the model or not. So the first requirement for any proper statistical inference is the use of an adequate and relevant model. The authors also believe *that the type of statistical inference used (be it classical, Bayesian or pure likelihood-based) may be less important to the conclusions than choosing a suitable model or models in the first place.* Of course different types of inferential approaches have their advantages and disadvantages. For example a classical model is faster to fit and provides good diagnostic tools, while the Bayesian approach provides a full posterior distribution (rather than estimates and standard errors) and is probably the best method for the elimination of nuisance parameters.

In this book and the current **gamlss** packages we concentrate on classical likelihood-based inferential approaches.

5.1.1   Asymptotic behaviour of a parametric GAMLSS model

We start by describing what properties we would expect asymptotically (as the sample size $n \to \infty$) from the coefficients of a parametric GAMLSS model. For such models, the estimation of the parameters is achieved through maximum likelihood estimation. (For models containing smoothing terms, penalized maximum likelihood estimation is used; see Chapter 3.) For a parametric GAMLSS model, if the model is correct, then in general, since maximum likelihood estimation is used, all the parameter estimators are consistent, with correct asymptotic standard errors, leading to asymptotically correct confidence interval coverage and test size [Cox and Hinkley, 1979, pages 279–311]. The GAMLSS parameter estimators are robust to outliers in specific cases. If the GAMLSS model is not correct then the parameter estimators may not be consistent. This is an important issue as it implies that, if a GAMLSS model is incorrect (e.g. if it fits poorly), it may be preferable to fit a generalized linear model (GLM) with robust standard errors or to estimate parameters (with robust standard errors) using generalized estimating equations (GEE) (see Section 5.5).

5.1.2   Types of inference in a GAMLSS model

Inference in simple parametric regression models is usually concentrated on (i) selecting (or testing) between different models, (ii) model fitting, i.e. estimating the parameters of a model, (iii) testing the values of a single or a group of parameters in the model, (iv) confidence interval(s) for parameter(s) in the model and (v) predicting the distribution of a future value of the response variable.

For additive regression models where smoothers are included as terms we need some additional inferential tools to deal with conclusions about the smoothing parameters and the behaviour of the fitted smoothing functions. The same type of problems arise in GAMLSS modelling, with the added burden of having to deal with up to four different models for the different distribution parameters. The following is a list of possible inferential questions we may require from a GAMLSS model:

a constant distribution parameter: Here we are drawing conclusions about a constant distribution parameter μ, σ, ν or τ, when no explanatory variables have been fitted for the parameter. It is possible to obtain approximate standard errors in this case by using `vcov()`, `summary()` and `predict()`. Confidence intervals can be obtained using the functions `confint()` and `prof.dev()`. All functions are described in this chapter.

coefficients from a parametric term in a predictor model: Here we are drawing conclusions about a parameter in a predictor model (for a distribution parameter). The parametric term can be a continuous or a categorical variable (factor). For a continuous variable the functions `vcov()` and `summary()` provide standard errors, while `confint()` and `prof.term()` provide confidence intervals. For factors, individual coefficients are seldom of interest but what is important is the contribution of the factor as a whole to the model. The function `drop1()` (described in Section 5.2.3) provides the generalized likelihood ratio test of whether a factor is significant or not. The function `pcat()` (described in Chapter 9) shows how the levels of a factor may be reduced. In the presence of smoothing terms in any of the models of a distribution parameter, special care has to be taken when making inferences about the parameters of linear terms.

smoothing curves: The shape of the fitted curve of a smoothing term and the pointwise standard errors attached to it are of special interest here. For the most common smoothers the standard errors of the fitted curves are a function of the trace of the smoothing matrix. Smooth functions are addressed in detail in Chapter 9.

smoothing parameters: The estimation of the smoothing parameter of a smoothing term is the main concern here because it determines the shape of the fitted curve. The selection of smoothing parameters is discussed in detail in Chapter 3, where distinction is made between 'global' and 'local' techniques. Standard errors of the smoothing parameters are seldom sought but may be approximated relatively easily (at extra computational cost) by bootstrapping, which is discussed in Section 5.1.4.

prediction of a future value for any of the distribution parameters: This involves prediction for any or all of the distribution parameters μ, σ, ν and τ. The functions `predict()` and `predictAll()`, discussed in Section 5.4, are appropriate here.

prediction of the distribution of a future value of the response variable: The fitted distribution of the response variable for a future observation can be obtained by substituting the predicted values of all the distribution parameters into the distribution (This is applied, for example, in centile estimation.) Note however the resulting distribution does not take account of the uncertainty in the predicted parameters.

selection between models Model selection in GAMLSS is covered in Chapter 11.

The inferential tools used to answer some of the above questions fall into two categories: (i) likelihood-based inference, and (ii) bootstrapping. We now consider both of these approaches.

5.1.3 Likelihood-based inference

Let us assume that we are dealing with a parametric GAMLSS model (3.4) and let $\boldsymbol{\theta}$ be the vector containing all the parameters. Note that $\boldsymbol{\theta}$ is a generic parameter vector, which in this context may be thought of as the collection of all the linear coefficient parameters for μ, σ, ν and τ, i.e. $(\boldsymbol{\beta}_1, \boldsymbol{\beta}_2, \boldsymbol{\beta}_3, \boldsymbol{\beta}_4)$. From classical maximum likelihood theory we have the result (stated informally) that asymptotically

$$\hat{\boldsymbol{\theta}} \sim \mathcal{N}(\boldsymbol{\theta}_T, i(\boldsymbol{\theta}_T)^{-1}), \tag{5.1}$$

where $\hat{\boldsymbol{\theta}}$ is the maximum likelihood estimator and

$$i(\boldsymbol{\theta}_T) = -\mathrm{E}\left[\frac{\partial^2 \ell(\boldsymbol{\theta})}{\partial \boldsymbol{\theta}\, \partial \boldsymbol{\theta}^\top}\right]_{\boldsymbol{\theta}_T}$$

is the (Fisher) expected information matrix evaluated at the assumed 'true' value $\boldsymbol{\theta}_T$. It is not always possible to derive the expected information $i(\boldsymbol{\theta}_T)$ analytically and therefore the *observed information* $\boldsymbol{I}(\boldsymbol{\theta}_T)$, defined as

$$\boldsymbol{I}(\boldsymbol{\theta}_T) = -\left[\frac{\partial^2 \ell(\boldsymbol{\theta})}{\partial \boldsymbol{\theta}\, \partial \boldsymbol{\theta}^\top}\right]_{\boldsymbol{\theta}_T}$$

is often used instead. Note $\boldsymbol{I}(\boldsymbol{\theta}_T)$ is equal to the negative of the Hessian matrix of the log-likelihood function at $\boldsymbol{\theta}_T$. The variance-covariance matrix of the asymptotic distribution of $\hat{\boldsymbol{\theta}}$ is then approximated by $\boldsymbol{I}(\boldsymbol{\theta}_T)^{-1}$ instead of $i(\boldsymbol{\theta}_T)^{-1}$. Of course since $\boldsymbol{\theta}_T$ is unknown and is estimated by $\hat{\boldsymbol{\theta}}$, we substitute $\hat{\boldsymbol{\theta}}$ for $\boldsymbol{\theta}_T$ in both the expected $i(\boldsymbol{\theta}_T)$ and observed $\boldsymbol{I}(\boldsymbol{\theta}_T)$ information, giving $i(\hat{\boldsymbol{\theta}})$ and $\boldsymbol{I}(\hat{\boldsymbol{\theta}})$.

In general, for parametric GAMLSS models the following asymptotic distribution for $\hat{\boldsymbol{\theta}}$ is used:

$$\hat{\boldsymbol{\theta}} \sim \mathcal{N}(\boldsymbol{\theta}_T, \boldsymbol{I}(\hat{\boldsymbol{\theta}})^{-1}) . \tag{5.2}$$

If the model is *incorrect* then, under certain conditions, the asymptotic distribution of $\hat{\boldsymbol{\theta}}$ may be approximated by

$$\hat{\boldsymbol{\theta}} \sim \mathcal{N}(\boldsymbol{\theta}_C, \boldsymbol{I}(\hat{\boldsymbol{\theta}})^{-1}K(\hat{\boldsymbol{\theta}})\boldsymbol{I}(\hat{\boldsymbol{\theta}})^{-1}) , \tag{5.3}$$

where $K(\hat{\boldsymbol{\theta}})$ is an estimate of the variance-covariance matrix of the first derivative of log-likelihood function with respect to the parameters and $\boldsymbol{\theta}_C$ is the value of $\boldsymbol{\theta}$ 'closest' to the true model as measured by a (weighted) Kullback–Leibler distance. For more information see Claeskens et al. [2008, Chapter 2]. Note in particular that if the model is incorrect then $\hat{\boldsymbol{\theta}}$ is not in general a consistent estimator of $\boldsymbol{\theta}_T$.

Under certain very specific conditions e.g. for the mean model parameters of a fitted GLM when the mean model is correct (but the distribution and/or dispersion could be incorrect), then $\boldsymbol{\theta}_C = \boldsymbol{\theta}_T$. Under such conditions the robust or sandwich estimator of the standard errors, given by the square root of the diagonal of the estimated variance-covariance matrix $\boldsymbol{I}(\hat{\boldsymbol{\theta}})^{-1}K(\hat{\boldsymbol{\theta}})\boldsymbol{I}(\hat{\boldsymbol{\theta}})^{-1}$, may be used.

The function `gen.likelihood()` obtains the Hessian of the log-likelihood, while `vcov()` inverts the negative of the Hessian and provides the variance-covariance matrix $\boldsymbol{I}(\hat{\boldsymbol{\theta}})^{-1}$ (or $\boldsymbol{I}(\hat{\boldsymbol{\theta}})^{-1}\boldsymbol{K}(\hat{\boldsymbol{\theta}})\boldsymbol{I}(\hat{\boldsymbol{\theta}})^{-1}$ for robust sandwich estimators). The usual standard errors of the parameter estimates are obtained by taking the square root of the diagonal of the variance-covariance matrix. An alternative method of obtaining standard errors for the estimate of a specific parameter (say β) when the observed information matrix is difficult to obtain (as for example in the finite mixtures of Chapter 7) is to use:

$$\text{s.e.}(\hat{\beta}) \approx \frac{|\hat{\beta}|}{\sqrt{\Delta\text{GDEV}}} \,, \tag{5.4}$$

where ΔGDEV is the difference in global deviance obtained by omitting the explanatory variable associated with β from the model; see Aitkin et al. [2009]. Such a procedure is justified by approximating the likelihood ratio test statistic by the Wald test statistic for testing the null hypothesis $\beta = 0$, i.e.

$$\left[\frac{\hat{\beta}}{\text{s.e.}(\hat{\beta})}\right]^2 \approx \Delta\text{GDEV} \,. \tag{5.5}$$

Bootstrapping is an alternative method of obtaining standard errors for the estimates of the parameters and will be considered next.

5.1.4 Bootstrapping

Bootstrapping was introduced by Efron [1979, 1992] as a way of estimating standard errors using computer power rather than formulae. More details about bootstrapping can be found in Efron and Tibshirani [1993] and Davison and Hinkley [1997]. Bootstrapping can be summarized as follows:

1. find an appropriate model and define which parameter(s) are of interest;

2. generate B new data sets (related somehow to the original data set);

3. fit the original model to the new data sets and save the estimates of the relevant parameter(s) of interest;

4. use the B realizations of the saved fitted parameter(s) of interest to draw inference.

In order to demonstrate how bootstrapping can be used for GAMLSS models we will need the following notation. Let M_0 be the appropriate fitted model where the fitted distribution is $\mathcal{D}(\mathbf{y} \mid \hat{\boldsymbol{\mu}}, \hat{\boldsymbol{\sigma}}, \hat{\boldsymbol{\nu}}, \hat{\boldsymbol{\tau}})$. The parameters of the distribution are fitted (if needed) using appropriate explanatory variables. Let $\mathcal{F}$, $\mathcal{F}^{-1}$ and $\mathcal{D}_r$ be the cdf, inverse cdf and the random generating function of $\mathcal{D}$, respectively. (That is, $\mathcal{D}$, $\mathcal{F}$, $\mathcal{F}^{-1}$ and $\mathcal{D}_r$ are the relevant $\mathbf{d}$, $\mathbf{p}$, $\mathbf{q}$ and $\mathbf{r}$, functions of a distribution in $\mathbf{R}$.) Let $\mathbf{y}_{s[b]}$ be the bth realization of a simulated response variable and $\mathbf{y}_{[b]}$ be the bth realization of a sample with replacement from the original response variable $\mathbf{y}$. Let $\mathbf{r}$ be the observed residuals of model M_0, and $\mathbf{r}_{[b]}$ the bth realization of a sample with replacement from the original residuals $\mathbf{r}$. Finally let D_0 be the original data frame and $D_{[b]}$ the bth realization of a sample with replacement from D_0. There are at least three distinct ways of performing bootstrapping, all of them involving different ways of generating new data. Here we describe them in terms of GAMLSS models. In all cases we are assuming that the fitted model, M_0, is a reasonable fit to the original data set.

The parametric bootstrap The new simulated realizations of $\mathbf{y}$ are created, i.e. $\mathbf{y}_{s[b]}$ from the fitted model M_0 using $\mathcal{D}_r(\hat{\boldsymbol{\mu}}, \hat{\boldsymbol{\sigma}}, \hat{\boldsymbol{\nu}}, \hat{\boldsymbol{\tau}})$. This is repeated B times, i.e. for $b = 1, \ldots, B$.

Bootstrapping the residuals The residuals, $\mathbf{r}$, of model M_0 are resampled with replacement, i.e. $\mathbf{r}_{[b]}$. The new values of the response are created using $\mathbf{y}_{[b]} = \mathcal{F}^{-1}(\hat{\mathbf{u}}_{[b]} \mid \hat{\boldsymbol{\mu}}, \hat{\boldsymbol{\sigma}}, \hat{\boldsymbol{\nu}}, \hat{\boldsymbol{\tau}})$ where $\hat{\mathbf{u}}_{[b]} = \Phi(\mathbf{r}_{[b]})$. (Remember that the residuals in GAMLSS are defined as $\mathbf{r} = \Phi^{-1}(\mathbf{u})$ where $\mathbf{u} = \mathcal{F}(\mathbf{y} \mid \hat{\boldsymbol{\mu}}, \hat{\boldsymbol{\sigma}}, \hat{\boldsymbol{\nu}}, \hat{\boldsymbol{\tau}})$.) This is repeated B times.

Nonparametric bootstrapping Here model M_0 is refitted to a data set $D_{[b]}$ sampled with replacement from the original data set D_0. This is repeated B times.

The use of bootstrapping to obtain standard errors and confidence intervals is demonstrated in Exercise 3 using the `abdom` data.

5.2   Functions to obtain standard errors

5.2.1   The `gen.likelihood()` function

We have seen in Chapter 3 that the fitting algorithms for GAMLSS models consist of repeatedly calling an iteratively reweighted least squares algorithm (possibly penalized). This process is repeated until convergence of the global deviance is achieved.

There is a problem in that the standard errors provided by each least squares fitting are not correct, because they assume that all the other parameters of the distribution are fixed at their current fitted values. The function gen.likelihood() was created to overcome this deficiency. Given a fitted gamlss model, gen.likelihood() generates the likelihood function of the model for the purpose of creating the Hessian matrix required for construction of the standard errors of the parameter estimates. The function gen.likelihood() is used by vcov() to obtain the correct Hessian matrix, after a model has been fitted. This is automatically used by summary() to obtain standard errors of the parameter estimates. The following example demonstrates gen.likelihood() using the aids data, which were described and plotted in Section 4.3.2:

```
data(aids)
m100 <- gamlss(y~x+qrt, data=aids, family=NBI, trace=FALSE)
logL<-gen.likelihood(m100)#generate the log-likelihood function
logL()# evaluate it at the final fitted values
```

[1] 246.3187

```
logL(c(coef(m100), coef(m100, "sigma")))#this code is equivalent
```

[1] 246.3187

```
optimHess(c(coef(m100), coef(m100, "sigma")), logL)#the Hessian
```

##	(Intercept)	x	qrt2
## (Intercept)	212.050893	4971.82091	51.5380919
## x	4971.820911	140205.85157	1198.0983500
## qrt2	51.538092	1198.09835	51.5380919
## qrt3	52.071512	1187.73218	0.0000000
## qrt4	51.791336	1237.55026	0.0000000
## (Intercept)	1.826129	-15.03447	0.6425665

##	qrt3	qrt4	(Intercept)
## (Intercept)	52.0715125	51.791336	1.8261293
## x	1187.7321846	1237.550255	-15.0344720
## qrt2	0.0000000	0.000000	0.6425665
## qrt3	52.0715125	0.000000	0.2461583
## qrt4	0.0000000	51.791336	0.2317890
## (Intercept)	0.2461583	0.231789	18.1635333

Note that the first intercept is in the predictor model for μ, while the second intercept is in the predictor model for σ in the negative binomial $NBI(\mu, \sigma)$ distribution.

When smoothing terms are fitted, gen.likelihood() considers them as fixed at their fitted values, so the Hessian in this case does not take into account the variability in the fitting of the smoothers.

```
m200 <- gamlss(y~pb(x)+qrt, data=aids, family=NBI, trace=FALSE)
logL2<-gen.likelihood(m200)#create the log-likelihood
#evaluate it at the final fitted values
logL2(c(coef(m200), coef(m200, "sigma")))
```

[1] 183.4629

```
optimHess(c(coef(m200), coef(m200, "sigma")), logL2)#Hessian matrix
```

```
##                 (Intercept)           pb(x)          qrt2
## (Intercept)   3.741335e+03   1.111319e+05    848.198810
## pb(x)         1.111319e+05   3.695284e+06  24596.919051
## qrt2          8.481988e+02   2.459692e+04    848.198810
## qrt3          9.470198e+02   2.750627e+04      0.000000
## qrt4          9.279800e+02   2.811189e+04      0.000000
## (Intercept)   1.183395e+00  -3.405392e+00      3.595562
##                       qrt3           qrt4 (Intercept)
## (Intercept)     947.019761     927.980000    1.183395
## pb(x)         27506.266553   28111.889104   -3.405392
## qrt2              0.000000       0.000000    3.595562
## qrt3            947.019761       0.000000   -7.024548
## qrt4              0.000000     927.980000    2.294323
## (Intercept)      -7.024548       2.294323    5.416412
```

The entry under the `pb(x)` column refers to the linear part of the smoother and therefore should not be interpreted on its own but only in combination with the fitted smoother.

5.2.2   The `vcov()` and `rvcov()` functions

The generic function `vcov()` within **gamlss** uses `gen.likelihood()` to construct the Hessian matrix numerically. Standard errors for the estimated coefficients are obtained from the observed information matrix. The standard errors obtained this way are more reliable than those obtained during the GAMLSS fitting algorithms since they take into account the information about the interrelationship between the distribution parameters. The function `rvcov()` creates the robust or sandwich standard errors; see Section 5.1.3. The following are the arguments of the generic function `vcov()`:

```
vcov(object, type = c("vcov", "cor", "se", "coef", "all"),
                robust = FALSE, hessian.fun = c("R", "PB"), ...)
```

where the arguments of the function in **gamlss** are

object   a **gamlss** fitted object;

type   what is required: (i) **type="vcov"**: variance-covariance matrix; (ii) type="cor": the correlation matrix; (iii) type="se": the standard errors, (iv) type="coef": the fitted coefficients; (v) type="all": all the above as a list;

robust   whether the usual (robust=FALSE) or robust or sandwich (robust=TRUE) standard errors are required;

hessian.fun   how to obtain the Hessian numerically: (i) hessian.fun="R": using optimHess(); (ii) hessian.fun="PB": using a function by Pinheiro and Bates taken from package **nlme**, which is the default;

...   extra arguments.

As an example we use the model fitted in the previous section:

```
vcov(m100, type="cor")#correlation betuwen the fitted parameters
```

```
##                (Intercept)            x         qrt2
## (Intercept)   1.00000000 -0.760891874 -0.462032219
## x            -0.76089187  1.000000000  0.018879191
## qrt2         -0.46203222  0.018879191  1.000000000
## qrt3         -0.47515350  0.034637199  0.477918686
## qrt4         -0.44671986 -0.002258329  0.476834481
## (Intercept) -0.08157924  0.088524725  0.001656598
##                      qrt3         qrt4  (Intercept)
## (Intercept) -0.47515350 -0.446719862 -0.081579245
## x            0.03463720 -0.002258329  0.088524725
## qrt2         0.47791869  0.476834481  0.001656598
## qrt3         1.00000000  0.477952149  0.012473668
## qrt4         0.47795215  1.000000000  0.009506204
## (Intercept)  0.01247367  0.009506204  1.000000000
```

```
vcov(m100, type="se")#the standard errors
```

```
## (Intercept)            x         qrt2         qrt3         qrt4
## 0.204805619 0.006535937 0.192532079 0.192104635 0.192260793
## (Intercept)
## 0.235686941
```

```
vcov(m100, type="se", robust=TRUE)#the sandwich standard errors
```

```
## (Intercept)            x         qrt2         qrt3         qrt4
## 0.291009501 0.008784397 0.212083065 0.210080980 0.198288097
## (Intercept)
## 0.250504613
```

The **gamlss** function rvcov() has the same arguments as the generic function vcov.gamlss(), apart from the default robust=TRUE. It provides the sandwich or robust standard errors. Robust standard errors, introduced by Huber [1967] and White [1980], may be more reliable than the usual standard errors when the variance model is suspected not to be correct (assuming the mean model is correct). The sandwich standard errors are usually (but not always) larger than the usual ones. The following is an example of simulated data from a gamma (GA) distribution with sigma=2, where an incorrect exponential distribution model is fitted instead. Therefore we would expect the robust standard errors to be greater than the standard ones and more reliable.

```
#set seed
set.seed(4321)
# generate from a gamma distribution with sigma=2
Y <- rGA(200, mu=1, sigma=2)
# fitting the wrong model i.e.   sigma=1
r1 <- gamlss(Y~1, family=EXP)

## GAMLSS-RS iteration 1: Global Deviance = 391.2369
## GAMLSS-RS iteration 2: Global Deviance = 391.2369

# the conventional se is too precise
vcov(r1, type="se")

## (Intercept)
##   0.07071067

# the sandwich se is larger
rvcov(r1, type="se")

## (Intercept)
##    0.1182156

# fitting the correct model
 r2 <- gamlss(Y~1, family=GA)

## GAMLSS-RS iteration 1: Global Deviance = 19.5225
## GAMLSS-RS iteration 2: Global Deviance = 19.5225

 # standard se's
 vcov(r2, type="se")

## (Intercept) (Intercept)
##   0.13170785   0.03935216

  # robust se's
 rvcov(r2, type="se")

## (Intercept) (Intercept)
##   0.11851375   0.03866188

# the same standard errors are obtained using
vcov(r2, type="se", robust=TRUE)

## (Intercept) (Intercept)
##   0.11851375   0.03866188
```

5.2.3 The summary() function

The generic function **summary()** within **gamlss** provides more detailed information about the fitted GAMLSS model than the method **print()**, which gives only limited information. The arguments of **summary()** in **gamlss** are as follows:

```
summary(object, type = c("vcov", "qr"), robust=FALSE, save = FALSE,
                hessian.fun = c("R", "PB"), ...)
```

where

object   a **gamlss** fitted object;

type   the default **type="vcov"** uses the **vcov()** method to get the variance-covariance matrix of the estimated beta coefficients. The alternative **type="qr"** produces standard errors from the individual least squares fits but is not reliable since it does not take into the account the correlation between the estimated distribution parameters μ, σ, ν and τ;

robust   whether the robust (or sandwich) standard errors are required, in which case the function **rvcov()** is called;

save   whether to save the environment of the function in order to have access to its values;

hessian.fun   how to obtain the Hessian numerically: (i) **hessian.fun="R"**, using **optimHess()** (ii) **hessian.fun="PB"**, using a function by Pinheiro and Bates taken from package **nlme**. The latter is used as the default;

...   extra arguments.

In the following example the environment of **summary()** is saved so the p-values, for example, can be accessed.

```
sm100 <-summary(m100, robust=TRUE, save=TRUE)

## ******************************************************************
## Family:  c("NBI", "Negative Binomial type I")
##
## Call:
## gamlss(formula = y ~ x + qrt, family = NBI, data = aids,
##      trace = FALSE)
##
## Fitting method: RS()
##
## --------------------------------------------------------------
## Mu link function:  log
## Mu Coefficients:
```

```
##               Estimate Std. Error t value Pr(>|t|)
## (Intercept)   2.885458   0.291010   9.915 3.25e-12 ***
## x             0.087433   0.008784   9.953 2.92e-12 ***
## qrt2         -0.120383   0.212083  -0.568   0.574
## qrt3          0.111753   0.210081   0.532   0.598
## qrt4         -0.075539   0.198288  -0.381   0.705
## ---
## Signif. codes:
## 0 '***' 0.001 '**' 0.01 '*' 0.05 '.' 0.1 ' ' 1
##
## -------------------------------------------------------------------
## Sigma link function:  log
## Sigma Coefficients:
##               Estimate Std. Error t value Pr(>|t|)
## (Intercept)   -1.6032     0.2505     -6.4 1.44e-07 ***
## ---
## Signif. codes:
## 0 '***' 0.001 '**' 0.01 '*' 0.05 '.' 0.1 ' ' 1
##
## -------------------------------------------------------------------
## No. of observations in the fit:  45
## Degrees of Freedom for the fit:  6
##         Residual Deg. of Freedom:  39
##                      at cycle:  3
##
## Global Deviance:    492.6373
##            AIC:     504.6373
##            SBC:     515.4773
## *******************************************************************
```

```
names(sm100)
```

```
##  [1] "ps"          "co"          "p1"          "pm"
##  [5] "est.disp"    "coef.table"  "pvalue"      "tvalue"
##  [9] "se"          "coef"        "ifWarning"   "covmat"
## [13] "object"      "type"        "robust"      "save"
## [17] "hessian.fun" "digits"
```

```
sm100$pvalue
```

```
##  (Intercept)            x         qrt2         qrt3
## 3.253778e-12 2.921666e-12 5.735457e-01 5.977766e-01
##        qrt4  (Intercept)
## 7.053023e-01 1.443488e-07
```

The fitted model is given by $Y \sim \text{NBI}(\hat{\mu}, \hat{\sigma})$ where

$$\begin{aligned}
\log(\hat{\mu}) =\ & 2.885 + 0.0874x - 0.1204(\text{if qrt} = 2) \\
& +0.1118(\text{if qrt} = 3) - 0.0755(\text{if qrt} = 4), \\
\log(\hat{\sigma}) =\ & -1.6032.
\end{aligned}$$

For testing the significance of individual terms, given all the rest of the terms in the model, it is generally better to use `drop1()` rather than relying on p-values from `summary()`. The `drop1()` function provides the generalized likelihood ratio test (GLRT) for dropping each term, which is generally much more reliable than the Wald test based on the standard error given by the p-value (i.e. `Pr(>|t|)` above). The GLRT has an asymptotic Chi-squared distribution with degrees of freedom equal to the number of parameters in the term dropped. This only applies if the model does not include smoothing terms. (In the presence of smoothing terms in the model, `drop1()` could be used as a rough guide to the significance of each of the parametric terms, with the smoothing degrees of freedom fixed at their values chosen from the model prior to `drop1()`.) Note that for complicated models with large data sets, `drop1()` can take a few minutes.

In the following example we first apply `drop1()` to the fully parametric model `m100`:

```
drop1(m100)

## Single term deletions for
## mu
##
## Model:
## y ~ x + qrt
##           Df    AIC     LRT  Pr(Chi)
## <none>        504.64
## x          1 576.91  74.271   <2e-16 ***
## qrt        3 500.23   1.593    0.661
## ---
## Signif. codes:
## 0 '***' 0.001 '**' 0.01 '*' 0.05 '.' 0.1 ' ' 1
```

The resulting test p-value for the parametric term `qrt` is 0.661. However this is unreliable because the linear dependence of $\log(\mu)$ on x is inadequate.

In model `m300` we fix the smoothing degrees of freedom in model `m200` (Section 5.2.1), which includes a smoothing term, and then apply `drop1(m300)`:

```
m300 <- gamlss(y~pb(x, df=m200$mu.nl.df)+qrt, data=aids, family=NBI,
               trace=FALSE)
drop1(m300)

## Single term deletions for
## mu
```

```
##
## Model:
## y ~ pb(x, df = m200$mu.nl.df) + qrt
##                                  Df    AIC      LRT    Pr(Chi)
## <none>                                390.10
## pb(x, df = m200$mu.nl.df)   6.5889 576.91  199.983 < 2.2e-16
## qrt                         3.0000 402.49   18.389 0.0003656
##
## <none>
## pb(x, df = m200$mu.nl.df) ***
## qrt                       ***
## ---
## Signif. codes:
## 0 '***' 0.001 '**' 0.01 '*' 0.05 '.' 0.1 ' ' 1
```

The resulting test p-value for the parametric term `qrt` is 0.00037. This gives a rough guide to testing `qrt`. The corresponding test not fixing the smoothing degrees of freedom is certainly may be unreliable. (Note also that the model `m300`, or equivalently `m200`, has the smallest value of AIC.)

5.3   Functions to obtain confidence intervals

The function `confint()` provides Wald standard error-based confidence intervals for parameters, while there are two functions providing profile likelihood confidence intervals for a single parameter of a GAMLSS model:

`prof.dev()`   produces a profile deviance plot for any constant distribution *parameter* μ, σ, ν or τ;

`prof.term()`   produces a profile deviance plot for any linear *term* parameter in a predictor model for μ, σ, ν or τ.

A profile deviance plot is a plot of the global deviance against a particular parameter in a model. It is obtained by fitting the model for each of a sequence of fixed values of the parameter.

5.3.1   The `confint()` function

The generic function `confint()` provides Wald standard error-based confidence intervals for the coefficients (i.e. parameters) in the predictor of a distribution parameter μ, σ, ν or τ. The Wald standard error-based confidence intervals are generally much less reliable than the profile likelihood confidence intervals, given in Section

5.3.2, if the model is correct, but may be preferred (with `robust=TRUE`) if the model is incorrect. `confint()` has the following arguments in **gamlss**:

```
confint(object, parm, level = 0.95,
                      what = c("mu", "sigma", "nu", "tau"),
                      parameter = NULL, robust = FALSE, ...)
```

where

object   a `gamlss` fitted object;

parm   which terms are to be given confidence intervals, either a vector of numbers or a vector of names. If missing, all terms are considered;

level   the confidence level required;

what or parameter
                which distribution parameter to consider;

robust   whether the usual (`robust=FALSE`) or robust or sandwich (`robust=TRUE`) standard errors should be used;

...   extra arguments.

The following is an example using `confint()`:

```
confint(m100)
```

```
##                   2.5 %      97.5 %
## (Intercept)   2.48404626  3.2868695
## x             0.07462303  0.1002434
## qrt2         -0.49773874  0.2569731
## qrt3         -0.26476513  0.4882712
## qrt4         -0.45236355  0.3012849
```

```
confint(m100,1, robust=TRUE)
```

```
##                 2.5 %    97.5 %
## (Intercept)   2.31509  3.455826
```

5.3.2   The `prof.dev()` function

The function `prof.dev()` produces a profile deviance plot for any of the distribution parameters μ, σ, ν or τ, and is useful for checking the reliability of models in which one (or more) of the parameters in the distribution are constant (and therefore have not been modelled as functions of explanatory variables). The `prof.dev()` also provides a $100(1-\alpha)\%$ profile likelihood confidence interval for the parameter which is, in general, much more reliable than a (Wald) standard error-based confidence interval if the model is correct.

```
prof.dev(object, which = NULL, min = NULL, max = NULL, step = NULL,
         length = 7, startlastfit = TRUE, plot = TRUE, perc = 95,...)
```

where

object	a **gamlss** fitted object;
which	which parameter to get the profile deviance for, e.g. `which="tau"`;
min	the minimum value for the parameter, e.g. `min=1`;
max	the maximum value for the parameter, e.g. `max=20`;
step	step length of the grid for the parameter, which determines how often to evaluate the global deviance, e.g. `step=1`;
length	if `step` is not set, how many times the global deviance has to be evaluated for the construction of the profile deviance. The default value is `length=7`, with equal spacing from `min` to `max`;
startlastfit	whether to start fitting from the last fit or not, with default `startlastfit=TRUE`;
plot	whether to plot (`plot=TRUE`) or save the results (`plot=FALSE`);
perc	what % confidence interval is required;
...	extra arguments.

As an example consider the abdominal circumference model fitted in Section 4.4, where a t-distribution is fitted to the data. Smooth terms are fitted for μ and σ and a constant for ν, which is the degrees of freedom parameter. It is of some interest to find a confidence interval for ν. Note that $\nu = 1$ corresponds to a Cauchy distribution while a large value of ν corresponds closely to a normal distribution. Usually it takes several attempts to select a suitable range for the parameter in order to produce a decent graph. As a default (if the argument **step** is not specified) the profile deviance is evaluated at only seven points and a cubic spline approximation of the function is computed. This can produce a wobbly function. The arguments **step** or **length** can then be used to improve the approximation. Our advice is to start with a sparse grid for the parameter (i.e. few points) and improve that when you see the resulting plot. This should include the full 95% confidence interval for the parameter within the horizontal axis scale, and the horizontal deviance bar representing the global deviance at the endpoints of the parameter interval should be roughly halfway up the vertical axis scale. Note that the procedure requires fitting the model repeatedly for a sequence of fixed values of the parameter of interest (ν in this example), so it can be slow.

Here we reproduce our first attempt (left panel of Figure 5.1) and our final attempt (right panel of Figure 5.1).

Fig.
5.1

```
### First attempt
### By default the global deviance is evaluated at seven values of nu
### equally spaced from min=5 to max=50.
h<-gamlss(y~pb(x),sigma.fo=~pb(x),family=TF,data=abdom,trace=FALSE)
h2<-gamlss(y~pb(x, df=h$mu.nl.df), sigma.fo=~pb(x, df=h$sigma.nl.df),
           family=TF, data=abdom, trace=FALSE)
pd1<-prof.dev(h2,"nu", min=5, max=50)
```

```
## ****************************************************************
## . . .
## ****************************************************************
## The Maximum Likelihood estimator is   13.88497
## with a Global Deviance equal to  4777.429
## A  95 % Confidence interval is: ( 7.342625 , 40.72531 )
## ****************************************************************
```

Now we increase the number of evaluations of the global deviance to 20 (i.e. 20 values of ν equally spaced from `min=5` to `max=50`). This provides a more accurate maximum likelihood estimate $\hat{\nu} = 11.61$ for ν and a more accurate 95% (profile likelihood) confidence interval for ν given by (6.18, 42.04):

Fig.
5.1

```
### Final attempt
pd2<-prof.dev(h2,"nu",min=5, max=50, length=20)
```

```
## ****************************************************************
## . . .
## ****************************************************************
## The Maximum Likelihood estimator is   11.61266
## with a Global Deviance equal to  4777.528
## A  95 % Confidence interval is: ( 6.179175 , 42.04074 )
## ****************************************************************
```

Note that the object **pd2** has several components saved, among them the profile deviance function under the name **fun**, which is a function of the distribution parameter (ν here). This function can be used for further evaluations at values of the distribution parameter (e.g. $\nu = 34$ here), as the following code shows. Beware of trying to evaluate outside the original range (e.g. 5 to 50 of ν here), since this can be very misleading.

```
names(pd2)
```

```
## [1] "values"    "fun"        "min"        "max"
## [5] "max.value" "CI"         "criterion"
```

```
pd2$fun(34)
```

```
## [1] 4780.655
```

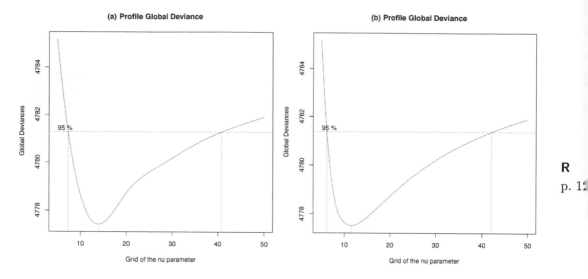

FIGURE 5.1: Profile deviance for ν from a t-family fitted model **h2** using **abdom** data. The left panel (a) has 7 evaluations of the function, while the right panel (b) has 20.

The profile deviance for ν (as in Figure 5.1(b)) can be plotted again by:

```
curve(pd2$fun(x), 5, 50)
```

For different levels of confidence interval, change the **perc** option, e.g. for 99% use **perc=99**.

5.3.3   The prof.term() function

The function **prof.term()** is similar to **prof.dev()** but it can provide a profile deviance for any parameter in the model, not just for a constant distribution parameter. That is, while **prof.dev()** can be applied to profile a (constant) parameter of the distribution (i.e. μ, σ, ν or τ), **prof.term()** can be applied to any parameter in the predictor model for μ, σ, ν or τ. In order to show how **prof.term()** works, consider again the **aids** data. Let us assume first that we are interested in fitting a linear term in time (**x**) plus a factor for the quarterly seasonal effect (**qrt**), using the negative binomial type I model. This model is fitted as

```
m100 <-gamlss(y~x + qrt, family = NBI, data = aids)
```

The coefficient for **x** has the value of 0.08743 and a t-value of 13.3773, which indicates that it is highly significantly different from zero. An approximate (Wald) standard error-based 95% confidence interval for this parameter can be obtained using the function **confint()**:

R
p. 1?

```
confint(m100, "x")
```

```
##          2.5 %     97.5 %
## x 0.07462303 0.1002434
```

We shall now use `prof.term()` to find, a (profile likelihood) 95% confidence interval (which is probably more accurate) for the linear term parameter for x. The parameter of interest is denoted `this` in the model formula and is the slope parameter of x in the predictor for μ. For a fixed value of `this`, the term `this*x` is fixed and known and so is offset in the predictor for μ:

```
mod<-quote(gamlss(y~offset(this*x) + qrt, data=aids, family=NBI))
prof.term(mod, min=0.06, max=0.11, length=20)
```

```
## GAMLSS-RS iteration 1: Global Deviance = 508.1867
## GAMLSS-RS iteration 2: Global Deviance = 508.1845
## GAMLSS-RS iteration 3: Global Deviance = 508.1845
## GAMLSS-RS iteration 1: Global Deviance = 500.8052
## . . .
## *************************************************************
## The Maximum Likelihood estimator is  0.08739487
## with a Global Deviance equal to   492.6384
## A   95 % Confidence interval is: ( 0.07458118 , 0.1008422 )
## *************************************************************
```

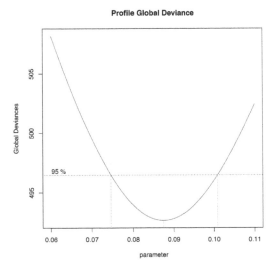

FIGURE 5.2: The profile deviance for the coefficient of x.

The profile deviance looks quadratic, so it is not a surprise that the approximate (Wald) standard error-based 95% confidence interval and the 95% profile interval are almost identical. In general this will not be the case if the likelihood is not nearly quadratic at the maximum. To obtain a 99% interval, use

```
prof.term(mod, min=0.06, max=0.11, length=20, perc=99)
```

Now we use **plof.term()** to find a breakpoint in the relationship between the response and one of the explanatory variables. Stasinopoulos and Rigby [1992] have shown that the **aids** data provide clear evidence of a breakpoint in the AIDS cases over time. Here we consider model **x+(x>break)*(x-break)**. This predictor model is linear in **x** up to **x=break**, then continues linearly in **x** but with a different slope, i.e. a piecewise linear model with continuity at the breakpoint. We are interested in estimating the breakpoint; see Stasinopoulos and Rigby [1992]. More detail on piecewise polynomials is given in Section 8.5.

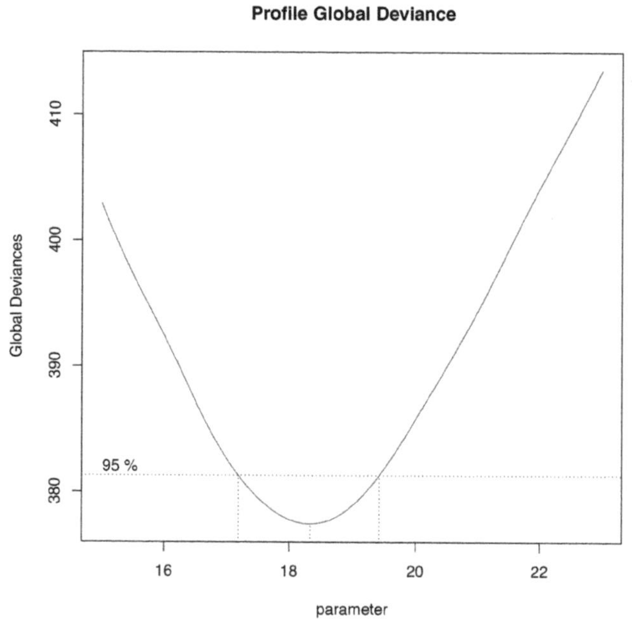

FIGURE 5.3: The profile deviance for the breakpoint parameter of **x**.

R

p. 132

```
aids.1 <- quote(gamlss(y~x+I((x>this)*(x-this))+qrt, family=NBI,
                 data=aids))
prof.term(aids.1, min=16, max=21, length=20,  criterion="GD")
```

```
## GAMLSS-RS iteration 1: Global Deviance = 403.8474
## GAMLSS-RS iteration 2: Global Deviance = 402.9138
## GAMLSS-RS iteration 3: Global Deviance = 402.913
## GAMLSS-RS iteration 1: Global Deviance = 396.6076
## . . .
## ***********************************************************************
## The Maximum Likelihood estimator is  18.33295
## with a Global Deviance equal to  377.4614
## A  95 % Confidence interval is: ( 17.19218 , 19.41849 )
```

Fig.
5.3

```
## **********************************************************************
```

The profile plot shown in Figure 5.3 suggests strong support for a breakpoint since the confidence interval for the breakpoint lies within the range of **x**. The estimated breakpoint is 18.33 with a 95% (profile likelihood) confidence interval given by (17.19, 19.42). Now we plot the **aids** data with the fitted breakpoint model:

```
aids.2<-gamlss(y~x+I((x>18.33)*(x-18.33))+qrt,family=NBI,data=aids)
plot(aids$x,aids$y)
lines(aids$x, fitted(aids.2), col="red")
```

Finally **prof.term()** can also be used as a way of determining a smoothing parameter in a model by plotting the GAIC(κ) (see Section 2.3.7). The penalty κ is specified by the **penalty** argument of **prof.term()**. Consider the model

```
gamlss(y~cs(x,df=??) + qrt, data = aids, family = NBI)
```

in which we would like to determine a reasonable value for the missing degrees of freedom of the cubic spline function **cs()**. (Note that in **pb()** the smoothing parameter and therefore the degrees of freedom are estimated automatically using a local maximum likelihood procedure, while here the estimation of effective degrees of freedom is done globally.) Models with different degrees of freedom can be fitted and their GAIC plotted against the degrees of freedom. This process can be automated using **prof.term()**:

Fig.
5.4

```
mod1<-quote(gamlss(y~cs(x,df=this) + qrt, data=aids, family=NBI))
prof.term(mod1, min=1, max=15, step=1, criterion="GAIC", penalty=2.5)
```

```
## GAMLSS-RS iteration 1: Global Deviance = 419.651
## GAMLSS-RS iteration 2: Global Deviance = 423.8293
## GAMLSS-RS iteration 3: Global Deviance = 425.0032
## GAMLSS-RS iteration 4: Global Deviance = 425.0032
## . . .
## GAMLSS-RS iteration 3: Global Deviance = 347.9341
## **********************************************************************
## The Mimimum is  8.240494
## with an an GAIC( 2.5 ) = 394.2163
## **********************************************************************
```

The profile GAIC($\kappa = 2.5$) plot, shown in Figure 5.4, suggests support for effective degrees of freedom for the cubic smoothing function in **x** of 8.24 (in addition to the constant and linear term). Alternative penalties values could be used, e.g. $\kappa = 2, 3, 4$ or $\log n$. (Note an alternative to the use of **prof.term()** here is to use the **find.hyper()** function.)

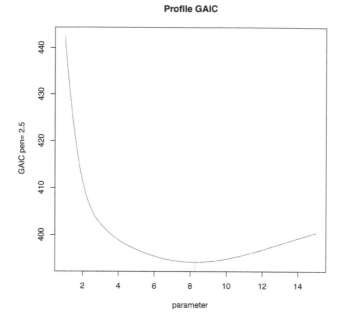

R

p. 13

FIGURE 5.4: Profile GAIC($\kappa = 2.5$) for the cubic spline degrees of freedom in the model
`gamlss(y ~ cs(x,df=this) + qrt, data = aids, family = NBI)`.

> **Important**: Profile deviance intervals of a parametric term coefficient should
> be used with care if smoothing or random effects are included in the model
> for any of the distribution parameters. In general they produce narrower inter-
> vals than the marginal likelihood approach; see Rigby and Stasinopoulos [2005,
> Section 6.2 and Appendix A.2]. The more accurate profile deviance intervals
> are obtained from the approximate marginal likelihoods which are model de-
> pendent. At present we do not provide a general function for calculating these
> intervals but see comments below.

Finding a confidence interval for a parametric term coefficient when the model contains smoothing terms is more difficult. One possible approach which may provide
a rough guide to the confidence interval is as follows:

(i) fit the full model including smoothing terms in which the smoothing parameter
or the effective degrees of freedom could be estimated, e.g. `pb(x)`;

(ii) fix each of the smoothing degrees of freedom to their values from the full fitted
model in (i);

(iii) profile a parametric term coefficient in model (ii).

5.4 Functions to obtain predictions

5.4.1 The predict() function

The generic function **predict()** in **gamlss** is the **gamlss** method which produces predictors for the current or a new data set for a specified parameter of a **gamlss** object. It can be used to extract predictors (**type="link"**), fitted values (**type="response"**) and contributions of terms in the predictor (**type="terms"**), for a specific parameter in the model, at the current or new values of the explanatory variables, in a similar way that **predict.lm()** and **predict.glm()** can be used for **lm** and **glm** objects. Certain problems that were associated with the above functions, see Venables and Ripley [2002, Section 6.4], are avoided here since **predict.gamlss()** is based on the **predict.gam()** S-PLUS function [Chambers and Hastie, 1992]. Note that the main difference between the **gamlss predict()** and the usual predictive functions in **R** is the fact that the **gamlss predict()** is distribution parameter specific, that is, predictions are for one of the distribution parameters **mu**, **sigma**, **nu** or **tau**.

Predictors, fitted values and specific terms for a specific distribution parameter in the model at the current data values of the explanatory variables can also be extracted using **lpred()** (which in fact is called from **predict()** if the **newdata** argument is NULL; see below).

> **Warning**: Extrapolation, i.e. prediction (of the response variable distribution) outside the range of the values of the explanatory variables in the original data set, is *dangerous* for any statistical model (because the relationship between the response variable and the explanatory variables may not extend outside the original data set). Extrapolation should be avoided or treated with great caution.

The **gamlss predict()** function is defined as

```
predict(object, what=c("mu", "sigma", "nu", "tau"), parameter=NULL,
            newdata=NULL, type = c("link", "response", "terms"),
            terms=NULL, se.fit=FALSE, data=NULL, ...)
```

where

object   a **gamlss** fitted object;

what or **parameter**
                what parameter of the distribution is required, **mu**, **sigma**, **nu** or **tau**. The default is **what="mu"**;

newdata   a data frame containing new values for the explanatory variables used in the model;

type	The default value `type="link"` gets the predictor for the specified distribution parameter. `type="response"` gets the fitted values for the response and `type="terms"` gets the contribution of fitted terms in the predictor for the specified parameter;
terms	if `type="terms"`, the specified term from the formula of the parameter at hand is selected. By default all terms are selected;
se.fit	if `se.fit=TRUE` the approximate standard errors of the appropriate type are extracted. Note that standard errors are not given for new data sets, i.e. when `newdata` is defined. The default is `se.fit=FALSE`;
data	the data frame used in the original fit if it is not defined in the call;
...	extra arguments.

The `lpred()` function of **gamlss** has identical arguments to the `predict()` function in **gamlss**, apart from the `newdata` argument which does not exist in `lpred()`. The functions `fitted()` and `fv()` are equivalent to using `lpred()` or `predict()` with the argument `type="response"`. The function `lp()` is equivalent to `lpred()` or `predict()` with the argument `type="link"`. We demonstrate some of these points using the **aids** data set.

```
data(aids)
 # fitting a negative binomial type I distribution
aids.1<-gamlss(y~poly(x,3)+qrt, family=NBI, data=aids,trace=FALSE)
```

First we plot the **aids** data with the fitted μ.

```
plot(aids$x, aids$y)
lines(aids$x, aids.1$mu.fv)
```

Fig.
5.5

```
#To see the first 6 values of the fitted predictor for mu
head(predict(aids.1))
```

```
##        1        2        3        4        5        6
## 1.322524 1.490931 1.996051 2.140244 2.540856 2.643345
```

```
#Alternative ways of obtaining the fitted predictor for mu
identical(predict(aids.1),predict(aids.1, parameter="mu"))
```

```
## [1] TRUE
```

```
identical(predict(aids.1),predict(aids.1, parameter="mu",
                              type="link"))
```

```
## [1] TRUE
```

```
#Obtaining the fitted parameter mu for the first 6 observations
head(predict(aids.1, type="response"))
```

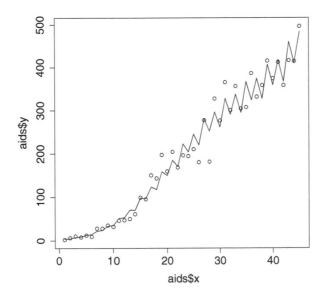

FIGURE 5.5: The `aids` data set with the fitted μ.

```
##         1        2        3        4         5          6
## 3.752880  4.441230  7.359933  8.501513  12.690525  14.060153
```

Standard errors for prediction

The argument `se.fit=TRUE` can be used to obtain approximate standard errors for
`predict()`. The result is a list containing two objects, `fit` and `se.fit`.

```
#Obtaining the fitted parameter mu and its approximate standard
#errors for the observations
paids.1 <- predict(aids.1, what="mu", se.fit=TRUE ,type="response")
names(paids.1)
```

```
## [1] "fit"    "se.fit"
```

```
head(paids.1$se.fit)
```

```
##         1        2        3        4         5          6
## 0.6739890  0.6939025  1.0019629  1.0183976  1.3176894  1.2834913
```

```
#Obtaining the fitted predictor for mu and its approximate standard
#errors for the observations
paids.2 <- predict(aids.1, what="mu", se.fit=TRUE, type="link")
head(paids.2$se.fit)
```

```
##          1          2          3          4          5
## 0.17959249  0.15624108  0.13613750  0.11979015  0.10383254
```

```
##              6
## 0.09128573
```

> **Important**: Standard errors should be used with caution. If the predictor contains only linear terms (i.e. no smoothing terms) then the standard errors of the predictor (using the option `type="link"`) are correctly calculated. Standard errors for fitted distribution parameters if the link function is not the identity function are calculated using the *delta method* which could be unreliable; see Chambers and Hastie [1992, page 240]. If additive (smoothing) terms are included in the model of a specific distribution parameter then the unreliability increases since the standard errors of the additive (smoothing) terms are crudely approximated using the method described in Chambers and Hastie [1992, Section 7.4.4].

The option `type="terms"` creates a matrix containing the contribution to the predictor from each of the terms in the model formula. If in addition the argument `se.fit=TRUE` is set, then a list of two objects is created, each containing a matrix. The first matrix contains the contribution of the terms to the predictor and the second their approximate standard errors. The number of columns in the matrices are the number of parameters in the model formula. The argument `terms` can be used in combination with `type="terms"` to select the contribution to the predictor of a specific term in the model. A typical use of the option `type="terms"` is for plotting the additive contribution of a specific term in modelling a distribution parameter, as in `term.plot()`.

```
#Obtaining the contributions to the fitted predictor of mu from its
#two model terms poly(x,3) and qrt
paids.2 <- predict(aids.1, what="mu", type="terms")
colnames(paids.2)
```

```
## [1] "poly(x, 3)" "qrt"
```

```
# now with se
paids.2 <- predict(aids.1, what="mu", type="terms", se.fit=TRUE)
names(paids.2)
```

```
## [1] "fit"    "se.fit"
```

```
colnames(paids.2$fit)
```

```
## [1] "poly(x, 3)" "qrt"
```

```
colnames(paids.2$se.fit)
```

```
## [1] "poly(x, 3)" "qrt"
```

```
# select only "qrt" to save
paids.2 <- predict(aids.1, what="mu", type="terms", se.fit=TRUE,
```

```
                        terms="qrt")
colnames(paids.2$fit)

## [1] "qrt"
```

The most common use of `predict()` is to obtain fitted values for a specific parameter at new values of the explanatory variables, for predictive purposes or for validation. To do so, the argument `newdata` should be set, and it needs to be a data frame containing the explanatory variables used in the model. This could cause problems if a transformed variable is used in the fitting of the original model (see below).

The `predict()` function for **gamlss** is based on the `predict.gam()` S-PLUS function; see Chambers and Hastie [1992, Section 7.3.3]. Note that `predict()` is fast but can be unreliable, especially for extrapolation. The steps used in the execution of the `predict()` function are described below. This may involve recalculating the model design matrix for the original data set (e.g. for the B-spline base, `bs()`, used in the P-spline, `pb()`). Especially for extrapolation, this may result in unreliable predictions, both for *extrapolated* and *interpolated* predictions at values of the exploratory variables *outside* and *inside* the original data set, respectively. For P-splines this is caused by the positions of the default 20 knots being potentially greatly displaced for an exploratory variable x when spread over the extrapolated x range, resulting in many fewer knots within the original x range. This distorting effect of extrapolation on the predictions within the original x range may be reduced by using more knots (e.g. `inter=50` instead of the default 20) in the **gamlss** fit before prediction is used, however this is *unlikely* to prevent the potentially unreliable extrapolated predictions. Note that the cubic splines function `cs()` does not have this problem of repositioned knots, since it has a knot for each distinct x value, but as always extrapolation is potentially unreliable.

Steps in `predict()`

We follow here the steps used in the execution of `predict()` taken from Chambers and Hastie [1992, Section 7.3.3]. Let $\mathbf{D}_{old}$ be the original data frame, with original design matrix $\mathbf{X}_{old}$, and $\mathbf{D}_{new}$ the new data frame (the new x-values where the fitted model has to be evaluated). We assume that both data frames contain the correct columns in the sense that the x-variables used in the model formula (for the specific distribution parameter) are present in both.

1. Construct a new data frame using the combined (old and new) data, with columns for the matching variables included in both data frames, i.e. $\mathbf{D}_n = \begin{bmatrix} \mathbf{D}_{old} \\ \mathbf{D}_{new} \end{bmatrix}$.

2. Construct the model frame and the corresponding new design matrix, $\mathbf{X}_n = \begin{bmatrix} \mathbf{X}_{old_2} \\ \mathbf{X}_{new} \end{bmatrix}$, using the combined data frame, $\mathbf{D}_n = \begin{bmatrix} \mathbf{D}_{old} \\ \mathbf{D}_{new} \end{bmatrix}$. Note that for certain

models (when the function used to construct the design matrix is data dependent) the submatrix $\mathbf{X}_{old_2}$ of the new design matrix $\mathbf{X}_n$, corresponding to the original observations in $\mathbf{D}_{old}$, may be different from the original design matrix $\mathbf{X}_{old}$. This can happen, for example, if the B-spline base (bs()) is used in the model.

3. The parametric part of the model for the specified parameter is refitted using only $\mathbf{X}_{old_2}$. If the difference between the old and the new fit is large, a warning is given.

4. The coefficients from the fit obtained using only $\mathbf{X}_{old_2}$ are used to obtain the new predictions.

5. If the gamlss object contains additive (smoothing) components an additional step is taken. The additive (smoothing) function is fitted to the old data but using the new basis, i.e. $\mathbf{X}_{old_2}$, (recalculating the smoothing parameters automatically if necessary). The additive (smoothing) function is then evaluated at the new data values using $\mathbf{X}_{new}$. (This requires that the additive function has a predict option.)

Here we use the aids data to fit a negative binomial model using a polynomial poly(), a B-spline base bs(), and a smoothing P-spline pb() to model time (x). The sigma parameter is a constant in the model. predict() is used first, to find fitted values for μ at new data values and finally for σ. Note that predict() gives a warning when bs is used in the μ model.

```
data(aids)
mod1 <- gamlss(y~poly(x,3)+qrt, family=NBI, data=aids, trace=FALSE)
mod2 <- gamlss(y~bs(x,5)+qrt, family=NBI, data=aids, trace=FALSE)
mod3 <- gamlss(y~pb(x)+qrt, family=NBI, data=aids, trace=FALSE)
```

```
# create a new data frame
newaids<-data.frame(x=c(46,47,48), qrt=c(2,3,4))
# predict "mu" at new values
(ap1 <- predict(mod1,what="mu",newdata=newaids,type="response"))
```

```
## [1] 430.3484 548.9467 498.9708
```

```
(ap2 <- predict(mod2,what="mu",newdata=newaids,type="response"))
```

```
## [1] 391.6007 474.3773 405.1320
```

```
(ap3 <- predict(mod3,what="mu",newdata=newaids,type="response"))
```

```
## new prediction
## [1] 426.3170 539.9632 491.5646
```

```
# get the term contributions to the predictor of mu
(ap4 <- predict(mod3,what="mu",newdata=newaids,type="terms"))
```

```
## new prediction
##      pb(x)          qrt
## 1 1.399057 -0.09869534
## 2 1.449094  0.08758643
## 3 1.501004 -0.05823180
## attr(,"constant")
## [1] 4.754821
```

```
# predict sigma at new values
(ap5 <- predict(mod3,what="sigma",newdata=newaids,type="response"))
```

```
## [1] 0.005131356 0.005131356 0.005131356
```

For example the predicted value of the predictor for μ for the first new case (i.e. case 46) is $\eta_1 = 4.7548 + 1.3991 - 0.09869 = 6.05521$ with corresponding predicted value for μ given by $\exp(\eta_1) = \exp(6.05521) = 426.3$. Note that the `se.fit` argument does not work with new data.

The following example is taken from Venables and Ripley [2002] (who used it to demonstrate that `predict.lm()` did not work properly for `lm` models). The data set `wtloss` (in the **MASS** library) is a data frame that contains the weight, in kilograms, of an obese patient at 52 time points over an eight-month period of a weight rehabilitation programme.

R data file: `wtloss` in package **MASS** of dimensions 52×2

variables

    Days : time in days since the start of the programme.

    Weights : weight in kilograms of the patient.

purpose: to demonstrate prediction in GAMLSS

Here we use the **gamlss** generic `predict()` function:

```
library(MASS)
data(wtloss)
# squaring Days
quad <-gamlss(Weight~Days+I(Days^2), data=wtloss)
```

```
## GAMLSS-RS iteration 1: Global Deviance = 137.8867
## GAMLSS-RS iteration 2: Global Deviance = 137.8867
```

```
# new data
new.x <-data.frame(Days=seq(250,300,10), row.names=seq(250,300,10))
# use predict
predict(quad, newdata=new.x)
```

```
## [1] 112.5061 111.4747 110.5819 109.8277 109.2121 108.7351
```

Prediction using a transformed explanatory variable

If a transformed variable is used in the fitting of the current data, some care has to be taken to ensure that the correct variables exist in the new data. Here we revisit the **abdom** data set. Let us assume that a transformation of age is needed in the model, e.g. square root of age. This could be fitted as

```
mod<-gamlss(y~cs(age^.5), data=mydata)
```

or by transforming age first:

```
mydata$nage<-(mydata$age)^.5
```

and then fitting

```
mod<-gamlss(y~cs(nage), data=mydata)
```

The latter fit is more efficient, particularly for a data set with a large number of cases. In the first case, the code

```
predict(mod, newdata=data.frame(age=c(34,56)))
```

would produce the prediction results. In the second case, a new data frame has to be created containing the old data plus any new transformed variable. This data frame has to be declared in the **data** argument of the **predict()** function. The option **newdata** should contain a **data.frame** with the transformed variable names and the transformed variable values for which prediction is required, as the following example demonstrates.

```
data(abdom)
# assume that a transformation x^.5 is required
aa<-gamlss(y~pb(x^.5), data=abdom, trace=FALSE)
# predict at old values, e.g.  case 610
predict(aa, what="mu")[610]
```

```
## [1] 371.4253
```

```
# predict at new data
predict(aa, newdata=data.frame(x=abdom$x[610]))
```

```
## new prediction
## [1] 371.4253
```

```
# now transform x first
nx<-abdom$x^.5
aaa<-gamlss(y~pb(nx), data=abdom, trace=FALSE)
# create a new data frame
newd<-data.frame(abdom, nx=abdom$x^.5)
# predict at old values
predict(aaa)[610]
```

```
## [1] 371.4253
```

```
# predict at new values
predict(aaa, newdata=data.frame(nx=abdom$x[610]^.5), data=newd)
```

```
## new prediction
## [1] 371.4253
```

5.4.2   The predictAll() function

The predictAll function produces predictors for the current or a new data set for all the distribution parameters of a gamlss object. (It is important to note that extrapolation should be avoided or treated with great caution, as discussed in Section 5.4.1.) By default the execution of predictAll() uses the same steps as predict() given in Section 5.4.1.

Alternatively, with the argument use.weights=TRUE, the execution of predictAll() is achieved by adding the new cases to the data set with weights given by set.to, e.g. set.to=0, and the gamlss model refitted to all (old and new) cases. The default value of set.to is the machine smallest non-zero normalized floating-point number, and so allows for approximate standard errors of predictions to be calculated. The user may need to change the starting values for the distribution parameters μ, σ, ν and τ (e.g. using argument mu.start). (Note that the same general problem of unreliable extrapolation still applies to predictAll(), as does the specific problem of extrapolation resulting in repositioned knots in pb(), as discussed in Section 5.4.1.)

By default in predictAll() the predicted values of parameters (type="response") are obtained, unlike the predict() function whose default is the predicted values of the predictors (i.e. type="link"). For example for the abdom data set:

```
h<-gamlss(y~pb(x),sigma.fo=~pb(x),family=TF,data=abdom,trace=FALSE)
hall<-predictAll(h)
```

This provides fitted (i.e. predicted) values (for all cases in the data set abdom) for the parameters of the distribution (TF), saved as hall$mu, hall$sigma and hall$nu. Also the dependent variable is saved as hall$y.

The fitted values of all the parameters can also be saved for new data cases, e.g. for gestational age x = 20, 25 and 30 weeks:

```
newabdom<-data.frame(x=c(20,25,30))
hall2<-predictAll(h,newdata=newabdom)
```

Hence the fitted distribution of the response variable can be plotted using the pdf for each of the three new cases by:

```
pdf.plot(mu=hall2$mu, sigma=hall2$sigma, nu=hall2$nu, family=TF,
        min=100, max= 350, step=1)
```

Fig.
5.6

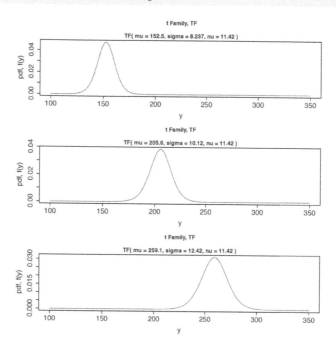

R

p. 14

FIGURE 5.6: **abdom** data: fitted pdf of **y** for ages **x** = 20, 25 and 30 weeks.

5.5   Appendix: Some theoretical properties of GLM and GAMLSS

The generalized linear model (GLM) of Nelder and Wedderburn [1972] (a fully parametric model) assumes an exponential family distribution for Y, denoted here by $\mathcal{E}(\mu, \sigma)$, with mean μ and scale σ (where $\phi = \sigma^2$ is called the dispersion parameter), and only allows modelling of the mean μ. The variance of Y, within the exponential family, is a function of both the mean parameter μ, (which depends, in general, on explanatory variables), and the scale parameter σ which is assumed to be constant (and therefore cannot be modelled explicitly by explanatory variables). The GLM mean model parameter estimators are **not** robust to outliers, so a single observation has potentially an unlimited effect on the mean model parameter estimators.

One of the important properties of a GLM (coming from the assumption of the exponential family) is that the estimators of the mean model parameters are consistent if the mean model is correct, even if the variance and/or distribution model is wrong; see Fahrmeir and Tutz [2001, page 57] and Gourieroux et al. [1984]. (This is an important property, not shared in general by GAMLSS models where the mean model parameters are consistent if the full GAMLSS model is correct.) However, for a GLM in which the true mean model is correct, but the variance and/or distribution model is wrong, then, although the GLM mean model parameter estimators are consistent, they are, in general, asymptotically inefficient (relative to the maximum likelihood parameter estimators from the true model). Also if the mean model is correct, but the variance model is wrong, then the usual estimated standard errors of the GLM mean model parameter estimates are, in general, asymptotically incorrect, leading to incorrect asymptotic confidence interval coverage and test size; see for example Fahrmeir and Tutz [2001, page 57]. In this case the usual estimated standard errors of the GLM parameter estimates should be replaced by robust standard errors (i.e. robust to mis-specification of the variance and/or distribution model), which are then asymptotically correct [Fahrmeir and Tutz, 2001, page 57]. However the mean model parameter estimators are still, in general, asymptotically inefficient and so asymptotically their mean squared errors and expected lengths of their robust confidence intervals will be higher than those of the maximum likelihood estimators of the true model.

Hence, for example, if σ in a gamma GLM depends on explanatory variables, then the GLM gamma variance ($\sigma^2\mu^2$ for fixed σ) must be incorrect, leading, in general, to asymptotically incorrect estimated standard errors (for the mean model parameter estimators), and incorrect confidence interval coverage and test size (for the mean model parameters). GLM robust estimated standard errors are preferred. However the true GAMLSS gamma model, with models for both μ and σ, provides mean model parameter estimators which are asymptotically most efficient (with asymptotically correct standard errors), provided the gamma distribution and the models for μ and σ are correct. (If, however, the GAMLSS model is incorrect, then a GLM with robust standard errors may be preferred, or generalized estimating equations (GEE) with robust standard errors, if interest lies in the mean model parameters. See also the normal family (NOF) distribution in **gamlss**.)

5.6 Bibliographic notes

An overview of classical inference is given by Cox and Hinkley [1979]. The use of profile likelihood to obtain a confidence interval for a parameter (which is generally preferred to the (Wald) standard error based confidence interval if the model is correct) is considered by Aitkin et al. [2009].

Robust standard errors which may be appropriate under certain conditions when the model is incorrect are considered by Huber [1967] and White [1980]. An outline derivation of the distribution of the maximum likelihood estimator of model parameters when the model is misspecified is given by Ripley [1996, page 32] and Claeskens and Hjort [2003, p.26-27], while a more rigorous derivation, with sufficient conditions, is given by White [1980]. Gourieroux et al. [1984] also consider parameter estimation under model misspecification. See also Fahrmeir and Tutz [2001, p.55-60]. The use of GEE for parameter estimation is discussed in Hardin and Hilbe [2012]. The use of bootstrapping for tests and confidence intervals for a parameter is discussed in Efron and Tibshirani [1993] and Davison and Hinkley [1997].

5.7   Exercises

1. **The salmonella assay data:** Margolin et al. [1981] present the following data from an Ames Salmonella assay. The objective is to analyze the number of revertant colonies observed on a plate, given a dose of quinoline.

> **R data file:** `margolin` in package **gamlss.data** of dimensions 18×2
>
> **variables**
>
>> y : the number of revertant colonies observed on a plate
>>
>> x : the dose of quinoline
>
> **purpose:** to demonstrate inference within GAMLSS.

The data were subsequently analyzed by Breslow [1984], Lawless [1987] and Saha and Paul [2005]. Margolin et al. [1981] suggested the following nonlinear model for μ, the mean number of colonies:

$$\log(\mu) = \beta_0 + \beta_1 x + \beta_2 \log(x + 10) \ .$$

(a) Input and plot the data.

(b) Fit the above mean model together with a constant sigma model, using a negative binomial type I ($\text{NBI}(\mu, \sigma)$) distribution for y. Use `summary()` to display the results. Use `confint()` to obtain (Wald) standard error based confidence intervals for the coefficients.

```
m1<-gamlss(y~x+log(x+10),sigma.fo=~1,data=margolin,family=NBI)
```

(c) Fit the linear model given by setting $\beta_2 = 0$. Hence test whether $\beta_2 = 0$.

```
m2 <- gamlss(y~x, sigma.fo=~1, data=margolin, family=NBI)
LR.test(m2,m1)
#models are nested, so this is better, i.e. GLRT
```

Alternatively use `drop1(m1)`.

(d) Compare a Poisson $\text{PO}(\mu)$ model with the $\text{NBI}(\mu, \sigma)$ model in (b). Note that this is equivalent to testing $H_0 : \sigma = 0$ against $H_1 : \sigma > 0$. This hypothesis can be tested formally using the generalized likelihood ratio test statistic (GLRT), given by the difference in global deviance between the PO and NBI models, which has an asymptotic distribution which is 0 with a point mass of 0.5 and a χ_1^2 distribution with probability 0.5 [Lawless, 1987].

```
m0 <- gamlss(y~x+log(x+10), data=margolin, family=PO)
LR.test(m0,m1)
```

Note this p-value is incorrect, since it is calculated from χ_1^2. The GLRT statistic and correct p-value are given by

```
glrt <- m0$G.deviance- m1$G.deviance
p <- 0.5*pchisq(glrt,1,lower.tail=FALSE)
```

(e) Fit a smooth curve for μ and then smooth curves for both μ and σ using the NBI distribution, and compare these models to the previous parametric models using AIC and SBC.

```
m3 <- gamlss(y~pb(x), data=margolin, family=NBI)
m4 <- gamlss(y~pb(x), sigma.fo=~pb(x), data=margolin,
             family=NBI, n.cyc=100, method=mixed(10,100))
```

(f) Check the residuals of the preferred model.

2. **Prediction using aids data** from the **gamlss.data** package. The data set **aids** is presented on page 99.

   (a) Fit the following models:
   ```
   # poly
   mod1<-gamlss(y~poly(x,3)+qrt,family=NBI,data=aids,trace=FALSE)
   # bs
   mod2<-gamlss(y~bs(x,5)+qrt,family=NBI,data=aids,trace=FALSE)
   # pb
   mod3<-gamlss(y~pb(x)+qrt,family=NBI,data=aids,trace=FALSE)
   ```

   (b) For the above three models, use the **predict()** function to predict the values for the response variable for the **newaids** data.
   ```
   newaids<-data.frame(x=c(46,47,48), qrt=c(2,3,4))
   (ap1<-predict(mod1,what="mu",newdata=newaids,type="response"))
   (ap2<-predict(mod2,what="mu",newdata=newaids,type="response"))
   (ap3<-predict(mod3,what="mu",newdata=newaids,type="response"))
   ```

   (c) Execute the followig code and comment on the results.
   ```
   plot(y~x, data=aids, xlim=c(0,55), ylim=c(0,600))
   lines(fitted(mod1)~aids$x, col=2)
   lines(fitted(mod2)~aids$x, col=3)
   lines(fitted(mod3)~aids$x, col=4)
   ```

```
lines(ap1~newaids$x, col=2)
points(ap1~newaids$x, col=2)
lines(ap2~newaids$x, col=3)
points(ap2~newaids$x, col=3)
lines(ap3~newaids$x, col=4)
points(ap3~newaids$x, col=4)
```

3. **Bootstrapping using** `boot`: The `abdom` data is used here to demonstrate how bootstrapping can be used in a GAMLSS model.

   Fit a logistic distribution with a parametric quadratic and cubic polynomial term for `x` in the μ model, and a linear term for `x` in the σ model. Use the GAIC and the likelihood ratio test to compare the two models.

```
fit2 <- gamlss(y~poly(x,2), sigma.fo=~x, data=abdom, trace=FALSE,
               family=LO)
fit3 <- gamlss(y~poly(x,3), sigma.fo=~x, data=abdom, trace=FALSE,
               family=LO)
AIC(fit2, fit3)
LR.test(fit2, fit3)
```

   The AIC and the likelihood ratio test indicate that the quadratic model is superior to the cubic one. Now confirm this result using bootstrapping methods.

   (a) **Parametric bootstrap (type 1):** simulating the response from a fitted model.

   Given model `fit3` and its fitted μ and σ values, simulate 999 bootstrapping samples for `y` from a logistic distribution. Refit those new `y`'s to the original `x`'s using the model given by `fit3` and check the distribution of the coefficients of the cubic term in the μ model.

```
library(boot)
# a new data.frame with fitted mu and sigma
abdomA <- data.frame(abdom, fmu=fitted(fit3),
                      fsigma=fitted(fit3, "sigma"))
# function to save the mu coefficients
abd.funA <- function(data, i)#i is the index for bootstrapping
  {
  d<-data # the data
  #omit the first
  d$y <- if ("original"%in%as.character(sys.call())) d$y
  #simulate y
          else rLO(dim(d)[1], mu=d$fmu, sigma=d$fsigma)
  coef(update(fit3, data=d)) # fit and get the coef for mu
}
```

   Do the simulation.

```
(abd.T1<-boot(abdomA,abd.funA, R=999))
abd.T1
```

Plot the cubic (4th) parameter and get confidence intervals.

```
plot(abd.T1, index=4)# parameter 4
boot.ci(abd.T1, index=c(4,1))
```

What do you conclude?

(b) **Semi-parametric bootstrap (type 2):** In type 2 situations you sample from the residuals of the fitted model. The residuals and the fitted values for μ and σ need to be saved. The new simulated y's are taken by permuting the residuals, transforming them using the cdf of the normal distribution (since they are normalized) to values between 0 and 1, and then using the inverse cdf of the logistic distribution.

```
abdomB <- data.frame(abdom, res=resid(fit3), fmu=fitted(fit3),
                     fsigma=fitted(fit3, "sigma"))
abd.funB <- function(data, i)
      {
          d <- data
    ## pNO is always the same for normalized residuals
    ## the q function changes according  to the family we fit
          d$y <- qLO(pNO(d$res[i]), mu=d$fmu, sigma=d$fsigma)
          coef(update(fit3, data=d))
          }
abd.T2<-boot(abdomB,abd.funB, R=999)
abd.T2
plot(abd.T2, index=4)
boot.ci(abd.T2, index=c(4,1))
```

Compare these confidence intervals with those obtained using type 1 bootstrapping.

(c) **Nonparametric bootstrap (type 3):** In the nonparametric bootstrap, the original y and x values are resampled with replacement, and the original model is refitted to the new data.

```
abd.funC <- function(data, i)
      {
      d<-data[i,]
      coef(update(fit3, data=d))
      }
abd.T3<-boot(abdom, abd.funC, R=999)
plot(abd.T3, index=4)
boot.ci(abd.T3, index=c(4,1))
```

Compare the confidence intervals with those obtained with type 1 and 2.

(d) You can use the function `histSmo()` to obtain probabilities from the fitted distribution of the cubic coefficient. For example, to obtain the probability of β_4 being greater than 20 or between -20 and 20 use, respectively:

```
FF<-histSmo(abd.T3$t[,4],plot=T)
1-FF$cdf(20)
FF$cdf(20)-FF$cdf(-20)
```

(e) Use the nonparametric bootstrap to calculate confidence intervals for smoothing parameters:

```
# fits the  model with P-spline for mu
fitG <- gamlss(y~pb(x),  sigma.fo=~x, data=abdom, trace=F,
               family=LO, gd.tol=Inf)
abd.Smo <- function(data, i)
    {
    d<-data[i,]
    getSmo(update(fitG, data=d))$lambda
    }
abd.Sim<-boot(abdom, abd.Smo, R=99)
plot(abd.Sim, index=1)
boot.ci(abd.Sim, type="perc")
```

Part III

Distributions

6

The GAMLSS family of distributions

CONTENTS

6.1 Introduction

This chapter describes:

1. the different types of distribution within **gamlss**;

2. how to visualize the different distributions;

3. how to create a new distribution; and

4. link functions within **gamlss**.

This chapter is essential for understanding the different types of distributions in GAMLSS and especially the need for more complex distributions.

Within the GAMLSS framework the probability (density) function of the response variable Y, $f(y \mid \boldsymbol{\theta})$, where $\boldsymbol{\theta} = (\mu, \sigma, \nu, \tau)$, is deliberately left general with no explicit distribution specified. Although μ, σ, ν and τ often represent parameters related to location, scale, skewness and kurtosis this is not always true in general. The only restriction that the **gamlss** package has for specifying the distribution of Y is that $f(y \mid \boldsymbol{\theta})$ and its first (and optionally expected second and cross) derivatives with respect to each of the parameters of $\boldsymbol{\theta}$ must be computable. Explicit derivatives are preferable, but numerical derivatives can be used (resulting in reduced computational speed). That is, the algorithm used for fitting the GAMLSS model needs only this information.

Here we introduce the available distributions within the **gamlss** package. We refer to this set of distributions as the GAMLSS family, to be consistent with **R** where the distributions are defined as `gamlss.family` objects. Note that a comprehensive review of all `gamlss.family` distributions can be found in the book *Distributions for Modelling Location, Scale, and Shape: Using GAMLSS in R* [Rigby et al., 2017].

Fitting a parametric distribution within the GAMLSS family can be achieved using the command `gamlss(y~1, family= )` where the argument `family` can take any `gamlss.family` distribution; see Tables 6.1, 6.2 and 6.3 for appropriate names. For example, in order to fit a negative binomial distribution to count data one can use `family=NBI`. Note also the following forms are acceptable: `family=NBI()`, `family="NBI"` or `family=NBI(mu.link=log, sigma.link=log)`. For example, Figure 6.1 shows a histogram of the Turkish stock exchange returns data `tse` (described and plotted in more detail in Exercise 3):

```
data(tse)
truehist(tse$ret, xlab="TSE")
```
Fig. 6.1

In the following code the t family (`TF`) distribution is fitted:

```
m1 <- gamlss(ret~1, data=tse, family=TF)
```

When no covariates are involved, (as above), the functions `gamlssML()`, `histDist()` and `fitDist()` can be used instead of `gamlss()`. `gamlssML()` uses numerical optimization techniques that are faster than the algorithms that `gamlss()` uses, which are designed for regression models.

```
m2 <- gamlssML(ret, data=tse, family=TF)
```

`histDist()` uses `gamlssML()` as a default algorithm to fit the model, but in addition displays the histogram together with the fitted distribution of the data; see Figure 6.2.

```
m3 <- histDist(ret, data=tse, family=TF, nbins=30, line.wd=2.5)
```
Fig. 6.2

The function `fitDist()` uses `gamlssML()` to fit a set of predetermined distributions to the data and chooses the 'best' according to the GAIC, with default penalty

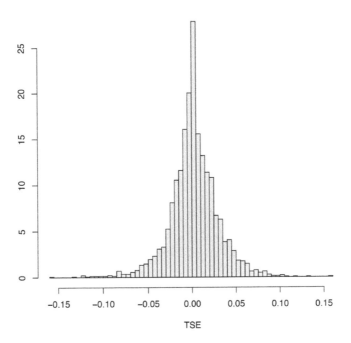

FIGURE 6.1: Histogram of the Turkish stock exchange returns.

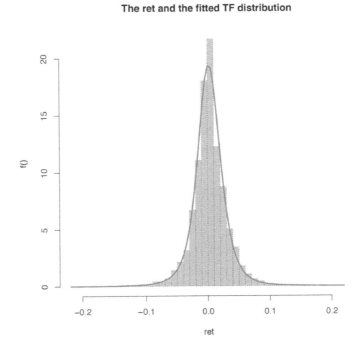

FIGURE 6.2: Histogram of the Turkish stock exchange returns together with fitted *t* distribution.

$\kappa = 2$. (This can be changed by specifying the argument k.) The order of the fitted models displayed in increasing order of GAIC, i.e. from 'best' to 'worst', as shown below:

```
m5 <- fitDist(ret, data=tse, type="realline")
m5$fits
```

```
##       SEP2        SEP1        SEP3        SEP4          PE         PE2
## -12879.01  -12876.88  -12876.14  -12865.04  -12862.09  -12862.09
##         GT      SHASHo     SHASHo2       SHASH        JSUo         JSU
## -12860.09  -12836.07  -12836.07  -12833.92  -12803.68  -12803.68
##        ST3          TF         ST2         ST5         ST1         ST4
## -12790.80  -12789.56  -12788.83  -12788.20  -12787.75  -12787.62
##        TF2         SST          LO      exGAUS          NO         SN2
## -12785.20  -12783.20  -12726.41  -12461.02  -12444.12  -12442.92
##        SN1          GU          RG
## -12442.12  -11578.60  -11305.11
```

All of the `gamlss.family` distributions on the real line have been considered (specified by the option `type="realline"`). The most preferred distribution is the skew exponential power type 2 (SEP2), and the least preferred is the reverse Gumbel (RG). Rigby et al. [2017] contains more examples demonstrating all of the above functions.

6.2   Types of distribution within the GAMLSS family

6.2.1   Explicit GAMLSS family distributions

The type of distribution to use depends on the type of response variable being modelled. Within the GAMLSS family there are three distinct types of distributions:

1. continuous distributions; see Figure 6.3(a),

2. discrete distributions; see Figure 6.3(b),

3. mixed distributions; see Figure 6.3(c).

Table 6.1 shows all the available continuous distributions within the `gamlss.family`. The columns of the table display the distribution name, the **R** implementation name, the range of Y (R_Y) and the default link functions of all available parameters. Link functions are discussed in Chapter 1 and Sections 6.4.1 and 6.5. Note that, for presentational reasons, the `identity` and the `logshiftto2` links are abbreviated to 'ident.' and 'log-2', respectively. Note also that 'o' at the end of a `gamlss` name indicates the original version of the distribution abbreviated to 'orig'. Also 'gen',

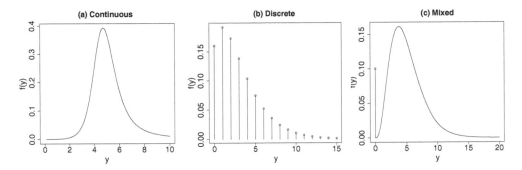

FIGURE 6.3: Different type of distributions in GAMLSS (a) continuous, (b) discrete, (c) mixed.

'inv', 'exp' and 'repar' are abbreviated from 'generalized', 'inverse', 'exponential' and 'reparameterized', respectively.

TABLE 6.1: Continuous distributions implemented within the **gamlss.dist** package, with default link functions

Distribution	gamlss name	Range R_Y	Parameter link functions			
			μ	σ	ν	τ
beta	BE	$(0,1)$	logit	logit	-	-
Box–Cox Cole–Green	BCCG	$(0,\infty)$	ident.	log	ident.	-
Box–Cox Cole–Green orig.	BCCGo	$(0,\infty)$	log	log	ident.	-
Box–Cox power exponential	BCPE	$(0,\infty)$	ident.	log	ident.	log
Box–Cox power expon. orig.	BCPEo	$(0,\infty)$	log	log	ident.	log
Box–Cox t	BCT	$(0,\infty)$	ident.	log	ident.	log
Box–Cox t orig.	BCTo	$(0,\infty)$	log	log	ident.	log
exponential	EXP	$(0,\infty)$	log	-	-	-
exponential Gaussian	exGAUS	$(-\infty,\infty)$	ident.	log	log	-
exponential gen. beta 2	EGB2()	$(-\infty,\infty)$	ident.	log	log	log
gamma	GA	$(0,\infty)$	log	log	-	-
generalized beta type 1	GB1	$(0,1)$	logit	logit	log	log
generalized beta type 2	GB2	$(0,\infty)$	log	log	log	log
generalized gamma	GG	$(0,\infty)$	log	log	ident.	-
generalized inv. Gaussian	GIG	$(0,\infty)$	log	log	ident.	-
generalized t	GT	$(-\infty,\infty)$	ident.	log	log	log
Gumbel	GU	$(-\infty,\infty)$	ident.	log	-	-
inverse Gamma	IGAMMA	$(0,\infty)$	log	log	-	-
inverse Gaussian	IG	$(0,\infty)$	log	log	-	-
Johnson's SU repar.	JSU	$(-\infty,\infty)$	ident.	log	ident.	log
Johnson's SU original	JSUo	$(-\infty,\infty)$	ident.	log	ident.	log
logistic	LO	$(-\infty,\infty)$	ident.	log	-	-
logit normal	LOGITNO	$(0,1)$	ident.	log	-	-
log normal	LOGNO	$(0,\infty)$	ident.	log	-	-

TABLE 6.1: Continuous distributions implemented within the **gamlss.dist** package, with default link functions (continued)

Distribution	gamlss name	Range R_Y	μ	σ	ν	τ
log normal 2	LOGNO2	$(0, \infty)$	log	log	-	-
log normal (Box–Cox)	LNO	$(0, \infty)$	ident.	log	fixed	-
NET	NET	$(-\infty, \infty)$	ident.	log	fixed	fixed
normal	NO, NO2	$(-\infty, \infty)$	ident.	log	-	-
normal family	NOF	$(-\infty, \infty)$	ident.	log	-	-
Pareto 2	PARETO2	$(0, \infty)$	log	log	-	-
Pareto 2 original	PARETO2o	$(0, \infty)$	log	log	-	-
Pareto 2 repar	GP	$(0, \infty)$	log	log	-	-
power exponential	PE	$(-\infty, \infty)$	ident.	log	log	-
reverse gen. extreme	RGE	$y > \mu - (\sigma/\nu)$	ident.	log	log	-
reverse Gumbel	RG	$(-\infty, \infty)$	ident.	log	-	-
sinh-arcsinh	SHASH	$(-\infty, \infty)$	ident.	log	log	log
sinh-arcsinh original	SHASHo	$(-\infty, \infty)$	ident.	log	ident.	log
sinh-arcsinh original 2	SHASHo2	$(-\infty, \infty)$	ident.	log	ident.	log
skew normal type 1	SN1	$(-\infty, \infty)$	ident.	log	ident.	-
skew normal type 2	SN2	$(-\infty, \infty)$	ident.	log	log	-
skew power exp. type 1	SEP1	$(-\infty, \infty)$	ident.	log	ident.	log
skew power exp. type 2	SEP2	$(-\infty, \infty)$	ident.	log	ident.	log
skew power exp. type 3	SEP3	$(-\infty, \infty)$	ident.	log	log	log
skew power exp. type 4	SEP4	$(-\infty, \infty)$	ident.	log	log	log
skew t type 1	ST1	$(-\infty, \infty)$	ident.	log	ident.	log
skew t type 2	ST2	$(-\infty, \infty)$	ident.	log	ident.	log
skew t type 3	ST3	$(-\infty, \infty)$	ident.	log	log	log
skew t type 3 repar	SST	$(-\infty, \infty)$	ident.	log	log	log-2
skew t type 4	ST4	$(-\infty, \infty)$	ident.	log	log	log
skew t type 5	ST5	$(-\infty, \infty)$	ident.	log	ident.	log
t Family	TF	$(-\infty, \infty)$	ident.	log	log	-
t Family repar	TF2	$(-\infty, \infty)$	ident.	log	log-2	-
Weibull	WEI	$(0, \infty)$	log	log	-	-
Weibull (PH)	WEI2	$(0, \infty)$	log	log	-	-
Weibull (μ the mean)	WEI3	$(0, \infty)$	log	log	-	-

The distributions in Table 6.1 (except for RGE) are defined on $(-\infty, +\infty)$, $(0, \infty)$ and $(0, 1)$, ranges. Users can restrict those ranges of the response variable Y by defining a truncated gamlss.family distribution using the package **gamlss.tr**.

Discrete distributions are usually defined on $y = 0, 1, 2, \ldots, n$, where n is either a known finite value, or infinite. Table 6.2 shows the available discrete `gamlss.family` distributions.

TABLE 6.2: Discrete distributions implemented within **gamlss.dist**, with default link functions

Distribution	gamlss name	Range R_Y	Parameter range		
			μ	σ	ν
beta binomial	BB	$\{0, 1, \ldots, n\}$	logit	log	-
binomial	BI	$\{0, 1, \ldots, n\}$	logit	-	-
Delaporte	DEL	$\{0, 1, 2, \ldots\}$	log	log	logit
geometric	GEOM	$\{0, 1, 2, \ldots\}$	log	-	-
logarithmic	LG	$\{1, 2, 3, \ldots\}$	logit	-	-
negative binomial type I	NBI	$\{0, 1, 2, \ldots\}$	log	log	-
negative binomial type II	NBII	$\{0, 1, 2, \ldots\}$	log	log	-
Poisson	PO	$\{0, 1, 2, \ldots\}$	log	-	-
Poisson inverse Gaussian	PIG	$\{0, 1, 2, \ldots\}$	log	log	-
Sichel	SI	$\{0, 1, 2, \ldots\}$	log	log	identity
Sichel (μ the mean)	SICHEL	$\{0, 1, 2, \ldots\}$	log	log	identity
Waring (μ the mean)	WARING	$\{0, 1, 2, \ldots\}$	log	log	-
Yule (μ the mean)	YULE	$\{0, 1, 2, \ldots\}$	log	-	-
zero altered beta binomial	ZABB	$\{0, 1, \ldots, n\}$	logit	log	logit
zero altered binomial	ZABI	$\{0, 1, \ldots, n\}$	logit	logit	-
zero altered logarithmic	ZALG	$\{0, 1, 2, \ldots\}$	logit	logit	-
zero altered neg. binomial	ZANBI	$\{0, 1, 2, \ldots\}$	log	log	logit
zero altered Poisson	ZAP	$\{0, 1, 2, \ldots\}$	log	logit	-
zero inflated beta binomial	ZIBB	$\{0, 1, 2, \ldots\}$	logit	log	logit
zero inflated binomial	ZIBI	$\{0, 1, \ldots, n\}$	logit	logit	-
zero inflated neg. binomial	ZINBI	$\{0, 1, 2, \ldots\}$	log	log	logit
zero inflated Poisson	ZIP	$\{0, 1, 2, \ldots\}$	log	logit	-
zero inflated Poisson (μ the mean)	ZIP2	$\{0, 1, 2, \ldots\}$	log	logit	-
zero inflated Poisson inv. Gaussian	ZIPIG	$\{0, 1, 2, \ldots\}$	log	log	logit

Mixed distributions are a special case of finite mixture distributions, described in Chapter 7, and are mixtures of continuous and discrete distributions, i.e. continuous distributions where the range of Y has been expanded to include some discrete values with non-zero probabilities. Table 6.3 shows the available mixed `gamlss.family` distributions. In the range R_Y a square bracket indicates that the endpoint is included in the range. Several distributions are defined on the interval from 0 to 1, including one or both endpoints, and are suitable for modelling a proportion or fractional response variable. The other distributions (**ZAGA** and **ZAIG**) are defined on the interval $[0, \infty)$ and are suitable for a response variable that is nonnegative with a positive probability at zero, e.g. insurance claim amount in a year on an insurance policy. The zero adjusted gamma distribution (ZAGA) shown in Figure 6.3(c) is a possible distribution for such data.

TABLE 6.3: Mixed distributions implemented within **gamlss.dist**, with default link functions

Distribution	gamlss name	Range R_Y	Parameter link functions			
			μ	σ	ν	τ
beta inflated (at 0)	BEOI	$[0, 1)$	logit	log	logit	-
beta inflated (at 0)	BEINFO	$[0, 1)$	logit	logit	log	-
beta inflated (at 1)	BEZI	$(0, 1]$	logit	log	logit	-
beta inflated (at 1)	BEINF1	$(0, 1]$	logit	logit	log	-
beta inflated (at 0 and 1)	BEINF	$[0, 1]$	logit	logit	log	log
zero adjusted GA	ZAGA	$[0, \infty)$	log	log	logit	-
zero adjusted IG	ZAIG	$[0, \infty)$	log	log	logit	-

All of the distributions in Tables 6.1, 6.2 and 6.3 have d, p, q and r functions corresponding, respectively, to the pdf, cdf, the quantile function (i.e. inverse cdf) and random number generating function. For example, the Box–Cox t (BCT) distribution has the functions dBCT, pBCT, qBCT and rBCT. In addition each distribution has a *fitting* function which helps the fitting procedure by providing link functions, first and (exact or approximate) expected second derivatives, starting values, etc. All fitting functions have as arguments the link functions for the distribution parameters. For example, the fitting function for the Box–Cox t distribution is called BCT with arguments mu.link and sigma.link. The function show.link() can be used to identify which are the available links for the distribution parameter. The following code displays the available links within the BCT distribution:

```
show.link(BCT)

## $mu
## c("inverse", "log", "identity", "own")
##
## $sigma
## c("inverse", "log", "identity", "own")
##
## $nu
## c("inverse", "log", "identity", "own")
##
## $tau
## c("inverse", "log", "identity", "own")
```

Available link functions are the usual glm() link functions plus some other functions such as logshiftto1, and own; see Section 6.4.

6.2.2 Extending GAMLSS family distributions

There are several ways to extend the `gamlss.family` distributions. This can be achieved by

- creating a new `gamlss.family` distribution;

- creating a *log* or *logit* version of a distribution from an existing continuous `gamlss.family` distribution on the real line;

- truncating an existing `gamlss.family`;

- using a censored version of an existing `gamlss.family`; and

- mixing different `gamlss.family` distributions to create a new finite mixture distribution.

These extensions are dealt with in the following sections.

New gamlss.family distribution

To create a new `gamlss.family` distribution is relatively simple if the probability (density) function of the distribution can be evaluated easily. To do that, a file of a current `gamlss.family` distribution, having the same number of distribution parameters, may be amended. Section 6.4 provides an example of this.

New log and logit versions from a continuous gamlss.family on $(-\infty, \infty)$

Any continuous random variable Z defined on $(-\infty, \infty)$ can be transformed by $Y = \exp(Z)$ to a random variable defined on $(0, \infty)$. A well-known example of this is the log-normal distribution, which is defined by $Y = \exp(Z)$ where Z is a normally distributed random variable. This is achieved in **gamlss** by using the function **gen.Family()** with the option **type="log"**. The following is an example in which we take a t family distribution, i.e. $Z \sim \text{TF}(\mu, \sigma, \nu)$ and apply an exponential transform $Y = \exp(Z)$ to give $Y \sim \log \text{TF}(\mu, \sigma, \nu)$, i.e. we create a log-$t$ family distribution on $(0, \infty)$. We then generate a random sample of 200 observations from the distribution and finally fit the distribution to the generated data.

```
# generate the distribution
gen.Family("TF", type="log")
```

Fig. 6.4

```
## A  log  family of distributions from TF has been generated
##  and saved under the names:
##  dlogTF plogTF qlogTF rlogTF logTF

#generate 200 observations with df=nu=10
#(and default mu=0 and sigma=1)
set.seed(134)
Y<- rlogTF(200, nu=10)
# fit the distribution
```

```
h1 <- histDist(Y, family=logTF, nbins=30, ylim=c(0,.65), line.wd=2.5)
```

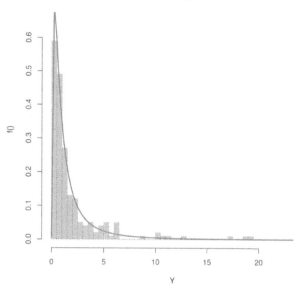

The Y and the fitted logTF distribution

R
p. 16

FIGURE 6.4: A fitted log-t distribution to 200 simulated observations.

Similarly take $Z \sim \text{TF}(\mu, \sigma, \nu)$ and apply the inverse logit transformation $Y = 1/(1 + \exp(-Z))$ to give $Y \sim \text{logitTF}(\mu, \sigma, \nu)$, i.e. a logit-$t$ family distribution on $(0, 1)$:

```
gen.Family("TF", type="logit")

## A  logit  family of distributions from TF has been generated
##  and saved under the names:
##  dlogitTF plogitTF qlogitTF rlogitTF logitTF
```

Truncating `gamlss.family` distributions

A truncated distribution is appropriate when the range of possible values of a variable Y is a subset of the range of an original distribution. Truncating existing **gamlss.family** distributions can be achieved using the package **gamlss.tr**. The function **gen.trun()** takes any **gamlss.family** distribution and generates the d, p, q, r and fitting functions for the specified truncated distribution. The truncation can be left, right or in both tails of the range of the response variable. For example, a t family distribution (**TF**) can be truncated in both tails, at and below 0 and at and above 100, as follows:

```
# generate the distribution
library(gamlss.tr)
gen.trun(par=c(0,100),family="TF", name="0to100", type="both")
```

Fig.
6.5

```
## A truncated family of distributions from TF has been generated
##   and saved under the names:
##   dTF0to100 pTF0to100 qTF0to100 rTF0to100 TF0to100
## The type of truncation is both
##   and the truncation parameter is 0 100

Y <-rTF0to100(1000, mu=80 ,sigma=20, nu=5)
h1 <- histDist(Y, family=TF0to100, nbins=30, xlim=c(0,100),
               line.col="darkblue",line.wd=2.5)
```

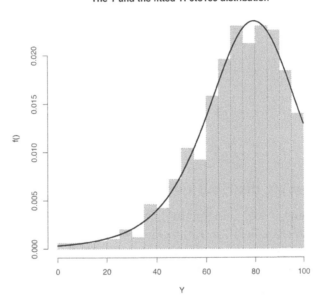

The Y and the fitted TF0to100 distribution

FIGURE 6.5: A truncated t distribution defined on $(0, 100)$, fitted to 1000 simulated observations.

The **gen.trun()** function has the following main arguments:

par: a vector with one element for left or right truncation, or two elements for truncating both sides;

family: a **gamlss.family** distribution;

name: the name for the new truncated distribution;

type: "left", "right" or "both" sides truncation.

For discrete count distributions,

- *left* truncation at the integer a means that the random variable can take values $\{a + 1, a + 2, \ldots\}$;

- *right* truncation at the integer b means that the random variable can take values up to but not including b, i.e. $\{\ldots, b - 2, b - 1\}$.

- *both* truncation at the integer interval (a, b) means the random variable can take values $\{a + 1, a + 2, \ldots, b - 1\}$.

For a continuous random variable Y, $\text{Prob}(Y = a) = 0$ for any constant a and therefore the inclusion or not of an endpoint in the truncated range is not an issue. Hence "left" truncation at value a results in variable $Y > a$, "right" truncation at b results in $Y < b$, and "both" truncation results in $a < Y < b$.

Assume Y is obtained from Z by left truncation at and below a, and right truncation at and above b. If Z is a discrete count variable with probability function $\text{Prob}(Z = z)$, cdf $F_Z(\cdot)$ and a and b are integers $(a < b)$, then

$$\text{Prob}(Y = y) = \frac{\text{Prob}(Z = y)}{F_Z(b - 1) - F_Z(a)} .$$

for $y = \{a + 1, a + 2, \ldots, b - 1\}$. If Z is continuous with pdf $f_Z(z)$ and a and b are any values with $a < b$,

$$f_Y(y) = \frac{f_Z(y)}{F_Z(b) - F_Z(a)}$$

for $a < y < b$.

Censored `gamlss.family` distributions

The package **gamlss.cens** is designed for the situation where one or more observations of the response variable are left- or right-censored (i.e. lie above or below known values), or more generally, they lie in a known interval. For example, in a survey, respondents may be asked to report their income in an interval, so the responses are in the form $[\$0, \$100]$, $(\$100, \$200]$, etc. The function `gen.cens()` will take any `gamlss.family` distribution and create a new function which fits a response variable with one or more observations of left-, right- or interval-censored type. Note that for independent interval response variable observations, the usual likelihood function defined as

$$L(\boldsymbol{\theta}) = \prod_{i=1}^{n} f(y_i \mid \boldsymbol{\theta})$$

changes to

$$L(\boldsymbol{\theta}) = \prod_{i=1}^{n} [F(y_{2i} \mid \boldsymbol{\theta}) - F(y_{1i} \mid \boldsymbol{\theta})] \tag{6.1}$$

where $F(\cdot)$ is the cdf and $(y_{1i}, y_{2i}]$ is the observed interval of the ith observation. For a left-censored observation, censored at and below y_{2i} set $y_{1i} = -\infty$, so $F(y_{1i} \mid \boldsymbol{\theta}) = 0$, in (6.1). For a right-censored observation, censored above y_{1i}, set $y_{2i} = \infty$, so $F(y_{2i} \mid \boldsymbol{\theta}) = 1$, in (6.1).

The following is an example of generating a Weibull distribution which allows an interval response variable to be fitted. The data are called `lip` and come from an experimental enzymology research project which attempted to develop a generic food spoilage model. The response variable `lip$y` is defined as an interval variable, while the explanatory variables are `Tem`, `pH`, `aw`.

> **R data file: `lip` in package gamlss.cens** of dimensions 120×14
>
> **var y** : a matrix with 3 columns, that is a `Surv()` object indicating the start, the finish and censored indicator as defined in function `Surv()` of the package **survival**,
>
> > **name** : a factor with levels for the different experiments
> >
> > **Tem** : the temperature
> >
> > **pH** : the PH
> >
> > **aw** : the water acidity
> >
> > **X0.d:X39.d** : vectors indicating whether enzyme reacted at specific days
>
> **purpose:** to demonstrate an interval response variable.

```
library(gamlss.cens)
data(lip)
head(lip$y, 10)
```

```
## [1] 1-       1-       1-       1-       [11, 18] 1-
## [7] 1-       1-       1-       [ 2,  4]
```

The value $1-$ indicates an interval $(1, \infty)$ not including 1, while $[11, 18]$ indicates the interval $(11, 18]$ not including 11 but including 18. This specific data set does not have left-censored observations. If it did then, for example, a value of -5 would indicate the interval $(-\infty, 5]$ including 5. For a continuous distribution the likelihood is unaffected by whether interval endpoints are included or not, but for (a discrete distribution this is very important. An interval-censored Weibull, $\text{WEI2}(\mu, \sigma)$, distribution is now generated which allows interval- (as well as left- and right-) censored response values. This is used to fit a model to the `lip` data.

```
gen.cens(WEI2,type="interval")
```

```
## A censored family of distributions from WEI2 has been generated
##  and saved under the names:
##  dWEI2ic pWEI2ic qWEI2ic WEI2ic
## The type of censoring is interval
```

```
WEI2ic()
```

```
##
## GAMLSS Family: WEI2ic interval censored Weibull type 2
## Link function for mu   : log
## Link function for sigma: log
```

```
weimi<- gamlss(y ~ poly(Tem,2)+poly(pH,2)+poly(aw,2), data=lip,
    family=WEI2ic, n.cyc=100, trace=FALSE)
summary(weimi)

## ****************************************************************
## Family:  c("WEI2ic", "interval censored Weibull type 2")
##
## Call:
## gamlss(formula = y ~ poly(Tem, 2) + poly(pH, 2) + poly(aw,
##     2), family = WEI2ic, data = lip, n.cyc = 100, trace = FALSE)
##
##
## Fitting method: RS()
##
## -----------------------------------------------------------------
## Mu link function:  log
## Mu Coefficients:
##               Estimate Std. Error t value Pr(>|t|)
## (Intercept)    -5.3059     0.7124  -7.447 2.12e-11 ***
## poly(Tem, 2)1  35.2577     4.5141   7.811 3.32e-12 ***
## poly(Tem, 2)2  -1.3722     2.0149  -0.681   0.4972
## poly(pH, 2)1   20.6126     3.2062   6.429 3.25e-09 ***
## poly(pH, 2)2   -4.5238     2.0601  -2.196   0.0302 *
## poly(aw, 2)1   31.5191     4.3209   7.295 4.58e-11 ***
## poly(aw, 2)2    1.2573     1.8689   0.673   0.5025
## ---
## Signif. codes:
## 0 '***' 0.001 '**' 0.01 '*' 0.05 '.' 0.1 ' ' 1
##
## -----------------------------------------------------------------
## Sigma link function:  log
## Sigma Coefficients:
##               Estimate Std. Error t value Pr(>|t|)
## (Intercept)     0.1159     0.1564   0.741     0.46
##
## -----------------------------------------------------------------
## No. of observations in the fit:  120
## Degrees of Freedom for the fit:  8
##         Residual Deg. of Freedom:  112
##                       at cycle:  43
##
## Global Deviance:     138.4933
##             AIC:     154.4933
##             SBC:     176.7932
## ****************************************************************
```

Finite mixtures of `gamlss.family` distributions

Finite mixtures, dealt with in detail in Chapter 7, can be fitted using the package **gamlss.mx**. A finite mixture of `gamlss.family` distributions has the form

$$f(y \mid \boldsymbol{\psi}) = \sum_{k=1}^{\mathcal{K}} \pi_k f_k(y \mid \boldsymbol{\theta}_k) \tag{6.2}$$

where $f_k(y \mid \boldsymbol{\theta}_k)$ is the probability (density) function of y for component k, and $0 \leq \pi_k \leq 1$ is the prior (or mixing) probability of component k, for $k = 1, 2, \ldots, \mathcal{K}$. Also $\sum_{k=1}^{\mathcal{K}} \pi_k = 1$ and $\boldsymbol{\psi} = (\boldsymbol{\theta}, \boldsymbol{\pi})$ where $\boldsymbol{\theta} = (\boldsymbol{\theta}_1, \boldsymbol{\theta}_2, \ldots, \boldsymbol{\theta}_{\mathcal{K}})$ and $\boldsymbol{\pi} = (\pi_1, \pi_2, \ldots, \pi_{\mathcal{K}})$. Any combination of (continuous or discrete) `gamlss.family` distributions can be used. The model in this case is fitted using the EM algorithm. The component probability (density) functions may have different parameters (fitted using the function `gamlssMX()`) or may have parameters in common (fitted using the function `gamlssNP()`). In the former case, the mixing probabilities may also be modelled using explanatory variables and the finite mixture may have a zero component (e.g. zero inflated negative binomial). Both `gamlssMX()` and `gamlssNP()` are in package **gamlss.mx**; see Chapter 7. Figure 6.6 shows an example of fitting a finite mixture of two reverse Gumbel distributions to the **enzyme** data. The finite mixture distribution (obtained by `getpdfMX()`) is compared with a nonparametric density estimate (obtained by `density()`).

> **R data file: enzyme** in package **gamlss.mx** of dimensions 245×1
>
> **variables**
>
>   act : enzyme activity in the blood of 245 individuals
>
> **purpose:** to demonstrate the fitting of a finite mixture distribution.

Fig. 6.6

```
library(gamlss.mx)
data(enzyme)
m3 <- gamlssMX(act ~ 1, data = enzyme, family = RG, K = 2)
truehist(enzyme$act, h = 0.1, xlab="y", ylab="f(y)")
fnRG <- getpdfMX(m3, observation=1)
lines(seq(0, 3, 0.01), fnRG(seq(0, 3, 0.01)), lty = 1, lwd=2)
lines(density(enzyme$act, width = "SJ-dpi"), lty = 2, lwd=2)
legend("topright", legend=c("Reverse Gumbel mixture",
      "nonparametric density estimate"), lty=1:2, lwd=2)
```

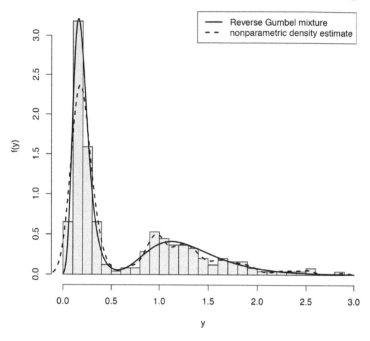

R

p. 16

FIGURE 6.6: Histogram of enzyme activity, with fitted two-component reverse Gumbel finite mixture distribution and nonparametric density estimate.

6.3   Displaying GAMLSS family distributions

6.3.1   Using the distribution demos

A `gamlss.family` distribution can be displayed interactively in **R** using the **gamlss.demo** package. The following commands will start the demos:

```
library(gamlss.demo)
gamlss.demo()
```

This will display a menu where, by choosing the option "Demos for gamlss.family distributions", the user can proceed to display the different distributions. Alternatively one can simply type `demo.NAME()` where `NAME` is a `gamlss.family` name, e.g. `demo.NO()` for the normal distribution. This allows any distribution in GAMLSS to be displayed graphically and its parameters adjusted interactively. A screenshot is given in Figure 6.7.

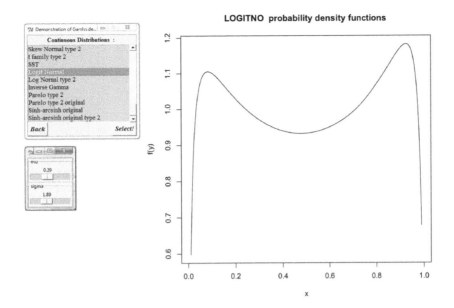

FIGURE 6.7: Screenshot demonstrating the logit Normal distribution (`LOGITNO`).

6.3.2 Using the `pdf.plot()` function

An alternative method of graphically displaying probability functions is to use the `pdf.plot()` function, shown in Figure 6.8, which plots a Poisson (`PO`(μ)) distribution for $\mu = 1, 2, 3, 4$.

Fig. 6.8
```
# generate the distribution
pdf.plot(family=PO(),mu=c(1,2,3,4), min=0, max=10,step=1)
```

This function is also useful for plotting different fitted distributions for specific observations. For example here we plot the fitted distribution for observations 100 and 200 after fitting a *t*-distribution to the **abdom** data, in which the response variable **y** is abdominal circumference, and the covariate **x** is gestational age. The result is shown in Figure 6.9.

Fig. 6.9
```
# fitted distribution for cases 100 and 200
m1<- gamlss(y~pb(x), sigma.fo=~pb(x), data=abdom, family=TF,
            trace=FALSE)
pdf.plot(m1,obs=c(100,200), min=50, max=250,step=.1)
```

The following code demonstrates the use of the **d**, **p** and **r** functions to plot the pdf, cdf and the histogram of a randomly generated sample from a continuous distribution (the gamma). The resulting plots are in Figure 6.10.

Fig. 6.10
```
mu=3
sigma=.5
curve(dGA(y, mu, sigma), 0.01, 10, xname="y",ylab="f(y)") # pdf
```

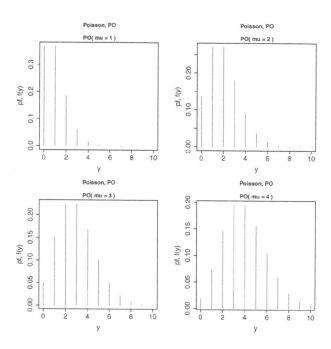

FIGURE 6.8: Plots of the Poisson distribution using the pdf.plot() function.

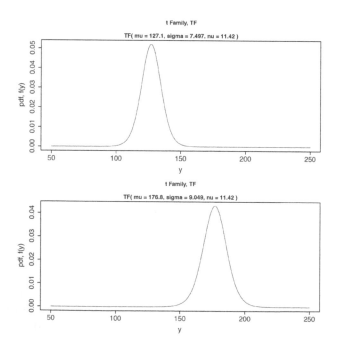

R
p. 16

R
p. 16

FIGURE 6.9: The fitted distribution for observations 100 and 200.

```
curve(pGA(y, mu, sigma), 0.01, 10, xname="y", ylab="F(y)")   # cdf
y<-rGA(1000, mu, sigma) # random sample
hist(y,col="lightgray",main="")
```

Similarly, to create Figure 6.11, the pf, cdf and histogram of a random sample of a discrete distribution (the negative binomial), use the following code:

Fig.
6.11

```
mu=8
sigma=.25
plot(function(y) dNBI(y, mu, sigma), from=0, to=30, n=30+1, type="h",
                 xlab="y",ylab="f(y)")
cdf <- stepfun(0:29, c(0,pNBI(0:29,mu, sigma)), f = 0)
plot(cdf, xlab="y", ylab="F(y)", verticals=FALSE, cex.points=.8,
     pch=16, main="")
Ni <- rNBI(1000, mu, sigma)
hist(Ni,breaks=seq(min(Ni)-0.5,max(Ni)+0.5,by=1),col="lightgray",
     main="")
```

R
p. 169

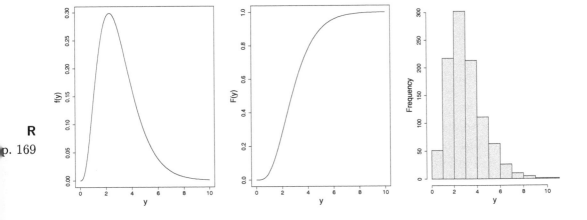

FIGURE 6.10: Probability density function, cumulative distribution function and histogram of a random sample of 1000 observations from a gamma distribution, using the **d**, **p** and **r** functions.

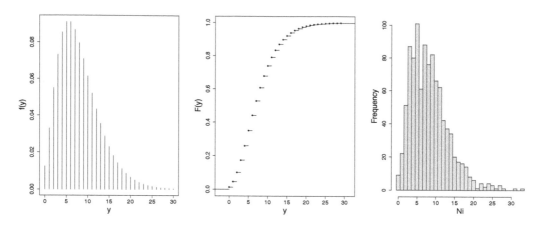

R

p. 17

FIGURE 6.11: Probability function, cumulative distribution function and histogram of a random sample of 1000 observations from a negative binomial distribution, using the d, p and r functions.

6.4   Amending an existing distribution and constructing a new distribution

This section can be omitted if the user does not plan to add a new distribution or amend an existing distribution.

In this section we describe the structure of a gamlss.family distribution and how it can be amended to produce a new distribution. As mentioned above, for each new distribution five different functions are required. Taking the normal distribution as an example, we have:

NO   the function used for fitting;

dNO   the pdf function;

pNO   the cdf function;

qNO   the inverse cdf function; and

rNO   the random number generation function.

The function NO provides information for fitting the normal distribution within **gamlss**. The function body for NO() has three fields:

- the definition of the link functions,

- the information needed for fitting the distribution, and

- the class definition of the fitted object.

The function NO() starts as:

```
NO<-function (mu.link = "identity", sigma.link = "log")
{
  mmstats<-checklink("mu.link", "Normal", substitute(mu.link),
      c("inverse", "log", "identity", "own"))
  dstats<-checklink("sigma.link", "Normal", substitute(sigma.link),
      c("inverse", "log", "identity", "own"))
```

The code shows that the available links for both μ and σ parameters of the $NO(\mu, \sigma)$ distribution are the identity, log, inverse and own links. The default links are identity and log for μ and σ, respectively.

6.4.1 Definition of the link functions

To define the link function of any of the parameters, the checklink() function is used as above. This takes four arguments:

which.link   which parameter the link is for, e.g. "mu.link";

which.dist   the current distribution, e.g. "Normal" (the name is only used to report an error in the specification of the link function);

link   which link is currently used, (the default value is the one given as the argument in the function definition, e.g. substitute(mu.link) will work);

link.List   the list of the possible links for the specific parameter, e.g. c("inverse", "log", "identity").

The available links in the link.List to choose from are currently the ones used by make.link.gamlss(). This list includes:
"identity", "log", "sqrt", "inverse", "1/mu^2", "mu^2", "logit", "probit", "cauchit", "cloglog", "logshiftto1", "logshiftto2", "logshiftto0" or "Slog", and "own". The parameter range and formula for each link function are given in Table 6.4.

This may change in future **gamlss** releases to incorporate more link functions. For the use of the user-defined **own** link function, see Section 6.5 or the help for make.link.gamlss(), where an example is given.

The object returned by checklink() contains the link function as a function of the current parameter, the inverse link function as a function of the current linear predictor and finally the first derivative of the inverse link function as a function of the linear predictor, i.e. d mu/d eta. These functions are used in the fitting of the distribution.

TABLE 6.4: Link functions available within the **gamlss** packages, according to the range of the distribution parameters

Parameter range	Link functions	Formula for $g(\theta)$
$-\infty$ to ∞	identity	θ
0 to ∞	log	$\log(\theta)$
	sqrt	$\sqrt{\theta}$
	inverse	$1/\theta$
	'1/mu$\wedge$2'	$1/\theta^2$
	'mu$\wedge$2'	θ^2
0 to 1	logit	$\log[\theta/(1-\theta)]$
	probit	$\Phi^{-1}(\theta)$
	cauchit	$\tan(\pi(\theta - 0.05))$
	cloglog	$\log(-\log(1-\theta))$
1 to ∞	logshiftto1	$\log(\theta - 1)$
2 to ∞	logshiftto2	$\log(\theta - 2)$
0.00001 to ∞	logshiftto0 or Slog[1]	$\log(\theta - 0.00001)$

[1] This function was created to avoid the value of positive parameters too close to zero.

6.4.2   The fitting information

The fitting algorithm uses the following information.

```
structure(list(family = c("NO", "Normal"),
        parameters = list(mu = TRUE, sigma = TRUE), nopar = 2,
            type = "Continuous",
          mu.link = as.character(substitute(mu.link)),
       sigma.link = as.character(substitute(sigma.link)),
        mu.linkfun = mstats$linkfun,
     sigma.linkfun = dstats$linkfun, mu.linkinv = mstats$linkinv,
     sigma.linkinv = dstats$linkinv, mu.dr = mstats$mu.eta,
         sigma.dr = dstats$mu.eta,
             dldm = function(y, mu, sigma) (1/sigma^2)*(y - mu),
           d2ldm2 = function(sigma) -(1/sigma^2),
             dldd = function(y, mu,sigma)
                    ((y - mu)^2 - sigma^2)/(sigma^3),
           d2ldd2 = function(sigma) -(2/(sigma^2)),
          d2ldmdd = function(y) rep(0, length(y)),
       G.dev.incr = function(y, mu, sigma, ...)
                    -2 * dNO(y, mu, sigma, log = TRUE),
            rqres = expression(rqres(pfun = "pNO", type =
                    "Continuous", y = y, mu = mu, sigma=sigma)),
       mu.initial = expression({ mu <- (y + mean(y))/2}),
    sigma.initial = expression({sigma <- rep(sd(y), length(y))}),
```

```
            mu.valid = function(mu) TRUE,
        sigma.valid = function(sigma) all(sigma > 0),
            y.valid = function(y) TRUE)
```

The meanings of the components are:

family   the name of the distribution, usually an abbreviated version and a more explicit one;

parameters   a list indicating whether the parameter will be fitted, e.g. mu=TRUE, or fixed at initial values, e.g. mu=FALSE;

nopar   the number of parameters;

type   the type of distribution, i.e. "Continuous", "Discrete" or "Mixed";

mu.link, sigma.link
         the current link functions as character strings;

mu.linkfun, sigma.linkfun
         the actual link functions returned from checklink();

mu.linkinv, sigma.linkinv
         the actual inverse link functions returned from checklink();

mu.dr, sigma.dr
         the actual first derivative of the inverse link functions returned from checklink();

dldm   the first derivative of the log-likelihood with respect to the location parameter mu. (Note that the log-likelihood here is regarded as a vector of contributions from the observations, rather than its sum which is a scalar);

d2ldm2   the expected second derivative of the log-likelihood with respect to mu;

dldd   the first derivative of the log-likelihood with respect to the scale parameter sigma;

d2ldd2   the expected second derivative of the log-likelihood with respect to sigma;

d2ldmddd   the expected cross derivative of the log-likelihood with respect to both mu and sigma;

G.dev.incr   the global deviance increment (equal to minus twice the log-likelihood);

rqres   the definition of the (normalized quantile) residuals. Note these are randomized for discrete distributions, this requires specification of the type of the distribution.

`mu.initial, sigma.initial`

>   the default initial starting values for `mu` and `sigma` (both vectors of
>   length n) for starting the algorithm;

`mu.valid, sigma.valid, y.valid`

>   valid range of values for the parameters (`mu` and `sigma`) and the
>   response variable.

The first and expected second and cross derivatives of the log-likelihood function
for a random observation from a normal distribution are obtained in Exercise 6.
Note that all of the above items are compulsory. The expected second derivatives
are replaced in some cases by the negative squared first derivatives. Similarly the
expected cross derivatives are replaced in some cases by the negative cross product
of the first derivatives.

6.4.3   The S3 class definition

Each family is defined as a `gamlss.family` object:

```
class = c("gamlss.family", "family")
}
```

The function `NO()` is now ended.

6.4.4   Definition of the d, p, q and r functions

```
dNO<-function(y, mu=0, sigma=1, log=FALSE)
{
 fy <- dnorm(y, mean=mu, sd=sigma, log=log)
 fy
}
pNO <- function(q, mu=0, sigma=1, lower.tail = TRUE, log.p = FALSE)
{
 if(any(sigma<=0)) stop(paste("sigma must be positive","\n",""))
 cdf<-pnorm(q,mean=mu,sd=sigma,lower.tail=lower.tail,log.p=log.p)
 cdf
}

qNO <- function(p, mu=0, sigma=1, lower.tail = TRUE, log.p = FALSE)
{
 if(any(sigma<=0))  stop(paste("sigma must be positive", "\n", ""))
 if(log.p==TRUE) p<-exp(p) else p <- p
 if(any(p<0)|any(p>1))
   stop(paste("p must be between 0 and 1", "\n", ""))
 q <- qnorm(p, mean=mu, sd=sigma, lower.tail = lower.tail )
```

```
  q
}

rNO <- function(n, mu=0, sigma=1)
{
 if(any(sigma<=0)) stop(paste("sigma must be positive", "\n", ""))
 r <- rnorm(n, mean=mu, sd=sigma)
 r
}
#-------------------------------------------------
```

The four functions `dNO`, `pNO`, `qNO` and `rNO` define in general the pdf, cdf, inverse cdf and random generating functions for the distribution. In the specific case of the normal distribution, these functions are not necessarily needed since **R** provides the equivalent functions `dnorm`, `pnorm`, `qnorm` and `rnorm`. We have included them here for convenience, and consistency with our parameterization of the distribution according to `mu` and `sigma`. Of the four functions, only the `d` function is usually used within the fitting function of a distribution, in the definition of global deviance, while the `p` function is needed in the definition of the normalized quantile residuals. The residuals are defined with the element `rqres` of the structure above, which uses the function `rqres()`. `rqres()` needs to know what type of `gamlss.family` distribution we are using. For example for the `NO` distribution above, we use the code

```
rqres(pfun="pNO", type="Continuous", y=y, mu=mu, sigma=sigma)
```

This in effect defines the residuals as `qnorm(pNO(y,mu,sigma))`. For discrete distributions, `rqres()` will randomize the residuals; see Chapter 12. For example the code for the Poisson distribution is

```
rqres(pfun="pPO", type="Discrete", ymin=0, y=y, mu=mu)
```

6.4.5 Example: reparameterizing the NO distribution

As an example in which a different parameterization of a distribution is required, consider the normal distribution in which `mu` is still the mean but `sigma` is now the variance of the distribution rather than the standard deviation. Only the changes from the previous definition of the function are reproduced here.

```
        O2 <- function (mu.link ="identity", sigma.link="log")
...

list(family = c("NO2","Normal with variance"),
...
        dldm = function(y,mu,sigma) (1/sigma)*(y-mu),
     d2ldm2 = function(sigma) -(1/sigma),
```

```
             dldd = function(y,mu,sigma)  0.5*((y-mu)^2-sigma)/(sigma^2),
          d2ldd2 = function(sigma)  -(1/(2*sigma^2)),
         d2ldmdd = function(y) rep(0,length(y)),
    G.dev.incr  = function(y,mu,sigma,...) -2*dNO2(y,mu,sigma,log=TRUE),
                            y=y, mu=mu, sigma=sigma)),
    ...
    }
```

The d, p, q and r functions have to be amended accordingly. Since **R** provides d, p, q and r functions for the normal distribution (dnorm, pnorm, qnorm and rnorm, respectively), the amendment can be done easily as follows:

```
dNO2<-function(y, mu=0, sigma=1, log=FALSE)
{
 if (any(sigma <= 0)) stop(paste("sigma must be positive", "\n", ""))
 fy <- dnorm(y, mean=mu, sd=sqrt(sigma), log=log)
 fy
}
pNO2 <- function(q, mu=0, sigma=1, lower.tail = TRUE, log.p = FALSE)
{
 if (any(sigma <= 0)) stop(paste("sigma must be positive", "\n", ""))
 cdf <- pnorm(q, mean=mu, sd=sqrt(sigma), lower.tail = lower.tail,
               log.p = log.p)
 cdf
}
qNO2 <- function(p, mu=0, sigma=1, lower.tail = TRUE, log.p = FALSE)
{
 if (any(sigma<=0))  stop(paste("sigma must be positive", "\n", ""))
 if (log.p==TRUE) p <- exp(p) else p <- p
 if (any(p < 0)|any(p > 1))
    stop(paste("p must be between 0 and 1", "\n", ""))
 q <- qnorm(p, mean=mu, sd=sqrt(sigma), lower.tail = lower.tail )
 q
}
rNO2 <- function(n, mu=0, sigma=1)
{
 if (any(sigma<=0))  stop(paste("sigma must be positive", "\n", ""))
 r <- rnorm(n, mean=mu, sd=sqrt(sigma))
 r
}
```

More generally, if an equivalent function does not exist already in **R** it has to be written explicitly. For example this is another version of dNO2:

```
dNO2<-function(y, mu=0, sigma=1, log=FALSE)
{
if (any(sigma<=0)) stop(paste("sigma must be positive", "\n", ""))
loglik <- -0.5*log(2*pi*sigma)-0.5*((y-mu)^2)/sigma
fy <- if(log==FALSE)  exp(loglik) else  loglik
fy
}
```

For users who would like to implement a different distribution from the ones in Tables 6.1, 6.2 and 6.3, the advice is to take one of the current distribution definition files (with the same number of parameters) and amend it. The GU(), TF() and BCT() distributions are good examples of 2-, 3- and 4-parameter continuous distributions, respectively. IG() provides a good example where the p and q functions are calculated using numerical methods. The PO(), NBI() and SI() are good examples of 1-, 2- and 3-parameter discrete distributions, respectively. The BB() provides a discrete distribution example where the p and q functions are calculated using numerical methods. The SICHEL() distribution is an example where numerical derivatives are implemented, using **numeric.deriv()**.

6.5 The link functions

A link function relates a distribution parameter to its predictor, e.g. $g_1(\mu) = \eta_1$ in equation (1.10). The function **make.link.gamlss()** creates all the standard link functions within the **gamlss.dist** package. Table 6.4 shows all the standard link functions currently available, (together with the usual range for the corresponding distribution parameter). Users can also create their own link function, as we show below.

6.5.1 How to display the available link functions

The default link functions for each parameter of a distribution in **gamlss** can be found by querying the name of the distribution. For example:

```
GA()

##
## GAMLSS Family: GA Gamma
## Link function for mu   : log
## Link function for sigma: log
```

indicates that the default link functions for both μ and σ are **log**.

6.5.2   Changing the default link function

The function `show.link()` displays all the available link functions for all the parameters of a distribution in `gamlss`.

```
show.link(GA)
```

```
## $mu
## c("inverse", "log", "identity", "own")
##
## $sigma
## c("inverse", "log", "identity", "own")
```

Hence the link functions `"inverse"` , `"log"`, `"identity"` are available for both the μ and σ parameters of the gamma distribution in `gamlss`. The `"own"` link is one way of defining a new link function for a parameter as described in Section 6.5.5.

The default link function for a parameter can be changed to one of its available link functions. This is demonstrated below, where the default link function for μ in the gamma distribution is changed to identity:

```
GA(mu.link=identity)
```

```
##
## GAMLSS Family: GA Gamma
## Link function for mu   : identity
## Link function for sigma: log
```

Note that such a change in the link function is usually applied using the `family` argument of the `gamlss()` function, e.g. `family=GA(mu.link=identity)`.

6.5.3   Defining a link function

Each link function requires the definition of four different functions, which are generated automatically for standard link functions, but must be created for new link functions:

`linkfun(mu)`   the link function defining the predictor η, (`eta`), as a function of the current distribution parameter (which is always referred to as `mu`), i.e. $\eta = g(\mu)$;

`linkinv(eta)`   the inverse of the link function as a function of the predictor η (`eta`), i.e. $\mu = g^{-1}(\eta)$;

`mu.eta(eta)`   the first derivative of the inverse link with respect to η (`eta`), i.e. $\frac{d\mu}{d\eta}$;

`validate(eta)`

        the range in which values of η (`eta`) are defined.

There are two ways for the user to generate a new link function within **gamlss**. The first is by creating a link function having the appropriate link information. This is the newest and recommended way. The other is by using the `own` link facility. That was the original way to generate a link function, but it is not as flexible, as it can only be used to change the link of one parameter of the current distribution.

6.5.4 Creating a link function

The `aep` data are described and analysed fully in Section 14.3. Here we demonstrate the creation of new and `own` link functions using these data.

We create the log-log link function

$$g(\mu) = \eta = -\log\left(-\log\mu\right)$$

with inverse link function

$$g^{-1}(\eta) = \mu = e^{-e^{-\eta}}$$

and derivative

$$\frac{d\mu}{d\eta} = e^{-e^{-\eta} - \eta} \ .$$

```
# creating a  log-log link
loglog  <- function()
{
linkfun<-function(mu) { -log(-log(mu))}
linkinv<-function(eta) {
 thresh<-log(-log(.Machine$double.eps))
    eta<-pmin(thresh, pmax(eta, -thresh))
       exp(-exp(-eta))}
 mu.eta<-function(eta) pmax(exp(-exp(-eta)-eta), .Machine$double.eps)
valideta<-function(eta) TRUE
   link<-"loglog"
structure(list(linkfun = linkfun, linkinv = linkinv, mu.eta = mu.eta,
        valideta = valideta, name = link), class = "link-gamlss")
}
# fitting a model
h1<-gamlss(y~ward+loglos+year, family=BI(mu.link=loglog()), data=aep,
                                 trace=FALSE)
```

6.5.5 Using the own link function

We use the `own` facility to create the complementary log-log link function:

$$g(\mu) = \eta = \log\left[-\log\left(1 - \mu\right)\right] \ ,$$

with inverse link function

$$g^{-1}(\eta) = \mu = 1 - e^{-e^{\eta}}$$

and derivative

$$\frac{d\mu}{d\eta} = e^{-e^{\eta} + \eta} \ .$$

We compare the results by using the existing `cloglog` link.

```
# The complementary log-log function using own link
own.linkfun<-function(mu) {log(-log(1-mu))}
own.linkinv<-function(eta) {
    thresh<--log(-log(1-.Machine$double.eps))
    eta<-pmin(thresh, pmax(eta, -thresh))
    1-exp(-exp(eta))}
own.mu.eta<-function(eta)pmax(exp(-exp(eta)+eta),.Machine$double.eps)
own.validate<-function(eta) TRUE

# fit model
h2<-gamlss(y~ward+loglos+year, family=BI(mu.link="own"), data=aep,
          trace=FALSE)
# h2 should be identical to cloglog in h3:
h3<-gamlss(y~ward+loglos+year, family=BI(mu.link="cloglog"),
          data=aep, trace=FALSE)
# compare the models
AIC(h1,h2,h3, k=0)

##     df      AIC
## h1   5 9439.480
## h2   5 9456.144
## h3   5 9456.144
```

6.6 Bibliographic notes

Continuous, discrete and mixed distributions used in **gamlss** are discussed in Rigby et al. [2017], where their probability (density) function and properties are given. For continuous distributions, classic references are Johnson et al. [1994] and Johnson et al. [1995].

Symmetric continuous distributions for $-\infty < Y < \infty$ include the two-parameter normal (NO), and the three-parameter t family (TF) [Lange et al., 1989], which is leptokurtic, and the power exponential (PE) [Nelson, 1991], which can be platykurtic or leptokurtic.

Flexible four-parameter continuous distributions for $-\infty < Y < \infty$ which can be positively or negatively skewed include the popular skew t type 3 (ST3) [Fernandez and Steel, 1998] and its reparameterized form (SST), which are both leptokurtic; the skew exponential power type 3 (SEP3) [Fernandez et al., 1995]; and the original sinh arc-sinh (SHASHo) [Jones and Pewsey, 2009], which can both be platykurtic or leptokurtic.

Continuous distributions for $0 < Y < \infty$ include the one-parameter exponential (EXP), two-parameter gamma (GA) and Weibull (WEI) and three-parameter generalized gamma (GG) [Lopatatzidis and Green, 2000] and generalized inverse Gaussian (GIG) [Jørgensen, 1982]. Popular continuous distributions for $0 < Y < \infty$ because of their flexibility are the three-parameter Box–Cox Cole and Green (BCCG, BCCGo) [Cole and Green, 1992] and the four-parameter Box–Cox power exponential (BCPE, BCPEo) [Rigby and Stasinopoulos, 2004] and Box–Cox t (BCT, BCTo) [Rigby and Stasinopoulos, 2006]. These are especially useful in centile estimation, when the values of the response are not close to zero.

Continuous distributions for $0 < Y < 1$ include the two-parameter beta (BE) and 'logit' distributions obtained by an inverse logit transformation of variable Z (see Section 6.2.2) or by truncation to range $(0, 1)$ (see Section 6.2.2). Mixed distributions which are continuous on the interval $(0, 1)$ and include 0, 1 or both endpoints, with point probabilities, include the beta inflated distributions [Ospina and Ferrari, 2010, 2012] (BEZI, BEOI, BEINF0, BEINF1 and BEINF), and inflated 'logit' distributions, inflated truncated distributions and generalized Tobit distributions [Hossain et al., 2016].

For discrete distributions the classic reference is Johnson et al. [2005]. A comprehensive thesaurus of discrete distributions is given by Wimmer and Altmann [1999]. The most popular discrete (count) distributions for $Y = 0, 1, 2, \ldots$ are the one-parameter Poisson (PO) and the two-parameter negative binomial (NBI, NBII) [Hilbe, 2011]. More flexible distributions, which can be more difficult to fit, are the two-parameter Poisson-inverse Gaussian (PIG) [Dean et al., 1989] and three-parameter Sichel (SI) and Delaporte (DEL) distributions; see Rigby et al. [2008]. Zero-inflated distributions for $Y = 0, 1, 2, \ldots$ include the zero-inflated Poisson and negative binomial (ZIP, ZINBI). Zero-adjusted distributions for $Y = 0, 1, 2, \ldots$, which allow for zero deflation as well as zero inflation, include the zero-altered (or adjusted) Poisson and negative binomial (ZAP, ZANBI) [Hilbe, 2011]. The most popular discrete distributions for $Y = 0, 1, 2, \ldots, N$ where N is known, are the one-parameter binomial (BI) and the two-parameter beta-binomial (BB).

6.7 Exercises

1. Use the **gamlss.demo** package to plot distributions.

```
library(gamlss.demo)
gamlss.demo()
```

Investigate how the following distributions change with their parameters:

(a) Continuous distributions

    i. Power exponential distribution (PE) for $-\infty < y < \infty$

    ii. Gamma distribution (GA) for $0 < y < \infty$

    iii. Beta distribution (BE) for $0 < y < 1$

(b) Discrete distributions

    i. Negative binomial type I (NBI) for $y = 0, 1, 2, 3, \ldots$

    ii. Beta binomial (BB) for $y = 0, 1, 2, 3, \ldots, n$

(c) Mixed distributions

    i. Zero adjusted gamma (ZAGA) for $0 \leq y < \infty$

    ii. Beta inflated (BEINF) for $0 \leq y \leq 1$

2. Plotting different distributions: The **gamlss.dist** package (which is downloaded automatically with **gamlss**) contains many distributions. Typing

```
?gamlss.family
```

will show all the available distributions in the **gamlss** packages. You can also explore the shape and other properties of the distributions. For example the following code will produce the pdf, cdf, inverse cdf and a histogram of a random sample generated from a gamma distribution:

```
PPP <- par(mfrow=c(2,2))
plot(function(y) dGA(y, mu=10 ,sigma=0.3),0.1, 25) # pdf
plot(function(y) pGA(y, mu=10 ,sigma=0.3), 0.1, 25) #cdf
plot(function(y) qGA(y, mu=10 ,sigma=0.3), 0, 1) # inverse cdf
hist(rGA(100,mu=10,sigma=.3)) # randomly generated values
par(PPP)
```

Note that the first three plots above can also be produced by using the function `curve()`, for example

```
curve(dGA(x=x, mu=10, sigma=.3),0, 25)
```

To explore discrete distributions use:

```
PPP <- par(mfrow=c(2,2))
plot(function(y) dNBI(y, mu = 10, sigma =0.5 ), from=0, to=40,
     n=40+1, type="h", main="pdf", ylab="pdf(x)")
cdf <- stepfun(0:39, c(0, pNBI(0:39, mu=10, sigma=0.5 )), f = 0)
```

```
plot(cdf,main="cdf", ylab="cdf(x)", do.points=FALSE )
invcdf <-stepfun(seq(0.01,.99,length=39), qNBI(seq(0.01,.99,
               length=40), mu=10, sigma=0.5 ), f = 0)
plot(invcdf,main="inverse cdf",ylab="inv-cdf(x)",do.points=FALSE)
tN <- table(Ni <- rNBI(1000,mu=5, sigma=0.5))
r <- barplot(tN, col='lightblue')
par(PPP)
```

Note that to find moments or to check if a distribution integrates or sums to one, the functions `integrate()` or `sum()` can be used. For example

```
integrate(function(y) dGA(y, mu=10, sigma=.1),0, Inf)
```

will check that the distribution integrates to one, and

```
integrate(function(y) y*dGA(y, mu=10, sigma=.1),0, Inf)
```

will give the mean of the distribution.

The pdf of a GAMLSS family distribution can also be plotted using the **gamlss** function `pdf.plot()`. For example

```
pdf.plot(family=GA, mu=10, sigma=c(.1,.5,1,2), min=0.01,max=20,
         step=.5)
```

will plot the pdf's of four gamma distributions $GA(\mu, \sigma)$, all with $\mu = 10$, but with $\sigma = 0.1, 0.5, 1$ and 2, respectively.

Try plotting other continuous distributions, e.g. `IG` (inverse Gaussian), `PE` (power exponential) and `BCT` (Box–Cox t); and discrete distributions, e.g. `NBI` (negative binomial type I) and `PIG` (Poisson inverse Gaussian). Make sure you define the values of all the parameters of the distribution.

3. **Turkish stock exchange: the `tse` data.** The data are for the eleven-year period 1 January 1988 to 31 December 1998. Continuously compounded returns in domestic currency were calculated as the first difference of the natural logarithm of the series. The objective is to fit a distribution to the Turkish stock exchange index.

> **R data file: `tse`** in package **gamlss.data** of dimensions 2868×6.
>
> **variables**
>
>> year
>>
>> month
>>
>> day
>>
>> ret : day returns `ret[t]=ln(currency[t])-ln(currency[t-1])`

> currency : the currency exchange rate
>
> tl : day return `ret[t]=log10(currency[t])-log10(currency[t-1])`
>
> **purpose:** to show the `gamlss` family of distributions.

(a) Input the data and plot the returns sequentially using

```
with(tse, plot(ret,type="l"))
```

(b) Fit continuous distributions on $(-\infty < y < \infty)$ to `ret`. Automatically choose the best fitting distribution according to AIC. Show the AIC for the different fitted distributions. Do any of the fits fail?

```
mbest<-fitDist(tse$ret,type="realline",k=2)
mbest
mbest$fits
mbest$fails
```

Repeat with `k=3.84` and `k=log(length(tse$ret))` (corresponding to criteria $\chi^2_{1,0.05}$ and SBC, respectively).

(c) For the chosen distribution, plot the fitted distribution using `histDist()`. Refit the model using `gamlss()` in order to output the parameter estimates using `summary()`.

(d) An alternative approach is to manually fit each of the following distributions for `ret` using `histDist()` (and using different model names for later comparison):

    i. two-parameter: normal $NO(\mu,\sigma)$,

```
mNO<-histDist(tse$ret,"NO",nbins=30, n.cyc=100)
```

    ii. three-parameter: t family $TF(\mu,\sigma,\nu)$ and power exponential $PE(\mu,\sigma,\nu)$

    iii. four-parameter: Johnson Su $JSU(\mu,\sigma,\nu,\tau)$, skew exponential power type 1 to 4, e.g. $SEP1(\mu,\sigma,\nu,\tau)$, skew t type 1 to 5, e.g. $ST1(\mu,\sigma,\nu,\tau)$ and sinh arc-sinh $SHASH(\mu,\sigma,\nu,\tau)$.

(Note that `histDist()` has as default `nbins=30`, to provide a detailed histogram.)

(e) Use `GAIC()` with each of the penalties $\kappa = 2, 3.84$ and $7.96 = \log(2868)$ (corresponding to criteria AIC, $\chi^2_{1,0.05}$ and SBC, respectively), in order to select a distribution model. Output the parameter estimates for your chosen model using the function `summary()`.

4. The stylometric data.

> **R data file:** `stylo` in package **gamlss.data** of dimensions 64×2
>
> **variables**
>
> > `word` : number of times a word appears in a single text
> >
> > `freq` : frequency of the number of times a word appears in a text
>
> **purpose:** to demonstrate the fitting of a truncated discrete distribution.

Note that the response variable `word` is (left) truncated at 0.

(a) Load the data and plot them.

(b) Create different truncated at zero count data distributions (`PO`, `NBII`, `DEL`, `SICHEL`), for example:

`gen.trun(par = 0, family = PO, type = "left")`

(c) Fit the different truncated distributions, for example:

```
mPO <- gamlss(word ~ 1, weights = freq, data = stylo,
              family = POtr, trace = FALSE)
```

(d) Compare the distributions using GAIC.

(e) Check the residuals of the chosen model using `plot()` and `wp()`.

(f) Plot the fitted distributions using `histDist`.

5. **Creating a new link function.** Create a general shifted log link function, taking as an argument a 'shifted from zero' value.

6. Let Y be a random observation from a `NO`(μ, σ) distribution with pdf

$$f(y \,|\, \mu, \sigma) = \frac{1}{\sqrt{2\pi}\sigma} \exp\left[-\frac{1}{2\sigma^2}(y - \mu)^2\right].$$

(a) Find the log-likelihood function, $\ell = \ell(\mu, \sigma; y) = \log f(y \,|\, \mu, \sigma)$.

(b) Find $\dfrac{\partial \ell}{\partial \mu}$ and $\dfrac{\partial \ell}{\partial \sigma}$.

(c) Find $\dfrac{\partial^2 \ell}{\partial \mu^2}$, $\dfrac{\partial^2 \ell}{\partial \sigma^2}$ and $\dfrac{\partial^2 \ell}{\partial \mu \partial \sigma}$.

(d) Find the expected values $\mathrm{E}_Y\left[\dfrac{\partial^2 \ell}{\partial \mu^2}\right]$, $\mathrm{E}_Y\left[\dfrac{\partial^2 \ell}{\partial \sigma^2}\right]$ and $\mathrm{E}_Y\left[\dfrac{\partial^2 \ell}{\partial \mu \partial \sigma}\right]$.

(e) Check that the results you obtained in (b) and (d) agree with the formulae as used in Section 6.4.2, by typing

7. Let Y be a random observation from a continuous distribution given below. For each distribution find (as in Exercise 6):

   (a) the log-likelihood function ℓ,

   (b) the first derivative of ℓ with respect to each parameter,

   (c) the second and cross derivatives of ℓ with respect to the parameters,

   (d) the expected second and cross derivatives of ℓ with respect to the parameters.

   (e) Check the results by typing the distribution name in **gamlss**.

   Continuous distributions

   (a) $Y \sim \text{EXP}(\mu)$ with pdf

   $$f(y) = \frac{1}{\mu} \exp\left(-\frac{y}{\mu}\right),$$

   for $y > 0$, where $\mu > 0$.

   (b) $Y \sim \text{GA}(\mu, \sigma)$ with pdf

   $$f(y) = \frac{1}{\Gamma\left(\frac{1}{\sigma^2}\right) y} \left(\frac{y}{\sigma^2 \mu}\right)^{\frac{1}{\sigma^2}} \exp\left(-\frac{y}{\sigma^2 \mu}\right),$$

   for $y > 0$, where $\mu > 0$ and $\sigma > 0$.

   (c) $Y \sim \text{GU}(\mu, \sigma)$ with pdf

   $$f(y) = \frac{1}{\sigma} \exp\left[\left(\frac{y-\mu}{\sigma}\right) - \exp\left(\frac{y-\mu}{\sigma}\right)\right],$$

   for $-\infty < y < \infty$, where $-\infty < \mu < \infty$ and $\sigma > 0$.

   (d) $Y \sim \text{WEI}(\mu, \sigma)$ with pdf

   $$f(y) = \frac{\sigma y^{\sigma-1}}{\mu^\sigma} \exp\left[-\left(\frac{y}{\mu}\right)^\sigma\right],$$

   for $y > 0$, where $\mu > 0$ and $\sigma > 0$.

8. Let Y be a random observation from a discrete distribution with probability function (pf) given below. Repeat Exercise 7 for each of the following:

   (a) $Y \sim \text{PO}(\mu)$ with pf

   $$\text{Prob}(Y = y) = \frac{\exp(-\mu)\mu^y}{y!},$$

   for $y = 0, 1, 2, \ldots$, where $\mu > 0$.

(b) $Y \sim \text{BI}(N, \mu)$ with pf

$$\text{Prob}(Y = y) = \frac{N!}{y!(N-y)!} \mu^y (1 - \mu)^{N-y},$$

for $y = 0, 1, 2, \ldots, N$ where $0 < \mu < 1$ and N is a known positive integer.

9. Write the **R** code needed to fit the original beta distribution $Y \sim \text{BEo}(\mu, \sigma)$ with pdf

$$f(y) = \frac{1}{B(\mu, \sigma)} y^{\mu-1} (1 - y)^{\sigma-1},$$

for $0 < y < 1$, where $\mu > 0$ and $\sigma > 0$.

7

Finite mixture distributions

CONTENTS

7.1   Introduction to finite mixtures

This chapter covers finite mixtures within GAMLSS, in particular:

1. finite mixtures with no parameters in common and

2. finite mixtures with parameters in common.

This chapter is important for fitting *multimodal* distributions to data or for fitting *latent* class models. Latent class models arise when we believe that the population under study is divided into distinct (but unknown) classes or categories and we would like to estimate the probabilities of individuals belonging to each of the classes.

Suppose that the random variable Y comes from a population which is partitioned into $\mathcal{K}$ subpopulations or components. In component k, the probability (density) function of Y is $f_k(y)$. The (marginal) probability (density) function of Y is given by

$$f(y) = \sum_{k=1}^{\mathcal{K}} \pi_k f_k(y), \tag{7.1}$$

where $0 \leq \pi_k \leq 1$ is the prior (or mixing) probability of Y coming from component k, for $k = 1, 2, \ldots, \mathcal{K}$ and $\sum_{k=1}^{\mathcal{K}} \pi_k = 1$. Note that the cumulative distribution function of Y has a similar form:

$$F(y) = \sum_{k=1}^{\mathcal{K}} \pi_k F_k(y) \, . \tag{7.2}$$

Using Bayes theorem, the conditional (posterior) probability of an observation belonging to component k given y is given by:

$$p_k = \frac{\pi_k f_k(y)}{f(y)} = \frac{\pi_k f_k(y)}{\sum_{h=1}^{\mathcal{K}} \pi_h f_h(y)} \, . \tag{7.3}$$

More generally the probability function $f_k(y)$ for component k may depend on parameters $\boldsymbol{\theta}_k$ (combining all the parameters in the predictor models for distribution parameters, $\mu, \sigma, \nu,$ and τ for mixture component k) which themselves can be a function of explanatory variables $\mathbf{x}_k$, i.e. $f_k(y) = f_k(y \,|\, \boldsymbol{\theta}_k)$. Hence the marginal distribution $f(y)$ depends on parameters $\boldsymbol{\psi} = (\boldsymbol{\theta}, \boldsymbol{\pi})$ where, in this chapter, $\boldsymbol{\theta} = (\boldsymbol{\theta}_1^\top, \ldots, \boldsymbol{\theta}_{\mathcal{K}}^\top)^\top$, and $\boldsymbol{\pi} = (\pi_1, \ldots, \pi_{\mathcal{K}})$, with explanatory variables $\mathbf{x} = (\mathbf{x}_1, \ldots, \mathbf{x}_{\mathcal{K}})$. This gives $f(y) = f(y \,|\, \boldsymbol{\psi})$, and

$$f(y \,|\, \boldsymbol{\psi}) = \sum_{k=1}^{\mathcal{K}} \pi_k f_k(y \,|\, \boldsymbol{\theta}_k) \, . \tag{7.4}$$

We now omit the conditioning on $\boldsymbol{\psi}$ and $\boldsymbol{\theta}_k$ to simplify the presentation. Note that in practical situations $\mathcal{K}$, the total number of components, is generally unknown and has to be estimated from the data.

Finite mixture distribution and their applications are extensively discussed in the literature; see Lindsay [1995], Böhning [2000], McLachlan and Peel [2004] and Aitkin et al. [2009, Chapter 7]. Here we discuss their implementation within GAMLSS. Finite mixture distributions are fitted within GAMLSS using the expectation-maximization (EM) algorithm [Dempster et al., 1977].

Specific mixed distributions are explicitly available in the **gamlss.dist** package, for example the zero adjusted gamma (ZAGA), the zero adjusted inverse Gaussian (ZAIG) and the zero-and-one inflated beta (BEINF). There are also a variety of zero inflated and zero adjusted discrete distributions (ZIP, ZIP2, ZAP, ZINBI, ZANBI, ZIPIG, ZIBI, ZIBB). Those specific distributions are not the subject of this chapter.

Sections 7.2 to 7.4 are concerned with finite mixtures with no parameters in common, while the remainder of the chapter is devoted to finite mixtures with common parameters. Section 7.2 is rather theoretical, explaining the justification of the fitting EM algorithm, and can be omitted by practitioners who are interested only in applying the models. Section 7.3 explains the function `gamlssMX()`. An example for fitting a finite mixture model with no parameters in common is given in Section 7.4. The theoretical justification of the algorithm of fitting finite mixtures with parameters in common is given in Section 7.5. The function `gamlssNP()` for fitting the model is explained in Section 7.6, and an example demonstrating its use is given in Section 7.7. Throughout this chapter we will assume that all $\mathcal{K}$ components of the mixture can be represented by GAMLSS models.

7.2 Finite mixtures with no parameters in common

Here the parameter sets $(\boldsymbol{\theta}_1, \boldsymbol{\theta}_2, \ldots, \boldsymbol{\theta}_{\mathcal{K}})$ are distinct, that is, there is no parameter common to two or more parameter sets. One implication that this has in practice is that the conditional distribution components $f_k(y)$ in (7.1) can also be different. Within GAMLSS this means that we can have different `gamlss.family` distributions, e.g. one can be `GA` and the other `IG`.

7.2.1 The likelihood function

Given n independent observations y_i for $i = 1, 2, \ldots, n$, from the finite mixture model (7.4), the likelihood function is given by

$$L = L(\boldsymbol{\psi} \,|\, \mathbf{y}) = \prod_{i=1}^{n} f(y_i) = \prod_{i=1}^{n} \left[\sum_{k=1}^{\mathcal{K}} \pi_k f_k(y_i) \right] \tag{7.5}$$

where $\boldsymbol{\psi} = (\boldsymbol{\theta}, \boldsymbol{\pi})$, $\mathbf{y} = (y_1, y_2, \ldots, y_n)^{\top}$ and $f_k(y_i) = f_k(y_i \,|\, \boldsymbol{\theta}_k)$. The log-likelihood function is given by

$$\ell = \ell(\boldsymbol{\psi} \,|\, \mathbf{y}) = \sum_{i=1}^{n} \log \left[\sum_{k=1}^{\mathcal{K}} \pi_k f_k(y_i) \right] . \tag{7.6}$$

We wish to maximize ℓ with respect to $\boldsymbol{\psi}$, i.e. with respect to $\boldsymbol{\theta}$ and $\boldsymbol{\pi}$. The problem is that the logarithm of the second summation in (7.6) makes the solution difficult. One possible way, especially for simple mixtures where no explanatory variables are involved, is to use a numerical maximization technique, e.g. the `optim()` function, to maximize ℓ numerically. (See for example Venables and Ripley [2002, Chapter 16].) In our case where we are interested in regression type models an EM algorithm can be used for the maximization.

7.2.2   Maximizing the likelihood function using the EM algorithm

Within the EM algorithm a *complete* log-likelihood is defined. The conditional expectation of the log-likelihood is taken (the E-step), and this conditional expectation is maximized with respect to the required parameters (the M-step). The EM algorithm alternates between the E-step and the M-step until convergence [Dempster et al., 1977]. In the case of finite mixtures, the EM algorithm will maximize (7.6) with respect to ψ by treating all the component indicator variables (δ, defined below) as missing variables.

Let

$$\delta_{ik} = \begin{cases} 1 & \text{if observation } i \text{ comes from component } k \\ 0 & \text{otherwise} \end{cases} \tag{7.7}$$

for $k = 1, 2, \ldots, K$ and $i = 1, 2, \ldots, n$. Let $\delta_i^\top = (\delta_{i1}, \delta_{i2}, \ldots, \delta_{iK})$ be the indicator vector for observation i. If observation i comes from component k then δ_i is a vector of zeros, except for $\delta_{ik} = 1$. For example, with $K = 4$, a typical δ_i will look like $(0, 0, 1, 0)$.

Let $\delta^\top = (\delta_1^\top, \delta_2^\top, \ldots, \delta_n^\top)$ combine all the indicator variable vectors. Then the complete data, i.e. observed $\mathbf{y}$ and unobserved δ, has a complete likelihood function given by

$$L_c = L_c(\psi, \mathbf{y}, \delta) = f(\mathbf{y}, \delta) = \prod_{i=1}^{n} f(y_i, \delta_i)$$

$$= \prod_{i=1}^{n} f(y_i \mid \delta_i) f(\delta_i)$$

$$= \prod_{i=1}^{n} \left\{ \prod_{k=1}^{K} \left[f_k(y_i)^{\delta_{ik}} \pi_k^{\delta_{ik}} \right] \right\}, \tag{7.8}$$

since if y_i comes from component k, then $\delta_{ik} = 1$ and $\delta_{ik'} = 0$ for $k' \neq k$, and

$$f(y_i \mid \delta_i) f(\delta_i) = f_k(y_i) \pi_k = f_k(y_i)^{\delta_{ik}} \pi_k^{\delta_{ik}} = \prod_{k=1}^{K} f_k(y_i)^{\delta_{ik}} \pi_k^{\delta_{ik}} .$$

From (7.8) the complete log-likelihood is given by

$$\ell_c = \ell_c(\psi, \mathbf{y}, \delta) = \sum_{i=1}^{n} \sum_{k=1}^{K} \delta_{ik} \log f_k(y_i) + \sum_{i=1}^{n} \sum_{k=1}^{K} \delta_{ik} \log \pi_k . \tag{7.9}$$

If δ were known then ℓ_c could be maximized over each θ_k separately, since $\theta_1, \theta_2, \ldots \theta_K$ have no parameters in common and the likelihood separates.

E-step: At the $(r+1)$st iteration, the E-step finds Q, the conditional expectation of the complete data log-likelihood (7.9), over the missing $\boldsymbol{\delta}$, given $\mathbf{y}$ and the current parameter estimates $\hat{\boldsymbol{\psi}}^{(r)}$ from iteration r. Following Aitkin [1999a], the Q function is defined as:

$$
\begin{aligned}
Q &= E_{\boldsymbol{\delta}}\left[\ell_c \,|\, \mathbf{y}, \hat{\boldsymbol{\psi}}^{(r)}\right] \\
&= \sum_{k=1}^{\mathcal{K}} \sum_{i=1}^{n} \hat{p}_{ik}^{(r+1)} \log f_k(y_i) + \sum_{k=1}^{\mathcal{K}} \sum_{i=1}^{n} \hat{p}_{ik}^{(r+1)} \log \pi_k \qquad (7.10)
\end{aligned}
$$

where

$$
\begin{aligned}
\hat{p}_{ik}^{(r+1)} &= E\left[\delta_{ik} \,|\, \mathbf{y}, \hat{\boldsymbol{\psi}}^{(r)}\right] \\
&= p\left(\delta_{ik} = 1 \,|\, y_i, \hat{\boldsymbol{\psi}}^{(r)}\right) \\
&= \frac{\hat{\pi}_k^{(r)} f_k(y_i \,|\, \hat{\boldsymbol{\theta}}_k^{(r)})}{\sum_{h=1}^{\mathcal{K}} \hat{\pi}_h^{(r)} f_h(y_i \,|\, \hat{\boldsymbol{\theta}}_h^{(r)})}. \qquad (7.11)
\end{aligned}
$$

using Bayes theorem. Note that $\hat{p}_{ik}^{(r+1)}$ is the current estimate of the posterior probability p_{ik} of the ith observation y_i belonging to the kth mixture component given the observed y_i value and given $\boldsymbol{\psi} = \hat{\boldsymbol{\psi}}^{(r)}$, while $\hat{\pi}_k^{(r)} = p(\delta_{ik} = 1 \,|\, \hat{\boldsymbol{\psi}}^{(r)})$ is the estimate of the prior (or mixing) probability that observation y_i comes from component k, given $\boldsymbol{\psi} = \hat{\boldsymbol{\psi}}^{(r)}$ only. (On convergence, i.e. $r = \infty$, $\hat{p}_{ik}^{(\infty)}$ and $\hat{\pi}_k^{(\infty)}$ are the **estimated** posterior and prior probabilities that the observation y_i comes from component k, respectively, since $\boldsymbol{\psi}$ is estimated by $\hat{\boldsymbol{\psi}}^{(\infty)}$.)

M-step: At the $(r+1)$st iteration, the M step maximizes Q with respect to $\boldsymbol{\psi}$, where Q is the conditional expected value from E-step, given by (7.10). Since the parameters $\boldsymbol{\theta}_k$ in $f_k(y_i)$ for $k = 1, 2, \ldots, \mathcal{K}$ are distinct, (i.e. there are no parameters in common to two or more $\boldsymbol{\theta}_k$'s), Q can be maximized with respect to each $\boldsymbol{\theta}_k$ by maximizing separately the kth part of the first term in (7.10), i.e. maximize $\sum_{i=1}^{n} \hat{p}_{ik}^{(r+1)} \log f_k(y_i)$ with respect to $\boldsymbol{\theta}_k$, for $k = 1, 2, \ldots, \mathcal{K}$. Assuming, for $k = 1, 2, \ldots, \mathcal{K}$, that component k follows a GAMLSS model, this is just a weighted log-likelihood for a GAMLSS model with weights $\hat{p}_{ik}^{(r+1)}$ for $i = 1, 2, \ldots, n$. Also the parameters $\boldsymbol{\pi}$ only occur in the second term in (7.10) and so can be estimated by maximizing the second term, subject to $\sum_{k=1}^{\mathcal{K}} \pi_k = 1$, i.e. where $\pi_k = 1 - \sum_{k=1}^{\mathcal{K}-1} \pi_k$, leading to $\hat{\pi}_k^{(r+1)} = \frac{1}{n} \sum_{i=1}^{n} \hat{p}_{ik}^{(r+1)}$ for $k = 1, 2, \ldots, \mathcal{K}$.

Summary of the $(r+1)$st iteration of the EM algorithm

E-step For $k = 1, 2, \ldots, \mathcal{K}$ and $i = 1, 2, \ldots, n$: replace δ_{ik} in (7.9) by $\hat{p}_{ik}^{(r+1)}$, its conditional expectation given $\mathbf{y}$ and given the current estimate $\hat{\boldsymbol{\psi}}^{(r)}$ from iteration r (obtained from (7.11)), to give (7.10).

M-step

(1) For each $k = 1, 2 \ldots, \mathcal{K}$, obtain $\hat{\boldsymbol{\theta}}_k^{(r+1)}$ by fitting the GAMLSS model for the kth component to response variable $\mathbf{y}$ with explanatory variables $\mathbf{x}_k$ using prior weights $\hat{\mathbf{p}}_k^{(r+1)}$, where $\mathbf{p}_k^\top = (p_{1k}, p_{2k}, \ldots, p_{nk})$,

(2) $\hat{\pi}_k^{(r+1)} = \frac{1}{n} \sum_{i=1}^n \hat{p}_{ik}^{(r+1)}$ for $k = 1, 2 \ldots, \mathcal{K}$,

(3) $\hat{\boldsymbol{\psi}}^{(r+1)} = \left[\hat{\boldsymbol{\theta}}^{(r+1)}, \hat{\boldsymbol{\pi}}^{(r+1)} \right]$ where $\boldsymbol{\theta} = (\boldsymbol{\theta}_1^\top, \boldsymbol{\theta}_2^\top, \ldots, \boldsymbol{\theta}_\mathcal{K}^\top)^\top$ and $\boldsymbol{\pi} = (\pi_1, \pi_2, \ldots, \pi_\mathcal{K})$.

One of the characteristics of the EM algorithm employed to find the ML estimates in a finite mixture is that it does **not** provide standard errors. The standard errors which are given by fitting each component of the finite mixture using a weighted likelihood do not take into account the uncertainty of estimating the parameters $\boldsymbol{\pi}$. One will need the information matrix for all the parameters, something that the **gamlss.mx** package does not currently provide. A simple solution suggested by Aitkin et al. [2009] is to use the approximation (5.4). Bootstrapping is an alternative method of obtaining standard errors for the parameters.

7.2.3   Modelling the mixing probabilities

The procedure for fitting the finite mixture model can be extended by assuming that the mixing probabilities π_k depend on p_1 explanatory variables $\mathbf{s} = (\mathbf{s}_1, \ldots, \mathbf{s}_{p_1})$, and hence on i. We assume that $\boldsymbol{\delta}_i$ is a single draw from a multinomial distribution with probability vector $\boldsymbol{\pi}$, i.e. $\boldsymbol{\delta}_i \sim \text{Multinomial}(1, \boldsymbol{\pi})$ and

$$\log \left(\frac{\pi_{ik}}{\pi_{i\mathcal{K}}} \right) = \boldsymbol{\alpha}_k^\top \mathbf{s}_i \qquad (7.12)$$

where $\boldsymbol{\alpha}_k^\top = (\alpha_{k1}, \ldots, \alpha_{k,p_1})$ and $\mathbf{s}_i = (s_{i1}, \ldots, s_{i,p_1})^\top$ for $k = 1, 2, \ldots, \mathcal{K} - 1$. Imposition of the constraint $\sum_{k=1}^\mathcal{K} \pi_{ik} = 1$ gives

$$\pi_{ik} = \frac{\exp \left(\boldsymbol{\alpha}_k^\top \mathbf{s}_i \right)}{\sum_{h=1}^\mathcal{K} \exp \left(\boldsymbol{\alpha}_h^\top \mathbf{s}_i \right)}, \qquad (7.13)$$

where $\boldsymbol{\alpha}_\mathcal{K} = \mathbf{0}$. Consequently the complete log-likelihood is given by replacing π_k by π_{ik} in equation (7.9) to give

$$\ell_c(\boldsymbol{\psi}, \mathbf{y}, \boldsymbol{\delta}) = \sum_{i=1}^n \sum_{k=1}^\mathcal{K} \delta_{ik} \log f_k(y_i) + \sum_{i=1}^n \sum_{k=1}^\mathcal{K} \delta_{ik} \log \pi_{ik} . \qquad (7.14)$$

7.2.4 Estimating $\mathcal{K}$: the total number of components

In the discussion so far we assumed that $\mathcal{K}$, the total number of components in the finite mixture, is known in advance, so we only have to estimate the $\boldsymbol{\theta}$ and $\boldsymbol{\pi}$ parameters. In practical applications $\mathcal{K}$ is usually unknown and has to be determined by the data. Unfortunately the estimation of $\mathcal{K}$ is an irregular statistical problem and the usual procedures such as the likelihood ratio test or even the GAIC do not apply. One approach is to apply a parametric bootstrap. The reader is referred to Aitkin et al. [2009] and the references therein for a detailed discussion on the problem. Here we take a pragmatic approach to selecting $\mathcal{K}$, relying on a GAIC with a strong penalty and on diagnostics such as worm plots.

7.2.5 Zero components

Zero-inflated and zero-adjusted distributions are special cases of the finite mixture models described above. For example, the zero-adjusted inverse Gaussian distribution (**ZAIG**) can be thought of as a two-component finite mixture in which the first component is identically zero, i.e. $y = 0$ with probability 1, and the second component is an inverse Gaussian distribution. Hence

$$ f_1(y) = \begin{cases} 1 & \text{if } y = 0 \\ 0 & \text{otherwise} , \end{cases} $$

and $f_2(y)$ is the inverse Gaussian pdf. Distributions of this type can be fitted with the EM algorithm described in the previous section but convergence is very slow. Specific zero-inflated and zero-adjusted distributions (e.g. **ZAIG**, **ZAGA**) are available explicitly in **gamlss**.

7.3 The gamlssMX() function

The function for fitting finite mixtures with no parameters in common is **gamlssMX()** in the **gamlss.mx** package. In this section we describe how it works, while examples of its use are given in Section 7.4. The function **gamlssMX()** has the following arguments:

formula   This argument is a single formula (or a list of formulae of length $\mathcal{K}$) for modelling the predictor for the μ parameter of the model. If a single formula is used then the $\mathcal{K}$ mixture components have the same predictor for μ, but different parameters in their predictors (since there are no parameters in common to two or more of the $\mathcal{K}$ components). Note that modelling the rest of the distributional parameters can be done by using the usual `gamlss()` formula arguments, e.g. `sigma.fo=~x`, which passes the arguments to `gamlss()`. Again either a single common formula or a list of formulae of length $\mathcal{K}$ is used;

`pi.formula`   This is a formula for modelling the predictor for prior (or mixing) probabilities as a function of explanatory variables in the multinomial model (7.12). The default model is constants for the prior (or mixing) probabilities. Note that no smoothing or other additive terms are allowed here, only the usual linear terms. The modelling here is done using the `multinom()` function from the **nnet** package;

`family`   This is a `gamlss.family` distribution (or a list of $\mathcal{K}$ distributions). Note that if different distributions are used here, it is preferable (but not essential) that their parameters are comparable for ease of interpretation;

`weights`   For declaring prior weights if needed;

`K`   the number of components $\mathcal{K}$ in the finite mixture with default K=2;

`prob`   starting values for the prior probabilities;

`data`   The data frame containing the variables in the fit. Note that this is compulsory if `pi.formula` is used for modelling the prior (or mixing) probabilities;

`control`   control parameters for the EM iterations algorithm. The default settings are given in the `MX.control` function;

`g.control`   This argument can be used to pass control parameters, as in `gamlss.control`;

`zero.component`
whether or not there is a zero component, i.e. y identically equal to zero, in the finite mixture;

`...`   For extra arguments to be passed to `gamlss()`.

Note that finite mixtures are notorious for having multi-modal likelihoods, therefore it is advisable to use the `gamlssMX()` function several times to ensure that a global maximum has been reached. The function `gamlssMXfits()` is designed for this purpose and has as its first argument **n**, the number of different fits to try. (The **n** fits are obtained using different seeds resulting in different random allocations of observations to components at the start of the EM-algorithm.) The model with the smallest global deviance is reported in the end.

A fitted `gamlssMX` object has the following components:

`models`   A list of the different fitted `gamlss` models which comprise the finite mixture. They can be accessed individually, e.g. `m1$models[1]`;

`model.pi`   The fitted multinomial model (if required);

`G.deviance`   The global deviance;

`df.fit, df.residual`
The fitted and residual degrees of freedom;

`post.prob`   The fitted posterior probabilities $\hat{p}_{ik}$ for $i = 1, \ldots, N$ and $k = 1, \ldots, \mathcal{K}$;

`prob`   The fitted prior probabilities, $\hat{\pi}_{ik}$ or $\hat{\pi}_k$ depending on whether the probabilities are modelled as function of explanatory variables or not, for $i = 1, \ldots, N$ and $k = 1, \ldots, \mathcal{K}$;

`family`   The different `gamlss.family` distributions fitted;

`call`   The function call;

`aic, sbc`   The information criterion;

`K`   The number of components $\mathcal{K}$ in the finite mixture;

`N`   Total number of observations;

`weights`   The prior weights;

`residuals`   The normalized (randomized) quantile residuals. The residuals from the fitted mixture distribution for each (continuous response) observation are obtained as $\hat{r}_i = \Phi^{-1}\left[\hat{F}(y_i)\right]$ where $\hat{F}(y_i) = \sum_{k=1}^{\mathcal{K}} \hat{\pi}_k \hat{F}_k(y_i)$ is the fitted cumulative distribution function of equation (7.2). For a discrete response a randomization, as described in Section 12.2, is performed;

`Presiduals`   These are alternative residuals using the estimated posterior probabilies (rather than prior probabilities), i.e. $\hat{F}(y_i) = \sum_{k=1}^{\mathcal{K}} \hat{p}_{ki} \hat{F}_k(y_i)$;

`seed`   The random seed used for fitting the model (useful if an identical model has to be refitted).

Several methods are available for fitted `gamlssMX` objects. The functions `fitted()`, `coef()`, `formula()`, `model.matrix()`, `terms()` and `predict()` all have an argument K which needs to be specified in order to pick up the right component of the finite mixture. The methods `print()` and `resid()` do not have this argument. The `summary()` method is not provided for `gamlssMX` objects.

7.4   Example using `gamlssMX()`: Reading glasses data

The age at which participants in the Blue Mountains Eye Study [Attebo et al., 1996] reported that they started wearing reading glasses (`ageread`) is analysed here. As a covariate we have the subjects' `sex`.

> **Data summary:** Reading glasses
>
> **R data file: `glasses`** in package **gamlss.data** of dimensions 1016×3
>
> **variables**
>
> > `age` : age of participants
> >
> > `sex` : 1=male, 2=female
> >
> > `ageread` : age at which participant started wearing reading glasses
>
> **purpose:** the variable `ageread` is used here to demonstrate the fitting of a finite mixture.
>
> **conclusion:** A two-component logistic distribution is found to be suitable

We firstly demonstrate the fitting of a finite mixture model to `ageread` without covariates, and then we model `ageread` with `sex` as an explanatory variable. The histogram and the two density estimates shown in Figure 7.1 are suggestive of two subpopulations of subjects: those who started wearing reading glasses in childhood and early adulthood, and those who started wearing them later in life. (The density estimates were created using the functions `histSmo()` and `histSmoC()`.)

```
data(glasses)
truehist(glasses$ageread,nbins=25,col="grey",xlab="Age",
        ylab="Density", ylim=c(0,0.05))
lines(histSmo(glasses$ageread), lty=1, lwd=2)
lines(histSmoC(glasses$ageread, df=7), lty=2, lwd=2)
legend("topleft",legend=c("local ML","fixed df"),lty=1:2, cex=1.5)
```

Fig. 7.1

We demonstrate how to fit a variety of two-component mixtures of continuous distributions and then select the 'best' using AIC. Two-component mixtures of normal, gamma, inverse Gaussian, Weibull, reverse Gumbel and logistic distributions are fitted using `gamlssMX()`, and then compared using `AIC()`:

```
library(gamlss.mx)
set.seed(3683)
readNO <- gamlssMX(ageread~1,family=NO,K=2,data=glasses)
readGA <- gamlssMX(ageread~1,family=GA,K=2,data=glasses)
readIG <- gamlssMX(ageread~1,family=IG,K=2,data=glasses)
```

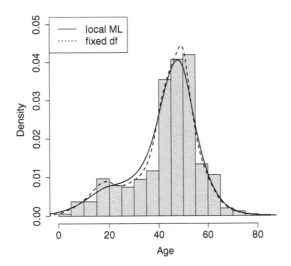

FIGURE 7.1: Histogram of age at which subjects started wearing reading glasses, with two different nonparametric density estimates, using the functions `histSmo()` and `histSmoC()`. In the solid line the smoothing parameter is estimated using local ML estimation, while in the dashed line the smoothing parameter is estimated by fixing the degrees of freedom at 7.

```
readWEI <- gamlssMX(ageread~1,family=WEI,K=2,data=glasses)
readRG <- gamlssMX(ageread~1,family=RG,K=2,data=glasses)
readLO <- gamlssMX(ageread~1,family=LO,K=2,data=glasses)
AIC(readNO,readGA,readIG,readWEI,readRG,readLO)

##            df      AIC
## readLO    5 7930.470
## readGA    5 7930.677
## readRG    5 7949.452
## readWEI   5 7956.933
## readNO    5 7958.670
## readIG    5 7971.757
```

(Note that we fix the seed using `set.seed()` so the results are reproducible.) The best model appears to be the two-component logistic (`LO`) model, although the gamma (`GA`) model gives a very comparable fit and may be preferred (because it has a positive response variable). We can verify that we have reached a global maximum using the function `gamlssMXfits()` and fitting five models with different starting values:

```
readLO1<-gamlssMXfits(n=5, ageread~1,family=LO,K=2,data=glasses)

## model= 1
```

```
## model= 2
## model= 3
## model= 4
## model= 5
```

The resulting fitted model `readLO1` is unchanged from `readLO`:

```
readLO
```

```
##
## Mixing Family:  c("LO", "LO")
##
## Fitting method: EM algorithm
##
## Call:
## gamlssMX(formula = ageread ~ 1, family = LO, K = 2,
##      data = glasses)
##
## Mu Coefficients for model: 1
## (Intercept)
##        18.86
## Sigma Coefficients for model: 1
## (Intercept)
##        1.492
## Mu Coefficients for model: 2
## (Intercept)
##        47.04
## Sigma Coefficients for model: 2
## (Intercept)
##        1.553
##
## Estimated probabilities: 0.1663841 0.8336159
##
## Degrees of Freedom for the fit: 5 Residual Deg. of Freedom    1011
## Global Deviance:      7920.47
##              AIC:      7930.47
##              SBC:      7955.09
```

As the logistic distribution has default identity link for μ and log link for σ, the estimated model is given by

$$f(y) = \hat{\pi}_i f_1(y) + \hat{\pi}_i f_2(y)$$
$$= 0.17 f_1(y) + 0.83 f_2(y)$$

where $f_1(y)$ is a logistic distribution pdf, $\text{LO}(\mu_1, \sigma_1)$ with $\hat{\mu}_1 = 18.86$ and $\hat{\sigma}_1 = \exp(1.492) = 4.45$ and $f_2(y)$ is a logistic distribution pdf, $\text{LO}(\mu_2, \sigma_2)$ with $\hat{\mu}_2 = 47.04$

and $\hat{\sigma}_2 = \exp(1.553) = 4.73$. Figure 7.2 shows the histogram of the data together with the fitted two-component LO model.

Fig.
7.2

```
### create a function to plot the fitted values
### of observation 1, i.e.   case 1
fnLO <- getpdfMX(readLO, observation=1)
fnGA <- getpdfMX(readGA, observation=1)
truehist(glasses$ageread, nbins=25, col="grey", xlab="Age",
        ylab="Density", ymax=0.05)
lines(seq(0.5,90.5,1),fnLO(seq(0.5,90.5,1)), lty=1, lwd=2)
lines(seq(0.5,90.5,1),fnGA(seq(0.5,90.5,1)), lty=2, lwd=2)
legend("topleft",legend=c("logistic","gamma"),lty=1:2, cex=1.5)
```

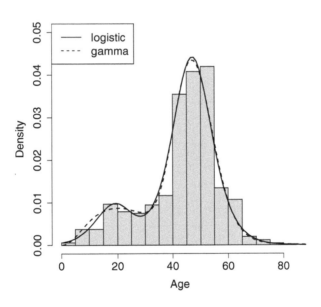

FIGURE 7.2: Histogram of age at which subjects started wearing reading glasses, with fitted two-component logistic distribution (black-solid line), and gamma distribution (dashed line).

Note the use of the `getpdfMX()` function in the above code. The command `getpdfMX(readLO, observation=1)` creates a new function which can then be used to plot the fitted pdf for a specified observation. Here we choose observation 1 since a single distribution is fitted with no explanatory variables, and the fitted mixture distribution applies to all observations.

The residual plot is shown in Figure 7.3. Note however that the top left plot should be ignored. This is a plot of the residuals against the fitted values for μ, a plot useful for regression models with continuous explanatory variables. Here we have fitted a constant model for μ, so all fitted values are the same and should lie in a vertical line in the plot. The fact that one of them does not appear to do so is because of a

small numerical discrepancy (see the horizontal scale). This feature appears often in
cases when no explanatory variables are involved in the fit, and should be ignored.

```
plot(readLO)
```
Fig.
7.3

```
## ***********************************************************************
##     Summary of the Randomised Quantile Residuals
##
##                          mean    =  -0.0001824348
##                      variance    =  0.9920201
##            coef. of skewness    =  0.02470025
##            coef. of kurtosis    =  2.922055
## Filliben correlation coefficient =  0.9968217
## ***********************************************************************
```

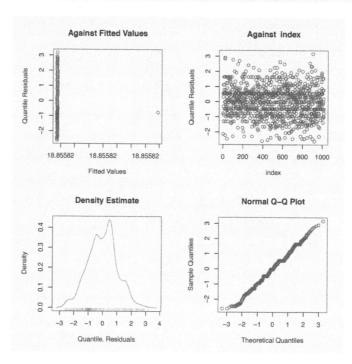

R
p. 20

FIGURE 7.3: Residual plots, two-component logistic mixture.

Whether the number of finite mixture components $\mathcal{K}$ should be increased or not is
left as an exercise for the reader.

We now consider using sex as a covariate. We compare models which include sex
in the model for μ, in the model for π, and in both:

```
readLO1 <- gamlssMX(ageread~sex,family=LO,K=2,data=glasses)
readLO2 <- gamlssMX(ageread~sex,pi.formula=~sex,family=LO,K=2,
                    data=glasses)

readLO3 <- gamlssMX(ageread~1,pi.formula=~sex,family=LO,K=2,
```

```
                    data=glasses)
AIC(readLO,readLO1,readLO2,readLO3)

##            df        AIC
## readLO2  8 7916.746
## readLO3  6 7918.782
## readLO1  7 7924.246
## readLO   5 7930.470
```

The preferred model is **readLO2**, which has **sex** as a covariate for both μ and π:

```
readLO2

##
## Mixing Family:  c("LO", "LO")
##
## Fitting method: EM algorithm
##
## Call:
## gamlssMX(formula = ageread ~ sex, pi.formula = ~sex,
##     family = LO, K = 2, data = glasses)
##
## Mu Coefficients for model: 1
## (Intercept)          sex2
##     18.8519        0.2559
## Sigma Coefficients for model: 1
## (Intercept)
##       1.497
## Mu Coefficients for model: 2
## (Intercept)          sex2
##      46.417         1.483
## Sigma Coefficients for model: 2
## (Intercept)
##       1.546
## model for pi:
##             (Intercept)       sex2
## fac.fit2      1.368604 0.6525015
##
## Estimated probabilities:
##          pi1        pi2
## 1 0.1170047 0.8829953
## 2 0.2028455 0.7971545
## 3 0.2028455 0.7971545
## ...
##
## Degrees of Freedom for the fit: 8 Residual Deg. of Freedom   1008
```

```
## Global Deviance:      7900.75
##            AIC:        7916.75
##            SBC:        7956.14
```

The estimates of μ_k, σ_k and π_k for the two components $k = 1, 2$ are obtained as:

Component 1: $\hat{\mu}_1 = 18.8532 + 0.2559 \, \text{if}(\text{sex} = 2)$

$$= \begin{cases} 18.8532 & \text{for males} \\ 19.1091 & \text{for females} \end{cases}$$

$$\hat{\sigma}_1 = \exp(1.497) = 4.468$$
$$\hat{\pi}_1 = 1 - \hat{\pi}_2$$

$$= \begin{cases} 1 - 0.7971 = 0.2029 & \text{for males} \\ 1 - 0.8830 = 0.1170 & \text{for females} \end{cases}$$

where $\hat{\pi}_2$ is obtained from component 2 below.

Component 2: $\hat{\mu}_2 = 46.417 + 1.483 \, \text{if}(\text{sex} = 2)$

$$= \begin{cases} 46.4170 & \text{for males} \\ 47.90 & \text{for females} \end{cases}$$

$$\hat{\sigma}_2 = \exp(1.546) = 4.6930$$

$$\log\left(\frac{\hat{\pi}_2}{1 - \hat{\pi}_2}\right) = 1.3686 + 0.6525 \, \text{if}(\text{sex} = 2)$$

$$= \begin{cases} 1.3686 & \text{for males} \\ 2.0211 & \text{for females} \end{cases}$$

$$\hat{\pi}_2 = \begin{cases} \frac{1}{1+\exp(-1.3686)} = 0.7972 & \text{for males} \\ \frac{1}{1+\exp(-2.0211)} = 0.8830 & \text{for females} \end{cases}$$

To summarize, the following is the fitted distribution for the age of starting to wear reading glasses:

Males: $f(y) = 0.20 \, f_L(y \,|\, \hat{\mu}_1 = 18.85, \hat{\sigma}_1 = 4.47) +$
$$0.80 \, f_L(y \,|\, \hat{\mu}_2 = 46.42, \hat{\sigma}_2 = 4.69)$$
Females: $f(y) = 0.12 \, f_L(y \,|\, \hat{\mu}_1 = 19.11, \hat{\sigma}_1 = 4.47) +$
$$0.88 \, f_L(y \,|\, \hat{\mu}_2 = 47.90, \hat{\sigma}_2 = 4.69)$$

where $f_L(\cdot \,|\, \mu, \sigma)$ is the pdf of the logistic (LO) distribution. Among wearers of glasses, males have a higher probability than females of starting to wear reading glasses in childhood or early adulthood. The fitted densities are shown in Figure 7.4.

Fig.
7.4

```
fnloFemale<-getpdfMX(readLO2, observation=1)#observation 1 is female
fnloMale<-getpdfMX(readLO2, observation=2)#observation 2 is male
truehist(glasses$ageread[glasses$sex==1],nbins=25,col="grey",
        xlab="Age",ylab="Density",ymax=0.05, main="Males")
lines(seq(0.5,90.5,1),fnloMale(seq(0.5,90.5,1)), lwd=2)
truehist(glasses$ageread[glasses$sex==2],nbins=25,col="grey",
        xlab="Age",ylab="Density",ymax=0.05, main="Females")
lines(seq(0.5,90.5,1),fnloFemale(seq(0.5,90.5,1)), lwd=2)
```

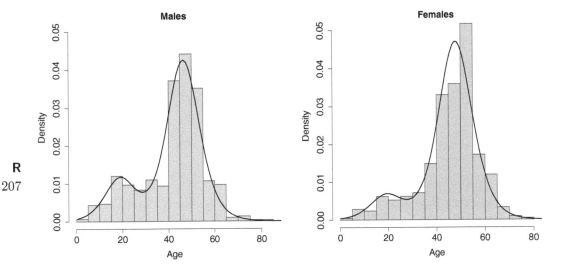

FIGURE 7.4: Histogram of age at which subjects started wearing reading glasses, by sex, with fitted two-component logistic distributions.

7.5   Finite mixtures with parameters in common

Here the $\mathcal{K}$ components of the mixture may have parameters in common, i.e. the parameter sets $(\boldsymbol{\theta}_1, \boldsymbol{\theta}_2, \ldots, \boldsymbol{\theta}_{\mathcal{K}})$ are not disjoint. The prior (or mixing) probabilities, in principle, can be either assumed to be constant or may depend on explanatory variables **s** and parameters $\boldsymbol{\alpha}$ through a multinomial logistic model, as in Section 7.2.3. Note, however, that in the implementation of the function `gamlssNP()`, which is used for fitting finite mixtures with common components, the probabilities are assumed to be constant, i.e. not dependent on explanatory variables.

It is assumed throughout that the $\mathcal{K}$ components $f_k(y) = f_k(y \mid \boldsymbol{\theta}_k)$ for $k = 1, 2, \ldots, \mathcal{K}$ can be represented by GAMLSS models. However, since some of the parameters may be common across components, the distribution used must be the same for all components. Similarly the link functions of the distribution parameters must be the same for all $\mathcal{K}$ components. In our notation in this chapter, the parameter vector $\boldsymbol{\theta}_k$ contains all the parameters in the predictor models for μ, σ, ν and τ for mixture component k. Here are some examples to clarify this.

Example 1. Mixture of $\mathcal{K}$ Poisson regression models: $f(y) = \sum_{k=1}^{\mathcal{K}} \pi_k f_k(y)$ where $f_k(y)$ is $\mathtt{PO}(\mu_k)$ for $k = 1, 2, \ldots, \mathcal{K}$, and $\log \mu_k = \beta_{0k} + \beta_1 x$. Here the coefficient β_1 is the same for all $\mathcal{K}$ components, but the intercept β_{0k} depends on k.

Example 2. Mixture of $\mathcal{K}$ negative binomial regression models: Let $f_k(y)$ be $\mathtt{NBI}(\mu_k, \sigma)$ for $k = 1, 2, \ldots, \mathcal{K}$, where $\log \mu_k = \beta_{10k} + \beta_{11}x$ and $\log \sigma = \beta_{20} + \beta_{21}x$. Here β_{11} (for μ_k) and both parameters for σ are the same for all $\mathcal{K}$ components, but the intercept parameter β_{10k} for μ_k depends on k.

Example 3. Mixture of $\mathcal{K}$ BCT models: Let $f_k(y)$ be $\mathtt{BCT}(\mu_k, \sigma_k, \nu, \tau)$, where

$$\log \mu_k = \beta_{10k} + \beta_{11k}x$$
$$\log \sigma_k = \beta_{20k} + \beta_{21k}x$$
$$\nu = \beta_{30}$$
$$\log \tau = \beta_{40} \ .$$

Here parameters β_{10k} and β_{11k} for μ and β_{20k} and β_{21k} for σ depend on k, but β_{30} for ν and β_{40} for τ are the same for all $\mathcal{K}$ components.

Maximizing the likelihood using the EM algorithm

As in Section 7.2.2 the complete log-likelihood is given by (7.9), and can be maximized using an EM algorithm. The M-step is achieved by expanding the data set $\mathcal{K}$ times as in Table 7.1. This method is identical to that used in Aitkin [1996, 1999b], described also in Aitkin et al. [2009, Chapter 7], but here we are not restricting ourselves to the exponential family. The column headed $\hat{\mathbf{p}}^{(r+1)}$ contains the weights (calculated internally) at the $(r+1)$st iteration. The column headed MASS identifies the $\mathcal{K}$ mixture components. This column is declared as a factor in the R implementation of the EM algorithm.

The posterior probability of membership of component k for observation i is estimated by the fitted weight $\hat{p}_{ik}$ at the last iteration of the EM algorithm. The fitted prior (or mixing) probabilities are estimated by

$$\hat{\pi}_k = \frac{1}{n} \sum_{i=1}^{n} \hat{p}_{ik} \ , \qquad \text{for } k = 1, 2, \ldots, \mathcal{K} \ . \tag{7.15}$$

TABLE 7.1: Table showing the expansion of data use in M-step of the EM algorithm for fitting the common parameter mixture model

i	MASS	y_e	$\mathbf{X}_e$	$\hat{\mathbf{p}}^{(r+1)}$
1	1			
2	1	$\mathbf{y}$	$\mathbf{X}$	$\hat{\mathbf{p}}_1^{(r+1)}$
⋮	⋮			
n	1			
1	2			
2	2	$\mathbf{y}$	$\mathbf{X}$	$\hat{\mathbf{p}}_2^{(r+1)}$
⋮	⋮			
n	2			
⋮	⋮	⋮	⋮	⋮
1	$\mathcal{K}$			
2	$\mathcal{K}$	$\mathbf{y}$	$\mathbf{X}$	$\hat{\mathbf{p}}_{\mathcal{K}}^{(r+1)}$
⋮	⋮			
n	$\mathcal{K}$			

7.6 The gamlssNP() function

The function for fitting finite mixtures with parameters in common is **gamlssNP()**, of the package **gamlss.mx**, with argument `mixture="np"`. This function was initially designed for fitting marginal likelihoods for random effect models; see Chapter 10. The function is based on `alldist()` and `allvc()` of the **npmlreg** package [Einbeck et al., 2014]. `gamlssNP()` has the following arguments:

formula — A formula defining the response variable and explanatory terms (fixed effects) for the μ parameter of the model. Note that modelling the rest of the distribution parameters can be done by using the usual formulae, e.g. `sigma.fo=~x`, which passes the arguments to `gamlss()`;

random — A formula defining the random part of the model. For random effect models it takes the form `random = ~1|f` or `random = ~ x|f` for (nonparametric) random intercept and for random intercept and random slopes models, respectively, where `f` represents a factor and `x` represents a continuous variable. This formula is also used for fixed effect finite mixture models to define interactions of the factor MASS with continuous explanatory variables x in the predictor for μ. If, for example, different coefficients in `x` in the predictor of μ

are needed for the $\mathcal{K}$ different components of the finite mixture, use random=~x;

family   A gamlss family distribution;

data   The data frame, which is mandatory for this function even if the data are attached. This is because the data frame is used to expand the data as in Table 7.1;

K   The number of mixture components (in fixed effects finite mixture models), or the number of mass points or integration quadrature points (for random effects models);

mixture   The type of mixing distribution: mixture="np" for nonparametric finite mixtures or mixture="gq" for Gaussian quadrature;

tol   The tolerance scalar, usually between zero and one, used for changing the starting values;

weights   Prior weights;

control   This sets the control parameters for the EM algorithm iterations. The default setting is the NP.control function;

g.control   This is for controlling the gamlss control function, gamlss.control, passed to the gamlss fit;

...   For extra arguments to pass to the gamlss() function.

The function gamlssNP() creates a gamlssNP object which has methods print(), fitted(), family(), plot() and residuals(). The saved components of the gamlssNP object are similar to those of a gamlss object with the following extra components:

prob   The fitted prior (or mixing) probabilities $\hat{\pi}_k$ for $k = 1, \ldots, \mathcal{K}$;

random   The random effect component;

post.prob   The fitted posterior probabilities $\hat{p}_{ik}$ for $i = 1, \ldots, n$ and $k = 1, \ldots, \mathcal{K}$;

mass.points   The mass points (appropriate for nonparametric random effects model; see Chapter 10);

ebp   The empirical Bayes predictors for the parameter μ defined as $\hat{\mu}_i = \sum_{k=1}^{\mathcal{K}} \hat{p}_{ik}\hat{\mu}_{ik}$ for $i = 1, \ldots, n$.

If MASS is included in the predictor for a distribution parameter (any of σ, ν or τ), then the predictor intercepts differ between the $\mathcal{K}$ components. If an interaction between MASS and an explanatory variable x is included in the predictor model for a distribution parameter (any of σ, ν or τ), then the coefficient of x differs across the $\mathcal{K}$ components. Note however that gamlssNP() automatically includes MASS in the predictor for μ and the syntax used in gamlssNP() for the interaction between

MASS and x in the predictor for μ is achieved using the random=~x argument (see Section 7.7 for an example).

7.7 Example using gamlssNP(): Animal brain data

Data summary: the animal brain data

R data file: brains in package **gamlss.mx** of dimensions 28×2 (identical to Animals in package **MASS**)

variables

brain : brain weight in g.

body : body weight in kg.

purpose: To fit a finite mixture model with different intercepts.

conclusion: A three component normal distribution mixture is found to be adequate

---

The average brain size (**brain**) and body weight (**body**) were recorded for 28 species of land animals. Since the distribution of both brain size and body weight are highly skewed, a log transformation was applied to both variables to give transformed variables **lbrain** and **lbody**. The resulting data are plotted in Figure 7.5.

Fig. 7.5
```
library(gamlss.mx)
data(brains)
brains<-transform(brains, lbrain = log(brain),lbody = log(body))
with(brains, plot(lbrain~lbody, ylab="log brain", xlab="log body"))
```

A normal error linear regression model of **lbrain** against **lbody** has a highly significant slope for **lbody** but it is believed that the data may represent different stages of evolution and so a mixture model is fitted to the data. In the mixture model, the evolution stage is represented by a shift in the intercept of the regression equation. Normal mixture models with 1, 2, 3 and 4 components are fitted below. Models **br.2, br.3** and **br.4** are models with different intercepts for the $\mathcal{K}$ components, i.e. $f_k(y)$ is $\text{NO}(\mu_k, \sigma)$, where for $\mathcal{K} = 1, 2, 3, 4$:

$$\mu_k = \beta_{0k} + \beta_1 x , \quad k = 1, \ldots, \mathcal{K} ,$$

and y is log brain size and x is log body size. As the slopes are the same for the $\mathcal{K}$ components, parallel lines are fitted (see later for how different slopes can be incorporated in the model). The plots of the EM trajectories in **gamlssNP()** are suppressed here.

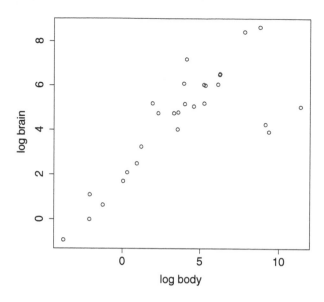

FIGURE 7.5: Log(brain size) plotted against log(body size).

R

p. 21

```
br.1<-gamlss(lbrain ~ lbody, data = brains,trace=FALSE)
br.2<-gamlssNP(formula=lbrain~lbody, mixture="np", K=2, tol=1,
    data=brains, family=NO, plot.opt=0)

## 1 ..2 ..3 ..4 ..5 ..6 ..7 ..8 ..9 ..10 ..11 ..12 ..
## 13 ..14 ..15 ..16 ..17 ..18 ..19 ..20 ..21 ..22 ..23 ..24 ..25 ..
## 26 ..27 ..28 ..29 ..30 ..31 ..32 ..33 ..
## EM algorithm met convergence criteria at iteration  33

br.3<-gamlssNP(formula=lbrain~lbody, mixture="np", K=3, tol=1,
    data=brains, family=NO, plot.opt=0)

## 1 ..2 ..3 ..4 ..5 ..6 ..7 ..8 ..9 ..10 ..11 ..12 ..
## 13 ..14 ..
## EM algorithm met convergence criteria at iteration  14

br.4<-gamlssNP(formula=lbrain~lbody, mixture="np", K=4, tol=1,
    data=brains, family=NO, plot.opt=0)

## 1 ..2 ..3 ..4 ..5 ..6 ..7 ..8 ..9 ..10 ..11 ..12 ..
## 13 ..14 ..15 ..16 ..17 ..18 ..19 ..20 ..21 ..22 ..23 ..24 ..25 ..
## 26 ..27 ..28 ..29 ..
## EM algorithm met convergence criteria at iteration  29
```

We compare the models using the criteria AIC and SBC:

```
GAIC(br.1, br.2, br.3, br.4)

##        df        AIC
## br.3   7   79.15079
## br.4   9   83.15613
## br.2   5   85.95938
## br.1   3  107.25779

GAIC(br.1, br.2, br.3, br.4, k = log(length(brains$body)))

##        df        AIC
## br.3   7   88.47622
## br.2   5   92.62040
## br.4   9   95.14598
## br.1   3  111.25440
```

The model `br.3` with three components (i.e. three parallel lines) is selected by both criteria. Changing the starting values by trying different values for `tol` (e.g. trying each of the values $0.1, 0.2, \ldots, 1$ in turn), for models `br.2`, `br.3` and `br.4`, did not change the values of AIC and SBC given by the two GAIC commands above (not shown). We now print model `br.3` and its estimated (fitted) posterior probabilities, $\hat{p}_{ik}$.

```
br.3

##
## Mixing Family:
## c("NO Mixture with NP", "Normal Mixture with NP")
##
## Fitting method: EM algorithm
##
## Call:
## gamlssNP(formula = lbrain ~ lbody, family = NO, data = brains,
##     K = 3, mixture = "np", tol = 1, plot.opt = 0)
##
## Mu Coefficients :
## (Intercept)         lbody          MASS2          MASS3
##     -3.0715        0.7499         4.9805         6.5530
## Sigma Coefficients :
## (Intercept)
##     -0.9387
##
## Estimated probabilities: 0.107 0.751 0.141
##
##
## Degrees of Freedom for the fit: 7 Residual Deg. of Freedom   21
## Global Deviance:       65.1508
```

```
##              AIC:        79.1508
##              SBC:        88.4762
```

```
head(br.3$post.prob[[1]])
```

```
##          [,1]        [,2]          [,3]
## [1,]      0 0.9999624 3.760045e-05
## [2,]      0 0.9999995 4.736429e-07
## [3,]      0 0.9996309 3.691210e-04
## [4,]      0 0.9979683 2.031733e-03
## [5,]      0 0.9999947 5.254125e-06
## [6,]      1 0.0000000 0.000000e+00
```

Model br.3 can be presented as $Y \sim \text{NO}(\hat{\mu}, \hat{\sigma})$ where

$$\hat{\mu} = \begin{cases} -3.072 + 0.750x & \text{with probability } 0.107 \\ 1.909 + 0.750x & \text{with probability } 0.751 \\ 3.482 + 0.750x & \text{with probability } 0.141 \end{cases} \qquad (7.16)$$

and $\hat{\sigma} = \exp(-0.9387) = 0.391$. (Note that the intercept for the second component in (7.16) is obtained from the estimated parameter coefficients for μ by $1.909 = -3.072 + 4.981$, since MASS2 gives the adjustment to the intercept for the second mixture component; similarly for MASS3.) The output given by br.3$post.prob contains the estimated posterior (conditional) probabilities $\hat{p}_{ik}$ of each of the observations in the data set belonging to each of the three components given that we observed the response variable lbrain.

A plot of the data together with the fitted values for the μ parameter of model br.3 is shown in Figure 7.6. Each observation of the data was allocated to the component for which it had the highest posterior probability and the observations are plotted using different symbols to represent allocation to each of the three components. Note that since the parameter μ in this (normal distribution) case is the mean of the distribution, the lines are the fitted means of the conditional distributions $f_k(y)$ for $k = 1, 2, 3$. Figure 7.6 is obtained by:

```
with(brains, plot(lbody, lbrain,
      pch = c(21, 22, 23)[max.col(br.3$post.prob[[1]])],
      bg = c("red", "green3", "blue")[max.col(br.3$post.prob[[1]])]))
for (k in 1:3){
with(brains,lines(fitted(br.3,K=k)[order(lbody)]~lbody[order(lbody)],
lty = k,lwd=2,col=c("red", "green3", "blue")[k]))
}
legend("topleft",legend=c("Component 1","Component 2","Component 3"),
      pch=c(21, 22, 23),pt.bg=c("red", "green3", "blue"),
      lty=1:3,lwd=2,col=c("red", "green3", "blue"))
```

Fig.
7.6

The weighted average for the (conditional) parameters $\hat{\mu}$ for the $\mathcal{K}$ components for each observation, i.e. $\sum_{k=1}^{\mathcal{K}} \hat{\pi}_k \hat{\mu}_{ik}$ are obtained as

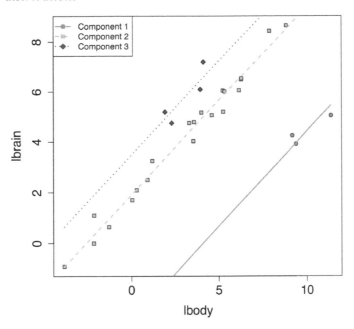

FIGURE 7.6: The brain size data with the three-component fitted means of log brain size (**lbrain**) against log body size (**lbody**).

```
fitted(br.3, K=0)
```

Note how the marginal mean, using the function **fitted()**, is obtained here compared to the conditional means. If the argument K of the **fitted()** function has any value in the range 1, 2 or 3 (that is the range of permissible values for the model **br.3**), then the conditional parameter μ_1, μ_2 or μ_3 is given. For any other value of K the average μ is given. This will be the marginal mean only if parameter μ is the mean of the conditional distribution for each component.

A residual plot of the finite mixture model is obtained in the usual way using the function **plot()**:

```
plot(br.3)
```

```
## ****************************************************************
##    Summary of the Randomised Quantile Residuals
##                          mean        =   -0.004003875
##                      variance        =   1.052469
##              coef. of skewness       =   0.1668313
##              coef. of kurtosis       =   2.739025
## Filliben correlation coefficient     =   0.9962244
## ****************************************************************
```

There are several other models that we could fit to these data, depending on which parameters are common to the three components in the model. The following table

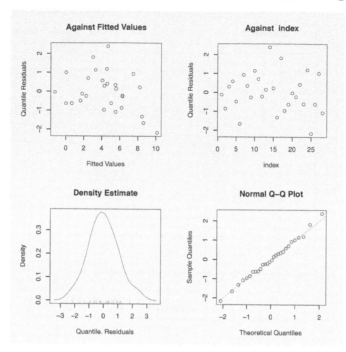

FIGURE 7.7: Residual plot of model `br.3` for the animal brain size data.

shows possible alternative models, and the code below fits them. Note however that
we have only 27 observations and therefore any fitted finite mixture model should
be treated with great caution.

model	μ intercept	μ slope	σ
br.3	different	same	same
br.31	different	same	different
br.32	different	different	same
br.33	different	different	different

```
br.31 <- gamlssNP(formula = lbrain ~ lbody, sigma.fo = ~MASS,
    mixture = "np", K = 3, tol = 1, data = brains, family = NO)
```

```
## 1 ..2 ..3 ..4 ..5 ..6 ..7 ..8 ..9 ..10 ..11 ..12 ..
## 13 ..14 ..15 ..16 ..17 ..18 ..19 ..20 ..21 ..22 ..23 ..24 ..25 ..
## 26 ..27 ..28 ..
## EM algorithm met convergence criteria at iteration    28
## Global deviance trend plotted.
## EM Trajectories plotted.
```

```
br.32 <- gamlssNP(formula = lbrain ~ lbody, random = ~lbody,
    sigma.fo = ~1, mixture = "np", K = 3, tol = 1, data = brains,
    family = NO)
```

```
## 1 ..2 ..3 ..4 ..5 ..6 ..7 ..8 ..9 ..10 ..11 ..12 ..
```

R
p. 21

```
## 13 ..14 ..15 ..16 ..
## EM algorithm met convergence criteria at iteration   16
## Global deviance trend plotted.
## EM Trajectories plotted.
```

```
br.33 <- gamlssNP(formula = lbrain ~ lbody, random = ~lbody,
    sigma.fo = ~MASS, mixture = "np", K = 3, tol = 1, data = brains,
    family = NO)
```

```
## 1 ..2 ..3 ..4 ..5 ..6 ..7 ..8 ..9 ..10 ..11 ..12 ..
## 13 ..14 ..15 ..16 ..17 ..
## EM algorithm met convergence criteria at iteration   17
## Global deviance trend plotted.
## EM Trajectories plotted.
```

We compare the models using AIC and SBC:

```
GAIC(br.3, br.31, br.32, br.33)
```

```
##        df      AIC
## br.32   9 77.31133
## br.3    7 79.15079
## br.33  11 80.26824
## br.31   9 81.93037
```

```
GAIC(br.3, br.31, br.32, br.33, k = log(length(brains$lbody)))
```

```
##        df      AIC
## br.3    7 88.47622
## br.32   9 89.30117
## br.31   9 93.92021
## br.33  11 94.92249
```

Model `br.3` has the smallest SBC value. (Note that model `br.32` has the smallest AIC value, however with so many parameters in the model and so few data points the model produces rather nonsensical results.) Note also that, in general, since model `br.33` has components with no parameters in common it could also be fitted using `gamlssMX()`. Again with only 27 observations any fitted model is too sensitive to starting values.

7.8 Bibliographic notes

Finite mixture distributions and their properties have been extensively discussed in the statistical literature in the past 40 years. There are several books on the subject, for example Everitt [1981], Lindsay [1995], Titterington et al. [1985], Böhning [2000] and McLachlan and Peel [2004].

The function `gamlssMX()` follows a similar approach to that adopted by Leisch [2004] and Grün and Leisch [2007]. `gamlssNP()` uses the methodology developed in Aitkin [1996, 1999b], also described in Aitkin et al. [2009, Chapter 7], but here we extend it to all `gamlss.family` distributions. This approach has its origin in a technique introduced to model overdispersion in GLMs by Hinde [1982]. The package **npmlreg** implements this methodology [Einbeck et al., 2014]. Both `gamlssMX()` and `gamlssNP()` functions estimate the finite mixture model with a fixed number of components using the EM algorithm in a likelihood-based framework [Dempster et al., 1977]. For a Bayesian approach using MCMC sampling, see Diebolt and Robert [1994]. Estimating the number of components of the finite mixture needs more thought. The reader can find more detail of the problem in Aitkin et al. [2009, Section 7.6].

There are several **R** packages featuring finite mixture models: **FlexMix** [Leisch, 2004]; **npmlreg** [Einbeck et al., 2014]; **mixtools** [Benaglia et al., 2009]; **mclust** for mixtures of multivariate Gaussian distributions [Fraley and Raftery, 2002] and **fpc** for mixtures of linear regression models [Hennig, 2015].

7.9   Exercises

1. **The acidity data:** This data set was analysed as a mixture of Gaussian distributions on the log scale by Crawford et al. [1992] and Crawford [1994].

   R data file: `acidity` in package **gamlss.data** of dimensions 155×1

   variable

   y : acidity index for 155 lakes in the northeastern United States

   purpose: to show an example of fitting a finite mixture distribution.

   (a) Load the `acidity` data, print the variable name and obtain a histogram of the acidity index values.

   ```
   data(acidity)
   names(acidity)
   with(acidity, truehist(y))
   ```

   (b) Fit a nonparametric density using the function `fitSmo()`, and save the fit.

```
h1 <- histSmo(acidity$y, plot=TRUE)
```

(c) Fit a mixture of two normal distributions (with different means and variances). (Note `gamlssMXfits()` fits the mixture from `n=10` different starting values and chooses the best mixture fit.)

```
mm <- gamlssMXfits(n=10,y~1, family=NO, K=2, data=acidity)
mm
```

(d) Calculate the pdf for the fitted mixture and plot it with the histogram of the acidity index values. Add the fitted density estimator and comment.

```
afn <- getpdfMX(mm)
with(acidity, truehist(y))
lines(seq(1,8,.01),afn(seq(1,8,.01)), col="red")
lines(h1, col="blue")
```

2. **Reading glasses data** revisited, `glasses` data from package **gamlss.data**.

(a) Use `gamlssMXfits()` with `n=10` to refit the data, using the formula `ageread~1` and the simple logistic finite mixture model with $K = 2$.

```
readLO2 <- gamlssMXfits(10, ageread~1,family=LO, K=2,
                        data=glasses)
```

(b) Refit the model using `gamlssNP()` and compare the results.

```
readnp2 <- gamlssNP(ageread~1,sigma.fo=~MASS, family=LO,K=2,
                    data=glasses)
```

(c) Now fit an extra model with $K = 3$.

```
readLO3 <- gamlssMXfits(10, ageread~1,family=LO, K=3,
                        data=glasses)
```

(d) Use GAIC (with a high penalty) to compare the two fitted models.

```
GAIC(readLO2, readLO3, k=log(1016))
```

(e) Plot the fitted distributions and compare the two critically.

```
fn2 <- getpdfMX(readLO2, observation=1)
fn3 <- getpdfMX(readLO3, observation=1)
truehist(glasses$ageread,nbins=25,col="grey",xlab="Age",
         ylab="Density", ymax=0.05)
lines(seq(0.5,90.5,1),fn2(seq(0.5,90.5,1)))
lines(seq(0.5,90.5,1),fn3(seq(0.5,90.5,1)), col="blue")
```

(f) Use worm plots to check their goodness of fit.

```
wp(readL02)
wp(readL03)
```

(g) Can you identify the reason for the parallel lines in the worm plots?

(h) Use **barplot()** to plot the original data and comment.

```
barplot(table(glasses$ageread))
```

Part IV

Model terms

8

Linear parametric additive terms

CONTENTS

8.1 Introduction to linear and additive terms

This chapter explains the types of linear terms within GAMLSS models, and how they can be used. In particular it explains:

1. linear terms and interactions for factors and numerical explanatory variables, and

2. different useful bases used for explanatory variables.

This chapter is essential for understanding the different types of additive terms in GAMLSS.

GAMLSS allows modelling of all the distribution parameters μ, σ, ν and τ as linear and/or nonlinear and/or 'nonparametric' smoothing functions of the explanatory variables. This allows the explanatory variables to affect the predictors of the specific parameters and consequently the parameters themselves. Therefore, while it is usual in regression modelling for the explanatory variables to affect the mean of the response variable, in GAMLSS the location, scale and shape of the response distribution could all be affected by the explanatory variables.

We shall refer to the explanatory variables as *terms* in the model. The relationships between a predictor η and the terms can be *linear* or *nonlinear*. A nonlinear relationship can be *parametric nonlinear* or a *smoother*. As an example of a parametric nonlinear relationship, consider the expression $\beta_1 x^{\beta_2}$ where β_1 and β_2 are parameters to be estimated. Smoothers are *nonparametric* techniques which allow the data to determine the relationship between the predictor and the explanatory variables. See Chapter 9 for more details about additive smoothing terms.

By *additive terms* we refer to the fact that in order to evaluate the effect of the explanatory variables on the predictor for a specific distribution parameter, we have to sum their individual effects. Additivity does not imply that there are no interactions in the model. Section 8.2.2 presents some examples.

As an example of what type of terms a GAMLSS model can take, let the x's represent continuous explanatory variables and f a factor (with three levels). (Note that throughout this chapter we refer to x as a continuous explanatory variable, although strictly speaking they are quantitative, i.e. continuous or discrete.) Let β_j for $j = 1, 2, \ldots, 7$ be constant (coefficient) parameters and s_j for $j = 0, 1, \ldots, 6$ be smooth functions. Then the following could be a typical predictor η for a parameter:

$$
\begin{aligned}
\eta = \ & \beta_0 + \beta_1 x_1 + \beta_2 x_2 + \beta_3(\text{if } f = 2) + \beta_4(\text{if } f = 3) + \beta_5 x_1 x_2 + \\
& \beta_6 x_1(\text{if } f = 2) + \beta_7 x_1(\text{if } f = 3) + s_0(x_1) + s_1(x_3) + s_2(x_4) + \\
& s_3(x_1)x_4 + s_4(x_1)(\text{if } f = 2) + s_5(x_1)(\text{if } f = 3) + s_6(x_3, x_5)
\end{aligned}
$$

where

β_0	is the constant term,
$\beta_1 x_1 + \beta_2 x_2$	are linear additive terms of x_1 and x_2,
$\beta_3(\text{if } f = 2) + \beta_4(\text{if } f = 3)$	is the main effect of the factor f,
$\beta_5 x_1 x_2$	is a linear interaction between x_1 and x_2,
$\beta_6 x_1(\text{if } f = 2) + \beta_7 x_1(\text{if } f = 3)$	is a linear interaction between x_1 and f,
$s_0(x_1) + s_1(x_3) + s_2(x_4)$	are smoothing additive terms for x_1, x_3 and x_4,
$s_3(x_1)x_4$	is a varying coefficient term for x_4 given x_1, i.e. the linear coefficient of x_4 varies according to x_1,
$s_4(x_1)(\text{if } f = 2) + s_5(x_1)(\text{if } f = 3)$	is an interaction of f with smoothing terms for x_1,
$s_6(x_3, x_5)$	is a smooth interaction between x_3 and x_5.

Figure 8.1 shows a classification of the different additive terms within the GAMLSS models. Additive terms can be parametric (left of the diagram), smoothing additive terms (centre) or random effects (right). Within the parametric additive terms we have linear parametric terms (the subject of this chapter) and nonlinear parametric terms. Smoothing terms are the subject of Chapter 9. Smoothing additive terms are divided into two distinct categories: (i) penalized additive terms, and (ii) other smoothing terms. Random effects, discussed in detail in Chapter 10, have a double arrow connection with penalized smoothers. This is because random effects can be seen and treated as penalized smoothers, but also we can gain more insight about penalized smoothers if we treat them as random effects. Random effects have their own history in the statistical literature with their own characteristics and peculiarities, which is the reason that we treat them separately.

Next we introduce the use of main effects and interactions for linear terms (based on the notation of Wilkinson and Rogers [1973]), and then show how some nonlinear relationships can be modelled using linear basis functions. In particular the following basis functions are introduced:

1. polynomials (Section 8.3),

2. fractional polynomials (Section 8.4),

3. piecewise polynomials (Section 8.5), and

4. B-splines (Section 8.6).

Section 8.2 describes how simple linear terms for continuous and categorical explanatory variables are accommodated within an additive model. It also describes linear interactions between terms.

8.2 Linear additive terms

The linear terms are declared by the use of formulae. A formula in **R** looks something like:

```
y ~ x1 + x2 + f1 + f2*x3
```

The x's denote *quantitative* explanatory variables (subsequently called *continuous*) and the f's denote *factors*. The symbols '+' and '*' have special meaning here derived from the Wilkinson and Rogers [1973] notation as applied in the S language by Chambers and Hastie [1992]. (It is the same notation used in **R** in the fitting of linear models (lm()) and GLMs (glm()); see for example Venables and Ripley [2002, Section 6.2].) The symbol '+' describes *additive* terms while '*' denotes *interactions* between terms. The purpose of the right-hand part of the formula is to create the design matrix **X** for fitting the linear part of the GAMLSS model.

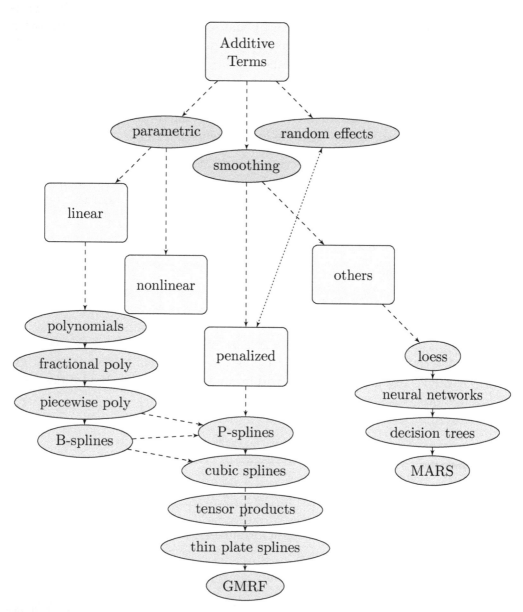

FIGURE 8.1: Diagram showing the different additive terms that can be fitted within GAMLSS.

8.2.1   Linear main effects

If the explanatory variable is continuous, the + sign will enter a single column in the design matrix **X**. If it is a factor it will enter a set of dummy variables. A *dummy variable* is a vector containing zeros and ones. As an example, if the factor f1 has four levels then f1 will be represented within **X** as a set of four dummy variables. Each dummy variable will have the value 1 if the observed value of f1 is at the appropriate level and zero otherwise (see example below).

Things are complicated somewhat by the presence of the constant in the model, represented in the design matrix **X** by a column of ones. This is automatically included in the design matrix (unless the user uses '-1' within the formula). The problem arises from the fact that by including both the constant and the factor (represented as a set of four dummy variables), the columns of the design matrix become linearly dependent. To avoid this, **R** drops the first dummy variable of the factor (corresponding to the first level of the factor). This type of coding of a factor is referred to as a 'dummy variable' coding with a reference category. The default reference category (the first level) can be changed using the function relevel(). The function contrasts() allows different coding systems such as 'effects' coding (not covered here).

To demonstrate we use the aids data first used in Chapter 4. The data were collected quarterly and the aids dataframe contains three variables: (i) y: the number of quarterly AIDS cases in England and Wales from January 1983 to March 1994, (ii) x: time in quarters from January 1983, and (iii) qrt: a factor for the quarterly seasonal effect. Here we input the data and output the first ten rows of the design matrix of the model with x and qrt as explanatory variables.

```
data(aids)
head(with(aids,model.matrix(formula(~x+qrt))), 10)
```

```
##    (Intercept)  x qrt2 qrt3 qrt4
## 1            1  1    0    0    0
## 2            1  2    1    0    0
## 3            1  3    0    1    0
## 4            1  4    0    0    1
## 5            1  5    0    0    0
## 6            1  6    1    0    0
## 7            1  7    0    1    0
## 8            1  8    0    0    1
## 9            1  9    0    0    0
## 10           1 10    1    0    0
```

8.2.2   Linear interactions

Interactions can be between:

- two or more continuous variables, or

- two or more factors, or

- one or more continuous variables and one or more factors.

The additive formula ~x1+x2 for two continuous variables initiates a linear plane fitting (by introducing two new columns in the design matrix $\mathbf{X}$). The interaction formula ~x1*x2 introduces a third column containing the element-wise multiplication of x1 by x2. This extra column makes the fitting surface a curvy one. This linear interaction surface is fitted globally and it is rather restrictive compared with surfaces fitted by nonparametric smoothers, where more flexibility locally is allowed. The formula ~x1*x2 is equivalent in $\mathbf{R}$ to ~x1+x2+x1:x2, which represents the main effect for x1, the main effect of x2 and the linear interaction between (i.e. the product of) x1 and x2. We refer to the coefficients of x1 and x2 as the *main effects* for x1 and x2, respectively; and the coefficient of x1:x2 as the *interaction* of x1 and x2. Depending on how many variables are involved we have 'two-way', i.e. x1*x2, 'three-way', i.e. x1*x2*x3 and up to say 'k-way' forms of interactions.

Interactions for categorical variables, say ~f1*f2, add columns in the design matrix to represent all combinations of the crossing of the levels of the two factors. For example if f1 has 2 levels $\{A, B\}$, and f2 has 3 levels $\{1, 2, 3\}$, then the interaction will have $2 \times 3 = 6$ levels reflecting all combinations $\{A1, A2, A3, B1, B2, B3\}$. The following is an example of how $\mathbf{R}$ creates the design matrix for factor interactions. First we create two factors f1 and f2 with 2 and 3 levels, respectively, both of length 24. We show the first twelve rows of the design matrix:

```
f1<-gl(2,1,24)
levels(f1) <- c("A", "B")
f1

## [1] A B A B A B A B A B A B A B A B A B A B A B A B
## Levels: A B

f2<-gl(3,2,24)
f2

## [1] 1 1 2 2 3 3 1 1 2 2 3 3 1 1 2 2 3 3 1 1 2 2 3 3
## Levels: 1 2 3

head(model.matrix(~f1*f2), 12)

##   (Intercept) f1B f22 f23 f1B:f22 f1B:f23
## 1           1   0   0   0       0       0
## 2           1   1   0   0       0       0
## 3           1   0   1   0       0       0
## 4           1   1   1   0       1       0
## 5           1   0   0   1       0       0
## 6           1   1   0   1       0       1
```

## 7	1	0	0	0	0	0
## 8	1	1	0	0	0	0
## 9	1	0	1	0	0	0
## 10	1	1	1	0	1	0
## 11	1	0	0	1	0	0
## 12	1	1	0	1	0	1

Note that, since the constant (intercept) is included by default in the design matrix, the main effect of f1 is represented by one column (number of levels of f1 minus 1) headed by f1B, the main effect of f2 by two columns $(3-1=2)$, and the interaction of the two factors by $(2-1) \times (3-1) = 2$ columns headed by f1B:f22 and f1B:f23. (For example, the column for f1B:f22 is obtained by a casewise multiplication of columns f1B and f22.) In general if f1 and f2 have k_1 and k_2 levels, respectively, then their interaction has $k = (k_1 - 1) \times (k_2 - 1)$ columns.

The following is an example of an interaction between a continuous variable and a factor.

```
data(aids)
head(with(aids,model.matrix(formula(~x*qrt))), 10)
```

##	(Intercept)	x	qrt2	qrt3	qrt4	x:qrt2	x:qrt3	x:qrt4
## 1	1	1	0	0	0	0	0	0
## 2	1	2	1	0	0	2	0	0
## 3	1	3	0	1	0	0	3	0
## 4	1	4	0	0	1	0	0	4
## 5	1	5	0	0	0	0	0	0
## 6	1	6	1	0	0	6	0	0
## 7	1	7	0	1	0	0	7	0
## 8	1	8	0	0	1	0	0	8
## 9	1	9	0	0	0	0	0	0
## 10	1	10	1	0	0	10	0	0

The continuous variable here is time which is coded as the integers 1 to 45. The design matrix in this case contains: (i) the constant as the first column, (ii) the x main effect as the second column, (iii) the main effect of the factor qrt as three dummy variables containing the last three levels of the factor, and (iv) the interaction x:qtr represented here by the last three columns. In order to show the interpretation of the model containing both a factor and a continuous variable, consider the simple analysis of covariance (ANCOVA), with a single covariate x and a single factor f. In the ANCOVA model we assume that the response variable is normally distributed and we are interested in modelling its mean. However within GAMLSS, we are interested in how the predictor η of a distribution parameter changes with the continuous variable x and the factor f. There are five different standard models of interest for η:

1. the *null* model in which neither **f** nor **x** are needed in the model for η, i.e. just the constant;

2. the *simple analysis of variance* model where only **f** is included in the model for η but not **x**;

3. the *simple regression model* where **x** is included in the model for η but not **f**;

4. the *additive model* where both **x** and **f** are included in the model for η, but with no interaction between them, so the slope for **x** is common for all levels of **f**, but the intercept varies according to the levels in **f**; and

5. the *interaction model* in which intercepts and slopes of η against x vary for different levels of **f**.

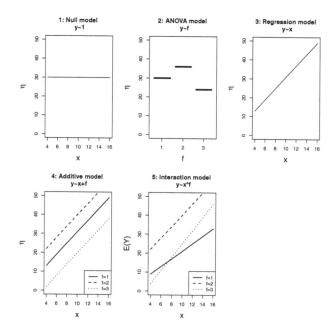

FIGURE 8.2: The five different models in the simple analysis of covariance.

Graphs for the different models involved are shown in Figure 8.2. The vertical axis represents $E(Y)$ in a normal error ANCOVA model, or more generally the predictor η for a distribution parameter in a GAMLSS model. The lines in the graph represent the different predictor values for each of the three levels of factor **f**. Model 1 has a constant predictor value. Model 2 has different values for the different levels of the factor **f**. The predictor of model 3 increases linearly with **x**, while the predictor of model 4 shows the same slope in the linear relationship between the predictor and **x** but with different intercept values. In model 5, both the intercepts and slopes of the predictor vary according to the levels of the factor **f**.

In the normal case where we model the mean, the choice between models is achieved by testing between nested models using an F-test. For example model 4 above is nested within model 5 so we can compare them. Models 2 and 3 are nested models within both models 4 and 5, while the null model 1 is nested within all the rest. Note model 2 is not nested within model 3. The appropriateness of the F-distribution comes as a result of the normal assumption with constant variance for the response variable, but in general it is not appropriate for a GAMLSS model. In this case selection between models can be achieved through the asymptotic χ^2_{df} distribution of the generalized likelihood ratio test statistic for nested models (where df is the difference in degrees of freedom between models), or GAIC for non-nested models.

8.3 Polynomials

Polynomials are the simplest way of modelling nonlinear relationships in regression. A polynomial of degree p has the form:

$$h(x) = \beta_0 + \beta_1 x + \beta_2 x^2 + \beta_3 x^3 + \ldots + \beta_p x^p \tag{8.1}$$

and the user has merely to add the appropriate columns in the design matrix $\mathbf{X}$. There are two ways of doing this in **R**: (i) by using the function I(), and (ii) through the orthogonal polynomial function poly(). I() allows the user to calculate an expression within a formula. For example

```
~ x + I(x^2) + I(x^3)
```

will fit a cubic polynomial for **x** by creating two extra columns containing x^2 and x^3. The columns of $\mathbf{X}$ form a *polynomial basis*. Higher-order polynomials can be added in the same way. The problem with defining polynomials this way is that it can lead to numerical inaccuracy. See for example Figure 8.3(a), where the basis functions for a 5th degree polynomial for the time variable (1 to 45) of the **aids** data are plotted. The values for x^p can easily become too big (as in Figure 8.3(a)) or too small (for small values of **x**). This problem is avoided if, instead of the standard basis, we use an orthogonal polynomial basis. Figure 8.3(b) shows a 5th order orthogonal polynomial basis for the same variable, computed using the poly() function. The curves are easily identified as linear, quadratic, cubic, etc. Furthermore the basis vectors in $\mathbf{X}$ are orthogonal.

The fitted values of the same order polynomial, irrespective of whether the standard or orthogonal basis is used, should be identical (unless something went numerically wrong). The fitted values are a linear function of the basis variables, weighted differently, according to the fitted coefficients. Figure 8.3(c) shows the fitted values of the model

```
y ~ poly(x,5)
```

for the **aids** data, as well as the orthogonal basis functions weighted by their fitted coefficients. Summing the weighted basis functions gives the fitted values. Note that in this specific case the constant (flat line) and the linear part of the basis play a major role in determining the shape of the fitted value, while the rest of the basis only adds a small curvature to it.

```
def.par <- par(no.readonly = TRUE)
lf<-layout(matrix(c(1,2,3,3), 2, 2, byrow = TRUE))
# simple
X <- with(aids,model.matrix(formula(~x+I(x^2)+I(x^3)+I(x^4)+
                                 I(x^5))))[,-1]
matplot(X, type="l", ylim=c(9, 1000), lwd=2, ylab="poly Basis",
        xlab="x",main="(a)",cex.lab=1.5)
legend("topright",legend=c("^1","^2","^3","^4","^5"),
       col=c(1,2,3,4,5), lty=c(1,2,3,4,5), ncol=1,
       bg="white", title="types")
# orthogonal
P <- with(aids,model.matrix(formula(~poly(x, 5)))[,-1])
matplot( P, type="l", lwd=2, ylab="poly Basis", xlab="x", main="(b)",
        cex.lab=1.5)
legend("bottomright", col=c(1,2,3,4,5), lty=c(1,2,3,4,5), ncol=1,
       legend=c("^1","^2","^3","^4","^5"), bg="white",title="types")
# fitting the model
m1 <- gamlss(y~poly(x, 5), data=aids)

## GAMLSS-RS iteration 1: Global Deviance = 430.2693
## GAMLSS-RS iteration 2: Global Deviance = 430.2693

P1 <- model.matrix(with(aids,formula(~poly(x, 5))))
b <- coef(m1)
b

## (Intercept) poly(x, 5)1 poly(x, 5)2 poly(x, 5)3 poly(x, 5)4
##   200.48889    961.16182    46.19070   -68.04244    46.24160
## poly(x, 5)5
##      22.13419

F <- t(rep(b,45) *t(P1))
Fit <- cbind(fitted(m1), F)
matplot(Fit, type="l", lwd=c(3,2,2,2,2,2,2), ylab="poly FIT",
        xlab="x", main="(c)",cex.lab=1.5)
legend("bottomright",  col=c(1,2,3,4,5,6,7),lty=c(1,2,3,4,5,6,7),
       legend=c("fitted","^0","^1","^2","^3","^4","^5"),
       ncol=1,bg="white")
par(def.par)
```

Fig. 8.3

Section 8.8.1 provides an example of the use of orthogonal polynomials. While orthogonal polynomials are more stable to fit, they suffer from the problem of in-

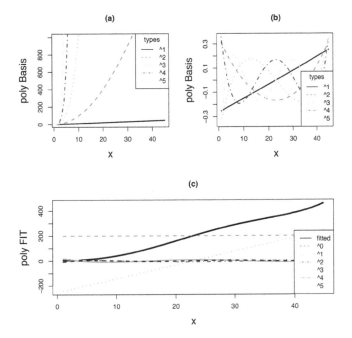

FIGURE 8.3: Polynomial for `aids` data: (a) standard polynomial basis, (b) orthogonal polynomial basis, (c) fitted values and the weighted orthogonal basis functions.

terpretability. Occasionally we may have to go back to the standard polynomial in order to explain the influence of an explanatory variable on the predictor.

The problem with polynomials in general is that, because their basis is defined globally in the full range of values for x, they can be highly influenced by a few observations in the data. This is a well-known phenomenon and is avoided by using smooth nonparametric functions.

8.4   Fractional polynomials

A polynomial x^p, for $x > 0$, is called *fractional* if the power p is not necessarily a positive integer, e.g. $x^{-1/2}$. Fractional polynomials were introduced by Royston and Altman [1994]. The idea is that a very flexible base to fit a parametric curve can be achieved with only a few fractional polynomials. The **gamlss** implementation uses the functions `fp()` and `bfp()`, which are loosely based on the fractional polynomial function `fracpoly()` for S-PLUS given by Ambler [1999]. The function `bfp()` generates the design matrix for fitting a fractional polynomial, while `fp()` works in `gamlss()` as an additive 'smoother' term.

The function `fp()` works as follows. Its argument `npoly` determines whether one, two or three terms in the fractional polynomial will be used in the fitting. For example with `npoly=3`, the following polynomial functions are fitted:

$$\beta_0 + \beta_1 x^{p_1} + \beta_2 x^{p_2} + \beta_3 x^{p_3} \ ,$$

where each p_j can take any value within the predetermined set $(-2, -1, -0.5, 0, 0.5, 1, 2, 3)$, with the value 0 interpreted as $\log(x)$. Figure 8.4 shows the shape of these basis functions, created with the following code:

Fig. 8.4

```
x<-seq(0,8,0.05)
plot(x, type="n",xlim=c(min(x), max(x)), ylim=c(-4,5),
        xlab="x", ylab="fractional polynomial",cex.lab=1.5)
fr<-c(-2, -1, -0.5,0,0.5, 1, 2, 3)
for(i in 1:8) {
if (fr[i]==0) lines(log(x)~x,lty=i+1, col=i+1, lwd=3)
else lines( I(x^fr[i])~x, lty=i+1, col=i+1, lwd=3)
}
legend("bottomright", legend=fr, col=2:9, lty=2:9, lwd=3)
```

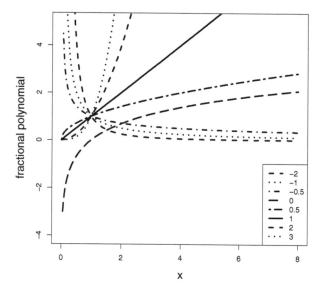

R
p. 23

FIGURE 8.4: Fractional polynomial basis used within `gamlss()`, i.e. polynomials with power $(-2, -1, -0.5, 0, 0.5, 1, 2, 3)$.

If two powers (p_j's) happen to be identical then the two terms $\beta_{1j}x^{p_j}$ and $\beta_{2j}x^{p_j}\log(x)$ are fitted instead. Similarly if three p_j's are identical, the terms fitted are $\beta_{1j}x^{p_j}$, $\beta_{2j}x^{p_j}\log(x)$ and $\beta_{3j}x^{p_j}[\log(x)]^2$. Note that `npoly=3` is rather slow since it fits all possible three-way combinations at each backfitting iteration.

Fractional polynomials can be fitted within `gamlss()` using the additive function fp(). It takes as arguments the x variable and npoly (the number of fractional polynomial terms) which takes the values 1, 2 or 3. An example of using fp() within gamlss() is given in Section 8.8.2.

8.5 Piecewise polynomials and regression splines

Piecewise polynomials are a useful tool in statistical modelling both on their own or in their penalized form. In fact fitting penalized piecewise polynomials is the most popular method of nonparametric smoothing because it usually works well in practice and is easy to implement. Penalized piecewise polynomials are smoothing functions and are covered in Chapter 9.

Sometimes we are confronted with data in which there is a sudden change in the relationship between the dependent variable and an independent variable. This type of data can be modelled using piecewise polynomials in the explanatory variable to describe the relationship. For simplicity, we will discuss the case where there is only one explanatory variable x. The values of x at which the piecewise polynomials change are called *breakpoints* or *knots*. These polynomials are known as splines if continuity restrictions are placed on them at the breakpoints. The name splines is derived from thin flexible rods that engineers have used to fit curves through given points. Smith [1979] referred to piecewise polynomials as regression splines and examined them as a tool in regression analysis.

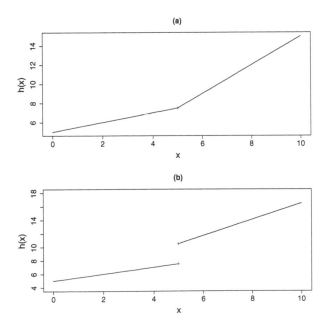

FIGURE 8.5: Piecewise linear (a) continuous and (b) discontinuous lines.

A simple example of a piecewise polynomial is the split line curve with a single breakpoint. There are two types of split line curve models, continuous and discontinuous split lines. The continuous split line has the form

$$h(x) = \beta_{00} + \beta_{01}x + \beta_{11}(x - b)H(x - b) \tag{8.2}$$

where $H(x - b)$ is the *Heaviside* function:

$$H(t) = \begin{cases} 1 & \text{if } t \geq 0 \\ 0 & \text{if } t < 0 \,. \end{cases}$$

(The notation $(x - b)_+$ is used often in the literature to denote $(x - b)H(x - b)$, i.e. $u_+ = u$ for $u > 0$.) β_{00}, β_{01}, and β_{11} are the linear parameters, and b is the breakpoint (or knot), being the nonlinear parameter. In a statistical modelling situation all four parameters need to be estimated. Figure 8.5(a) shows a continuous split line with parameters $\beta_{00} = 5$, $\beta_{01} = 0.5$, $\beta_{11} = 1$ and $b = 5$. With a discontinuous split line, the function has the form

$$h(x) = \beta_{00} + \beta_{01}x + (\beta_{10} + \beta_{11}(x - b)) H(x - b) \tag{8.3}$$

with β_{00}, β_{10}, β_{01}, β_{11}, and b being the parameters which need to be estimated. In a data set exhibiting piecewise linear behaviour, the breakpoint parameter b is usually the parameter of interest. Figure 8.5(b) shows a discontinuous split line curve with parameters $\beta_{00} = 5$, $\beta_{01} = 0.5$, $\beta_{10} = 3$, $\beta_{11} = 0.7$ and $b = 5$.

A quadratic piecewise polynomial with one breakpoint has the form

$$h(x) = \beta_{00} + \beta_{01}x + \beta_{02}x^2 + \left(\beta_{10} + \beta_{11}(x - b) + \beta_{12}(x - b)^2\right) H(x - b) \,. \tag{8.4}$$

The function is discontinuous at the breakpoint, with its first and second derivatives discontinuous (see Figure 8.6(a)). By dropping β_{10} from the equation, the function becomes continuous but still has discontinuous first and second derivatives at the breakpoint; see Figure 8.6(b). The function becomes continuous and with a continuous first derivative when the term $\beta_{11}(x - b)$ is dropped; see Figure 8.6(c). To create Figure 8.6 the following values for the parameters were used: $\beta_{00} = 5$, $\beta_{01} = -0.1$, $\beta_{02} = 0.1$, $\beta_{10} = 2$, $\beta_{11} = 1$, $\beta_{12} = -0.4$ and $b = 5$.

More general piecewise polynomials are defined as

$$h(x) = \sum_{j=0}^{D} \beta_{0j}x^j + \sum_{k=1}^{K} \sum_{j=0}^{D} \beta_{kj}(x - b_k)^j H(x - b_k) \tag{8.5}$$

where D is the degree of the polynomial in x and K is the number of breakpoints. The presence or absence of the term $\beta_{kj}(x - b_k)^j$ in (8.5) allows a discontinuity or continuity, respectively, at the breakpoint b_k in the jth derivative of the function. If continuity is required in the jth derivative of the function at a particular breakpoint b_k, it would probably be required also in all the lower-order derivatives at b_k. This

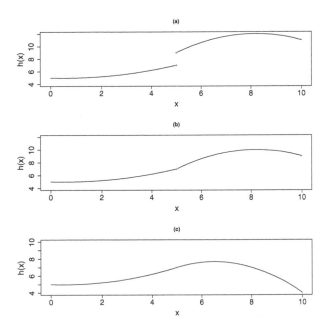

FIGURE 8.6: Piecewise quadratic, (a) discontinuous with discontinuous first derivative, (b) continuous with discontinuous first derivative and (c) continuous with continuous first derivative.

is achieved by removing all the terms $\beta_{km}(x - b_k)^m$, for $m = 0, 1, \ldots k$. The name *spline* is usually applied to piecewise polynomials with all the derivatives lower than D continuous at b_k. For example

$$h(x) = \sum_{j=0}^{D} \beta_{0j}x^j + \sum_{k=1}^{K} \beta_k (x - b_k)^D H(x - b_k) \qquad (8.6)$$

is a spline function of degree D. For $D = 3$ we have the *cubic spline*

$$h(x) = \beta_{00} + \beta_{01}x + \beta_{02}x^2 + \beta_{03}x^3 + \sum_{k=1}^{K} \beta_k (x - b_k)^3 H(x - b_k) \ . \qquad (8.7)$$

Cubic splines are, because of their continuous first and second derivatives at the breakpoints, very smooth curves and therefore ideal for smoothing techniques.

In order to fit a spline curve as in equation (8.7) within a regression model, one would need K nonlinear b_k breakpoint parameters and $K + 4$ linear β parameters to completely specify the equation. A design matrix basis based on equation (8.7) is called a *truncated piecewise polynomial* basis. Figure 8.7 shows a truncated piecewise polynomial basis for degrees of polynomials equal to constant, linear, quadratic and cubic. x is in the range $(0, 1)$ and there are five knots at $(0.2, 0.3, 0.5, 0.7, 0.8)$. For degree 0 (Figure 8.7(a)) the basis functions comprise six dummy variables corresponding to the intervals $(0 < x \leq 1)$, $(.20 < x \leq 1)$, $(0.3 < x \leq 1)$, $(0.5 < x \leq 1)$,

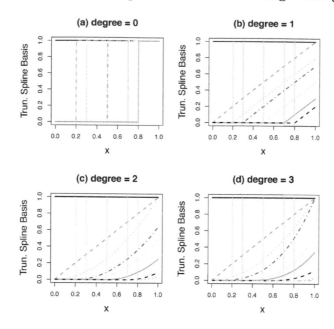

FIGURE 8.7: Truncated piecewise polynomial basis functions for different degrees: (a) constant, (b) linear, (c) quadratic and (d) cubic. x is defined from zero to one with breakpoints at $(0.2, 0.3, 0.5, 0.7, 0.8)$.

$(0.7 < x \leq 1)$ and $(0.8 < x \leq 1)$, having ones if x is in the interval and zero otherwise. For degree 1 the basis functions, in Figure 8.7(b), comprise the constant plus six extra linear functions. The first linear function is defined on the whole range of x and the remaining five only on a limited range, e.g. the second on the range $(0.2 < x < 1)$. Figure 8.7(c) shows the basis functions for degree 2. Here we have the constant, linear, plus six quadratic functions. The constant, linear and first quadratic functions are defined on the whole range of x, while the other five quadratic functions are defined on a limited range of x depending on the breakpoints. The same pattern appears in Figure 8.7(d) for degree 3, where the basis functions comprise the constant, linear, quadratic and cubic functions defined on the whole range of x, while the other five cubic functions are defined only on a limited range depending on the breakpoints.

While the truncated basis of a spline function is intuitively simple, it suffers from the same problem as a polynomial basis in that it is not numerically stable. The B-splines introduced in the next section are numerically superior.

8.6 B-splines

B-splines are to truncated piecewise polynomials what orthogonal polynomials are to polynomials, that is, a B-spline basis provides a superior numerical basis to equation (8.7). The basis functions in B-splines are defined only locally in the sense that they are non-zero only on the domain spanned by at most $2 + D$ knots, where D is the degree of the piecewise polynomial. See de Boor [2001] for further details. The term 'B-spline' is short for basis spline. The important point here is that any function given by equation (8.7) of a given degree and for a given x range can be uniquely represented as a linear combination of B-splines of the same degree within the same range. There are several properties of B-splines worth noting:

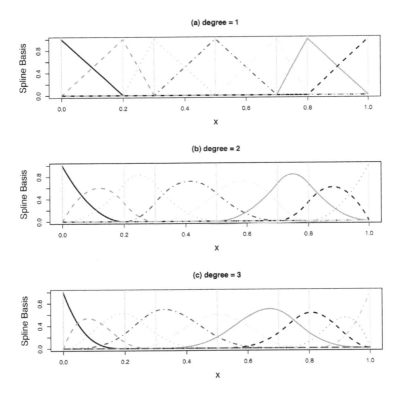

FIGURE 8.8: B-spline basis for degrees (a) linear, (b) quadratic and (c) cubic. x is defined from zero to one, with unequally spaced knots (breakpoints) at $(0.2, 0.3, 0.5, 0.7, 0.8)$.

- The B-splines are defined by local functions which have their domain within $2 + D$ knots of the x range. For example for cubic splines with $D = 3$, each basis function is defined within 5 knots.

- Depending on the degree of the piecewise polynomial, the B-splines could be

 – local constants ($D = 0$);

 – local (two piece) linear function of x ($D = 1$);

 – local (three piece) quadratic function of x ($D = 2$);

 – local (four piece) cubic function of x ($D = 3$);

 – local ($D + 1$ piece) higher level polynomial ($D \geq 4$).

Figure 8.8 shows an example for degree $D = 1, 2, 3$. Note that the basis functions of a cubic B-spline are mostly very similar in shape to the normal distribution.

- The knots do not have to be equally spaced, so general patterns of knots are possible.

- The number of knots determines the size of the B-spline basis which makes up the piecewise polynomial function.

- B-splines are columns of a basis matrix $\mathbf{B}$, which can be used in a regression framework as the design matrix. The fitted coefficients in such a regression ($\hat{\mathbf{y}} = \mathbf{B}\hat{\boldsymbol{\beta}}$) produce a flexible relationship between y and x. Figure 8.9 shows a regression fit using the **aids** data. Figure 8.9(a) shows the 11 B-spline basis functions with $D = 3$ for fitting x (time), generated with 9 equally spaced knots (7 internal knots and 2 boundary knots). Figure 8.9(b) shows the fitted spline (solid line) and the basis functions weighted by their fitted coefficients.

```
library(splines)
data(aids)
def.par <- par(no.readonly = TRUE)
# simple
X<-bs(aids$x, df=11, intercept=T)
knots<-c(1, attr(X, "knots"),45)
nf<-layout(matrix(c(1,2),2,1,byrow=TRUE))
matplot(X, type="l",lwd=2,col=2:12,lty=2:12,ylab="B-spline basis",
        xlab="x",main="(a)",cex.lab=1.5,cex.main=1.5)
abline(v=knots, col="gray")
# fitting the model
m1<-gamlss(y~X-1,data=aids,trace=FALSE)
P1<-model.matrix(with(aids,
                    formula(~bs(aids$x, df=11,intercept=T)-1)))
b<-coef(m1)
F<-t(rep(b,45) *t(P1))
Fit<-cbind(fitted(m1), F)
matplot(Fit, type="l", lwd=c(3,rep(2,11)),lty=1:12,col=1:12,
  ylab="B-spline fit", xlab="x", main="(b)",cex.lab=1.5,cex.main=1.5)
abline(v=knots, col="gray")
par(def.par)
```

Fig.
8.9

Models in which the breakpoints (or knots) b_k are determined in advance, so only the linear parameters β have to be estimated, are called *regression spline* models. In *cardinal spline* models the positions of the knots are chosen uniformly over the range of x. Another method of regression splines uses positions determined from the quantile values of x.

The number of knots affects the degrees of freedom for the fitted model and therefore the complexity of the model. Parameters which determine the complexity of the model are usually referred to as *smoothing parameters*. The degrees of freedom for the fitted model in this case are $1 + D + K$: one for the intercept, D for the degree of the polynomial in x and K for the number of internal knots. This design matrix **B** has $1 + D + K$ independent columns. To create a B-spline basis, the functions `bs()` or `ns()` (**splines** package) can be used. The former generates the B-spline basis matrix for a polynomial spline of any degree, and the latter the B-spline basis matrix for a *natural* cubic spline. Natural cubic splines are cubic splines having the added condition that the behaviour of the function outside the range of x is linear. An example of the use of `bs()` is given in Section 8.8.3.

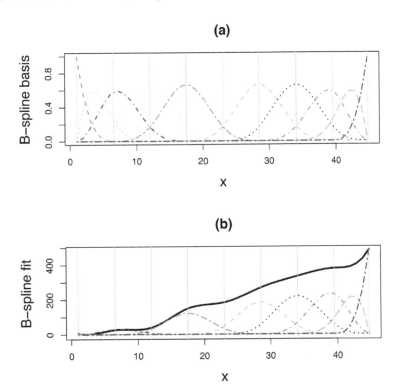

FIGURE 8.9: B-splines fit of y (number of aids cases) against x (time) for the `aids` data, using 9 equally spaced knots. (a) B-spline basis for x, and (b) fitted values for y in black plus the B-splines basis functions weighted by their coefficients $\hat{\beta}$.

Models in which the breakpoints are estimated are called *free knot* models and they are examined in the next section.

R

240

8.7   Free knot models

Free knot models are piecewise polynomial models in which the position and number of knots are estimated from the data. These models are useful if it is believed that there are structural changes in the relationship between y and x, and that the time(s) or point(s) at which the relationship changes are unknown. Estimation of the knots is a highly nonlinear problem. Here we focus on models with a fixed known number of knots. The likelihood function of the knot parameters is notorious for its multiple maxima. In this section we concentrate on cases where there are relatively few knots. The positions of the knots can be fixed (i.e. known) or free (i.e. unknown). Figure 8.5(a) is a typical example where at some point the linear relationship between y and x changes.

The **gamlss.add** package provides the following functions for knot modelling:

fitFixedKnots()

        for fitting a univariate regression model using piecewise polynomials with known knots;

fitFreeKnots()

        for fitting a univariate regression model using piecewise polynomials with unknown knots; and

fk()   for fitting a regression additive terms model using piecewise polynomials with unknown (or known) knots.

The arguments for `fitFixedKnots()` and `fitFreeKnots()` are:

x   the explanatory variable;

y   the response variable;

weights   the prior weights;

knots   the position of the interior knots for `fitFixedKnots()` or starting values for `fitFreeKnots()`;

data   the dataframe;

degree   the degree of the piecewise polynomials;

base   the basis functions for the piecewise polynomials: `base="trun"` for truncated (default), and `base="Bbase"` for B-spline basis piecewise polynomials

trace   controlling the trace of `optim()`, only used for `fitFreeKnots()`;

...   for extra arguments.

These two functions return S3 class objects `"FreeBreakPointsReg"` and

"FixBreakPointsReg", respectively. These objects have methods print(), fitted(), residuals(), coef(), knots() and predict().

The "FixBreakPointsReg" objects also have vcov() and summary().

The function fk() provides an interface so that fitFixedKnots() and fitFreeKnots() can be utilised within gamlss(). The main arguments of fk() are:

x   the explanatory variable;

start   starting values for the breakpoints. The number of breakpoints is also determined by the length of start;

all.fixed   fixes all knots at their start values, with default FALSE, i.e. free knots.

Other arguments for fitFreeKnots() or fitFixedKnots() can be passed through control. An example of the use of fk() is given in Section 8.8.4.

8.8 Example: the CD4 data

Data summary: the CD4 data

R data file: CD4 in package **MASS** of dimensions 609×2

variables

cd4 : CD4 counts from uninfected children born to HIV-1 mothers.

age : child's age in years

purpose: to demonstrate the use of linear parametric terms

This example illustrates the use of some of the techniques described in the previous section. The data are given by Wade and Ades [1994] and they refer to **cd4** counts from uninfected children born to HIV-1 mothers and the **age** in years of the child. We input the data, and plot it in Figure 8.10.

```
data("CD4")
plot(cd4 ~ age, data = CD4)
```
Fig. 8.10

This is a simple regression example with only one explanatory variable, **age**, which is continuous. The response, while strictly speaking a count, is sufficiently large for us to treat it at this stage as a continuous response variable.

There are several striking features in this data.

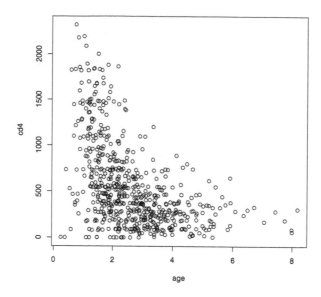

R
p. 24

FIGURE 8.10: The cd4 data.

1. The first has to do with the relationship between the mean of **cd4** and **age**. It is hard to see from the plot whether this relationship is linear or not.

2. The second has to do with the heterogeneity of variance in the response variable **cd4**. It appears that the variation in **cd4** is decreasing with age.

3. The final problem has to do with the distribution of **cd4** given **age**. Is this distribution normal? It is hard to tell from the figure but probably we will need a more flexible distribution.

Traditionally, problems of this kind were dealt with by a transformation of the response variable, or of both the response and the explanatory variable(s). One could hope that this would possibly correct for some or all of the above problems simultaneously. Figure 8.11 (produced with the following code) shows plots in which several transformations for **cd4** and **age** were tried. It is hard to see how we can improve the situation by transformations.

Fig.
8.11

```
op <- par(mfrow = c(3, 4), mar = par("mar") + c(0, 1, 0, 0),
     pch = "+", cex = 0.45, cex.lab = 2, cex.axis = 1.6)
page <- c("age^-0.5", "log(age)", "age^.5", "age")
pcd4 <- c("cd4^-0.5", "log(cd4+1)", "cd4^.5")
for (i in 1:3) {
    yy <- with(CD4, eval(parse(text = pcd4[i])))
    for (j in 1:4) {
        xx <- with(CD4, eval(parse(text = page[j])))
        plot(yy ~ xx, xlab = page[j], ylab = pcd4[i])
```

```
        }
    }
par(op)
```

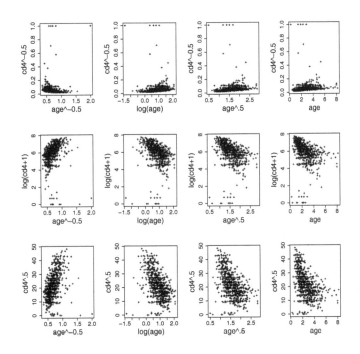

FIGURE 8.11: The CD4 data with various transformations for `cd4` and `age`.

Within the GAMLSS framework we can deal with these problems one at a time. Firstly we start with the relationship between the mean of `cd4` and `age`.

8.8.1 Orthogonal polynomials

We fit orthogonal polynomials of different orders, and choose the best using the GAIC. For now we fit a constant variance and the default normal distribution.

```
m1<-gamlss(cd4~age, sigma.fo=~1, data=CD4, trace=FALSE)
m2<-gamlss(cd4~poly(age,2), sigma.fo=~1, data=CD4, trace=FALSE)
m3<-gamlss(cd4~poly(age,3), sigma.fo=~1, data=CD4, trace=FALSE)
m4<-gamlss(cd4~poly(age,4), sigma.fo=~1, data=CD4, trace=FALSE)
m5<-gamlss(cd4~poly(age,5), sigma.fo=~1, data=CD4, trace=FALSE)
m6<-gamlss(cd4~poly(age,6), sigma.fo=~1, data=CD4, trace=FALSE)
m7<-gamlss(cd4~poly(age,7), sigma.fo=~1, data=CD4, trace=FALSE)
m8<-gamlss(cd4~poly(age,8), sigma.fo=~1, data=CD4, trace=FALSE)
```

We compare the models using the AIC:

```
GAIC(m1, m2, m3, m4, m5, m7, m8)
```

```
##    df     AIC
## m7  9 8963.263
## m8 10 8963.874
## m5  7 8977.383
## m4  6 8988.105
## m3  5 8993.351
## m2  4 8995.636
## m1  3 9044.145
```

and also using the SBC:

```
GAIC(m1, m2, m3, m4, m5, m7, m8, k = log(length(CD4$age)))
```

```
##    df     AIC
## m7  9 9002.969
## m8 10 9007.992
## m5  7 9008.266
## m2  4 9013.284
## m4  6 9014.576
## m3  5 9015.410
## m1  3 9057.380
```

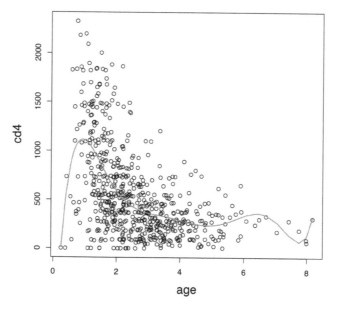

FIGURE 8.12: CD4 data and fitted values using polynomial of degree 7 in age.

Remarkably both AIC and SBC select model m7, i.e. a polynomial of degree 7. Unfortunately the fitted values for the mean of cd4 shown together with the data in Figure 8.12 look rather unconvincing. The line is too wobbly at the ends of the range of **age**, trying to be very close to the data. This is typical behaviour of polynomial fitting.

8.8.2 Fractional polynomials

Next we demonstrate fractional polynomials. The function **fp()** works in **gamlss()** as an additive smoother term; see Section 9.7. It can be used to fit the best (fractional) polynomial within a specific set of possible power values. Its argument **npoly** determines whether one, two or three terms in the fractional polynomial will be used in the fitting. Here we fit fractional polynomials with one, two and three terms and choose the best using GAIC:

```
m1f<-gamlss(cd4~fp(age,1), sigma.fo=~1, data=CD4, trace=FALSE)
m2f<-gamlss(cd4~fp(age,2), sigma.fo=~1, data=CD4, trace=FALSE)
m3f<-gamlss(cd4~fp(age,3), sigma.fo=~1, data=CD4, trace=FALSE)
GAIC(m1f, m2f, m3f)
```

```
##      df      AIC
## m3f  8 8966.375
## m2f  6 8978.469
## m1f  4 9015.321
```

```
GAIC(m1f, m2f, m3f, k = log(length(CD4$age)))
```

```
##      df      AIC
## m3f  8 9001.669
## m2f  6 9004.940
## m1f  4 9032.968
```

```
# to get the fitted GAMLSS model
m3f
```

```
##
## Family:  c("NO", "Normal")
## Fitting method: RS()
##
## Call:
## gamlss(formula = cd4 ~ fp(age, 3), sigma.formula = ~1,
##     data = CD4, trace = FALSE)
##
## Mu Coefficients:
## (Intercept)    fp(age, 3)
##        557.5           NA
## Sigma Coefficients:
## (Intercept)
##        5.929
##
```

```
## Degrees of Freedom for the fit: 8 Residual Deg. of Freedom   601
## Global Deviance:      8950.37
##            AIC:       8966.37
##            SBC:       9001.67
```

```
# to get the fitted fractional polynomial (lm class object)
getSmo(m3f)
```

```
##
## Call:
## lm(formula = y ~ x.fp, weights = w)
##
## Coefficients:
## (Intercept)        x.fp1        x.fp2        x.fp3
##      -599.3       1116.8       1776.2        698.6
```

```
# to get the power parameters
getSmo(m3f)$power
```

```
## [1] -2 -2 -2
```

```
plot(cd4 ~ age, data = CD4, cex.lab=1.5)
lines(CD4$age[order(CD4$age)], fitted(m1f)[order(CD4$age)],
    lty=1, lwd=2.5, col = "green")
lines(CD4$age[order(CD4$age)], fitted(m2f)[order(CD4$age)],
    lty=2, lwd=2.5, col = "blue")
lines(CD4$age[order(CD4$age)], fitted(m3f)[order(CD4$age)],
    lty=3, lwd=2.5, col = "red")
legend("topright",legend=c("1","2","3"),lty=1:3,
    col=c("green","blue","red"), lwd=2, title="degree")
```

Fig. 8.13

Both AIC and SBC favour the model m3f with a fractional polynomial with three terms. Note that printing m3f the model for μ gives a value of 557.5 for the Intercept and NA for the coefficient for fp(age, 3). This is because within the backfitting the constant is fitted first and then the fractional polynomial is fitted to the partial residuals of the constant model. As a consequence the constant is fitted twice. The coefficients and the power transformations of the fractional polynomials can be obtained using the getSmo() function of the gamlss fitted object. For the CD4 data all powers are -2, indicating that the following terms are fitted in the model: age^{-2}, $age^{-2} \log(age)$ and $age^{-2} [\log(age)]^2$. Hence the fitted model m3f is given by

$$cd4 \sim \mathcal{N}(\hat{\mu}, \hat{\sigma})$$
$$\hat{\mu} = 557.5 - 599.3 + 1116.8\, age^{-2} + 1776.2\, age^{-2} \log(age) +$$
$$698.6\, age^{-2} [\log(age)]^2$$
$$\hat{\sigma} = \exp(5.929) = 375.8 \,.$$

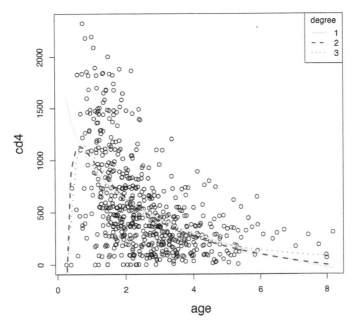

FIGURE 8.13: CD4 data and fitted values using fractional polynomials of degree 1, 2 and 3 in `age`.

Figure 8.13 shows the best fitted models using one, two or three fractional polynomial terms. The situation remains unconvincing. None of the models seems to fit particularly well.

8.8.3 Piecewise polynomials

Next we fit piecewise polynomials using `bs()`. We try different degrees of freedom (effectively different number of knots) and choose the best model using AIC and SBC:

```
m2b <- gamlss(cd4 ~ bs(age), data = CD4, trace = FALSE)
m3b <- gamlss(cd4 ~ bs(age, df = 3), data = CD4, trace = FALSE)
m4b <- gamlss(cd4 ~ bs(age, df = 4), data = CD4, trace = FALSE)
m5b <- gamlss(cd4 ~ bs(age, df = 5), data = CD4, trace = FALSE)
m6b <- gamlss(cd4 ~ bs(age, df = 6), data = CD4, trace = FALSE)
m7b <- gamlss(cd4 ~ bs(age, df = 7), data = CD4, trace = FALSE)
m8b <- gamlss(cd4 ~ bs(age, df = 8), data = CD4, trace = FALSE)
GAIC(m2b, m3b, m4b, m5b, m6b, m7b, m8b)

##      df     AIC
## m7b   9 8959.519
## m6b   8 8960.353
## m8b  10 8961.073
```

```
## m5b  7 8964.022
## m4b  6 8977.475
## m2b  5 8993.351
## m3b  5 8993.351
```

```
GAIC(m2b, m3b, m4b, m5b, m6b, m7b, m8b, k = log(length(CD4$age)))
```

```
##       df      AIC
## m5b   7 8994.904
## m6b   8 8995.648
## m7b   9 8999.225
## m4b   6 9003.946
## m8b  10 9005.191
## m2b   5 9015.410
## m3b   5 9015.410
```

Note that the model degrees of freedom (given in column df) include 1 for the σ parameter of the normal distribution. Note that m2b uses the default df = 3. The best model with AIC uses df = 7 in bs, while the best model with SBC uses df = 5. Figure 8.14 shows the resulting fitted models using df = 5 and df = 7 within the function bs(), for the parameter μ.

```
plot(cd4 ~ age, data = CD4, cex.lab=1.5)
lines(CD4$age[order(CD4$age)], fitted(m7b)[order(CD4$age)],
      col = "blue", lty=1, lwd=2.5)
lines(CD4$age[order(CD4$age)], fitted(m5b )[order(CD4$age)],
      col = "red", lty=2, lwd=2.5)
legend("topright",legend=c("5","7"),lty=1:2,col=c("blue","red"),
      lwd=2, title="degrees of freedom")
```

Fig. 8.14

8.8.4 Free knot models

Here we model the relationship between cd4 and age using simple piecewise polynomials. We fit four different models using linear (degree=1) and quadratic (degree=2) functions, and using one and two breakpoints, respectively. Note that starting values have to be specified for the breakpoints. The number of breakpoints to fit is taken from the number of starting values.

```
library(gamlss.add)
 f1<-gamlss(cd4~fk(age,degree=1,start=2),data=CD4,trace=FALSE)
 f2<-gamlss(cd4~fk(age,degree=1,start=c(2,5)),data=CD4,trace=FALSE)
 f3<-gamlss(cd4~fk(age,degree=2,start=2),data=CD4,trace=FALSE)
 f4<-gamlss(cd4~fk(age,degree=2,start=c(2,5)),data=CD4,trace=FALSE)
GAIC(f1, f2, f3, f4)
```

```
##      df      AIC
```

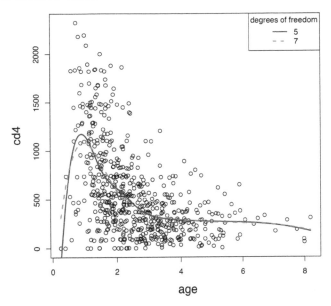

FIGURE 8.14: CD4 data and fitted values using piecewise polynomials with 5 and 7 degrees of freedom in **age** using the function **bs()**.

```
## f1   5 8984.558
## f3   5 8984.558
## f2   7 8988.357
## f4   7 8988.357

GAIC(f1, f2, f3, f4, k = log(length(CD4$age)))

##      df      AIC
## f1   5 9006.617
## f3   5 9006.617
## f2   7 9019.239
## f4   7 9019.239
```

Note the model degrees of freedom (**df**) includes 1 for σ. From the GAIC it can be seen that there is no support for the quadratic models **f3** and **f4**, and that the data support only one breakpoint parameter. To get the two different slopes and the breakpoint parameter, use the **getSmo()** function:

```
f1

##
## Family:  c("NO", "Normal")
## Fitting method: RS()
##
## Call:
## gamlss(formula = cd4 ~ fk(age, degree = 1, start = 2),
```

```
##      data = CD4, trace = FALSE)
##
## Mu Coefficients:
##                      (Intercept)
##                            557.5
## fk(age, degree = 1, start = 2)
##                               NA
## Sigma Coefficients:
## (Intercept)
##       5.949
##
##  Degrees of Freedom for the fit: 5 Residual Deg. of Freedom   604
## Global Deviance:    8974.56
##            AIC:     8984.56
##            SBC:     9006.62
```

getSmo(f1)

```
##
## Call:
## fitFreeKnots(y = y, x = xvar, weights = w, degree = degree,
##     knots = lambda, fixed = control$fixed, base = control$base)
##
##
## Coefficients:
## (Intercept)          x        XatBP1
##       809.1      -361.5        334.9
## Estimated Knots:
##   BP1
## 2.87
```

The breakpoint can also be found simply using

knots(getSmo(f1))

```
##        BP1
## 2.869966
```

The fitted linear plus linear model for μ is given by

$$\mu = (557.5 + 809.1) - 361.5\, \text{age} + 334.9\, (\text{age} - 2.87) H(\text{age} - 2.87)$$
$$= \begin{cases} 1366.6 - 361.5\, \text{age} & \text{if age} \le 2.87 \\ 329.1 - 26.6\, \text{age} & \text{if age} > 2.87 \,. \end{cases}$$

The plot of the fitted model is given in Figure 8.15.

```
plot(cd4 ~ age, data = CD4)
lines(CD4$age[order(CD4$age)], fitted(f1)[order(CD4$age)],
      col="blue", lty=1,lwd=2.5)
```

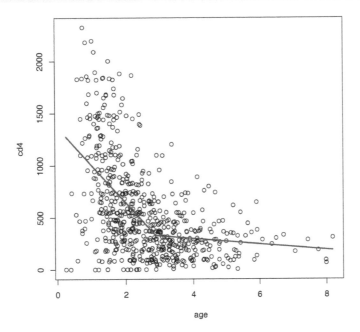

FIGURE 8.15: CD4 data and fitted values using piecewise linear fit, with the knot estimated from the data.

8.9 Bibliographic notes

Wilkinson and Rogers [1973] present symbolic notation and syntax for specifying factorial models for analysis of variance. Aiken and West [1991] interpreted interactions for multiple regression. Stigler [1974] shows that Gergonne [1815] was the first to present a design of an experiment for polynomial regression. Stasinopoulos and Rigby [1992] presented breakpoint models. Dierckx [1995] showed breakpoint models and more generally showed curve and surface fitting with splines. Royston and Altman [1994] introduced fractional polynomials. Schoenberg [1946b,a] was the first to introduce the terminology of the spline function. Schumaker [2007] presents the basic theory of spline functions, and de Boor [2001] discusses splines and B-splines in detail.

8.10   Exercises

1. **Fitting linear models to `whiteside` data:** For the gas consumption data first
   analysed on page 101 and exercise 1 of Chapter 4, fit the five different models
   described in Figure 8.2. Use GAIC to choose the best model. Find the structure
   of the design matrix for each model.

2. **The CD4 data:** Familiarize yourself with the different linear terms in **R** by
   repeating the commands given in Section 8.8.

3. **Fitting different distributions to the `cd4` data:** For the CD4 count data,
   fit the SEP3 model and check it using the following:

   (a) Load the `cd4` data and print the variable names.

   (b) Fit the SEP3 model by first fitting a PE (power exponential) model and using
       its fitted values as starting values for fitting the SEP3 (skew exponential
       power type 3) model. Use the `start.from` argument of `gamlss()`.

   (c) For the SEP3 model, print the total effective degrees of freedom and plot
       the fitted parameters against `age`. Note that the fitted μ's have different
       interpretations in the two distributions. For PE the distribution is symmetric
       about μ, which is the mean, median and mode. For SEP3, μ is the mode.

   (d) For the SEP3 model, look at the

       • residual and worm plots and Q statistics;

       • centile curves.

9

Additive smoothing terms

CONTENTS

This chapter provides an introduction to smoothing techniques and how to use those techniques within a GAMLSS model. In particular:

- univariate penalized smoothing techniques are introduced,

- the penalized approach to smoothing is explained, and

- the GAMLSS smoothing additive terms (including multivariate terms) are described.

9.1   Introduction

This chapter is dedicated to univariate and multivariate smoothing techniques and how they are applied within the GAMLSS framework. We can think of the univariate smoothers as the *main* (nonlinear) effects of the explanatory variables on the predictor of a distribution parameter, while the multivariate smoothers are their nonlinear *interaction* effects.

In particular, when only one explanatory variable x is used, this is a univariate smoother $s(x)$. A multivariate smoother, e.g. $s(x_1, x_2)$, is defined where two or more explanatory variables are involved in the fitting. Both univariate and multivariate smoothers can be used as additive terms within a GAMLSS model formula.

We classify all smoothers used within GAMLSS into two main categories:

penalized smoothers: which use quadratic penalties on the smooth model parameters to control the amount of smoothing, and

all other smoothers: which use different ideas (e.g. locality) or non-quadratic penalties to achieve the resulting smooth functions.

The distinction is illustrated in Figure 9.1, where the difference between univariate and multivariate penalized smoothers is also highlighted. The 'other' smoothers are all multivariate smoothers since they can take multiple explanatory variables. Figure 9.1 shows all smoothers discussed in this chapter. More specifically the univariate penalized smoothers include: (i) P-splines (together with their monotonic and cyclic versions), (ii) cubic splines, (iii) ridge and lasso regression (which are multivariate smoothers since they use more than one explanatory variable but it is convenient to be introduced here), (iv) penalized categorical variables (as a way of reducing the level of a factor) and (v) Gaussian Markov random field (GMRF) for spatial factors.

The structure of this chapter is as follows. Section 9.2 is an introduction to smoothing techniques in general. Section 9.3 describes local regression smoothers and serves as an introduction to basic ideas in smoothing, such as smoothing parameters and locality of the estimates. Sections 9.4 and 9.5 explain how the univariate and multivariate penalized smoothers can be used within GAMLSS, respectively. The 'other' smoothers are explained in Section 9.6. Table 9.1 provides a summary of the smooth functions available in the GAMLSS packages.

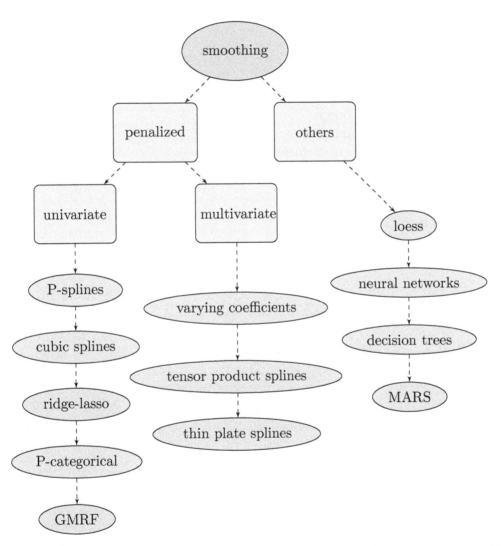

FIGURE 9.1: Diagram showing the different additive smoothing terms covered in this chapter.

TABLE 9.1: Additive terms implemented within the **gamlss** packages

Additive term	**gamlss** function name
cubic splines	`cs()`, `scs()`
decision trees	`tr()`
fractional and power polynomials	`fp()`, `pp()`
free knots (break points)	`fk()`
`loess`	`lo()`
neural networks	`nn()`
nonlinear fit	`nl()`
P-splines	`pb()`, `pb0()`, `ps()`,
P-splines cyclic	`pbc()`, `cy()`
P-splines monotonic	`pbm()`
P-splines shrinking to zero	`pbz()`
P-splines varying coefficient	`pvc()`
penalized categorical	`pcat()`
random effects	`random()`, `re()`
ridge regression	`ri()`
Simon Wood's **gam**	`ga()`
Stephen Milborrow's `earth`	`ma()`

9.2   What is a scatterplot smoother?

Suppose we have n measurements of a response variable $\mathbf{y} = (y_1, y_2, \ldots, y_n)^\top$ and a single explanatory variable $\mathbf{x} = (x_1, x_2, \ldots, x_n)^\top$, and we want to study their relationship. The first thing we should do is to plot $\mathbf{y}$ against $\mathbf{x}$. A curve fitted through the data shows the kind of relationship existing between the two variables.

For demonstration purposes, we return to the Munich 1990s rent data first introduced in Chapter 1. Figure 9.2(a) is a plot of the rent R against floor space (`Fl`), and 9.2(b) is a plot of R against the age (i.e. year of construction) of the building (`A`). If we ignore the fitted line smoothers shown in the plots for the moment, the left plot shows a clear positive relationship between rent and floor space but the relationship between rent and age is not clear-cut.

First we consider a simple scatterplot smoother to summarize the relationship between the mean of a response variable and a single explanatory variable. A simple *scatterplot smoother* is a statistical device which fits a curve as a way of exploring plausible relationships between a response (e.g. rent) and a predictor (e.g. floor space). It is this help in interpreting relationships which makes scatterplot smoothers important in statistics.

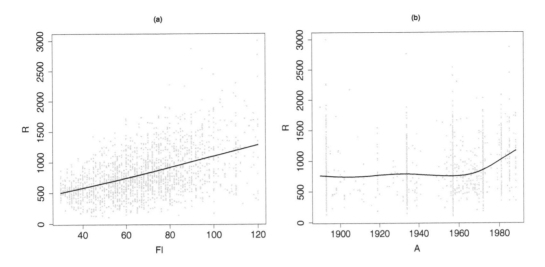

FIGURE 9.2: The Munich rent data set: (a) rent against floor space (b) rent against age of the building, with smooth curves fitted.

> *Definition:* A simple *scatterplot smoother*, or for convenience a *smoother*, summarizes the trend of the mean of a response variable y as a function of x by not assuming any parametric functional form for the dependence of the mean of y on x.

For example, the fitted smoother in Figure 9.2(a) confirms our belief of a positive (almost linear) relationship between the mean of rent and the floor space. From the smoother in Figure 9.2(b), we can conclude that for flats built between 1900 and 1960 the mean rent values are relatively constant, while there is a strong positive relationship between the mean rent and the age of the building after 1960.

Smoothing is also helpful in cases where the response variable is binary. Consider the data in Figure 9.3, kindly provided by Professor Brian Francis of Lancaster University. Here we have a scatterplot of $n = 10,590$ observations. The response variable is whether a particular crime was reported ($y = 1$) or not ($y = 0$) in the media. The explanatory variable is the age of the victim of the crime. The scatterplot in this case (ignoring for the moment the fitted smoothing curve) is uninformative due to the binary nature of the data. The fitted smoother shows the fitted or estimated probability of reporting a crime in the media according to age. It shows that the estimated probability that a crime is reported is higher when the crime is committed on a young person, with a peak at around age ten. The estimated probability then declines until the victim reaches the age of twenty. From then on, the estimated probability, remains constant until the age of sixty after which the reporting probability rises steadily with age.

A simple univariate smoother is a generalization of the simple linear regression

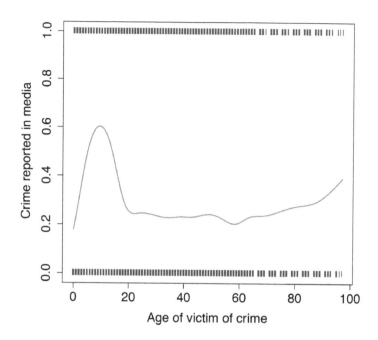

FIGURE 9.3: Whether crime was reported in the media (1=yes, 0=no) against the age of the victim, together with smooth curve of the probability that crime was reported.

model and can be written formally as a statistical model as:

$$E(Y_i) = \alpha + s(x_i) , \qquad \text{for } i = 1, \ldots, n \qquad (9.1)$$

where $s(\cdot)$ is an arbitrary function which we assume to exist, α is a constant which most of the time we absorb into the function $s(\cdot)$, say $E(Y_i) = s(x_i)$, and $s(x_i)$ is the trend that we would like to estimate. $s(\cdot)$ is arbitrarily defined but we may assume that it has some well-known properties. For example, a cubic spline smoother assumes that $s(\cdot)$ has continuous first and second derivatives.

There was an explosion of statistical smoothing techniques in the late 1980s and early 1990s. The reader is referred to the texts of Hastie and Tibshirani [1990], Green and Silverman [1994], Fahrmeir and Tutz [2001], Ruppert et al. [2003], Wang [2011] and Fahrmeir et al. [2013] for more details. Originally, the smoother was used to estimate the conditional expected value (or mean) of $\mathbf{y}$ given $\mathbf{x}$ without assuming a parametric functional form for this dependence, allowing the data to indicate the functional form. Later, this was extended to any location parameter (e.g. median) of the distribution of $\mathbf{y}$, and more generally, smoothers are used for the estimation of quantiles or expectiles of the distribution of $\mathbf{y}$ given $\mathbf{x}$ [Schnabel and Eilers, 2013a,b].

There are several ways to fit a smoother: for example to rely on local estimators or to put penalties on the behaviour of the smoother. The fact that they are called nonparametric is somewhat misleading, since all smoothers do estimate parameters and also contain a dominant parameter which determines the amount of smoothing for the data (the *smoothing parameter*). An advantage of a smoother over a parametric function is its local behaviour. This means that smoothers are affected by local observations more than by observations far away. Some of the basic ideas of smoothing can be introduced through local regression models and this is what the next section describes.

9.3   Local regression smoothers

The idea with the local smoother is that instead of using all available data to obtain a suitable estimate of the current value, only part of the data is used at a time. This part is determined by a *window*, which is an interval of the explanatory variable x. The window allows only observations that fall within it to be used in the calculation of the current smoothed value. Except near the ends of the range of x, the window is a 'symmetric' two-sided neighbourhood around the target x with an equal number of observations on each side of the target value. (Note that in time series analysis, one-sided windows are used extensively.)

The size of the window in the unweighted smoothers (see below for the definition of weighted smoothers) plays the role of the smoothing parameter. Large windows produce estimates of the trend that are smoother and hence high in bias, but low in variance. Small windows, conversely, produce estimates that are more wiggly, i.e. high in variance but low in bias. Hence there is always a trade-off between bias and variance. The windows are controlled in unweighted local regression by the *span*, defined as

$$\text{span} = (2k + 1)/n ,$$

where k is the number of observations in the window to the left and right of the target (middle) value. The span can take values from 0 to 2. For a very small value close to zero, the window will contain only one observation, while at a value of 2 it will contain all data points. The span for local unweighted polynomial regression is the smoothing parameter.

Definition: A *smoothing parameter* determines how smooth the fitted curve is and is effectively used for striking a balance between bias and variance in the estimation of the curve. We denote the smoothing parameter as λ, in general.

How to choose λ is one of the most important topics in the literature of smoothing techniques.

In any local regression scatterplot smoother, there are three main decisions that need to be made: (i) the choice of the smoothing parameter, (ii) the degree of polynomial and (iii) how the response values are averaged. The second is dealt with by fitting different degree polynomial functions in x to the data; and the third by deciding whether to use unweighted or weighted polynomial regression.

For weighted local regression, the specification of a *kernel* function and its smoothing parameter λ is required. Kernels are positive symmetric functions, usually looking similar to the normal distribution, having as smoothing parameter a scaling parameter which makes the shape of the function narrower or wider. For example, if the normal distribution is chosen as kernel, then the standard deviation σ is used as the smoothing parameter, i.e. $\lambda = \sigma$. It is well known in the smoothing literature that to determine the smoothness of the fitted curve, it is the smoothing parameter λ rather than the choice of kernel that matters more.

The following describes how local regression smoothing with a symmetric window works.

- Start by ordering the pair of values (y_i, x_i) for $i = 1, 2, \ldots, n$ with respect to x.

- Use the smoothing parameter to select the size of the window or the scale of the kernel function.

- Focus on a single observation (with the target x value) and fit a polynomial regression model only to observations falling into the current window (for unweighted local regression) or weight the observations according to the kernel function (for weighted local regression).

- Use the fitted values $\hat{y}$ for the x value of the target observation (x, y) as the fitted value for the smoother (at the target x value).

- Repeat this for all observations.

Figure 9.4 demonstrates some aspects of this process. Each plot shows:

- The current target observation (x, y) in bold and with a pointed arrow,

- The fitted polynomial. For example, Figure 9.4(a) shows a constant fit (or moving average), 9.4(b) a linear fit, 9.4(c) a quadratic fit and finally 9.4(d) a cubic fit.

The plots in Figure 9.4 (a) and (b) use an unweighted fit and therefore show the chosen windows as shaded areas. Note that the symmetry of the window (that is, containing an equal number k of observations on the left and on the right of the target value) breaks down at the two extreme ends of the range of x. For example, if there are fewer than k observations on the left of the target value, the window will contain fewer observations on the left of the target value than on the right, as demonstrated in Figure 9.4(a). The opposite behaviour will happen on the right part of the data. Note also that a span value close to zero will interpolate the data. A value of span equal to 2 will fit a global polynomial to all the data, while a value equal to 1 will fit a global polynomial for the middle part of the ordered data but not for the rest of the target points. Usually span = 0.5 is a good starting point and this is the value used in Figures 9.4 (a) and (b). The plots in Figures 9.4 (c) and (d) use a weighted fit using a normal kernel with smoothing parameter $\sigma = 0.25$. The

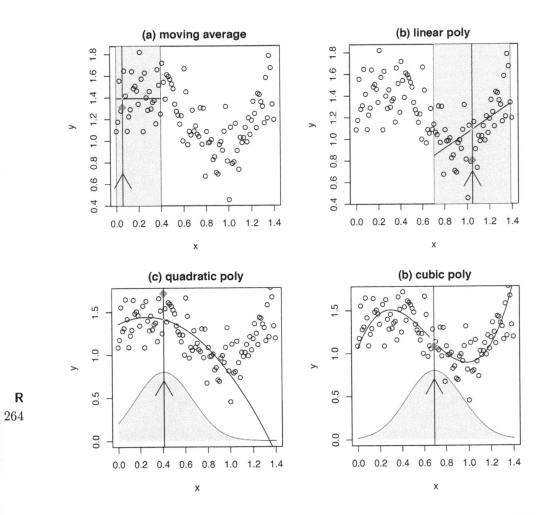

FIGURE 9.4: Different aspects of fitted local polynomial regression: Plots (a) and (b) show unweighted local regression fits (at the target values indicated by an arrow) with span = 0.5 while plots (c) and (d) show a weighted fit using a normal kernel with smoothing parameter $\sigma = 0.25$. Plot (a) uses a constant fit (i.e. a moving average), plot (b) uses a local linear fit, plot (c) a local quadratic fit and plot (d) a local cubic fit.

shaded area in the plot shows how much weight an observation has when fitting the local polynomial to determine the fitted value of y at the target observation. Observations far from the target x value have negligible effect since their weights are close to zero. The **R** code to produce Figure 9.4 is given below.

```
library(gamlss.demo)
n <- 100
x <- seq(0, 1, length = n)*1.4
set.seed(123)
y <- 1.2 + .3*sin(5*x) + rnorm(n) * 0.2
op <- par(mfrow=c(2,2))
LPOL(y,x, deg=0, position=5)
title("(a) moving average")
LPOL(y,x, deg=1,  position=75)
title("(b) linear poly")
WLPOL(y,x, deg=2, position=30)
title("(c) quadratic poly")
WLPOL(y,x, deg=3, position= 50)
title("(b) cubic poly")
par(op)
```

Fig. 9.4

There are several demos within the package **gamlss.demo** which show how different local regression models work. Readers with less experience in smoothing techniques are encouraged to use them in order to understand the basic ideas behind local regression smoothing. Table 9.2 gives the names of these functions and shows their functionality. The demos can be obtained by typing the name of the **R** function, e.g. `demo.Locpoly()`, or can be accessed using the function **gamlss.demo()** and following the menu for 'Demos for local polynomial smoothing'.

TABLE 9.2: Different ways of using local regression smoothers

	unweighted	weighted
mean	moving average	kernel smoothers
demos	`demo.Locmean()`	`demo.WLocmean()`
linear	simple regression	weighted linear regression
demos	`demo.LocalRegression()`	`demo.WLocpoly()`
polynomial	polynomial regression	weighted polynomial regression
demos	`demo.Locpoly()`	`demo.WLocpoly()`

The following are general comments on local polynomial smoothers:

1. Weighted local polynomial smoothers using a kernel function as weights produce much smoother fits than unweighted local regression using a window. The latter produce rather wiggly functions.

2. The running-mean smoothers (that is, moving average and kernel smoothers shown in the top row of Table 9.2) tend to flatten out the trends at the endpoints

of the x variable and thus the fitted values produced are biased at the endpoints. They can also be biased when the values of the x variable are unevenly spread out.

3. Every univariate smoother has a smoothing parameter which controls the amount of smoothing done to the data, i.e. the span or scale λ.

4. All the local polynomial regression smoothers discussed up to now are 'linear' in the sense that we can write the vector of fitted values as $\hat{\mathbf{y}} = \mathbf{S}\mathbf{y}$.

5. Smoothers are affected differently in sparse regions of the x-values. For example in the local weighted window, the values of the weights are affected by the sparseness of the x-values. Also the kernel smoother can misbehave at the end of the range of x-values where data are sparse, and more generally when the x-values are unevenly spaced.

6. The smoother takes its values locally since only local observations are used in calculating the fit. As a consequence, the smoother at non-extreme x-values is generally robust to extreme x-values since those values will only contribute locally to the fit. This is contrary to, for example, polynomial regression fits where extreme x-values can have a great influence on the whole fitted curve.

7. Outlier observations in y affect the smoother in the locality of their x values. That is why, for example, one of the most successful algorithms for weighted local regression, the loess algorithm, also provides a robust version. The implementation of the loess function within GAMLSS is discussed in Section 9.6.3.

9.4 Penalized smoothers: Univariate

Penalized smoothers are the most important smoothers within the GAMLSS family of smoothers because of their flexibility and the fact that they can be applied in a variety of different situations. All of the penalized smoothers considered in this section can be thought of as the solution to the following least squares minimization problem, where a quadratic penalty applies to the parameters.

Let $\mathbf{Z}$ be a $n \times p$ basis matrix (bases are defined in Chapter 8), $\boldsymbol{\gamma}$ a $p \times 1$ vector of parameters, $\mathbf{W}$ a $n \times n$ diagonal matrix of weights, $\mathbf{G}$ a $p \times p$ penalty matrix, λ a smoothing parameter and $\mathbf{y}$ the variable of interest. Penalized smoothers are the solution to minimizing the quantity Q with respect to $\boldsymbol{\gamma}$:

$$Q = (\mathbf{y} - \mathbf{Z}\boldsymbol{\gamma})^\top \mathbf{W}(\mathbf{y} - \mathbf{Z}\boldsymbol{\gamma}) + \lambda \boldsymbol{\gamma}^\top \mathbf{G}\boldsymbol{\gamma} . \tag{9.2}$$

The solution to equation (9.2) is given by:

$$\hat{\boldsymbol{\gamma}} = \left(\mathbf{Z}^\top \mathbf{W}\mathbf{Z} + \lambda \mathbf{G}\right)^{-1} \mathbf{Z}^\top \mathbf{W}\mathbf{y} . \tag{9.3}$$

It is important to emphasise here the fact that by slightly modifying the different bases $\mathbf{Z}$, and the different penalty matrices $\mathbf{G}$, different types of smoothers can be produced having certain desired properties. (In GAMLSS different weights $\mathbf{W}$ and different partial residuals replacing y are used iteratively within the backfitting algorithm to fit different distributions for the response variable. For details see; Sections 3.2.1.3 and 3.2.2.3 and also Rigby and Stasinopoulos [2005, Appendix C].)

Fitted values: The fitted values for y are given by:

$$\hat{\mathbf{y}} = \mathbf{Z}\left(\mathbf{Z}^\top \mathbf{W}\mathbf{Z} + \lambda \mathbf{G}\right)^{-1}\mathbf{Z}^\top \mathbf{W}\mathbf{y}$$
$$= \mathbf{S}\mathbf{y} \tag{9.4}$$

where $\mathbf{S}$ is the *smoothing matrix*. A smoothing matrix plays a similar role to the hat matrix $\mathbf{H}$ in least squares estimation.

Effective degrees of freedom: Another quantity of interest using penalized smoothers is the trace of $\mathbf{S}$ since it is used as the *effective degrees of freedom* of the smoother:

$$\text{tr}(\mathbf{S}) = \text{tr}\left[\mathbf{Z}\left(\mathbf{Z}^\top \mathbf{W}\mathbf{Z} + \lambda \mathbf{G}\right)^{-1}\mathbf{Z}^\top \mathbf{W}\right]$$
$$= \text{tr}\left[\left(\mathbf{Z}^\top \mathbf{W}\mathbf{Z} + \lambda \mathbf{G}\right)^{-1}\mathbf{Z}^\top \mathbf{W}\mathbf{Z}\right]. \tag{9.5}$$

The penalty matrix: The form of the penalty matrix $\mathbf{G}$ is also of great interest. Very often the penalty matrix is defined as $\mathbf{G} = \mathbf{D}_k^\top \mathbf{D}_k$, where the matrix $\mathbf{D}_k$ is a $(p-k) \times p$ difference matrix of order k. For example $\mathbf{D}_1$ and $\mathbf{D}_2$ matrices of order 1 and 2, respectively, look like:

$$\mathbf{D}_1 = \begin{bmatrix} -1 & 1 & 0 & \cdots & 0 \\ 0 & -1 & 1 & \cdots & 0 \\ \cdots & \cdots & \cdots & \cdots & \cdots \\ 0 & 0 & \cdots & -1 & 1 \end{bmatrix} \quad \text{and} \quad \mathbf{D}_2 = \begin{bmatrix} 1 & -2 & 1 & 0 & \cdots & 0 \\ 0 & 1 & -2 & 1 & \cdots & 0 \\ \cdots & \cdots & \cdots & \cdots & \cdots & \cdots \\ 0 & 0 & \cdots & 1 & -2 & 1 \end{bmatrix}.$$

$\mathbf{D}_1$ and $\mathbf{D}_2$, and the corresponding penalty matrices $\mathbf{G}_1$ and $\mathbf{G}_2$, can be generated easily in R using the `diff()` function:

```
D1 <- diff(diag(10), diff=1)
D2 <- diff(diag(10), diff=2)
G1 <- t(D1)%*%D1
G2 <- t(D2)%*%D2
```

Penalties are a powerful tool for imposing different properties into fitted curves. By introducing more than one penalty or by modifying an existing one, curves with desired properties can be produced. For example when the smoothing parameter $\lambda \to \infty$, then for polynomial degree ≥ 1 the usual penalty matrix $\mathbf{D}_2$ will shrink the fitted curve to a line. In this case the resulting curve is identical to a least squares fit. By adding a second penalty of order $k=1$, the shrinkage can be changed from a line to a constant; see Section 9.4.3.

The order k of the penalty is also important when fitted values are required inside or outside the actual values of the x-variable, that is, when we need to *interpolate* or *extrapolate*, respectively. The gamlss demo function `demo.interpolateSmo()`, described in Section 9.4.1, shows the behaviour the penalty imposes on the fitted values in the situation when we interpolate or extrapolate the data. It uses the simple smoother, known as the Whittaker smoother, where the x-variable is equally spaced. Similar behaviour is observed in more general situations.

Random effects and penalized smoothers: The connection of penalized smoothers to random effects is worth noting here. Assume that

$$\begin{aligned}
\mathbf{y} &= \mathbf{Z}\boldsymbol{\gamma} + \mathbf{e} \\
\mathbf{e} &\sim \mathcal{N}(\mathbf{0}, \sigma_e^2 \mathbf{W}^{-1}) \\
\boldsymbol{\gamma} &\sim \mathcal{N}(\mathbf{0}, \sigma_b^2 \mathbf{G}^{-1})
\end{aligned} \tag{9.6}$$

where $\mathbf{G}^{-1}$ is a (generalized) inverse of matrix $\mathbf{G}$, which is similar to equation (3.21). Model (9.6) assumes that $\mathbf{y}$ has two sources of variation, the random effects $\boldsymbol{\gamma}$ and the error term $\mathbf{e}$. The matrices $\mathbf{W}$ and $\mathbf{G}$ are assumed to be known, but the parameters $\boldsymbol{\gamma}$, σ_e^2 and σ_b^2 have to be estimated from the data. Since both random variables are normally distributed, the extended or hierarchical likelihood (see Pawitan [2001] and Lee et al. [2006]), which is proportional to the posterior density for $\boldsymbol{\gamma}$ (given σ_e^2 and σ_b^2), is given by:

$$\begin{aligned}
L_h &= f(\mathbf{y} \,|\, \boldsymbol{\gamma}, \sigma_e^2) f(\boldsymbol{\gamma} \,|\, \sigma_b^2) \\
&= \left|2\pi\sigma_e^2 \mathbf{W}\right|^{-1/2} \exp\left\{\frac{-1}{2\sigma_e^2}\mathbf{e}^\top \mathbf{W}\mathbf{e}\right\} \left|2\pi\sigma_b^2 \mathbf{G}\right|^{-1/2} \exp\left\{\frac{-1}{2\sigma_b^2}\boldsymbol{\gamma}^\top \mathbf{G}\boldsymbol{\gamma}\right\}
\end{aligned} \tag{9.7}$$

where $|\mathbf{A}|$ denotes the determinant of the matrix $\mathbf{A}$. Taking logs we have

$$\begin{aligned}
\ell_h &= \log f(\mathbf{y} \,|\, \boldsymbol{\gamma}, \sigma_e^2) + \log f(\boldsymbol{\gamma} \,|\, \sigma_b^2) \\
&= -\frac{1}{2}\log\left|2\pi\sigma_e^2\mathbf{W}\right| - \frac{1}{2\sigma_e^2}\mathbf{e}^\top\mathbf{W}\mathbf{e} - \frac{1}{2}\log\left|2\pi\sigma_b^2\mathbf{G}\right| - \frac{1}{2\sigma_b^2}\boldsymbol{\gamma}^\top\mathbf{G}\boldsymbol{\gamma} \\
&= -\frac{1}{2\sigma_e^2}\mathbf{e}^\top\mathbf{W}\mathbf{e} - \frac{1}{2\sigma_b^2}\boldsymbol{\gamma}^\top\mathbf{G}\boldsymbol{\gamma} + \text{terms not depending on } \boldsymbol{\gamma} \\
&= -\frac{1}{2\sigma_e^2}(\mathbf{y} - \mathbf{Z}\boldsymbol{\gamma})^\top\mathbf{W}(\mathbf{y} - \mathbf{Z}\boldsymbol{\gamma}) - \frac{1}{2\sigma_b^2}\boldsymbol{\gamma}^\top\mathbf{G}\boldsymbol{\gamma} + \ldots
\end{aligned} \tag{9.8}$$

Multiplying equation (9.8) by $-2\sigma_e^2$ and letting $\lambda = \sigma_e^2/\sigma_b^2$ yields equation (9.2). There are several points to be made here:

- The penalized least squares estimator of $\boldsymbol{\gamma}$ in equation (9.3) obtained by minimizing (9.2) is the same as the posterior mode or MAP estimator if we had assumed the random effect model in (9.6).

- The MAP estimators are also known in the statistical literature as BLUP (Best Linear Unbiased Predictors); see Robinson [1991] for more details.

- The smoothing parameter is the ratio of two variances, the error variance divided by the random effect variance. This is the inverse of the signal-to-noise ratio (SNR), a measure used in science and engineering that compares the level of a desired signal to the level of background noise. Signal-to-noise ratios are also used in time series analysis.

- The classical random effect methodology provides two ways of estimating the two variances σ_e^2 and σ_b^2, the maximum likelihood (ML) and the restricted maximum likelihood (REML) estimators; see Pinheiro and Bates [2000] for definitions. Those methods can be also used to provide estimates for σ_e^2 and σ_b^2 and therefore λ.

The γ's as random walks: Note that

$$\boldsymbol{\gamma} \sim \mathcal{N}(\mathbf{0}, \sigma_b^2 \mathbf{G}^{-1})$$

implies that

$$\mathbf{D}\boldsymbol{\gamma} \sim \mathcal{N}(\mathbf{0}, \sigma_b^2 \mathbf{I})$$

since $\mathbf{G} = \mathbf{D}^\top \mathbf{D}$ and

$$\mathrm{Var}(\mathbf{D}\boldsymbol{\gamma}) = \mathbf{D}\,\mathrm{Var}(\boldsymbol{\gamma})\mathbf{D}^\top = \sigma_b^2 \mathbf{D} \left(\mathbf{D}^\top \mathbf{D}\right)^{-1} \mathbf{D}^\top = \sigma_b^2 \mathbf{I}\,.$$

$\mathbf{D}$ is a difference matrix depending on the order k, i.e. $\mathbf{D}_k$ with dimension $(p-k) \times p$. Therefore an important feature of the order k is that it introduces different types of stochastic dependency on the coefficients $\boldsymbol{\gamma}$. For example, for $k = 0$ we have $\mathbf{D}_0 = \mathbf{I}$ so $\boldsymbol{\gamma}$ is a simple random effect. For $k = 1$ we have the penalty matrix $\mathbf{D}_1$ in (9.6) so

$$\gamma_j - \gamma_{j-1} \sim \mathcal{N}(0, \sigma_b^2)$$

which is a random walk of order 1. For $k = 2$ we have penalty matrix $\mathbf{D}_2$ in (9.6) and therefore

$$\gamma_j - 2\gamma_{j-1} - \gamma_{j-2} \sim \mathcal{N}(0, \sigma_b^2)$$

which is a random walk of order 2 and so on.

Fitting penalized smoothers using augmented least squares: For readers familiar with simple least squares estimation, it is worth pointing out that penalized least squares can be solved by expanding the original data and then using standard least squares software to do the analysis. This can be demonstrated as follows.

Let $\tilde{\mathbf{y}} = \begin{pmatrix} \mathbf{y} \\ \mathbf{0} \end{pmatrix}$, $\tilde{\mathbf{X}} = \begin{pmatrix} \mathbf{Z} \\ \sqrt{\lambda}\mathbf{D} \end{pmatrix}$ and $\tilde{\mathbf{W}} = \begin{pmatrix} \mathbf{W} & \mathbf{0} \\ \mathbf{0} & \mathbf{I}_p \end{pmatrix}$, where $\mathbf{0}$ in $\tilde{\mathbf{y}}$ is of length p. Then it is easy to show that minimizing the quantity

$$Q_1 = \left(\tilde{\mathbf{y}} - \tilde{\mathbf{X}}\boldsymbol{\gamma}\right)^\top \tilde{\mathbf{W}} \left(\tilde{\mathbf{y}} - \tilde{\mathbf{X}}\boldsymbol{\gamma}\right)$$

leads to the same solution as in (9.3),

$$\hat{\boldsymbol{\gamma}} = \left(\tilde{\mathbf{X}}^\top \tilde{\mathbf{W}} \tilde{\mathbf{X}}\right)^{-1} \tilde{\mathbf{X}}^\top \tilde{\mathbf{W}} \tilde{\mathbf{y}}$$

$$= \left(\mathbf{Z}^\top \mathbf{W} \mathbf{Z} + \lambda \mathbf{G}\right)^{-1} \mathbf{Z}^\top \mathbf{W} \mathbf{y}$$

since

$$\tilde{\mathbf{X}}^\top \tilde{\mathbf{W}} \tilde{\mathbf{X}} = (\mathbf{Z} \quad \sqrt{\lambda}\mathbf{D}) \begin{pmatrix} \mathbf{W} & \mathbf{0} \\ \mathbf{0} & \mathbf{I}_p \end{pmatrix} \begin{pmatrix} \mathbf{Z} \\ \sqrt{\lambda}\mathbf{D} \end{pmatrix} = \mathbf{Z}^\top \mathbf{W} \mathbf{Z} + \lambda \mathbf{D}^\top \mathbf{D}$$

and

$$\tilde{\mathbf{X}}^\top \tilde{\mathbf{W}} \tilde{\mathbf{y}} = (\mathbf{Z} \quad \sqrt{\lambda}\mathbf{D}) \begin{pmatrix} \mathbf{W} & \mathbf{0} \\ \mathbf{0} & \mathbf{I}_p \end{pmatrix} \begin{pmatrix} \mathbf{y} \\ \mathbf{0} \end{pmatrix} = \mathbf{Z} \mathbf{W} \mathbf{y},$$

and $\mathbf{G} = \mathbf{D}^\top \mathbf{D}$.

Estimating the smoothing parameter: There are several ways of estimating the smoothing parameter within GAMLSS and they are described in Section 3.4, with GAMLSS functions given in Section 11.2. The methods used here are *local* and they are:

- minimize local generalized cross validation (GCV) [Wood, 2006a],

- minimize local GAIC,

- maximum local likelihood method (as earlier in this section).

See Section 3.4.2 for more details on these local methods and Section 9.4.2 for their implementation using the argument `method` of the smoothing function `pb()`.

9.4.1 Demos on penalized smoothers

There are several demos within the **gamlss.demo** package for helping to understand how penalized smoothers work. These are based on **R** functions provided by Professor Paul Eilers. They are accessed by typing `gamlss.demo()` and then clicking on the menu item 'Demos for smoothing techniques', or by typing the names of the functions below:

`demo.BSplines()`: This function is designed for exploring the B-splines basis ideas. The user can control the number of knots for the basis and also the degree of the B-spline. The demo also shows, by clicking the button 'random', different shapes of curves that can be generated from a linear combination of such a B-spline basis.

`demo.RandomWalk()`: This function demonstrates the most basic penalized smoother, the random walk. Random walks are appropriate for time series data when the observations are equally spaced in time and there is no explicit explanatory variable, i.e.

$$\mathbf{y} = \boldsymbol{\mu} + \mathbf{e} , \quad \mathbf{e} \sim \mathcal{N}(\mathbf{0}, \sigma_e^2 \mathbf{I}) , \quad \mathbf{D}\boldsymbol{\mu} \sim \mathcal{N}(\mathbf{0}, \sigma_b^2 \mathbf{I}) .$$

It can be seen as the solution of the problem of minimizing the quantity Q with respect to $\boldsymbol{\mu}$ where $Q = (\mathbf{y} - \boldsymbol{\mu})^\top (\mathbf{y} - \boldsymbol{\mu}) + \lambda \boldsymbol{\mu} \mathbf{D}^\top \mathbf{D} \boldsymbol{\mu}$ and $\lambda = \sigma_e^2 / \sigma_b^2$. The solution is $\hat{\boldsymbol{\mu}} = (I + \lambda \mathbf{D}^\top \mathbf{D})^{-1} \mathbf{y}$ where $\mathbf{D}$ is usually a difference matrix of order 1. These smoothers are also called the Whittaker smoothers [Whittaker, 1922, Eilers, 2003].

`demo.interpolateSmo()`: This function explores how the fitted values of a random walk behave in the case of *interpolation* and *extrapolation*, that is when data are missing or when we are trying to predict outside the current range of x, respectively. The user will find that the interpolation is done by a polynomial of degree $2k - 1$, while extrapolation is done by a polynomial of degree $k - 1$, where k is the order of the penalty matrix $\mathbf{D}_k$. Both are done by introducing extra data (at the missing or extrapolation time points) with weights zero.

`demo.histSmo()`: This function shows the power of penalties when we are trying to smooth a histogram. It is using an old trick within the GLM literature of treating the counts within the histogram bin as Poisson distributed observations [Eilers and Marx, 2010].

`demo.PSplines()`: This demo shows the effect on the fitted P-spline curve of changing (i) the number of knots in the B-spline basis, (ii) the degree D of the polynomial used, (iii) the order k of the penalty matrix $\mathbf{D}_k$, and (iv) most importantly the smoothing parameter λ.

Next we will consider the univariate penalized smoothers implemented within GAMLSS.

9.4.2   The `pb()`, `pbo()` and `ps()` functions for fitting a P-splines smoother

`pb()` stands for penalized B-splines and it is a GAMLSS implementation of the Eilers and Marx [1996] P-splines methodology. The functions `pb()` and its earlier implementation `pbo()` give identical results but `pb()` is faster than `pbo()`. P-splines use $\mathbf{Z} = \mathbf{B}$ in equation (9.2), where $\mathbf{B}$ is a B-spline basis [de Boor, 2001] of a piecewise polynomial of degree d with *equally spaced* knots over the x range. We emphasise the 'equally spaced' knots since, according to Eilers et al. [2015], this is a very important aspect of the P-splines. The coefficients γ are penalized using the penalty matrix $\mathbf{G} = \mathbf{D}_k^\top \mathbf{D}_k$ of appropriate order k.

The function `pb()` has only three main arguments, while the rest of the arguments can be passed through the `control` option:

x	the single explanatory variable;
df	the desired effective degrees of freedom (trace of the smoother matrix minus two for the constant and linear part). This does not need to be an integer but must be positive. There is a check after fitting that the desired and fitted effective degrees of freedom are the same, and a warning is given if there is a discrepancy (with advice to try reducing the value of `inter` defined below);
lambda	the smoothing parameter;
control	the function `pb.control()` which sets the smoothing parameters.

If both `df` and `lambda` are set to NULL, then the smoothing parameter is estimated using one of the methods described below. If `df` is set, then the smoother will have fixed degrees of freedom. If `lambda` is set, then its value is used for smoothing. If both `df` and `lambda` are set, `lambda` takes priority.

The `pb.control()` function has the following options which can be specified within `pb()` or set prior to using `pb()`:

`inter`	the number of equally spaced intervals between the minimum and maximum values of x used for the creation of a B-splines basis **B**. The default value is `inter=20`. The resulting number of basis columns is `inter+degree`;
`degree`	the degree of the piecewise polynomial used for the basis **B**. The default is `degree=3`;
`order`	the required difference k in the difference matrix $\mathbf{D}_k$ with default `order=2`;
`start`	the starting value for the smoothing parameter `lambda` if it is estimated;
`quantiles`	if `quantiles=TRUE` the quantile values in x are used to determine the knots, rather than equally spaced knots. The default is `quantiles=FALSE`;
`method`	The method used in the (local) estimation of the smoothing parameter. Available methods are `"ML"`, `"GAIC"` and `"GCV"`. The older version `pbo()` has in addition the methods `"ML-1"` and `"EM"`. The `"ML"` method is described in Section 3.4.2.1 and Rigby and Stasinopoulos [2013]; `"ML-1"` is an experimental method identical to `"ML"` with the exception that the σ_e parameter is set to 1. This seems to make the algorithm unstable, so it is not recommended. `"EM"` is based on the method described by Fahrmeir and Wagenpfeil [1997], which should give identical results to `"ML"` but is generally slower. `"GAIC"` and `"GCV"` use the local generalized Akaike information criterion and the local generalized cross validation, respectively; see Sections 3.4.2.2 and 3.4.2.3, respectively.

The `ps()` function is an earlier version of `pb()` and there is no option for estimation of the smoothing parameters. It is based on an original **Splus** function of Professor Brian Marx. `ps()` uses as default fit a smooth function in x using three extra degrees of freedom, i.e. five degrees of freedom overall, three for smoothing, one for the linear part and one for the constant.

The Dutch boys data

For illustration we will use data from the Fourth Dutch Growth Study [Fredriks et al., 2000a,b], which was a cross-sectional study that measured growth and development of the Dutch population between the ages 0 and 21 years. The data were kindly provided by Professor Stef van Buuren. The study measured, amongst other variables, height, weight, head circumference and age. Here we use the BMI (y) and age (x) of the boys.

> **R data file:** `dbbmi` in package **gamlss.data** of dimensions 7294×2
>
> **var age** : age (years)
>
>   **bmi** : Body Mass Index (BMI)
>
> **purpose:** to demonstrate P-splines estimation

The data are plotted in Figure 9.5.

```
data(dbbmi)
plot(bmi~age, data=dbbmi, pch = 15, col = gray(0.5))
```

Fig.
9.5

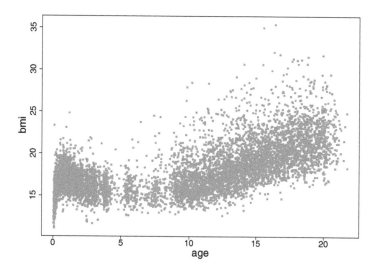

R
p. 272

FIGURE 9.5: BMI against age, Dutch boys data.

The following code demonstrates that different methods could lead to slightly different fitted curves. In general for large data and for smoother fitted curves, use a local SBC as a selection procedure for the smoothing parameter, by selecting `method="GAIC"` with `k=log(n)`, as in model `p4` below. Figure 9.6 plots the P-spline smoothers for four different methods of estimating the smoothing parameter.

```
p1<-gamlss(bmi~pb(age, method="ML"), data=dbbmi,trace=FALSE)
p2<-gamlss(bmi~pb(age, method="GCV") , data=dbbmi,trace=FALSE)
p3<-gamlss(bmi~pb(age, method="GAIC", k=2) , data=dbbmi,trace=FALSE)
p4<-gamlss(bmi~pb(age, method="GAIC", k=log(length(dbbmi$bmi))),
          data=dbbmi,trace=FALSE)
```

Fig.
9.6

```
plot(bmi~age, data=dbbmi, cex=.2, col="gray",main="ML")
lines(fitted(p1)~dbbmi$age, lwd=2)
plot(bmi~age, data=dbbmi, cex=.2, col="gray",main="GCV")
lines(fitted(p2)~dbbmi$age, lwd=2 )
plot(bmi~age, data=dbbmi, cex=.2, col="gray",main="AIC")
lines(fitted(p3)~dbbmi$age, lwd=2)
plot(bmi~age, data=dbbmi, cex=.2, col="gray",main="SBC")
lines(fitted(p4)~dbbmi$age, lwd=2)
```

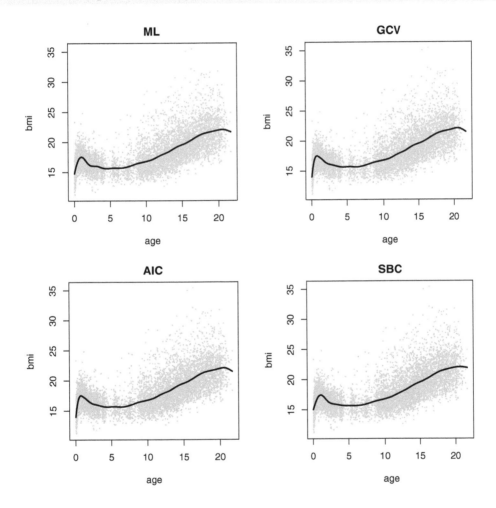

FIGURE 9.6: Fitted curves using different methods of estimating the smoothing parameters in `pb()`.

Important: For large data, use the local SBC method for a smoother fitted curve.

The fitted smoother is saved as an object of class `pb`. The methods `plot()`, `coef()`, `fitted()` and `print()` are applied to the class `pb`. The object can be retrieved from a `gamlss` fitted object using the function `getSmo()`.

9.4.3 The `pbz()` function for fitting smooth curves which can shrink to a constant

The selection of explanatory variables is important in any statistical modelling situation. Chapter 11 provides a general framework and ideas on how this can be achieved. The function `pbz()` provides one of the solutions to the variable selection problem using penalized least squares methodology. Let us first explain the problem.

The P-splines function `pb()` uses a default penalty $\mathbf{G}_2 = \mathbf{D}_2^\top \mathbf{D}_2$. If the estimation of λ leads to $\lambda \to \infty$ then for polynomial degree ≥ 1 the fitted curve is linear. Fitting a linear curve requires two degrees of freedom, one for the constant and one for the linear part. If the linear relationship is supported by the data, then `pb()` usually leads us to the right representation of the data and there is no problem. If, on the other hand, the linear part is not required (that is, there is no linear relationship between x and the current y), then we have unnecessarily used an extra degree of freedom for the linear part.

To demonstrate the problem we use the **abdom** data, introduced in Chapter 4, which has two variables `x` and `y`. Consider what will happen if we add an extra variable `x1`, which is not related to the response. The model `m0` below fits a model with the correct explanatory variable `x` while models `m1` and `m2` add the bogus `x1` variable using `pb()` and `pbz()`, respectively.

```
data(abdom)
# add a nuisance variable
abdom$x1 <- rNO(610, mu=5, sigma=5)
# fitting the original x
m0 <- gamlss(y~pb(x), data=abdom, trace=FALSE)
# fitting extra x1 with pb()
m1 <- gamlss(y~pb(x)+pb(x1), data=abdom, trace=FALSE)
# fitting extra x1 with pbz()
m2 <- gamlss(y~pbz(x)+pbz(x1), data=abdom, trace=FALSE)
# smaller deviance but liitle reduction in deviance
AIC(m0,m1,m2, k=0)

##           df      AIC
## m1 7.511102 4935.509
## m2 6.508276 4935.853
## m0 6.508274 4935.853
```

Models `m0` and `m2` give identical results as far as the degrees of freedom and deviance are concerned, while model `m1`, which uses `pb()`, gives an extra degree of freedom (for the linear part in `x1` of the fit even though it is not needed) and a slightly different deviance. The fitted values for m1 and m2 are almost identical as Figure 9.7 shows. In conclusion, if a continuous explanatory variable is not needed at all as a term in the model, pbz() provides a way of eliminating it. The way that the function pbz() works is by using a single penalty $\lambda \mathbf{D}_2^\top \mathbf{D}_2$ if the effective degrees of freedom for the smoother are greater than 3, and switching to a double penalty $\lambda_1 \mathbf{D}_2^\top \mathbf{D}_2 + \lambda_2 \mathbf{D}_1^\top \mathbf{D}_1$ if the effective degrees of freedom fall below 3.

Fig.
9.7

```
term.plot(m1)
term.plot(m2)
```

R
275

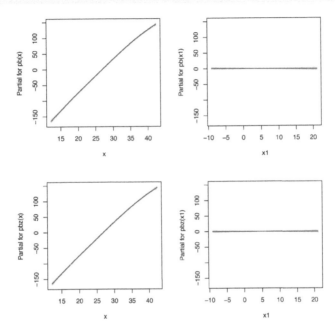

FIGURE 9.7: Fitted terms using the functions pb() (top) and pbz() (bottom), respectively. The fitted functions look identical, but model m2 using pbz() has degrees of freedom reduced by 1.

> **Important:** The function pbz() behaves identically to the function pb(), but when the effective degrees of freedom are small allows the fitted term to shrink to a constant.

The fitted smoothers are saved as objects of class pbz. The methods that apply to a pb class also apply to pbz. Note that the practical effects of using pbz, and in particular the automatic selection of smoothing parameter(s), are under investigation.

9.4.4 The pbm() function for fitting monotonic smooth functions

The function pbm() can be used to fit monotone curves to the data. It is a modified P-splines fit, so like pb() it uses $\mathbf{Z} = \mathbf{B}$ in equation (9.2), where $\mathbf{B}$ is a B-spline basis of a piecewise polynomial of degree d with equally spaced knots over the x range. The modification is in the penalty part of the fitting. The coefficients γ are penalized using two penalty matrices: (i) one for the smoothness of γ, $\mathbf{G} = \mathbf{D}_k^\top \mathbf{D}_k$ and (ii) one which penalizes the γ's if the monotonic property of the fitted function is violated. The latter penalty has the form $\mathbf{P}_k^\top \mathbf{W}_P \mathbf{P}_k$ where the matrix $\mathbf{P}_k$ is a difference matrix defined similarly to the penalty matrix $\mathbf{D}_k$. $\mathbf{W}_P$ is a diagonal matrix of weights which take the value 1 if the monotonic property of the function is violated and 0 otherwise. Note that the resulting monotonic function is achieved after several iterations of the penalized least squares algorithm.

The arguments of the function are almost identical to pb() apart from the argument mono, which can take the values "up" (the default) or "down".

Fig.
9.8

```
set.seed(1334)
x = seq(0, 1, length = 1000)
p = 0.4
y = sin(2 * pi * p * x) + rnorm(1000) * 0.1
plot(y~x, cex=.2, col="gray")
m1 <- gamlss(y~pbm(x), trace=FALSE)
lines(fitted(m1)~x, col="red")
yy <- -y
plot(yy~x, cex=.2, col="gray")
m2 <- gamlss(yy~pbm(x, mono="down"), trace=FALSE)
lines(fitted(m2)~x, col="red")
```

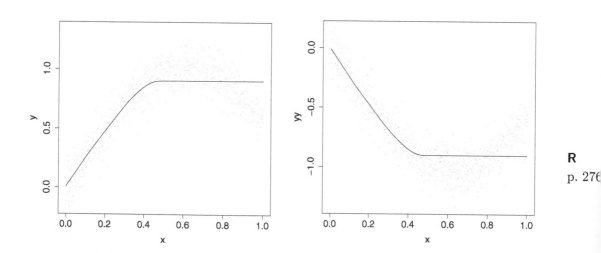

R
p. 276

FIGURE 9.8: Monotone fitted curves using the function pbm().

The fitted smoothers are saved as objects of class pbm. The methods that apply to a pb class also apply to pbm.

9.4.5 The pbc() and cy() functions for fitting cyclic smooth functions

The function pbc() produces smooth fitted curves with the property that the two ends (left and right) of the fitted smooth function have identical values. This is achieved by the following features added to the standard P-splines:

1. A circular penalty which connects the first and last elements of the coefficient vector using differences. For example for orders 1 and 2 (the only orders supported by pbc()) the penalty matrices are of the form:

$$\mathbf{D}_1 = \begin{bmatrix} -1 & 0 & 0 & \cdots & 1 \\ 1 & -1 & 0 & \cdots & 0 \\ 0 & 1 & -1 & \cdots & 0 \\ \cdots & \cdots & \cdots & \cdots & \cdots \\ 0 & 0 & \cdots & 1 & -1 \end{bmatrix} \quad \text{and} \quad \mathbf{D}_2 = \begin{bmatrix} 2 & -1 & 0 & 0 & \cdots & -1 \\ -1 & 2 & -1 & 0 & \cdots & 0 \\ 0 & -1 & 2 & -1 & \cdots & 0 \\ \cdots & \cdots & \cdots & \cdots & \cdots & \cdots \\ -1 & 0 & \cdots & 0 & -1 & 2 \end{bmatrix}.$$

2. A circular B-basis.

3. A different second-order penalty leading to a different fitted function when the smoothing parameter goes to infinity. For example, penalty of order 2, $\gamma_j - 2\gamma_{j-1} + \gamma_{j-2}$, leads to a straight line when $\lambda \to \infty$. The penalty can be modified slightly to $\gamma_j - 2\phi\gamma_{j-1} + \gamma_{j-2}$ where $\phi = \cos(2\pi/p)$. This penalty will lead, when $\lambda \to \infty$, to a (co)sine fitted function with period p. The option for this to happen is sin=TRUE.

The cyclic behaviour of the fitted function pbc() is ideal for fitting periodic functions for which the start of a new period is not expected to vary considerably from the end of the last one.

Next, we give two examples, one using simulated data to show an example of a continuous explanatory variable, while the second, a time series example, where we use pbc() to model a seasonal factor. For factors, the behaviour of pbc() is different in that the knots of the basis are chosen as the factor levels and the order of penalty switches to 1.

Fig. 9.9

```
set.seed(555)
x = seq(0, 1, length = 1000)
y<-cos(1 * x * 2 * pi + pi / 2)+rnorm(length(x)) * 0.2
plot(y~x, cex=.2, col="gray")
m1<-gamlss(y~pbc(x), sin=TRUE, trace=FALSE)
lines(fitted(m1)~x, col=2)
```

The second example uses the **R** time series data set accdeaths, which describes monthly accidental deaths in the US during the period 1973–1978. First we extract the month as a factor cc (with 12 levels) and time as a numeric vector ti. We then fit a smooth curve for ti and a cyclic smooth function for cc.

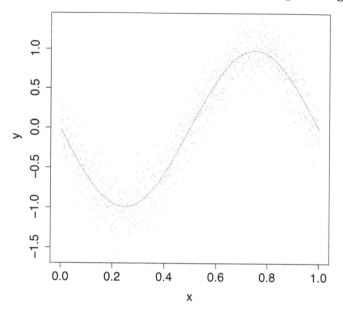

FIGURE 9.9: Fitted cyclic curves using the function pbc().

R
p. 2?

```
cc <- factor(cycle(accdeaths))
ti <- as.numeric(time((accdeaths)))
m1 <- gamlss(accdeaths~pb(ti)+pbc(cc), family=NBI)

## GAMLSS-RS iteration 1: Global Deviance = 1022.284
## GAMLSS-RS iteration 2: Global Deviance = 1022.285
## GAMLSS-RS iteration 3: Global Deviance = 1022.286

plot(m1$mu.s[,2]~ti, type="l", ylab="seasonal effect",
    xlab="Time", main="(a)")
plot(accdeaths, main="(b)", lty=3, lwd=2.5)
lines(fitted(m1)~ti, col="red")
legend("topright",legend=c("observed","fitted"),
    lty=c(3,1),col=c("black","red"))
```

Fig.
9.10

The fitted smoothers are saved as objects of class pbc. The methods which apply to a pb class also apply to pbz.

9.4.6   The cs() and scs() functions for fitting cubic splines

Both cubic spline functions are based on the smooth.spline() function of **R** and can be used for univariate smoothing. They fit a cubic smoothing spline function; see for example Hastie and Tibshirani [1990], Green and Silverman [1994, page 46] or Wood [2006a, pages 124 and 149].

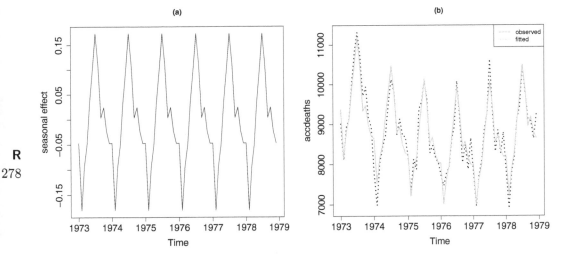

FIGURE 9.10: (a) The fitted seasonal effect using the function `pbc()`. (b) The observed and fitted values.

Cubic splines are the solution to the following minimization problem. Let $g(x)$ be a twice differentiable function of x in the interval $[a, b]$ and λ a smoothing parameter. Define the penalized sum of squares function:

$$Q_2(g) = \sum_{i=1}^{n} w_i \left(y_i - g(x_i)\right)^2 + \lambda \int_a^b \left\{g''(t)\right\}^2 dt$$

where the w_i are prior weights and $g''(\cdot)$ is the second derivative of $g(\cdot)$. It turns out that the minimizer of $Q_2(g)$ over the class of all twice-differentiable functions g is a cubic spline. Also $Q_2(g)$ can be written as

$$Q_2 = (\mathbf{y} - \mathbf{g})^\top \mathbf{W} (\mathbf{y} - \mathbf{g}) + \lambda \mathbf{g} \mathbf{K} \mathbf{g}$$

where $\mathbf{g}^\top = (g(x_1), g(x_2), \ldots, g(x_n))$, for a suitably defined $\mathbf{K}$ matrix. For details see Green and Silverman [1994].

There are two main differences between cubic spline smoothers and P-splines. The basis functions used for cubic smoothing splines fitting are similar to the P-splines. It is a B-spline basis $\mathbf{B}$ of a piecewise polynomial of degree 3. But while in P-splines we take equidistant knots in the x-axis, in cubic smoothing splines the knots are at the distinct x-variable values. The second difference is in the penalty. P-splines achieve smoothness in the fitted function by penalizing the parameters $\boldsymbol{\gamma}$. Cubic smoothing splines achieve smoothness by penalizing the second derivative of the function. The resulting smoothing curves are usually very similar for the same effective degrees of freedom.

The functions `cs()` and `scs()` behave differently at their default values when the degrees of freedom `df` and `lambda` are not specified. For example `cs(x)` by default will use three extra degrees of freedom for smoothing for x (five in total, if you include the linear and the constant). However scs(x) by default will estimate λ (and therefore the degrees of freedom) automatically using generalized cross validation (GCV). Note however that for small data sets the GCV can create instability in the algorithm.

The cs() function has the following arguments:

x   the univariate vector of an explanatory variable;

df   the desired equivalent number of degrees of freedom for smoothing on top of 2 (for the constant and linear terms). (This is the trace of the smoother matrix minus two). The real smoothing parameter (spar below) is found such that df=tr(S)-2, where S is the smoother matrix which depends on spar. Values for df should be positive, with zero implying a linear fit. The default is df=3, i.e. 3 degrees of freedom for smoothing x on top of a linear and constant term in x giving a total of 5 degrees of freedom;

spar   smoothing parameter, typically (but not necessarily) in the default range for spar (-1.5,2]. The coefficient λ of the integral of the squared second derivative in the fitted (penalized log-likelihood) criterion is a monotone function of spar. See the details in the help file of smooth.spline;

c.spar   This specifies minimum and maximum limits for the smoothing parameter, the default limits being -1.5 to 2. This is an option to be used when the degrees of freedom of the fitted gamlss object are different from those given as input in the option df, which is caused by the default limits for the smoothing parameter being too narrow to obtain the required degrees of freedom. The default values used are the ones given by the option control.spar in smooth.spline(), i.e. c.spar=c(-1.5, 2). For very large data sets, e.g. $n = 10,000$ observations, the upper limit may have to increase for example to c.spar=c(-1.5, 2.5). Use this option if you have received the warning 'The output df are different from the input, change the control.spar'. c.spar can take both vectors or lists of length 2; for example c.spar=c(-1.5, 2.5) or c.spar=list(-1.5, 2.5) would have the same effect.

The scs() function has identical arguments plus arguments which can be passed to the smooth.spline() function.

As an example, we apply the smoothing cubic spline functions cs() and scs() to the Munich rent data. We fit an additive model for floor space Fl and year of construction A.

```
# fitting cubic splines with fixed default degrees of freedom df=3
rcs1<-gamlss(R~cs(Fl)+cs(A), data=rent, family=GA, trace=FALSE)
# fitting cubic splines by estimating the smoothing parameter
rcs2<-gamlss(R~scs(Fl)+scs(A), data=rent, family=GA, trace=FALSE)
term.plot(rcs1)
```

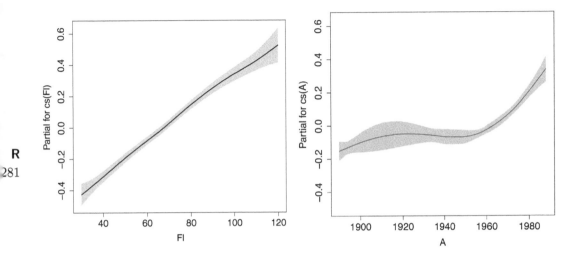

FIGURE 9.11: Fitted curves using the function `cs()` (cubic splines).

Next we use **predict** to create the fitted mean rent surfaces and plot them as contour plots.

```
newrent<-data.frame(expand.grid(Fl=seq(30,120,5),A=seq(1890,1990,5)))
pred1<-predict(rcs1,  newdata=newrent, type="response")

## new prediction

pred2<-predict(rcs2,  newdata=newrent, type="response")

## new prediction

Fln<-seq(30,120,5)
An<-seq(1890,1990,5)
contour(Fln,An,matrix(pred1,nrow=length(Fln)),nlevels=30,
ylab="year of construction", xlab="floor space", main="cs()")
contour(Fln,An,matrix(pred2,nrow=length(Fln)),nlevels=30,
ylab="year of construction", xlab="floor space",  main="scs()")
```

Three-dimensional plots of the fitted surfaces are created as follows:

```
library(lattice)
p1<-wireframe(pred1~Fl*A, newrent, aspect=c(1,0.5), drape=TRUE,
colorkey=list(space="right", height=0.6))
```

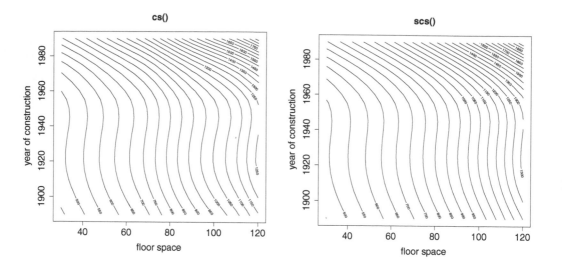

R

p. 28

FIGURE 9.12: Fitted additive curves surface using `cs()` and `scs()` for the rent data.

```
p2<-wireframe(pred2~Fl*A, newrent, aspect=c(1,0.5), drape=TRUE,
colorkey=list(space="right", height=0.6))
print(p1, split = c(1, 1, 2, 1),more = TRUE)
print(p2, split = c(2, 1, 2, 1))
```

9.4.7   The `ri()` function for fitting ridge and lasso regression terms

Shrinkage methods are methodologies for estimating parameters of explanatory variables which enter linearly in the predictor equation; see Hastie et al. [2009, page 61]. Ridge regression is a simple example of how these methods can be used for estimating parameters in a linear regression model. The shrinkage methods impose a penalty on the size of the linear coefficients of the explanatory variables. Assume $\mathbf{y}$ is mean centered to $\tilde{\mathbf{y}}$ and the explanatory variables are standardized to mean 0 and standard deviation 1, so the design matrix $\mathbf{X}$ is amended to $\tilde{\mathbf{X}}$. The standard ridge regression considers the simple least squares regression problem and minimizes the quantity $(\tilde{\mathbf{y}} - \tilde{\mathbf{X}}\boldsymbol{\beta})^\top(\tilde{\mathbf{y}} - \tilde{\mathbf{X}}\boldsymbol{\beta})$ with respect to $\boldsymbol{\beta}$. The solution is given by $\hat{\boldsymbol{\beta}} = (\tilde{\mathbf{X}}^\top\tilde{\mathbf{X}})^{-1}\tilde{\mathbf{X}}^\top\tilde{\mathbf{y}}$.

Ridge regression coefficients are the solution to the penalized least squares problem of minimizing Q_R with respect to $\boldsymbol{\beta}$:

$$Q_R = (\tilde{\mathbf{y}} - \tilde{\mathbf{X}}\boldsymbol{\beta})^\top(\tilde{\mathbf{y}} - \tilde{\mathbf{X}}\boldsymbol{\beta}) + \lambda\boldsymbol{\beta}^\top\boldsymbol{\beta} \tag{9.9}$$

with solution

$$\hat{\boldsymbol{\beta}} = (\tilde{\mathbf{X}}^\top\tilde{\mathbf{X}} + \lambda\mathbf{I})^{-1}\tilde{\mathbf{X}}^\top\tilde{\mathbf{y}} . \tag{9.10}$$

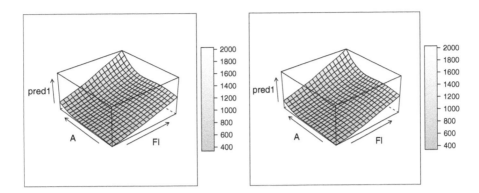

R
281

FIGURE 9.13: Three-dimensional additive surfaces using `cs()` (left) and `scs()` (right) for the rent data.

The effect of the quadratic penalty $\lambda\boldsymbol{\beta}^{\top}\boldsymbol{\beta}$ is to shrink the least squares coefficients towards zero. Equation (9.9) is similar to equation (9.2) with $\mathbf{Z}$, $\boldsymbol{\gamma}$, $\mathbf{W}$ and $\mathbf{G}$ replaced by $\tilde{\mathbf{X}}$, $\boldsymbol{\beta}$, $\mathbf{I}$ and $\mathbf{I}$, respectively, and equation (9.10) is similar to (9.3).

The penalty in equation (9.9) can be replaced with a more general penalty:

$$(\tilde{\mathbf{y}} - \tilde{\mathbf{X}}\boldsymbol{\beta})^{\top}(\tilde{\mathbf{y}} - \tilde{\mathbf{X}}\boldsymbol{\beta}) + \lambda||\boldsymbol{\beta}||_p{}^p \qquad (9.11)$$

where $\boldsymbol{\beta}^{\top} = (\beta_1, \beta_2, \ldots, \beta_J)$ and $||\boldsymbol{\beta}||_p{}^p = \sum_{j=1}^{J}|\beta_j|^p$ and where the summation is over all the elements of $\boldsymbol{\beta}$.[1] Different values of p are associated with different types of penalties: e.g. $p = 2$, which is associated with the L_2 or Euclidean norm and has quadratic penalties; $p = 1$, which is associated with the L_1 (or Manhattan) norm and has absolute values as penalties, etc. Note that $||\boldsymbol{\beta}||_1$ is also known as the *lasso* penalty. Different penalties shrink the least squares coefficients towards zero in different ways. The way the least squares coefficients are shrunk towards zero for different values of p is shown in Figure 9.14, which shows contour plots of the penalty $||\boldsymbol{\beta}||_p{}^p$ for two parameters β_1 and β_2, (i.e. for $J = 2$). This is also described in Hastie et al. [2009, pages 69-73].

Fig.
9.14

```
normPlot <- function(p=2)
{
  beta1 <- beta2 <- seq(-1, 1, len = 100)
  r <- (outer( abs(beta1)^p, abs(beta2)^p, "+"))
  image(beta1,beta2,r, main=paste("p=",p,sep=""),
        ylab=expression(beta[2]), xlab=expression(beta[1]))
  contour(beta1,beta2,r, add=T)
```

---

[1]Note that the mathematical definition of a p-norm is $||\boldsymbol{\beta}||_p = \left(\sum_{j=1}^{J}|\beta_j|^p\right)^{1/p}$ for $p > 1$.

```
}
normPlot(p=10)
normPlot(p=5)
normPlot(p=2)
normPlot(p=1.5)
normPlot(p=1.2)
normPlot(p=1)
normPlot(p=.8)
normPlot(p=.5)
normPlot(p=.001)
```

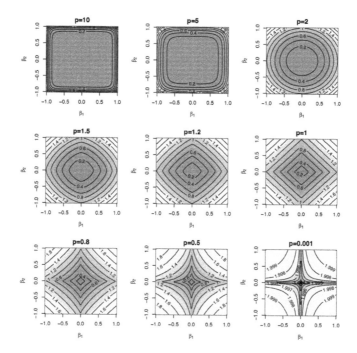

R
p. 28

FIGURE 9.14: How the type of penalty changes with different values of p.

The way that **gamlss** fits nonstandard p is by approximating the $||\boldsymbol{\beta}||_p^p$ with $\boldsymbol{\beta}^\top \boldsymbol{\Omega}_p \boldsymbol{\beta}$, where $\boldsymbol{\Omega}_p$ is a diagonal matrix with elements $1/\left(|\beta_j|^{2-p} + \kappa^2\right)$ for $j = 1, 2, \ldots, J$, and κ takes a very small value. The effective degrees of freedom used in the shrinkage are given by the trace of the resulting smoothing matrix.

The general ridge regression based on (9.11) can be fitted within **gamlss** using the `ri()` function. The most important arguments of `ri()` are:

X   the design matrix of the explanatory variables (whose columns are standardized automatically to mean zero and standard deviation one).

df, lambda   effective degrees of freedom and smoothing parameter which act the same way as in all other smoothers method   supporting methods for automatic selection of the smoothing parameter λ. Only local "ML" and "GAIC" are available.

Lp   the type of shrinkage penalty with Lp=2, i.e. ridge regression, as default.

In the example below, we use the US pollution data usair [Hand et al., 1994], which has six explanatory variables x1:x6, and $n = 41$ observations.

Data summary: US pollution data

R data file: usair in package **gamlss** of dimensions 41×7

variables

    y : sulphur dioxide concentration in air in mg per cubic metre

    x1 : average annual temperature in degrees F

    x2 : number of manufacturers employing more than 20 workers

    x3 : population size in thousands

    x4 : average annual wind speed in miles per hour

    x5 : average annual rainfall in inches

    x6 : average number of days rainfall per year

purpose: to demonstrate ridge regression techniques

We create the matrix X containing all the explanatory variables and then fit four different models. The first is the least squares model, and the rest are different shrinkage approaches: ridge, lasso and 'best subset':

```
X<-with(usair, cbind(x1,x2,x3,x4,x5,x6))
m0<-gamlss(y~scale(X), data=usair, trace=FALSE)    #least squares
m1<-gamlss(y~ri(X), data=usair, trace=FALSE)        #ridge
m2<-gamlss(y~ri(X, Lp=1), data=usair, trace=FALSE) #lasso
m3<-gamlss(y~ri(X, Lp=0), data=usair, trace=FALSE) #best subset
AIC(m0,m1,m2,m3)

##           df       AIC
## m2 5.336309 341.2492
## m0 8.000000 344.7232
## m1 5.884452 345.6097
## m3 2.838310 350.1807
```

Model m2, the lasso, seems most appropriate here. The different coefficients of the four fitted models are displayed below.

```
cbind(
zapsmall(coef(m0)[-1], digits=4),
zapsmall(coef(getSmo(m1)), digits=3),
zapsmall(coef(getSmo(m2)), digits=3),
zapsmall(coef(getSmo(m3)), digits=3))
```

```
##                [,1]   [,2]    [,3]   [,4]
## scale(X)x1   -9.16  -9.44   -8.76  -6.69
## scale(X)x2   36.58  18.54   26.54  13.39
## scale(X)x3  -22.75  -5.43  -12.95   0.00
## scale(X)x4   -4.55  -4.16   -3.94   0.00
## scale(X)x5    6.03   4.71    4.73   0.00
## scale(X)x6   -1.38   0.70    0.00   0.00
```

In the above output, the ridge regression, Lp=2, shrinks most of the least squares coefficients towards zero. The lasso, Lp=1, does the same, but also sets the coefficient of x6 to zero. The best subset, Lp=0, sets four coefficients to zero leaving only x1 and x2 whose coefficients are substantially shrunk towards zero. The fitted coefficients from the three different shrinkage methods are plotted in Figure 9.15.

```
plot(getSmo(m1))
plot(getSmo(m2))
plot(getSmo(m3))
```

Fig.
9.15

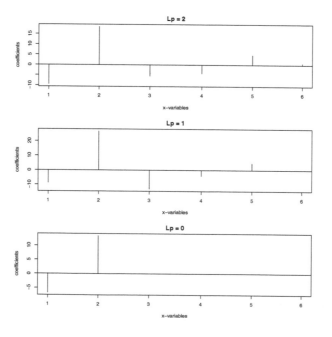

R
p. 28

FIGURE 9.15: Fitted linear coefficients using three different shrinkage approaches. Top: ridge; middle: lasso; bottom: best subset.

The fitted smoothers are saved as object of class `ri`. The methods `plot()`, `coef()` and `fitted()` apply to the class `ri`.

9.4.8   The `pcat()` function for reducing levels of a factor

The function `pcat()` is one of the three smoothers which apply exclusively to categorical variables. The most common 'smoothing' function for categorical variables is the random effect function `random()`, which is described in detail in Chapter 10. The main functionality of `random()` is to take the estimates for each factor level and shrink them towards the overall mean. That is, the estimates for each level of the factor may still be different from each other but the difference between them is reduced because of the shrinkage effect. The amount of shrinkage depends on the smoothing parameter λ, and the effective degrees of freedom used are in general less than the number of levels of the factor.

The function `pcat()` operates differently in that its main objective is to bring levels with similar values together rather than to shrink them towards the overall mean. The idea is to try to merge the different levels of a factor into a set in which similar levels are combined. This can be seen as a way of classifying the different levels of a factor into groups of levels with similar characteristics. The resulting fits will in general have used fewer degrees of freedom than the actual number of levels of a factor. Again the amount of shrinkage of the levels towards each other will depend on the smoothing parameter λ. A review and further discussion about penalties for categorical variables can be found in Gertheiss and Tutz [2016a,b].

The penalized categorical function can be seen as the solution to the penalized least squares problem:

$$(\mathbf{y} - \mathbf{Z}\boldsymbol{\gamma})^{\top}(\mathbf{y} - \mathbf{Z}\boldsymbol{\gamma}) + \lambda\|\mathbf{D}\boldsymbol{\gamma}\|_{p}^{p} \tag{9.12}$$

where $\mathbf{Z}$ is the incidence matrix (a dummy variable matrix) defined by the levels of the factor, and $\|\mathbf{D}\boldsymbol{\gamma}\|_{p}^{p}$ is a general penalty. This penalty is similar to the one defined in ridge regression. That is, different values of p are associated with different types of norms, e.g. $p = 1$ is associated with the L_1 norm and is the summation of the absolute values of the elements of vector $\mathbf{D}\boldsymbol{\gamma}$. Different penalties shrink the levels of the factor towards each other in different ways and for this type of problem it seems that both L_0 [Oelker et al., 2015] and L_1 [Gertheiss and Tutz, 2016a] norms produce reasonable results. Each row of the matrix $\mathbf{D}$ takes one value of 1, one of -1, and the rest zeroes. For example a typical $\mathbf{D}$ matrix of a factor with five levels is given by:

$$\mathbf{D} = \begin{bmatrix} -1 & 1 & 0 & 0 & 0 \\ -1 & 0 & 1 & 0 & 0 \\ 0 & -1 & 1 & 0 & 0 \\ -1 & 0 & 0 & 1 & 0 \\ 0 & -1 & 0 & 1 & 0 \\ 0 & 0 & -1 & 1 & 0 \\ -1 & 0 & 0 & 0 & 1 \\ 0 & -1 & 0 & 0 & 1 \\ 0 & 0 & -1 & 0 & 1 \\ 0 & 0 & 0 & -1 & 1 \end{bmatrix}$$

with $\mathbf{G} = \mathbf{D}^\top \mathbf{D}$ given by:

$$\mathbf{G} = \begin{bmatrix} 4 & -1 & -1 & -1 & -1 \\ -1 & 4 & -1 & -1 & -1 \\ -1 & -1 & 4 & -1 & -1 \\ -1 & -1 & -1 & 4 & -1 \\ -1 & -1 & -1 & -1 & 4 \end{bmatrix}.$$

The way that the **gamlss** function `pcat()` fits non-standard p is by approximating the penalty $||\mathbf{D}\boldsymbol{\gamma}||_p{}^p$ with $\boldsymbol{\gamma}^\top \mathbf{D}^\top \boldsymbol{\Omega} \mathbf{D}\boldsymbol{\gamma}$, where $\boldsymbol{\Omega}$ is a diagonal matrix with elements $1/\left(|(\mathbf{D}\boldsymbol{\gamma})_j|^{2-p} + \kappa^2\right)$ for $j = 1, 2, \ldots, J$, with κ taking a very small value. The solution is given by:

$$\hat{\boldsymbol{\gamma}} = (\mathbf{Z}^\top \mathbf{Z} + \lambda \mathbf{D}^\top \boldsymbol{\Omega} \mathbf{D})^{-1} \mathbf{Z}^\top \mathbf{y}. \tag{9.13}$$

The effective degrees of freedom used in the shrinkage is given by the trace of the smoothing matrix, $\mathbf{Z}(\mathbf{Z}^\top \mathbf{Z} + \lambda \mathbf{D}^\top \boldsymbol{\Omega} \mathbf{D})^{-1} \mathbf{Z}^\top$. The default value of p in `pcat()` is 0.

We demonstrate the use of `pcat()` using the **rent** data. A new factor is created from the continuous variable age of the building (A). The new factor cA has 10 levels and the idea here is to try to merge those levels into a smaller more compact set. Figure 9.16 plots the rent against the new created factor cA and it shows a general increase in the rent prices for newer flats.

```
#creating a new factor with levels: "1890" "1900" "1910" "1920"
#"1930" "1940" "1950" "1960" "1970" "1980"
rent <- transform(rent, cA=cut(rent$A,c(1880, seq(1900,1990, 10)),
                    labels=as.character(seq(1890,1980, 10))))
nlevels(rent$cA)

## [1] 10

plot(R~cA, data=rent)
```

Fig. 9.16

To demonstrate the use of `pcat()` we firstly consider cA on its own, then in combination with other explanatory variables. We fit four different models for cA: (i)

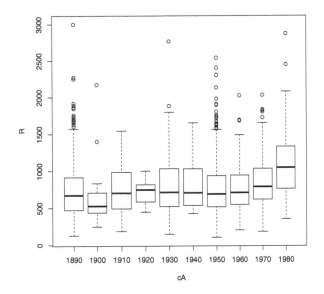

FIGURE 9.16: Monthly net rent (R) against categorized age of the building (cA).

the null model, (ii) the saturated model, i.e. cA as a factor, (iii) penalized cA by selecting the smoothing parameter using local maximum likelihood (ML) and (iv) penalized cA where the smoothing parameter is selected using a local GAIC.

```
rn<-gamlss(R~1,data=rent,family=GA,trace=F)#null

## GAMLSS-RS iteration 1: Global Deviance = 28611.58
## GAMLSS-RS iteration 2: Global Deviance = 28611.58

rs<-gamlss(R~cA,data=rent,family=GA,trace=F)#saturated

## GAMLSS-RS iteration 1: Global Deviance = 28490.11
## GAMLSS-RS iteration 2: Global Deviance = 28490.11

r1<-gamlss(R~pcat(cA), data=rent, family=GA,trace=F)#ML

## GAMLSS-RS iteration 1: Global Deviance = 28494.39
## GAMLSS-RS iteration 2: Global Deviance = 28494.39
## GAMLSS-RS iteration 3: Global Deviance = 28494.39

r2<-gamlss(R~pcat(cA,method="GAIC",k=log(1968)),data=rent,#GAIC
           family=GA, trace=F)

## GAMLSS-RS iteration 1: Global Deviance = 28611.58
## GAMLSS-RS iteration 2: Global Deviance = 28495.19
## GAMLSS-RS iteration 3: Global Deviance = 28495.1
## GAMLSS-RS iteration 4: Global Deviance = 28495.1
```

```
GAIC(rn,rs,r1,r2)
```

```
##           df       AIC
## r1  3.899228 28502.19
## r2  3.767306 28502.64
## rs 11.000000 28512.11
## rn  2.000000 28615.58
```

By using pcat() we improve the fit as judged by the GAIC. We can find the amount of smoothing required by getting the effective degrees of freedom:

```
r1$mu.df
```

```
## [1] 2.899228
```

```
r2$mu.df
```

```
## [1] 2.767306
```

It appears that we can move from ten degrees of freedom (saturated model) to three degrees of freedom without a great loss of information. The pcat() function automatically saves a new factor with the new classification of the original levels. Here we obtain the new factor:

```
f1 <- getSmo(r1)$factor
levels(f1)
```

```
## [1] "-0.0552" "0.0467"  "0.2829"
```

```
table(f1,rent$cA)
```

```
##
## f1         1890 1900 1910 1920 1930 1940 1950 1960 1970 1980
##   -0.0552   327   16   27   13  257    8  612  110    0    0
##   0.0467      0    0    0    0    0    0    0    0  397    0
##   0.2829      0    0    0    0    0    0    0    0    0  202
```

We see that the levels of the new factor f1 correspond to the periods: up to 1960, 1970 and 1980. We refit the model using f1:

```
r3 <- gamlss(R~f1, data=rent, family=GA, trace=F)
```

```
## GAMLSS-RS iteration 1: Global Deviance = 28494.23
## GAMLSS-RS iteration 2: Global Deviance = 28494.23
```

```
AIC(r1, r3)
```

```
##          df      AIC
## r1 3.899228 28502.19
## r3 4.000000 28502.23
```

The package **gamlss** provides two extra plots for helping to understand how pcat()

works. They are (i) `plotLambda()` and (ii) `plotDF()` for plotting the path of the merging levels against log λ and against the effective degrees of freedom, respectively.

Fig.
9.17
```
plotLambda(R,factor=cA,data=rent,family=GA,along=seq(-10,2,.2))
abline(v=log(getSmo(r1)$lambda),col='gray',lwd=2.5)#ML
abline(v=log(getSmo(r2)$lambda),col='darkblue',lty=3,lwd=2.5)#GAIC
```

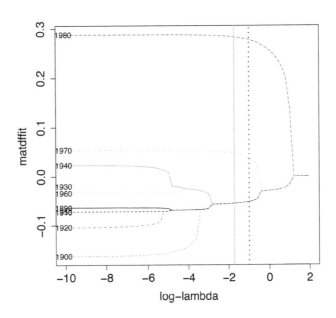

FIGURE 9.17: Plot of the merging levels of the factor `cA` for different values of log λ.

A similar plot (but somehow clearer) is produced using:

Fig.
9.18
```
plotDF(R,factor=cA,data=rent,family=GA)
abline(v=getSmo(r1)$edf,col='gray',lwd=2.5)# ML
abline(v=getSmo(r2)$edf,col='darkblue',lty=3,lwd=2.5)#GAIC
```

When other explanatory variables exist in the model we can still use `pcat()` to reduce the levels of a factor, as the following example shows:

```
rs<-gamlss(R~pb(Fl)+cA+B+H+loc,data=rent,family=GA,
            trace=F) #satutated

## GAMLSS-RS iteration 1: Global Deviance = 27651.05
## GAMLSS-RS iteration 2: Global Deviance = 27651.05

r1<-gamlss(R~pb(Fl)+pcat(cA)+B+H+loc,data=rent,family=GA,
            trace=F) #ML

## GAMLSS-RS iteration 1: Global Deviance = 27657.64
## GAMLSS-RS iteration 2: Global Deviance = 27657.64
## GAMLSS-RS iteration 3: Global Deviance = 27657.64
```

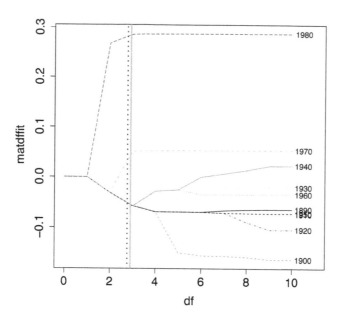

FIGURE 9.18: Plot of the merging levels of the factor `cA` for different effective degrees of freedom.

R

p. 29

```
AIC(rs,r1)

##          df       AIC
## r1  9.709107  27677.05
## rs 16.845471  27684.74

getSmo(r1, which=2)$edf

## [1] 2.87092

nf <- getSmo(r1, which=2)$factor
levels(nf)

## [1] "-0.0673" "-0.0012" "0.2149"

r2 <- gamlss(R~pb(Fl)+nf+B+H+loc, data=rent, family=GA, trace=F)

## GAMLSS-RS iteration 1: Global Deviance = 27657.36
## GAMLSS-RS iteration 2: Global Deviance = 27657.36

AIC(r1,r2)

##          df       AIC
## r1 9.709107  27677.05
## r2 9.858274  27677.08
```

The functions `plotLambda()` and `plotDF()` can also be used when more explanatory variables exist in the model. For example to get the path for different degrees of freedom use:

```
plotDF(formula=R~pb(F1)+B+H+loc, factor=cA, data=rent, family=GA)
abline(v=getSmo(r1, which=2)$edf, col='gray', lwd=2.5)
```

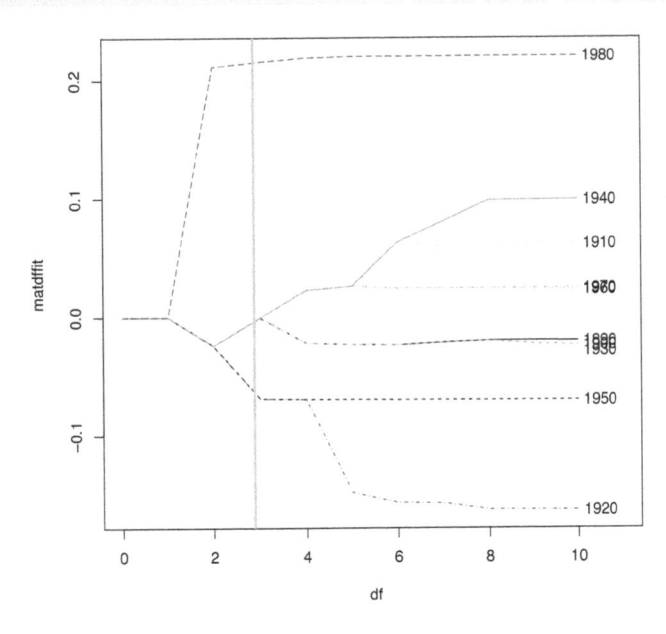

FIGURE 9.19: Plot of the merging levels of the factor `cA` fitted with explanatory variables `pb(F1)+B+H+loc` for different effective degrees of freedom.

Note that `pcat()` creates `pcat` class objects which have `print()`, `fitted()`, `coef()`, and `plot()` as methods.

9.4.9 The `gmrf()` function for fitting Gaussian Markov random fields

Gaussian Markov random fields (GMRF) model discrete spatial variation, where the variables are defined on discrete domains, such as regions, regular grids or lattices and can be applied to GLMs, GAMs, spatial and spatio-temporal models and other structured additive models. The explanatory variable is a factor but further spatial information is needed to identify neighbourhood information. The function to fit a GMRF term in GAMLSS is `gmrf()`, in the **gamlss.spatial** package. For more details about GMRF in GAMLSS see De Bastiani et al. [2016]. A GMRF model can be seen as a penalized smoother defined by equations (9.2) and (9.3), where the basis matrix $\mathbf{Z}$ is an index matrix defining which observation belongs to which spatial area and $\mathbf{G}$ provides spatial neighbourhood information. In order to be more specific about the nature of the $\mathbf{G}$ matrix, consider the following simple example.

Consider five areas. Area 1 is a neighbour of areas 2, 3 and 4, area 3 is a neighbour of areas 1, 4 and 5, and area 4 is a neighbour of areas 1, 3 and 5. The relationship between the areas can be described by the undirected graph shown in Figure 9.20.

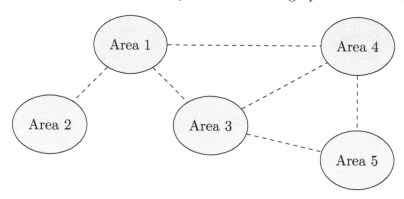

FIGURE 9.20: Diagram showing the relationship between the five regions.

Typically the $\mathbf{D}$ matrix will be of dimension $k \times l$, where k is the number of edges in the graph and l is the number of areas. For our example $\mathbf{D}$ is given by:

$$\mathbf{D} = \begin{bmatrix} 1 & -1 & 0 & 0 & 0 \\ 1 & 0 & -1 & 0 & 0 \\ 1 & 0 & 0 & -1 & 0 \\ 0 & 0 & 1 & -1 & 0 \\ 0 & 0 & 1 & 0 & -1 \\ 0 & 0 & 0 & 1 & -1 \end{bmatrix}$$

where, for example, in row 1 the values 1 and -1 indicate that areas 1 and 2 are connected. The matrix $\mathbf{G} = \mathbf{D}^\top \mathbf{D}$ is given by:

$$\mathbf{G} = \begin{bmatrix} 3 & -1 & -1 & -1 & 0 \\ -1 & 1 & 0 & 0 & 0 \\ -1 & 0 & 3 & -1 & -1 \\ -1 & 0 & -1 & 3 & -1 \\ 0 & 0 & -1 & -1 & 2 \end{bmatrix}.$$

The effect of the $\mathbf{G}$ matrix is to bring fitted values from neighbouring regions closer together, rather than to shrink them towards the overall mean as, for example, `random()` does. The neighbourhood information can be provided in three different ways using the arguments of `gmrf()`:

`precision`   by specifying the matrix $\mathbf{G}$,

`polys`   by giving the polygon information of the regions, or

`neighbour`   by the neighbour information (that is, which region is connected to which other region).

Here we use a reduced version of the Columbus OH, crime data taken from the package **mgcv** to demonstrate the use of `gmrf()`.

Data summary: District crime data from Columbus OH.

R data file: `columb` in package **mgcv** of dimensions 49×8

variables

 `area` : land area of district

 `home.value` : housing value in 1000USD

 `income` : household income in 1000USD

 `crime` : residential burglaries and auto thefts per 1000 households

 `open.space` : measure of open space in district

 `district` : code identifying district, and matching names for the columb.polys data set

purpose: to demonstrate the function `gmrf()`

`columb.polys` is a data set containing the polygon definitions for the areas.

Fig. 9.21

```
library(gamlss.spatial)
data(columb)
data(columb.polys)
m1 <- gamlss(crime~ gmrf(district, polys=columb.polys), data=columb)
draw.polys(columb.polys, getSmo(m1), scheme="topo")
```

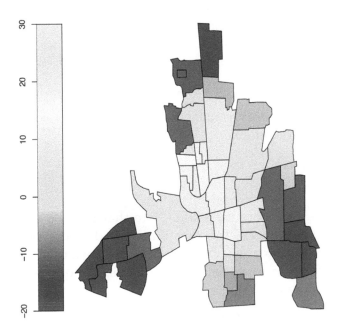

FIGURE 9.21: Fitted values from the Columbus OH, crime data, using `gmrf()`.

The functions `polys2nb()`, and `nb2prec()` are designed to transfer information from polygons to neighbour and from neighbour to precision, respectively. The following code shows how the functions are used:

```
# get the neighbour information from polygons
vizinhos <- polys2nb(columb.polys)
m2<-gamlss(crime~ gmrf(district, neighbour=vizinhos), data=columb)
# transform the neighbour information to a 'precision' matrix
precisionC <- nb2prec(vizinhos,x=columb$district)
m3<-gamlss(crime~ gmrf(district, precision=precisionC), data=columb)
```

The models `m2` and `m3` are identical to `m1` above. For data with a large number of regions the argument `precision` should be used in the fitting. The function `gmrf()` creates GMRF objects with methods `fitted()`, `coef()`, `AIC()`, `deviance()`, `print()`, `summary()`, `plot()`, `residuals()` and `predict()`.

9.5   Penalized smoothers: Multivariate

9.5.1   The `pvc()` function for fitting varying coefficient models

Varying coefficient terms were introduced by Hastie and Tibshirani [1993] to accommodate a special type of interaction between explanatory variables. This interaction takes the form of $\beta(x)z$; that is, the linear coefficient of the explanatory variable z changes smoothly according to another explanatory variable x. In time series data x can be time, so the linear coefficient of z varies over time. More generally, x should be a continuous variable, while z can be either continuous or categorical.

The `pvc()` function for varying coefficient terms has the following arguments:

x   the explanatory variable x which affects the coefficient of the explanatory variable z;

df   as in `pb()`;

lambda   as in `pb()`;

by   the explanatory variable z. Note that if z is a continuous variable (rather than a factor) then it is centred automatically, i.e. $z - \bar{z}$, by `pvc()` due to the invariance of varying coefficient models to location shifts in z; see comments of Green in the discussion of Hastie and Tibshirani [1993];

control   options for controlling the P-splines fitting.

9.5.1.1   Continuous z

As an example of using the function `pvc()` where the `by` argument is a continuous variable, consider the model in which smooth functions for age `A` and floor space `Fl` are fitted but in which we also investigate whether a linear or varying coefficients interaction exists between the two variables.

```
# main smoothing effects for Fl and A
m0<-gamlss(R~pb(Fl)+pb(A), data=rent, family=GA, trace=FALSE)
# linear interaction between A and Fl
m1<-gamlss(R~pb(Fl)+pb(A)+A:Fl, data=rent, family=GA, trace=FALSE)
# varying coefficients interaction b(A)Fl
m2<-gamlss(R~pb(Fl)+pb(A)+pvc(A, by=Fl), data=rent, family=GA,
           trace=FALSE)
# varying coefficients interaction b(Fl)A
m3<-gamlss(R~pb(Fl)+pb(A)+pvc(Fl, by=A), data=rent, family=GA,
           trace=FALSE)
# linear interaction plus varying coefficients interaction b(A)Fl
m4<-gamlss(R~pb(Fl)+pb(A)+A:Fl+pvc(A, by=Fl), data=rent, family=GA,
           trace=FALSE)
AIC(m0, m1,m2,m3, m4)
```

```
##             df        AIC
## m2 10.925246  27927.36
## m4 11.925247  27929.36
## m1  8.059831  27938.16
## m0  7.371373  27938.35
## m3 10.315233  27938.37
```

Model `m2` with the varying coefficient interaction $\beta(\mathtt{A})\mathtt{Fl}$ seems the most appropriate here. Note however that the most general interaction model is the one in which a 'smooth' surface is fitted for both explanatory variables, as discussed in Section 9.5.2.2. The `term.plot()` function for continuous z in the varying coefficient situation plots the relationship of the estimated smooth coefficient $\beta(x)$ against x, as shown in the 'varying coef' plot in Figure 9.22.

Fig.
9.22

```
term.plot(m2, pages=1)
```

The relationship of the estimated smoothed coefficient $\beta(x)$ against x of Figure 9.22 provides some information on how β is changing, but as with all two-way interactions of continuous variables a more informative plot is a contour plot or a three-dimensional plot of the relationship between $\hat{\mu}$ (the fitted or predicted mean of the rent `R`) and the two explanatory variables `A` and `Fl`, as below.

Fig.
9.23

```
newrent<-data.frame(expand.grid(Fl=seq(30,120,5),A=seq(1890,1990,5)))
newrent$pred<-predict(m2,newdata=newrent, type="response", data=rent)

## new prediction
```

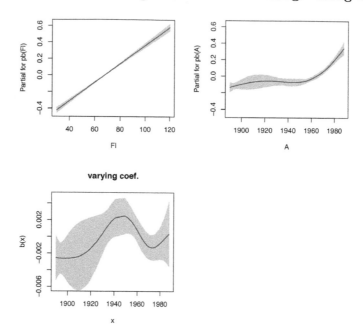

R

p. 29

FIGURE 9.22: The term plot for the varying coefficient interaction model m2.

```
Fln<-seq(30,120,5)
An<-seq(1890,1990,5)
op <- par(mfrow=c(1,1))
contour(Fln,An,matrix(newrent$pred,nrow=length(Fln)),nlevels=30,
        ylab="year of construction", xlab="floor space")
```

9.5.1.2   Categorical z

When the z variable used in the argument by is a factor, the varying coefficient function fits separate smooth curves against x for each level of the factor z. In the next example we use the factor loc, which identifies different locations (below, average or above average), to demonstrate the point.

```
g1<-gamlss(R~pb(Fl)+pb(A)+pvc(Fl,by=loc), data=rent, family=GA,
        trace=FALSE)
g2<-gamlss(R~pb(Fl)+pb(A)+pvc(A,by=loc), data=rent, family=GA,
        trace=FALSE)
g3<-gamlss(R~pb(Fl)+pb(A)+pvc(Fl,by=loc)+pvc(A,by=loc), data=rent,
        family=GA, trace=FALSE)
AIC(g1,g2,g3)

##          df      AIC
## g1 11.38144 27858.10
## g2 11.97336 27859.57
```

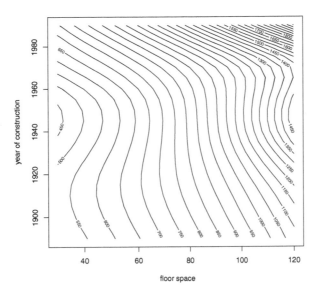

FIGURE 9.23: The fitted surface plot of the varying coefficient interaction model m2.

g3 15.88057 27859.64

The resulting best model according to AIC is **g1**, with a much reduced AIC relative to model **m0**. Figure 9.24 shows the results of using **term.plot()** for model **g1**.

Fig.
9.24

```
term.plot(g1, pages=1)
```

The three different smooth fits (for the three levels of **loc**) are cramped within the third panel of Figure 9.24. Individual fitted smooth curves can be shown using the following commands:

Fig.
9.25

```
plot(getSmo(g1, para="mu", which=3),  factor.plots = TRUE)
```

The three smooth fits appear linear, so a linear interaction **Fl*loc** may be adequate

```
g4<-gamlss(R~pb(Fl)+pb(A)+Fl*loc, data=rent, family=GA, trace=FALSE)
```

```
AIC(g4)
```

```
## [1] 27858.09
```

R
297

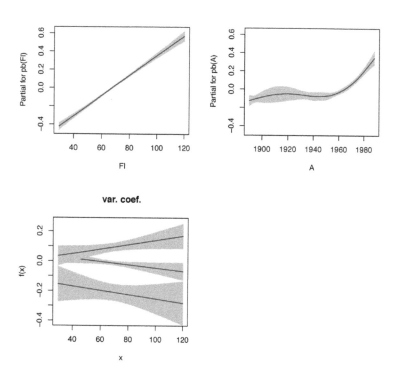

FIGURE 9.24: Term plot figures from model **g1**.

R
p. 29

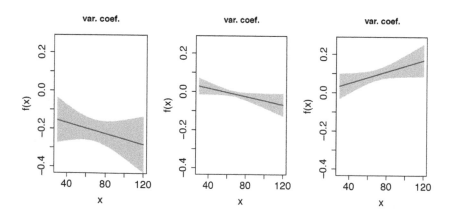

FIGURE 9.25: Individual fitted smooth curves from model **g1**.

R
p. 299

9.5.2 Interfacing with gam(): The ga() function

The ga() function is an additive smoothing function which can be used within a GAMLSS model. It is an interface to the gam() function in the **mgcv** package [Wood, 2001] in which several smoothers are available. (See the help function for formula.gam for information on the smoothers available in gam().) ga() is in the package **gamlss.add**, where other interfaces of this kind are also located. For simple one-dimensional smoothers, using pb() or the ga() interface usually makes little difference to the resulting fitted smoothing terms. In our experience, for exponential family models, the fitted smoothing curves using a single smoother in gam() within **mgcv**, or pb() within **gamlss** produce very similar results. The great advantage of the interface function ga() is the access to more than one-dimensional smoothers, such as thin plate cubic splines (s()) and tensor products (tp()), which are efficiently implemented within **mgcv**, in combination with a response variable distribution outside the exponential family, and the capability for any or all parameters of the distribution to be modelled.

The function ga() has two arguments. The first is a gam() type formula and the second is the gam() control. Here we demonstrate the use of ga() by fitting different smoothing models for floor (Fl) and age of construction (A) to the Munich rent data. Firstly we use smooth additive terms for Fl and A (main effects) and later we fit a smooth surface which explores the interaction between them.

9.5.2.1 Additive terms

We use the normal and the gamma error distributions as examples. Three different models are fitted first using gam(), and then using gamlss(), first calling the ga() interface with gam() and second the smoothing function pb(). The resulting deviances and effective degrees of freedom of the fitted models are displayed using GAIC().

```
library(gamlss.add)
data(rent)
# additive fits
# normal distribution
ga1<-gam(R~s(Fl)+s(A),method="REML",data=rent)
gn1<-gamlss(R~ga(~s(Fl)+s(A),method="REML"),data=rent,trace=FALSE)
gb1<-gamlss(R~pb(Fl)+pb(A),data=rent,trace=FALSE)
AIC(ga1, gn1, gb1, k=0)

##            df      AIC
## ga1 9.653262 28264.38
## gn1 8.275100 28264.58
## gb1 8.372701 28264.19
```

```
# gamma distribution
ga2<-gam(R~s(Fl)+s(A),method="REML",data=rent,family=Gamma(log))
gn2<-gamlss(R~ga(~s(Fl)+s(A),method="REML"),data=rent,family=GA,
            trace=FALSE)
gb2<-gamlss(R~pb(Fl)+pb(A), data=rent, family=GA, trace=FALSE)
AIC(ga2, gn2, gb2, k=0)

##              df       AIC
## ga2 8.295446 27924.42
## gn2 7.016779 27924.26
## gb2 7.371373 27923.61
```

For the normal models the fitted deviances are identical, but with slightly different
degrees of freedom for the gam() model. For the gamma models, the gamlss()
fit using pb() gives slightly different results to the other two models. Figure 9.26
shows the three resulting plots of the fitted terms for the different models using the
gamma response distribution. The top row shows the model fitted using gam(), the
middle the model fitted using ga() within gamlss(), while the bottom row shows
gamlss() using the pb() function. For all practical purposes the three plots lead
to identical conclusions.

```
plot(ga2, scheme=1)
term.plot(gn2)
term.plot(gb2)
```

Fig.
9.26

9.5.2.2 Smooth surface fitting

For surface fitting, **mgcv** provides several options, for example s() for thin plate
splines, te() for tensor products and ti() which is a variant of tensor product
designed to be used for interaction terms when the main effects (and any lower
order interactions) are present. Thin plate splines are rotationally invariant and are
well suited to explanatory variables which are spatial coordinates in two (or three)
dimensions. Tensor product splines are scale invariant and are suited to explanatory
variables measured in different units. First, to illustrate, we use thin plate splines:

```
ga4 <-gam(R~s(Fl,A), method="REML", data=rent, family=Gamma(log))
gn4 <- gamlss(R~ga(~s(Fl,A), method="REML"), data=rent, family=GA)

## GAMLSS-RS iteration 1: Global Deviance = 27891.31
## GAMLSS-RS iteration 2: Global Deviance = 27891.36
## GAMLSS-RS iteration 3: Global Deviance = 27891.36

AIC(ga4,gn4, k=0)

##              df       AIC
## ga4 17.61822 27893.40
```

Fig.
9.27

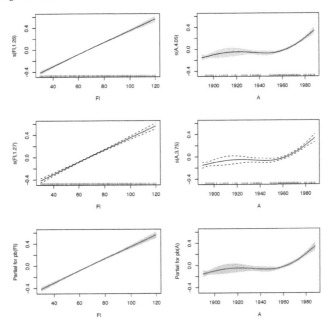

FIGURE 9.26: The plotting of terms of a gamma distribution model fitted using alternative methods. Top row: `gam()`; middle row: `ga()` within `gamlss()`; bottom row: `pb()` within `gamlss()`.

```
## gn4 17.60032 27891.36

vis.gam(ga4)
vis.gam(getSmo(gn4))
```

Note that `term.plot()` produces a contour plot similar to the `plot(getSmo(gn4))` command.

```
term.plot(gn4)
```

Fig.
9.28

For tensor product smoothers the `gam()` function `te()` is used:

```
ga5 <- gam(R~te(Fl,A), data=rent, family=Gamma(log))
gn5 <- gamlss(R~ga(~te(Fl,A)), data=rent, family=GA)

## GAMLSS-RS iteration 1: Global Deviance = 27887.83
## GAMLSS-RS iteration 2: Global Deviance = 27887.83

AIC(ga5,gn5, k=0)

##               df       AIC
## ga5 18.14628 27889.38
## gn5 18.61222 27887.83
```

R
302

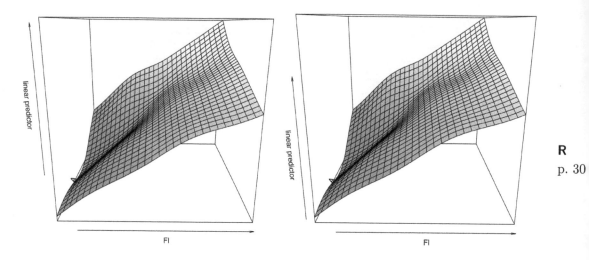

FIGURE 9.27: Surface fitting of the gamma model. Left: `gam()`; right: `ga()` within `gamlss()`.

R

p. 30

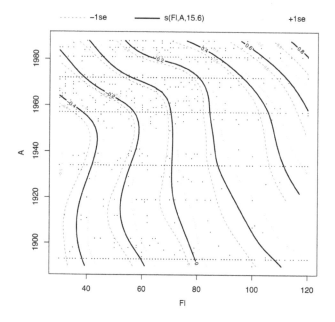

R

p. 303

FIGURE 9.28: Contour plot for a `gam()` model fitted within `gamlss()`.

```
vis.gam(getSmo(gn5))
```

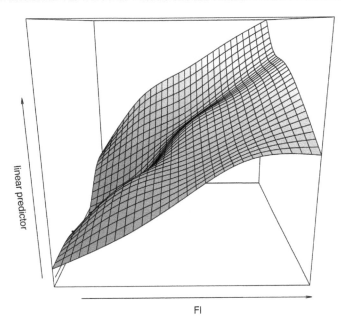

FIGURE 9.29: Surface plot for tensor product model fitted within `gamlss()`.

9.6 Other smoothers

9.6.1 Interfacing with nnet(): nn()

Neural networks provide a flexible way of fitting nonlinear regression models; see Bishop [1995] and Ripley [1993, 1996]. They are overparameterized nonlinear statistical models, which allows them to be very flexible and therefore approximate any smooth function. Because of the overparameterization of the neural network, they are difficult models to interpret compared to the more traditional smoothing models. They may be thought of as *black box* models, meaning that they can work in practice, but are not transparent and therefore not easy to interpret. On the other hand, they can pick up high level interactions among the explanatory variables that are difficult to find through the more classical regression approach.

The **gamlss.add** package provides an interface with the function **nnet()** from the **nnet** package. More information about the use of **nnet()** can be found in Venables and Ripley [2002]. The **gamlss** interface function is **nn()** and it takes as arguments a formula plus other control arguments needed for **nnet()**.

Note that, because the neural network models are overparameterized, different ini-

tial values can result in a different final fitted model. This is apparent within a
`gamlss` model where the `nnet()` function is called within the backfitting algorithm.
Setting the random generating seed at the beginning of the `gamlss` fit will ensure
that the same fitted model can be recreated later. Also one way to help the optimization process and possibly to avoid over-fitting is the use of the argument `decay`,
which is a smoothing parameter within `nnet`. The `size` argument is the number of
hidden variables.

Here we demonstrate the use of `nn()` by fitting three different models to the Munich
rent data. Following the advice of Ripley [1996, page 157], the continuous variables
`Fl` and `A` are rescaled to the interval $[0, 1]$. Categorical variables are automatically
coded 0 or 1 in the design matrix. In the first model, we fit surface models for
`Fl` and `A`. In the second model we add the main effects of the factors `B` (whether
there is a premium bathroom), `H` (whether there is central heating), `L` (whether the
kitchen equipment is above average), and `loc` (three different locations). In the last
model and in order to explore the interaction facilities of the neural network, we fit
a neural network model with all explanatory continuous and categorical variables.

```
library(gamlss.add)
set.seed(1432)
rent$Fls<-(rent$Fl-min(rent$Fl))/(max(rent$Fl)-min(rent$Fl))#scale Fl
rent$As<-(rent$A-min(rent$A))/(max(rent$A)-min(rent$A))#scale A
mr1<-gamlss(R~nn(~Fls+As, size=5, decay=0.01), data=rent, family=GA)

## GAMLSS-RS iteration 1: Global Deviance = 27902.55
## GAMLSS-RS iteration 2: Global Deviance = 27890.03
## GAMLSS-RS iteration 3: Global Deviance = 27890.03

mr2<-gamlss(R~nn(~Fls+As, size=5, decay=0.01) +H+B+loc, data=rent,
            family=GA)

## GAMLSS-RS iteration 1: Global Deviance = 27643.57
## GAMLSS-RS iteration 2: Global Deviance = 27634.4
## GAMLSS-RS iteration 3: Global Deviance = 27634.4

mr3<-gamlss(R~nn(~Fls+As+H+B+loc, size=5, decay=0.01), data=rent,
            family=GA)

## GAMLSS-RS iteration 1: Global Deviance = 27612.51
## GAMLSS-RS iteration 2: Global Deviance = 27597.81
## GAMLSS-RS iteration 3: Global Deviance = 27597.81

AIC(mr1, mr2, mr3)

##      df      AIC
## mr3 43 27683.81
## mr2 27 27688.40
## mr1 23 27936.03
```

```
AIC(mr1, mr2, mr3, k=log(1969))
```

```
##       df        AIC
## mr2  27  27839.20
## mr3  43  27923.98
## mr1  23  28064.49
```

Note that the results may depend on the chosen seed. The AIC selects the `mr3` model while SBC selects `mr2`. The **nnet** fitted object can be retrieved using the function `getSmo()`. For example to get the fitted coefficients use:

```
summary(getSmo(mr3))
```

```
## a 6-5-1 network with 41 weights
## options were - linear output units  decay=0.01
##    b->h1 i1->h1 i2->h1 i3->h1 i4->h1 i5->h1 i6->h1
##    -8.20   5.93   2.05  -0.93  -1.98  -3.34   6.33
##    b->h2 i1->h2 i2->h2 i3->h2 i4->h2 i5->h2 i6->h2
##     4.81   0.89  -5.14  -4.89   5.71  -7.73   2.29
##    b->h3 i1->h3 i2->h3 i3->h3 i4->h3 i5->h3 i6->h3
##   -10.73   1.69  10.01   1.68   0.30  -0.45   1.24
##    b->h4 i1->h4 i2->h4 i3->h4 i4->h4 i5->h4 i6->h4
##     1.85   3.15  -6.50   1.49  -4.35  -0.92  -0.04
##    b->h5 i1->h5 i2->h5 i3->h5 i4->h5 i5->h5 i6->h5
##     2.05  -0.87  -1.02   0.61   0.01  -1.09   0.53
##     b->o  h1->o  h2->o  h3->o  h4->o  h5->o
##     1.22   0.45   0.53   0.60   0.69  -2.82
```

In the summary b represents a constant intercept (or 'bias'), h represents a hidden variable and o represents the output. A visual presentation of the fitted neural network model is obtained as follows:

Fig. 9.30

```
plot(getSmo(mr3), y.lab=expression(g[1](mu)))
```

Thicker lines represent coefficients with high values, while 'black' and 'grey' colours represent positive and negative values, respectively.

Next we fit a neural network model to the σ parameter of the gamma distribution and compare the model with the previously fitted models, in which only μ was modelled as a function of the explanatory variables.

```
mr4 <- gamlss(R~nn(~Fls+As+H+B+loc, size=5, decay=0.1),
   sigma.fo=~nn(~Fls+As+H+B+loc, size=5, decay=0.1),data=rent,
   family=GA, gd.tol=1000,trace=FALSE)
AIC(mr2, mr3, mr4)
```

```
##       df        AIC
## mr4  84  27614.32
```

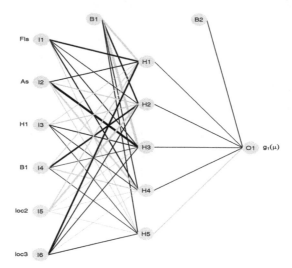

FIGURE 9.30: Visual representation of `mr3`, the neural network model for μ.

R
p. 30

```
## mr3 43 27683.81
## mr2 27 27688.40

AIC(mr2, mr3, mr4, k=log(1969))

##      df      AIC
## mr2 27 27839.20
## mr3 43 27923.98
## mr4 84 28083.49
```

Again the AIC selects the more complicated model `mr4`, while the SBC selects the simpler model `mr2`. The graphical representation of the neural network model for each of μ and σ in model `mr4` is displayed using:

```
plot(getSmo(mr4), y.lab=expression(g[1](mu)))
plot(getSmo(mr4, what="sigma"), y.lab=expression(g[2](sigma)))
```

9.6.2 Interfacing with rpart(): tr()

The `rpart()` function of package **rpart** fits decision trees, and the **gamlss.add** function `tr()` provides an interface to it. Hence, decision trees can be fitted as additive terms within GAMLSS. Here we give a small example of the use of this function. Note that the function is rather experimental and may be unstable, i.e. convergence may be dependent on the starting values.

```
r1 <- gamlss(R ~ tr(~Fl+A+H+B+loc), data=rent, family=GA, gd.tol=100)

## GAMLSS-RS iteration 1: Global Deviance = 27824.91
```

```
## GAMLSS-RS iteration 2: Global Deviance = 27833.9
## GAMLSS-RS iteration 3: Global Deviance = 27833.9

r2 <- gamlss(R ~ tr(~Fl+A+H+B+loc), sigma.fo=~tr(~Fl+A+H+B+loc),
          data=rent, family=GA, gd.tol=100, c.crit=0.1)

## GAMLSS-RS iteration 1: Global Deviance = 27779.77
## GAMLSS-RS iteration 2: Global Deviance = 27754.62
## GAMLSS-RS iteration 3: Global Deviance = 27754.53

AIC(r1,r2)

##     df       AIC
## r2 12 27778.53
## r1  9 27851.90
```

Note that for the second model we have increased the convergence criterion to 0.1. The plotting of the fitted decision trees for the predictor of a distribution parameter can be achieved by using `term.plot()`, or by individually plotting the fitted objects as demonstrated below:

```
term.plot(r2, parameter="mu", pages=1)
```

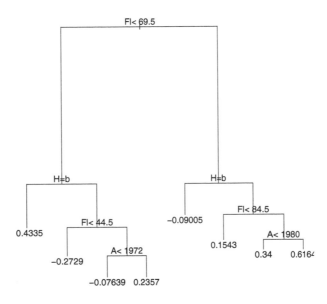

FIGURE 9.31: Visual representation for the μ parameter of the decision tree model r2.

```
plot(getSmo(r2, parameter="sigma"))
text(getSmo(r2, parameter="sigma"))
```

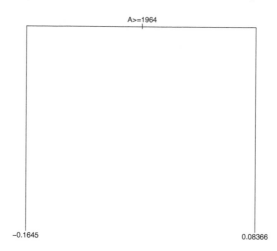

R

p. 3(

FIGURE 9.32: Visual representation for the σ parameter of the decision tree model r2.

9.6.3 Interfacing with `loess()`: `lo()`

loess (or 'lowess', locally weighted scatterplot smoothing) is a local regression smoother, as described in Section 9.3, created by Cleveland [1979]. loess fits smooth curves or surfaces for a dependent variable using one or more continuous explanatory variable, respectively. The function `loess()` and its older version `lowess()` are **R** implementations of the original Cleveland [1979] algorithm. `loess()` also provides an outlier resistant smoothing method specified by the argument `family="symmetric"` rather than the default `family="gaussian"` (see `?loess`). The default algorithm used for a single explanatory variable can be described in the following steps:

1. Define the window width or span for x.

2. Order the x values and focus on each x value one at a time.

3. Assign weights to the observations, using a kernel function to give greater weight to observations with x values that are closer to the x value of interest.

4. Fit a polynomial regression by weighted least squares and obtain the fitted dependent variable value at the x value of interest.

5. Repeat steps 3 and 4 for each x value in the data and plot the line connecting the fitted values of the dependent variable against x.

The **gamlss** package provides an interface to `loess()` through the smoothing term function `lo()`. This fits a local polynomial line or surface determined by one or more continuous predictors, within `gamlss()`, and is similar to `lo()` in the `gam()` implementation within the **gam** package [Chambers and Hastie, 1992]. Since there is no plotting function for fitted **loess** objects, a purpose built function vis.lo() has been created within **gamlss.add**.

The function lo() needs a formula for the declaration of the (continuous) explanatory variables. This formula is then passed to loess() for the actual fit. The formula accepts both the additive (+) and the interaction (*) symbols and the result will be fitted smooth curves and/or surfaces. Note that in loess(), it is recommended that no more than four continuous variables should be used for fitting surfaces since results may be unreliable for more variables. The authors only have experience with two-dimensional surfaces and the function vis.lo() only works for up to two explanatory variables.

The Munich rent data are used to demonstrate the use of lo() within **gamlss**. We fit two models: the first is an additive model for floor space (Fl) and year of construction (A) while the second is a smooth surface interaction model for Fl and A.

```
r1 <- gamlss(R~lo(~Fl)+lo(~A), data=rent, family=GA)# additive model

## GAMLSS-RS iteration 1: Global Deviance = 27921.71
## GAMLSS-RS iteration 2: Global Deviance = 27921.75
## GAMLSS-RS iteration 3: Global Deviance = 27921.75

r2 <- gamlss(R~lo(~Fl*A), data=rent, family=GA)   # interaction model

## GAMLSS-RS iteration 1: Global Deviance = 27910.92
## GAMLSS-RS iteration 2: Global Deviance = 27911.12
## GAMLSS-RS iteration 3: Global Deviance = 27911.12
```

We use **term.plot()** (which calls vis.lo()) to plot the fitted model for the predictor of the distribution parameter (in this case μ):

Fig. 9.33
```
term.plot(r1)
```

For more options in the plot, the function vis.lo()) can be used directly as we demonstrate below:

Fig. 9.34
```
vis.lo(getSmo(r1, which=1), partial=T)
vis.lo(getSmo(r1, which=2), partial=T)
```

The fitted interaction smooth surface is plotted using either the **term.plot()** function:

Fig. 9.35
```
term.plot(r2, pages=1)
```

or (for more options) using the vis.lo() function:

Fig. 9.36
```
vis.lo(getSmo(r2, which=1), se=1.97, main="(a) 95% CI")
vis.lo(getSmo(r2, which=1), partial.res=T,
        main="(b) partial residuals")
```

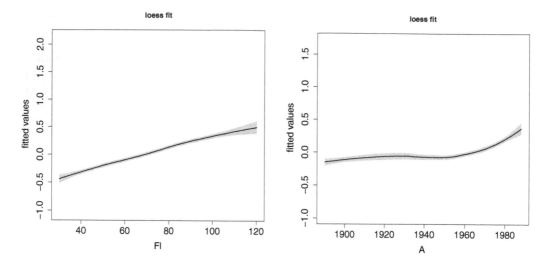

R
p. 31

FIGURE 9.33: Fitted terms for the additive `r1` model using `term.plot()`.

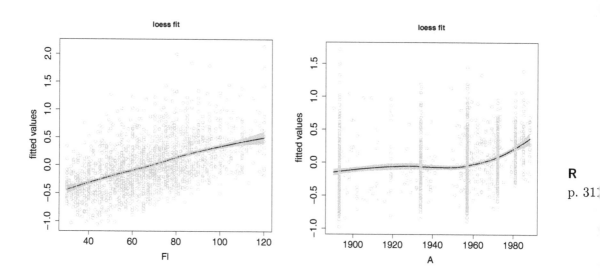

R
p. 31

FIGURE 9.34: The fitted additive `r1` model using the function `vis.lo()`.

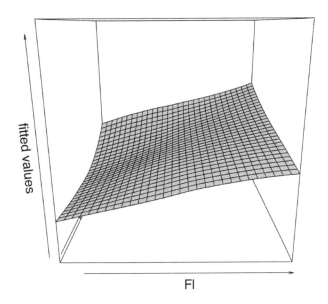

FIGURE 9.35: Fitted interaction surface for model `r2` using `term.plot()`.

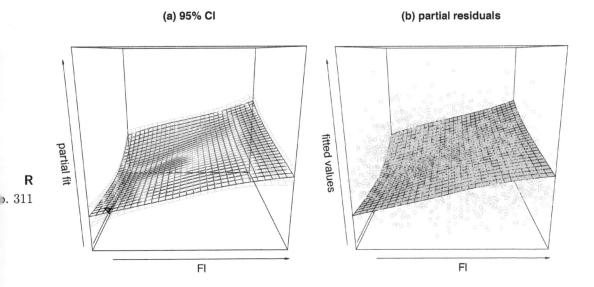

FIGURE 9.36: Fitted interaction surface of model `r2` with (a) an approximate 95% CI surface (b) partial residuals.

9.6.4   Interfacing with earth(): ma()

The **earth** package in **R** builds regression models using the techniques in the paper "Multivariate Adaptive Regression Splines" [Friedman, 1991], known as MARS. **earth** is an acronym for "Enhanced Adaptive Regression Through Hinges", used because the term MARS is trademarked.

The **gamlss.add2** package provides an interface with the function **earth()** through the smoothing terms function **ma()**, which fits nonlinear (relatively smooth) functions (through hinges) based upon one or more predictors (including interaction terms). In other words, the **ma()** function allows univariate and multivariate nonlinear functions within **gamlss**. The function **ma()** needs a formula for the declaration of the explanatory variables, which is passed to **earth()** for the fitting. To allow for interactions, the **degree** parameter controls the maximum degree of interactions to be considered. The default is **degree=1** which builds an additive model (i.e. no interactions).

The Munich rent data are used to demonstrate the use of **ma()**. We fit an additive model (with **degree=2**) to explain rent (**R**) as a function of the plausible predictors **Fl**, **A**, **loc**, **Sp** and **H**, without assuming all predictors and their interactions are necessarily needed to explain rent.

```
library(gamlss.add2)
ma1<-gamlss(R~ma(~Fl+A+loc+H, control=ma.control(degree=2,
            pmethod='exhaustive')),data=rent,family=GA,
            method=CG(20),trace=FALSE)
```

We use the **summary()** and **plotmo()** functions from **earth** to investigate the fitted model.

```
#get the earth model for the mu parameter.
ma.earth=getSmo(ma1,parameter = 'mu')
#print a summary of the earth model  for the mu parameter.
summary(ma.earth)
```

Fig.
9.37

```
## Call: earth(formula=form, data=Data, weights=W.var,
##             wp=control$wp, pmethod=control$pmethod,
##             keepxy=control$keepxy, trace=control$trace,
##             degree=control$degree, nprune=control$nprune,
##             ncross=control$ncross, nfold=control$nfold,
##             stratify=control$stratify,
##             Scale.y=control$Scale.y)
##
##             coefficients
## (Intercept)  0.061654578
## loc2         0.192792626
## loc3         0.274195750
## H1          -0.295646492
## h(90-Fl)    -0.010752490
## h(Fl-90)     0.007872509
```

```
## h(A-1968)      0.017617842
##
## Selected 7 of 8 terms, and 5 of 5 predictors
## Termination condition: RSq changed by less than 0.001 at 8 terms
## Importance: Fl, H1, A, loc3, loc2
## Number of terms at each degree of interaction: 1 6 (additive model)
## GCV 0.1344713    RSS 260.4884    GRSq 0.3668454    RSq 0.3764604
```

#plot the earth model for the mu parameter.
```
plotmo(ma.earth)
```

```
## plotmo grid:    Fl    A loc H
##                 67 1957   1 0
```

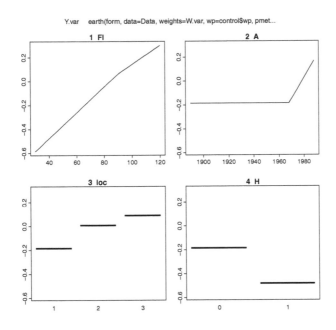

FIGURE 9.37: Visual representation for the μ parameter of the **earth** (MARS) model **r2**.

One of the advantages of the **ma()** function is the 'built-in' model selection technique. (See also Chapter 11 for model selection techniques in GAMLSS.) This means that the final model will only include the selected predictors. For example, the final model **ma.1** does not include interaction terms.

9.7   Bibliographic notes

Whittaker [1922] was the first to present a penalized smoothing technique now known as 'Whittaker smoothing' in a time series setting. The Whittaker technique can be seen in terms of equations (9.2) and (9.3) where the basis $\mathbf{Z}$ is an identity matrix; see for example Eilers [2003] and Eilers and Marx [1996].

Most of the current smoothing techniques have their genesis in the 1980s. Explanation about local regression models based on kernels is given in Härdle [1990]. Wand and Jones [1999] provide information on kernel smoothing. For robust loess scatterplot smoothing, see Cleveland [1979]. For local likelihood methods in smoothing see Tibshirani and Hastie [1987] and Staniswalis [1989]. Wahba [1990] gives a rigorous treatment of smoothing splines. Penalization and spline models in a variety of settings are discussed in Green and Silverman [1994]. Hastie and Tibshirani [1990] discuss the procedures concerning linear smoothers in detail. For applications in medicine, see Hastie and Herman [1990]. Ridge regression was introduced by Hoerl and Kennard [1970], and Lasso regression was proposed by Tibshirani [1996]. For a formulation of neural network using smoothing splines, see Roosen and Hastie [1994] and Venables and Ripley [2002]. Breiman et al. [1984] present regression trees and Li and Goel [2007] explain additive regression trees and smoothing splines.

For piecewise polynomials and splines, smoothing splines, multidimensional splines and wavelet smoothing see Hastie et al. [2009]. Ruppert et al. [2003] and Ruppert et al. [2009] review semiparametric regression. Wand and Ormerod [2008] present semiparametric regression with O'Sullivan penalized splines. Wang [2011] explores smoothing splines methods and applications. Wand [2003] presents smoothing and mixed models. Currie et al. [2006] present generalized linear array models with applications to multidimensional smoothing. Eilers and Marx [2010] compare numerical stability, quality of the fit, interpolation/extrapolation, derivative estimation, visual presentation and extension to multidimensional smoothing of different approaches of splines, and also discuss the mixed model. Fahrmeir and Tutz [2001] explain semi- and nonparametric approaches to regression analysis including smoothing splines, local regression and varying coefficient models. Park et al. [2015] give a review of varying coefficient models.

For smoothing procedures in spatial statistics see Fahrmeir et al. [2004]. For thin plate splines see for instance Wahba [1990], Wood [2006a], Fahrmeir et al. [2013]. Nychka [2000] explores the relationship between kriging and smoothing procedures that are based on radial basis functions. Eilers and Marx [2003] discuss bivariate polynomial splines in detail. Wood [2006a] presents a variety of smoothing bases such as cubic regression splines, P-splines, thin plate splines and tensor product bases and also shows smoothers as mixed model components. Kneib and Fahrmeir [2006] present a mixed model approach to structured additive regression for multi-categorical space-time data. Fahrmeir et al. [2013] give an overview of nonparametric regression for univariate and bivariate smoothing. Eilers et al. [2015] cover the history of P-splines and discuss how it can be seen as a mixed model component for correlated and uncorrelated data. For Bayesian P-splines see Lang and Brezger [2004]. Eilers and Marx [2010] discuss Bayesian parallels to penalized regression.

9.8 Exercises

1. **Victims of crime: smoothing with binary data.** The `VictimsOfCrime` data were introduced on page 259.

> **R data file:** `VictimsOfCrime` in package **gamlss.data** of dimensions 10590×2
>
> **variables**
>
> > `reported` : whether the crime was reported in local media (0 =no, 1 =yes)
> >
> > `age` : age of the victim
>
> **purpose:** to demonstrate binary data smoothing.

(a) Load the data and plot `reported` against `age`.

```
data(VictimsOfCrime)
plot(reported~age, data=VictimsOfCrime,  pch="|")
```

(b) Now use the different smoothers investigated in this chapter to fit smooth curves for `age`. Note that the response is binary and therefore the binomial distribution (BI) is used in the `family` argument. For example:

```
# P-splines
m1<- gamlss(reported~pb(age), data=VictimsOfCrime, family=BI)
```

The smoothers include **pb**, **pbm**, **cy**, **scs**, **lo**, **nn** and **tr**.

(c) Compare the results using AIC and SBC.

(d) Plot the different fitted μ (probability of a crime being reported in local media) for comparison. First study the behaviour of the P-spline based curves, i.e. pb(), pbm() and cy(), e.g.

```
plot(reported~age, data=VictimsOfCrime,  type="n")
with(VictimsOfCrime, lines(fitted(m1)[order(age)]~
                    age[order(age)],col="red", lwd=2))
```

(e) Compare the fitted curves of the P-splines and cubic splines.

(f) Compare the fitted curves of the P-splines and the neural network.

(g) Compare the P-splines with the decision trees fitted curves.

(h) Check the residuals of model `m1`. Note that for binary responses, the function `rqres.plot()` returns multiple realizations of the residuals.

```
rqres.plot(m1, ylin.all=.6)
```

(i) Obtain a multiple worm plot of the residuals.

```
wp(m1, xvar=age, n.inter=9)
```

2. **EU 15:** The eu15 data in **gamlss.data** package was first presented on page 110.

   (a) Load the data and transform it, as in Exercise 2 of Chapter 4, and plot the data to visually explore plausible relationships.

   ```
   #Create a function to generate a continuous colour palette
   rbPal <- colorRampPalette(c('red','blue'))
   #This adds a column of colour values based on the year values
   col <- rbPal(5)[as.numeric(cut(eu15$Year,breaks = 5))]
   with(eu15, plot(lLabor,lGDP,pch=21,col=col,main='Labor'))
   with(eu15, plot(lCapital,lGDP,pch=21,col=col,main='Capital'))
   with(eu15, plot(lUsefulEnergy,lGDP,pch=21,col=col,
                   main='Capital'))
   ```

   (b) The "Labor" plot shows clearly that high values are associated with different levels of GDP and time period (dark red indicates around 1960, dark blue around 2009), while there is a clearer relationship with Capital and UsefulEnergy (all on the log scale). To estimate the effects, we fit a GAMLSS model using the TF2 distribution (see Chapter 6 for the different choices of distributions).

   ```
   pbc=pb.control(method="GAIC",k=2.5)
   mod1<-gamlss(lGDP~pb(lCapital,control=pbc)+pb(lUsefulEnergy,
        control=pbc)+pb(lLabor,control=pbc),data=eu15,family=TF2)
   ```

   (c) Plot the model terms using **term.plot()**.

   (d) Plot the fitted mean of lGDP with the data.

   (e) Estimate the derivative of the smoothing functions.

   The function **getPEF()** can be used to calculate the partial effect and the elasticity of a continuous explanatory variable x.

   ```
   der.capital<-getPEF(mod1,data=eu15,term='lCapital',plot=T)
   der.labor<-getPEF(mod1,data=eu15,term='lLabor',plot=T)
   der.energy<-getPEF(mod1,data=eu15,term='lUsefulEnergy',plot=T)
   #estimate the average derivative, which gives us
   #the effect on (log) GDP
   der.capital(mean(eu15$lCapital), deriv=1)
   der.labor(mean(eu15$lLabor), deriv=1)
   der.energy(mean(eu15$lUsefulEnergy), deriv=1)
   ```

(f) Update `mod1` by fitting smooth functions for the σ and ν distribution parameters.

(g) Using the updated model, estimate the derivatives for μ, σ and ν. Do you observe any changes? Why?

(h) Using the updated model, fit a model with different smoothers (e.g. cubic splines) and compare the new model with `mod1`.

(i) Fit and plot a varying coefficient model whereby the coefficients of capital, useful energy and labor change smoothly according to year.

```
mod.pvc<-gamlss(lGDP~pvc(Year,by=lCapital)+
        pvc(Year,by=lUsefulEnergy)+pvc(Year,by=lLabor),
        data=eu15, family=TF2, n.cyc=100)
term.plot(mod.pvc,rug=T,ylim='free',ask=F,scheme='lines',
        pages=1)
```

(j) Compare the varying coefficient model `mod.pvc` with `mod1`.

3. **The Tokyo rainfall data: cycle smoothing with binomial data.** The data are taken from Kitagawa [1987] and contain observations from the years 1983 and 1984. They record whether there is more than 1 mm rainfall in Tokyo. The data consists of 366 observations of the variable `trd`, which takes values 0, 1, 2 on the number of times there was rain on the specific day of the year (during the two year period). Observation number 60 corresponds to 29 February, and therefore only one day is observed during the two years.

R data file: `trd` in package **gamlss.data** of dimensions 366×1

variable

　　`trd` : values 0, 1 or 2

purpose: to demonstrate cycle smoothing.

The data can be analysed using a binomial distribution with a binomial denominator equal to 2 (apart from 29 February which has 1). The data were analysed by Rue and Held [2005] and Fahrmeir and Tutz [2001].

(a) Load the data and plot them.

```
data(trd)
plot(trd)
```

(b) Define the binomial response variable (making sure that 29 February was observed only once) and a variable for the different days of the year.

```
NY <- 2-trd
NY[60] <- 1-trd[60]
y <- cbind(trd,NY) # response
ti <- 1:366 # time
```

(c) Fit a standard and a cycle P-spline smoother to your data and compare them using the GAIC.

```
t1 <- gamlss(y~pb(ti, inter=30), family=BI)
t2 <- gamlss(y~pbc(ti, inter=30), family=BI)
AIC(t1,t2)
```

(d) Plot the data and the fitted μ (i.e. fitted probability of rainfall) from the two fitted models. (You may divide y by 2 so the scale of the y-axis is compatible.)

```
plot(trd/2)
lines(fitted(t1)~ti, col=2, lty=3)
lines(fitted(t2)~ti, col=3,  lwd=2 )
```

(e) Use `term.plot()` to compare the fitted predictors for μ and comment.

(f) Use `getSmo()` to get information about the smoothing parameter.

(g) Fit different link functions and compare the results using GAIC. Which model and link function seems to fit best?

(h) Check the residuals for the different models. Since the fitted distribution is binomial, you may like to use `rqres.plot()` to obtain multiple realizations of the residuals. Obtain worm plots of the residuals. Comment on the plots.

4. Demonstrate that the penalized least squares quantity in (9.2) is equal to the following augmented least squares quantity:

$$\left\| \begin{bmatrix} \sqrt{\mathbf{W}} & \mathbf{0} \\ \mathbf{0} & \mathbf{I} \end{bmatrix} \left(\begin{bmatrix} \mathbf{y} \\ \mathbf{0} \end{bmatrix} - \begin{bmatrix} \mathbf{Z} \\ \sqrt{\lambda}\mathbf{D} \end{bmatrix} \gamma \right) \right\|^2$$

where $\|\mathbf{z}\|^2 = \mathbf{z}^\top \mathbf{z}$, $\mathbf{D}$ is any square root of the matrix $\mathbf{G}$ such that $\mathbf{G} = \mathbf{D}^\top \mathbf{D}$ and $\mathbf{W}$ is a diagonal matrix of weights. Discuss the implications of this result.

5. Let

$$\widetilde{\mathbf{Z}} = \begin{bmatrix} \mathbf{Z} \\ \sqrt{\lambda}\mathbf{D} \end{bmatrix}$$

be the augmented model matrix as above. Show that the sum of the first n elements of the diagonal of $\widetilde{\mathbf{Z}} \left(\widetilde{\mathbf{Z}}^\top \widetilde{\mathbf{Z}} \right)^{-1} \widetilde{\mathbf{Z}}^\top$ is $\mathrm{tr} \left(\mathbf{Z} \left(\mathbf{Z}^\top \mathbf{Z} + \lambda \mathbf{G} \right)^{-1} \mathbf{Z}^\top \right)$.

10

Random effects

CONTENTS

10.1 Introduction

This chapter explains how random effects models can be used within GAMLSS. In particular:

- it introduces the different ways in which random effects can be fitted within GAMLSS models;

- it explains the advantages and disadvantages of such modelling;

- it gives examples demonstrating the different approaches.

Random effect methodology is a powerful tool in statistics which was originally introduced to deal with *group* data. By group data we mean data in which the individual observations fall naturally into distinct groups in such a way that observations within a group are likely to be correlated. Therefore the basic assumption that the individual observations are independent of each other does not hold. Random effects can be used in the analysis of longitudinal studies, repeated measurements, block designs, multilevel data, time series, spatial and spatio-temporal analysis and also overdispersion. While GAMLSS was originally designed for independent observations, the fact that smoothing can be represented as a random effect model facilitates the idea of introducing random effects within GAMLSS.

Random effects models deal with correlation between individual observations by introducing different sources of variation in the data. This is achieved by assuming that there are random effects variables γ providing an additional source of variation, which have pdf $f(\gamma \mid \lambda)$, where λ is a vector of hyperparameters to be estimated from the data. We can think of γ as a vector of *latent* or unobserved random variables, which exist in the model to account for inter-dependence or overdispersion within the response variable. Under the GAMLSS model (3.2), given the random effects γ, the response variables $Y_1, Y_2, \ldots, Y_n$ are assumed to be independently distributed with pdf $f(y_i \mid \beta, \gamma)$. Under these assumptions the marginal probability (density) function of $\mathbf{Y} = (Y_1, Y_2, \ldots, Y_n)^\top$ is given by

$$\underbrace{f(\mathbf{Y} \mid \beta, \lambda)}_{\text{marginal } \mathbf{Y}} = \int_\gamma \overbrace{\underbrace{f(\mathbf{Y} \mid \beta, \gamma)}_{\text{conditional } \mathbf{Y} \mid \gamma} \underbrace{f(\gamma \mid \lambda)}_{\text{marginal } \gamma}}^{\text{joint } \mathbf{Y} \text{ and } \gamma} \, d\gamma \qquad (10.1)$$

where, given β and λ, $f(\mathbf{Y} \mid \beta, \lambda)$ denotes the marginal distribution of $\mathbf{Y}$, $f(\mathbf{Y} \mid \beta, \gamma)$ is the conditional distribution of $\mathbf{Y}$ given γ, and $f(\gamma \mid \lambda)$ is the marginal distribution for the random effects γ. The main obstacle to modelling the marginal distribution of $\mathbf{Y}$ is the integral over the random variables γ. Integrals over one or more dimensions are in general not easy to evaluate, and very seldom have explicit solutions. Under the assumption that both the conditional distribution of $\mathbf{Y}$ and the marginal distribution of γ are normal, the marginal distribution for $\mathbf{Y}$ may also be normal and therefore (10.1) has an explicit solution. Unfortunately with models with a conditional distribution within the exponential family (GLMs), and even more so within the `gamlss.family`, the marginal distribution of the response is, in general, not normal. To overcome the problem, there are several tricks to evaluate (10.1) implemented in the literature, including the EM algorithm, Laplace approximation, Markov chain Monte Carlo techniques (MCMC) or replacing the integral with a summation (e.g. Gaussian quadrature). All of these techniques come with some extra computational cost.

Within the GAMLSS models (3.2) an extra complication arises from the fact that there are potentially four different distribution parameters, each with its own predictor model, and therefore many ways to introduce random effects. In this chapter we examine how random effects are currently implemented within GAMLSS and how the hyperparameters $\boldsymbol{\lambda}$ are estimated. The following points are worth emphasizing here and will be considered in this chapter:

- the distinction between random effects at the *observational* level and random effects at the *factor* level;

- the difference between using the marginal likelihood of the response or the joint likelihood of the response and the random effects as an inferential tool [Pawitan, 2001, Lee et al., 2006, Wood, 2006a]; and

- the **gamlss** functions used to estimate random effects.

10.1.1 Random effects at the observational and at the factor level

When modelling random effects at the observational level, there are as many random effects as the number of observations, i.e. γ_i and Y_i for $i = 1, 2, \ldots, n$. The main application of random effects of this type is to deal with overdispersion, that is, when the variability in the data cannot be explained purely by the response distribution. Random effects at the observational level are used to generalize the distribution of Y to a more flexible, more general distribution. For example if $Y \,|\, \gamma \sim \texttt{PO}(\gamma\mu)$, and $\gamma \sim \texttt{GA}(1, \sigma^{1/2})$ then the marginal distribution of Y is $\texttt{NBI}(\mu, \sigma)$, the negative binomial distribution. (Provided that $\mathrm{E}(\gamma) = 1$, μ is the marginal mean of Y, a desirable property for interpretation of the fitted model.) Random effects at the observational level which provide an explicit marginal for Y are not the subject of this chapter, since if an explicit marginal distribution exists then, within GAMLSS, we can concentrate on modelling this marginal directly. Here, we are interested in random effects introduced in the predictors of distribution parameters. For example, assume for simplicity that we are dealing with only one set of random effects, $\boldsymbol{\gamma}_k$, for parameter $\boldsymbol{\theta}_k$. A random effect model of this parameter can be written as

$$g_k(\boldsymbol{\theta}_k) = \mathbf{X}_k\boldsymbol{\beta}_k + \mathbf{Z}_k\boldsymbol{\gamma}_k \,, \qquad \text{for } k = 1, 2, 3 \text{ or } 4.$$

Random effects at the observational level will have $\mathbf{Z}_k = \mathbf{I}_n$, the identity matrix, and therefore can be simplified as

$$g_k(\boldsymbol{\theta}_k) = \mathbf{X}_k\boldsymbol{\beta}_k + \boldsymbol{\gamma}_k \ .$$

Random effects at the factor level are the usual 'group' random effects. In this case one or more factors classify the observations into different categories, so $\mathbf{Z}_k$ in this case will be an incidence matrix with one if the observation belongs to a specific level of the factor and zero otherwise, i.e. a matrix of dummy variables.

The types of distribution for the random effect $\boldsymbol{\gamma}$ which we will consider in this chapter are: (i) nonparametric (discrete, see Section 10.2) and (ii) multivariate normal, i.e. $\boldsymbol{\gamma} \sim \mathcal{N}(\mathbf{0}, \boldsymbol{\Psi})$. The variance-covariance matrix $\boldsymbol{\Psi}$ of the multivariate normal involves the variance of the random effect σ_b^2, so for independent random effects we have $\boldsymbol{\Psi} = \sigma_b^2\mathbf{I}$.

10.1.2 Marginal and joint likelihood

Equation (3.2) gives the GAMLSS model as a random effects model. Section 3.4 gives a discussion on the different methods for estimating the random effects (or smoothing) hyperparameters $\boldsymbol{\lambda}$ which we called 'local' and 'global', depending on whether they are applied internally within the GAMLSS algorithm or externally.

The function `gamlssNP()` uses 'global' estimation and can apply Gaussian quadrature to approximate the marginal likelihood of the response variable Y by replacing the integration over a normally distributed random effects variable in the predictor for μ by a summation. This allows (global) estimation of the corresponding single random effects hyperparameter λ by maximizing the marginal likelihood of the data. (`gamlssNP()` can also fit (nonparametric) discrete random effects.)

In contrast the functions `random()` and `re()` use a 'local' normal approximation to the likelihood, known in the literature as penalized quasi likelihood (PQL) [Breslow and Clayton, 1993]. For these models the fitted values of the model are derived from the joint likelihood function, while inference for the behaviour of the response variable is based on the conditional likelihood given the random effects.

10.1.3 Functions available for fitting random effects

The function `gamlssNP()`, first introduced in Chapter 7 for fitting finite mixtures with common parameters, provides a way of fitting the marginal distribution of the response variable. It is, however, rather limited in the type of random effect models used. Figure 10.1 describes the capabilities of `gamlssNP()`. The left of the diagram shows that the function can be used to fit a normal random intercept model in the predictor for the location parameter μ. This is achieved by approximating the integral in (10.1) using Gaussian quadrature, and is discussed in Section 10.3. On the right side of the diagram it is shown that `gamlssNP()` can also fit nonparametric

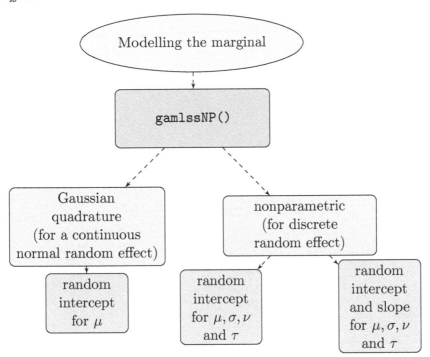

FIGURE 10.1: Diagram showing the different random effects that `gamlssNP()` can estimate using the marginal likelihood.

random effect models, discussed in Section 10.2. Random intercepts and random slopes models for μ, σ, ν and τ can be fitted using this method.

The functions `random()` and `re()` are based on PQL methodology, and can be used like all other additive smoothing terms within GAMLSS to model any or all of the parameters of the response distribution, using normal random effects. Figure 10.2 shows the different functionalities of these two functions.

The function `random()` (left of Figure 10.2), was originally based on the function with the same name in the package **gam**, but has been amended here to allow the estimation of the smoothing parameter using the local ML method described in Sections 3.4.2.1 and 9.4 and also Rigby and Stasinopoulos [2013]. It can be used for fitting normal random intercept models in the predictors of one or more of the parameters of the response distribution.

The function `re()` is an interface to the widely used function `lme()` of the **nlme** package. The different capabilities of `re()` are shown on the right of Figure 10.2. `re()` can be used for normally distributed random intercepts and slopes, repeated measurements, multilevel modelling and all other models implemented in `lme()`, as described by Pinheiro and Bates [2000].

Section 10.2 introduces nonparametric random effects models, while Section 10.3 focuses on a normal random effects model in the predictor for μ. Section 10.4 explains the `gamlssNP()` function for fitting the models in Sections 10.2 and 10.3 by 'global'

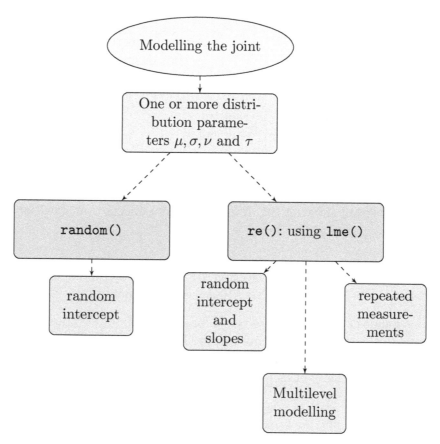

FIGURE 10.2: Diagram showing different types of normally distributed random effects that can be fitted using the functions **random()** and **re()**, which use PQL (joint likelihood).

estimation using the marginal likelihood and Section 10.5 provides examples. Sections 10.6 and 10.8 explain the `random()` and `re()` functions, with examples in Sections 10.7 and 10.9, respectively.

10.2 Nonparametric random effect models

In this section we explain the nonparametric random effect model and how it can be fitted using the EM algorithm. The nonparametric random effect model can be applied to one or more of the distribution parameters. Readers who are familiar with the model, or who are interested only in practical aspects, can move to Section 10.3.

10.2.1 Nonparametric random intercept model for μ at the factor level

Nonparametric random effect models for μ may be specified at the factor level or observational level. Here we consider the factor level. The observational level is a special case where there is one observation in each group.

Let y_{ij} be observation i on subject (or group) j for $i = 1, 2, \ldots, n_j$ and $j = 1, 2, \ldots, J$. (For example the ith student at the jth school, or the ith measurement of the jth person.) Let

$$Y_{ij} \mid \gamma_j \sim \mathcal{D}(\mu_{ij}, \sigma_{ij}, \nu_{ij}, \tau_{ij})$$

where

$$\begin{aligned}
g_1(\mu_{ij}) &= \mathbf{x}_{1ij}^\top \boldsymbol{\beta}_1 + \gamma_j \\
g_2(\sigma_{ij}) &= \mathbf{x}_{2ij}^\top \boldsymbol{\beta}_2 \\
g_3(\nu_{ij}) &= \mathbf{x}_{3ij}^\top \boldsymbol{\beta}_3 \\
g_4(\tau_{ij}) &= \mathbf{x}_{4ij}^\top \boldsymbol{\beta}_4 \,,
\end{aligned} \tag{10.2}$$

and γ_j is the random intercept for subject j. We write (10.2) in matrix notation as

$$\begin{aligned}
g_1(\boldsymbol{\mu}) &= \mathbf{X}_1 \boldsymbol{\beta}_1 + \mathbf{Z}_1 \boldsymbol{\gamma} \\
g_2(\boldsymbol{\sigma}) &= \mathbf{X}_2 \boldsymbol{\beta}_2 \\
g_3(\boldsymbol{\nu}) &= \mathbf{X}_3 \boldsymbol{\beta}_3 \\
g_4(\boldsymbol{\tau}) &= \mathbf{X}_4 \boldsymbol{\beta}_4
\end{aligned} \tag{10.3}$$

where

$$
\mathbf{Z}_1 = \begin{pmatrix}
1 & 0 & \cdots & 0 \\
\vdots & \vdots & \vdots & \vdots \\
1 & 0 & \cdots & 0 \\
0 & 1 & \cdots & 0 \\
\vdots & \vdots & \vdots & \vdots \\
0 & 1 & \cdots & 0 \\
\vdots & \vdots & \vdots & \vdots \\
0 & 0 & \cdots & 1 \\
\vdots & \vdots & \vdots & \vdots \\
0 & 0 & \cdots & 1
\end{pmatrix}
\tag{10.4}
$$

is an $n \times J$ matrix of dummy variables, where $n = \sum_{j=1}^{J} n_j$ and $\boldsymbol{\gamma}^\top = (\gamma_1, \gamma_2, \ldots, \gamma_J)$.

The random effects γ_j are assumed to be independent and coming from a *nonparametric (discrete) random effect* distribution given by

$$
\mathrm{Prob}(\gamma_j = u_k) = \pi_k, \qquad \text{for } k = 1, \ldots, \mathcal{K} .
\tag{10.5}
$$

Figure 10.3 shows an example of a nonparametric distribution with $\mathcal{K} = 5$. The u_k's and π_k's are assumed to be fixed but unknown constants and are estimated from the data. The u_k's are called the *mass points* for the nonparametric distribution while the π_k's are their probabilities. Note that if $n_j = 1$ for $j = 1, 2, \ldots, J$ then model (10.3) is a nonparametric random intercept model for μ at the observation level.

10.2.2   Fitting the nonparametric random intercept model for μ at the factor level

We denote the mass points as $\mathbf{u} = (u_1, u_2, \ldots, u_{\mathcal{K}})^\top$, their probabilities as $\boldsymbol{\pi} = (\pi_1, \pi_2, \ldots, \pi_{\mathcal{K}})^\top$ and the parameters of model (10.3), (10.4) and (10.5) as $\boldsymbol{\psi} = (\boldsymbol{\theta}, \boldsymbol{\pi})$ where, in this chapter, $\boldsymbol{\theta} = (\boldsymbol{\beta}_1, \boldsymbol{\beta}_2, \boldsymbol{\beta}_3, \boldsymbol{\beta}_4, \mathbf{u})$.

In order to fit model (10.3), (10.4) and (10.5) using the EM algorithm we need to define indicator variables $\boldsymbol{\delta}$. Let $\boldsymbol{\delta}_j = (\delta_{j1}, \delta_{j2}, \ldots, \delta_{j\mathcal{K}})^\top$ be the indicator variable vector for subject j, where

$$
\delta_{jk} = \begin{cases} 1 & \text{if subject } j \text{ has random effect value } u_k, \text{ i.e. } \gamma_j = u_k \\ 0 & \text{otherwise.} \end{cases}
\tag{10.6}
$$

Let $\boldsymbol{\delta}^\top = (\boldsymbol{\delta}_1^\top, \boldsymbol{\delta}_2^\top, \ldots, \boldsymbol{\delta}_J^\top)$ combine all indicator variable vectors. If the values for $\boldsymbol{\delta}$

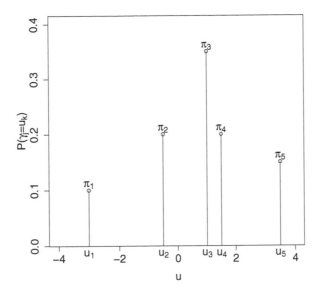

FIGURE 10.3: An example of a nonparametric distribution, with $\mathcal{K} = 5$.

were observed then we could write down the complete likelihood for the sample as:

$$
\begin{aligned}
L_c = L_c(\boldsymbol{\psi}, \mathbf{y}, \boldsymbol{\delta}) &= f(\mathbf{y}, \boldsymbol{\delta}) = f(\mathbf{y} \mid \boldsymbol{\delta}) f(\boldsymbol{\delta}) \\
&= \prod_{j=1}^{J} \left\{ \left[\prod_{i=1}^{n_j} f(y_{ij} \mid \boldsymbol{\delta}_j) \right] f(\boldsymbol{\delta}_j) \right\} \\
&= \prod_{j=1}^{J} \left\{ \left[\prod_{i=1}^{n_j} \prod_{k=1}^{\mathcal{K}} f_k(y_{ij})^{\delta_{jk}} \right] \prod_{k=1}^{\mathcal{K}} \pi_k^{\delta_{jk}} \right\} \\
&= \prod_{j=1}^{J} \prod_{k=1}^{\mathcal{K}} \left\{ \left[\prod_{i=1}^{n_j} f_k(y_{ij})^{\delta_{jk}} \right] \pi_k^{\delta_{jk}} \right\} .
\end{aligned} \tag{10.7}
$$

Hence the complete log-likelihood is given by:

$$
\ell_c = \ell_c(\boldsymbol{\psi}, \mathbf{y}, \boldsymbol{\delta}) = \sum_{j=1}^{J} \sum_{k=1}^{\mathcal{K}} \sum_{i=1}^{n_j} \delta_{jk} \log f_k(y_{ij}) + \sum_{j=1}^{J} \sum_{k=1}^{\mathcal{K}} \delta_{jk} \log \pi_k. \tag{10.8}
$$

However $\boldsymbol{\delta}$ is not observed, so is regarded as variables with missing values. The EM algorithm maximizes the marginal likelihood $L(\boldsymbol{\psi}, \mathbf{y})$ with respect to the parameters $\boldsymbol{\psi}$ using two steps. In the (expectation) E-step, the function $\mathbf{Q}$ is defined by replacing the missing values $\boldsymbol{\delta}$ in the complete log-likelihood (10.8) by their conditional expectations given $\mathbf{y}$ and the current value of $\boldsymbol{\psi}$. In the M-step the $\mathbf{Q}$ function is maximized with respect to $\boldsymbol{\psi}$.

The Q function

Let us first define the **Q** function for model (10.3), (10.4) and (10.5). The conditional expectation of the complete log-likelihood given $\mathbf{y}$ and the current parameter estimates $\hat{\boldsymbol{\psi}}^{(r)}$ at iteration r is:

$$\mathbf{Q} = \sum_{k=1}^{\mathcal{K}} \sum_{j=1}^{J} \sum_{i=1}^{n_j} \hat{w}_{jk}^{(r+1)} \log f_k(y_{ij}) + \sum_{k=1}^{\mathcal{K}} \sum_{j=1}^{J} \hat{w}_{jk}^{(r+1)} \log \pi_k \qquad (10.9)$$

where

$$
\begin{aligned}
\hat{w}_{jk}^{(r+1)} &= \mathrm{E}\left[\delta_{jk} \,|\, \mathbf{y}, \hat{\boldsymbol{\psi}}^{(r)}\right] \\[2mm]
&= \mathrm{Prob}\left(\delta_{jk} = 1 \,|\, \mathbf{y}, \hat{\boldsymbol{\psi}}^{(r)}\right) \\[2mm]
&= \frac{f\left(\mathbf{y} \,|\, \delta_{jk} = 1, \hat{\boldsymbol{\psi}}^{(r)}\right) \; \mathrm{Prob}\left(\delta_{jk} = 1 \,|\, \hat{\boldsymbol{\psi}}^{(r)}\right)}{f\left(\mathbf{y} \,|\, \hat{\boldsymbol{\psi}}^{(r)}\right)} \\[2mm]
&= \frac{\hat{\pi}_k^{(r)} \prod_{i=1}^{n_j} f_k\left(y_{ij} \,|\, \hat{\boldsymbol{\theta}}^{(r)}\right)}{\sum_{h=1}^{\mathcal{K}} \left[\hat{\pi}_h^{(r)} \prod_{i=1}^{n_j} f_h(y_{ij} \,|\, \hat{\boldsymbol{\theta}}^{(r)})\right]} \\[2mm]
&= \frac{\hat{\pi}_k^{(r)} m_{jk}}{\sum_{h=1}^{\mathcal{K}} \hat{\pi}_h^{(r)} m_{jh}} \qquad (10.10)
\end{aligned}
$$

where $f_k(y_{ij} \,|\, \hat{\boldsymbol{\theta}}^{(r)}) = f(y_{ij} \,|\, \delta_{jk} = 1, \hat{\boldsymbol{\theta}}^{(r)})$ and $m_{jk} = \prod_{i=1}^{n_j} f_k(y_{ij} \,|\, \hat{\boldsymbol{\theta}}^{(r)})$. Note that $\hat{w}_{jk}$ in (10.10) can be interpreted as the posterior probability for the jth individual to have mass point u_k.

The first term in (10.9) is just a weighted log-likelihood for a GAMLSS model for $n \times \mathcal{K}$ observations (where $n = \sum_{j=1}^{J} n_j$), with dependent variable $y_{ijk} = y_{ij}$ and predictor for μ given by

$$g_1(\mu_{ijk}) = \mathbf{x}_{1ij}^{\top} \boldsymbol{\beta}_1 + u_k$$

with weights $w_{ijk} = \hat{w}_{jk}^{(r+1)}$ for $i = 1, 2, \ldots, n_j$, $j = 1, 2, \ldots, J$ and $k = 1, 2, \ldots, \mathcal{K}$. An easy way to fit this model is to expand the data $\mathcal{K}$ times and then fit a weighted GAMLSS model. Table 10.1 shows how this expansion is achieved. (It is assumed that the cases are ordered by i within j, for ease of presentation.) Each data point is replicated $\mathcal{K}$ times and a factor created with $\mathcal{K}$ levels to indicate the different replications. In the `gamlssNP()` implementation, this factor is called `MASS`. Note that only the weights are changing in Table 10.1 in each iteration of the EM algorithm. We now describe the algorithm.

Summary of the $(r+1)$st iteration of the EM algorithm

E-step For $k = 1, 2, \ldots, \mathcal{K}$ and $j = 1, 2, \ldots, J$:

    Estimate δ_{jk} by $\hat{w}_{jk}^{(r+1)}$ (its current conditional expectation given $\mathbf{y}$ and the current estimate $\hat{\psi}^{(r)}$ at iteration r) obtained from (10.10) to give (10.9), where $\psi = (\theta, \pi)$ and $\theta = (\beta_1, \beta_2, \beta_3, \beta_4, \mathbf{u})$. This step just updates the weights $\hat{w}^{(r+1)}$ in Table 10.1.

M-step

(1) To estimate $\theta = (\beta_1, \beta_2, \beta_3, \beta_4, \mathbf{u})$ fit a single GAMLSS model to the expanded data given in Table 10.1 using the expanded dependent variable $\mathbf{y}_e$ with the expanded explanatory variables $\mathbf{x}_{1e}$ and factor MASS (i.e. column headed k) in the predictor for μ and the expanded explanatory variables $\mathbf{x}_{2e}$, $\mathbf{x}_{3e}$ and $\mathbf{x}_{4e}$ in the predictors for σ, ν and τ, respectively, using prior weights $\hat{\mathbf{w}}^{(r+1)}$.

(2) To estimate the (prior) probabilities π use

$$\hat{\pi}_k^{(r+1)} = \frac{1}{J} \sum_{j=1}^{J} \hat{w}_{jk}^{(r+1)} \qquad \text{for } k = 1, 2, \ldots, \mathcal{K} \, .$$

Usually several steps of the EM algorithm are required for the algorithm to converge. Convergence of the EM algorithm provides maximum (marginal) likelihood estimates of the parameters $\hat{\psi} = (\hat{\theta}, \hat{\pi})$. The maximized (marginal) likelihood is given by

$$L = \prod_{j=1}^{J} \left[\prod_{i=1}^{n_j} f(y_{ij} \mid \hat{\psi}) \right] = \prod_{j=1}^{J} \prod_{i=1}^{n_j} \sum_{k=1}^{\mathcal{K}} \hat{\pi}_k f_k(y_{ij} \mid \hat{\theta}) \, .$$

10.2.3 Nonparametric random intercept and slopes model for μ

In order to accommodate random slopes (in a variable Z) as well as random intercepts, model (10.2) is now modified to:

$$\begin{aligned}
g_1(\mu_{ij}) &= \mathbf{x}_{1ij}^\top \beta_1 + \gamma_{0j} + \gamma_{1j} z_{ij} \\
g_2(\sigma_{ij}) &= \mathbf{x}_{2ij}^\top \beta_2 \\
g_3(\nu_{ij}) &= \mathbf{x}_{3ij}^\top \beta_3 \\
g_4(\tau_{ij}) &= \mathbf{x}_{4ij}^\top \beta_4 \, ,
\end{aligned} \qquad (10.11)$$

for $i = 1, 2, \ldots, n_j$ and $j = 1, 2, \ldots, J$. In matrix form this model can still be written as (10.3), but the design matrix $\mathbf{Z}_1$ now includes an interaction matrix of

TABLE 10.1: Table showing the $\mathcal{K}$-fold expansion of the data for fitting a nonparametric random effect model at the M-step of the EM algorithm

k	i	j	y_e	x_{1e}	x_{2e}	x_{3e}	x_{4e}	$\hat{\mathbf{w}}^{(r+1)}$
1	1	1						$\hat{\mathbf{w}}_{11}^{(r+1)}$
1	2	1	$\mathbf{y_1}$	$\mathbf{x}_{11}$	$\mathbf{x}_{21}$	$\mathbf{x}_{31}$	$\mathbf{x}_{41}$	$\hat{\mathbf{w}}_{11}^{(r+1)}$
$\vdots$	$\vdots$	$\vdots$						$\vdots$
1	n_1	1						$\hat{\mathbf{w}}_{11}^{(r+1)}$
1	1	2						$\hat{\mathbf{w}}_{21}^{(r+1)}$
1	2	2	$\mathbf{y_2}$	$\mathbf{x}_{12}$	$\mathbf{x}_{22}$	$\mathbf{x}_{32}$	$\mathbf{x}_{42}$	$\hat{\mathbf{w}}_{21}^{(r+1)}$
$\vdots$	$\vdots$	$\vdots$						$\vdots$
1	n_2	2						$\hat{\mathbf{w}}_{21}^{(r+1)}$
$\vdots$	$\vdots$	$\vdots$	$\vdots$	$\vdots$	$\vdots$	$\vdots$	$\vdots$	$\vdots$
1	1	J						$\hat{\mathbf{w}}_{J1}^{(r+1)}$
1	2	J	$\mathbf{y}_J$	$\mathbf{x}_{1J}$	$\mathbf{x}_{2J}$	$\mathbf{x}_{3J}$	$\mathbf{x}_{4J}$	$\hat{\mathbf{w}}_{J1}^{(r+1)}$
$\vdots$	$\vdots$	$\vdots$						$\vdots$
1	n_J	J						$\hat{\mathbf{w}}_{J1}^{(r+1)}$
$\vdots$	$\vdots$	$\vdots$	$\vdots$	$\vdots$	$\vdots$	$\vdots$	$\vdots$	$\vdots$
$\mathcal{K}$	1	1						$\hat{\mathbf{w}}_{1\mathcal{K}}^{(r+1)}$
$\mathcal{K}$	2	1	$\mathbf{y_1}$	$\mathbf{x}_{11}$	$\mathbf{x}_{21}$	$\mathbf{x}_{31}$	$\mathbf{x}_{41}$	$\hat{\mathbf{w}}_{1\mathcal{K}}^{(r+1)}$
$\vdots$	$\vdots$	$\vdots$						$\vdots$
$\mathcal{K}$	n_1	1						$\hat{\mathbf{w}}_{1\mathcal{K}}^{(r+1)}$
$\mathcal{K}$	1	2						$\hat{\mathbf{w}}_{2K}^{(r+1)}$
$\mathcal{K}$	2	2	$\mathbf{y_2}$	$\mathbf{x}_{12}$	$\mathbf{x}_{22}$	$\mathbf{x}_{32}$	$\mathbf{x}_{42}$	$\hat{\mathbf{w}}_{2\mathcal{K}}^{(r+1)}$
$\vdots$	$\vdots$	$\vdots$						$\vdots$
$\mathcal{K}$	n_2	2						$\hat{\mathbf{w}}_{2\mathcal{K}}^{(r+1)}$
$\vdots$	$\vdots$	$\vdots$	$\vdots$	$\vdots$	$\vdots$	$\vdots$	$\vdots$	$\vdots$
$\mathcal{K}$	1	J						$\hat{\mathbf{w}}_{J\mathcal{K}}^{(r+1)}$
$\mathcal{K}$	2	J	$\mathbf{y}_J$	$\mathbf{x}_{1J}$	$\mathbf{x}_{2J}$	$\mathbf{x}_{3J}$	$\mathbf{x}_{4J}$	$\hat{\mathbf{w}}_{J\mathcal{K}}^{(r+1)}$
$\vdots$	$\vdots$	$\vdots$						$\vdots$
$\mathcal{K}$	n_J	J						$\hat{\mathbf{w}}_{J\mathcal{K}}^{(r+1)}$

Note: Here we assume, for simplicity of presentation, that there are only four explanatory variable vectors $\mathbf{x}_1$, $\mathbf{x}_3$, $\mathbf{x}_3$ and $\mathbf{x}_4$ (one for each distribution parameter μ, σ, ν and τ) which combine the values for the J levels of the factor, e.g. $\mathbf{x}_1^\top = (\mathbf{x}_{11}^\top, \mathbf{x}_{12}^\top, \ldots, \mathbf{x}_{1J}^\top)$ where, e.g. $\mathbf{x}_{1J}^\top = (x_{11J}, x_{12J}, \ldots, x_{1n_J J})$. The column headed by k indicates the levels of a factor which in the `gamlssNP()` implementation is called `MASS`. The only variable above which changes in each iteration of the EM algorithm is the weights, $\mathbf{w}$.

the random effect factor with the continuous variable Z, i.e.

$$\mathbf{Z}_1 = \begin{pmatrix} 1 & z_{11} & 0 & 0 & \ldots & 0 & 0 \\ \vdots & \vdots & \vdots & \vdots & \vdots & \vdots & \vdots \\ 1 & z_{n_1 1} & 0 & 0 & \ldots & 0 & 0 \\ 0 & 0 & 1 & z_{12} & \ldots & 0 & 0 \\ \vdots & \vdots & \vdots & \vdots & \vdots & \vdots & \vdots \\ 0 & 0 & 1 & z_{n_2 2} & \ldots & 0 & 0 \\ \vdots & \vdots & \vdots & \vdots & \vdots & \vdots & \vdots \\ 0 & 0 & 0 & 0 & \ldots & 1 & z_{1J} \\ \vdots & \vdots & \vdots & \vdots & \vdots & \vdots & \vdots \\ 0 & 0 & 0 & 0 & \ldots & 1 & z_{n_J J} \end{pmatrix}.$$

The random effects are now double indexed as

$$\boldsymbol{\gamma} = (\gamma_{01}, \gamma_{11}, \gamma_{02}, \gamma_{12}, \ldots, \gamma_{0J}, \gamma_{1J})^{\top}$$

where the first subscript indicates whether the parameter is for the intercept (0) or the slopes (1), and the second subscript is for the level of the random factor. Each pair $(\gamma_{0j}, \gamma_{1j})$ is assumed to be a random observation from a bivariate 'nonparametric' distribution, taking values (u_{0k}, u_{1k}) with probability π_k for $k = 1, 2, \ldots, \mathcal{K}$. That is,

$$\text{Prob}(\gamma_{0j} = u_{0k}, \gamma_{1j} = u_{1k}) = \pi_k, \qquad \text{for } k = 1, \ldots, \mathcal{K}, \qquad (10.12)$$

independently for $j = 1, 2, \ldots, J$. In Figure 10.4 a hypothetical two dimensional nonparametric distribution with $\mathcal{K} = 10$ is plotted. Note that in practice an important point in any nonparametric random effect modelling is determination of the number of mass points $\mathcal{K}$.

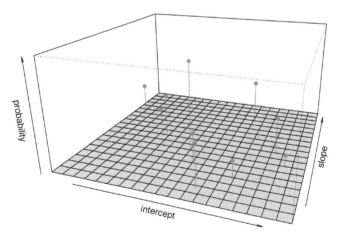

FIGURE 10.4: A nonparametric distribution in two dimensions with $\mathcal{K} = 10$.

The fitting of the random intercept and slopes model for μ can be achieved with the same EM algorithm described in this section for the fitting of the random intercept model. In Section 10.4 we describe the use of `gamlssNP()` for fitting the nonparametric random effects.

10.3   Normal random effect models

Here we include a normal random intercept in the predictor model for μ in model (10.3), i.e. assume

$$\gamma_j \sim \mathcal{N}(0, \sigma_b^2) \quad \text{independently for } j = 1, 2, \ldots, J. \tag{10.13}$$

The model defined by equations (10.3), (10.4) and (10.13) may be fitted using Gaussian quadrature to approximate the integral of equation (10.1).[1] Gaussian quadrature is a numerical integration method in which the integral is approximated by a summation. Gaussian quadrature replaces the continuous distribution $f(\gamma)$ with an approximating discrete distribution.

Hence, from (10.13), $\boldsymbol{\gamma}$ has a normal distribution,

$$\boldsymbol{\gamma} \sim \mathcal{N}(\mathbf{0}, \sigma_b^2 \mathbf{I}) \ .$$

Then by letting

$$\gamma_j = \sigma_b U_j \ ,$$

where

$$U_j \overset{\text{ind}}{\sim} \mathcal{N}(0, 1) \ ,$$

it follows that

$$\mathbf{U} = \frac{\boldsymbol{\gamma}}{\sigma_b} \sim \mathcal{N}(\mathbf{0}, \mathbf{1}) \ ,$$

where $\mathbf{U} = (U_1, U_2, \ldots, U_J)^\top$. Gaussian quadrature effectively approximates the continuous $\mathcal{N}(0, 1)$ distribution for each U_j by a discrete distribution:

$$\text{Prob}(U_j = u_k) = \pi_k \ , \qquad \text{for } k = 1, \ldots, \mathcal{K} \ , \tag{10.14}$$

where the u_k's and the π_k's are fixed and known for a fixed total number of discrete points $\mathcal{K}$ used for the Gaussian quadrature approximation. Figure 10.5 gives a visual explanation of the Gaussian quadrature discrete distribution approximation to $\mathcal{N}(0, 1)$ with $\mathcal{K} = 10$. The model (10.3), (10.4) and (10.14) can now be considered as a finite mixture of $\mathcal{K}$ components in which the prior (or mixing) probabilities π_k's

---

[1] "Gaussian" refers to the originator of the method [Gauss, 1815] and not to any specific (e.g. normal) distribution for $f(\gamma)$.

are fixed and known and the u_k's are also fixed and known (once the total number of quadrature points $\mathcal{K}$ has been chosen). Hence the predictor for μ_{ijk} is

$$g_1(\mu_{ijk}) = \mathbf{x}_{1ij}^\top \boldsymbol{\beta}_1 + \sigma_b u_k$$

with (component) prior probability π_k, for $k = 1, 2, \ldots, \mathcal{K}$. Note that the estimated coefficient of the explanatory variable U ($\mathbf{Z}$ in the gamlss output) in the predictor for μ is an estimate of σ_b.

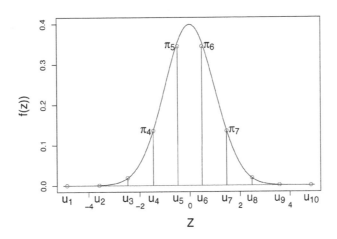

FIGURE 10.5: Plot showing how the $\mathcal{N}(0, 1)$ distribution is approximated by Gaussian quadrature with $\mathcal{K} = 10$.

10.3.1 Summary of the $(r + 1)$st iteration of the EM algorithm

The model is fitted using the following EM algorithm (amended from the algorithm in Section 10.2.2):

E-step Update the weights $\hat{w}^{(r+1)}$ in Table 10.1. For $k = 1, 2, \ldots, \mathcal{K}$ and $j = 1, 2, \ldots, J$: replace δ_{jk} in (10.8) by weights $\hat{w}_{jk}^{(r+1)}$ (its current conditional expectation given y and given the current estimate $\hat{\psi}^r$ at iteration r), obtained from (10.10), where $\hat{\pi}_k^{(r)} = \pi_k$. This gives (10.9), where $\psi = (\boldsymbol{\beta}_1, \boldsymbol{\beta}_2, \boldsymbol{\beta}_3, \boldsymbol{\beta}_4, \sigma_b)$.

M-step To estimate $\psi = (\boldsymbol{\beta}_1, \boldsymbol{\beta}_2, \boldsymbol{\beta}_3, \boldsymbol{\beta}_4, \sigma_b)$, fit a single GAMLSS model to the data given in Table 10.1 (extended to include an extra column $\mathbf{u}$ with values $u_{ijk} = u_k$ for $k = 1, 2, \ldots, \mathcal{K}$; $i = 1, 2, \ldots, n_j$ and $j = 1, 2, \ldots, J$) using the expanded dependent variable $\mathbf{y}_e$ with the expanded explanatory variables $\mathbf{x}_{1e}$ and $\mathbf{u}$ in the predictor for μ and the expanded explanatory variables $\mathbf{x}_{2e}, \mathbf{x}_{3e}$ and $\mathbf{x}_{4e}$ in the predictors for σ, ν and τ, respectively, using prior weights $\hat{\mathbf{w}}^{(r+1)}$.

10.4   The function `gamlssNP()` for random effects

The function `gamlssNP()`, of package **gamlss.mx**, was first introduced in Chapter 7 as a way to fit finite mixtures with parameters in common; see Section 7.5. It is based on the work of Einbeck et al. [2014] and uses an EM algorithm which requires expanding the data $\mathcal{K}$ times, as shown in general in Table 7.1 where the iterative weights are denoted by $\hat{\mathbf{p}}$ rather than $\hat{\mathbf{w}}$ in this chapter (as for example in Table 10.1). The function `gamlssNP()` can be used to fit:

1. normal random intercept models for μ, or

2. nonparametric random effect models (which can be random intercept, or random intercept and slope models) for μ, σ, ν and/or τ.

The arguments for the function were shown in Section 7.6, but here we would like to emphasize three of them which are crucial for fitting the random effect models.

`random`   A formula defining the random part of the model.

> **Important**: For random intercept and random intercept and slope models the argument `random` of the `gamlssNP()` function has to be set:
>
> - for a normal or nonparametric random intercept model at a factor level say (**f**), set `random =~1|f` ;
>
> - for a nonparametric random intercept and slope model at a factor level (**f**) with explanatory variable **x**, set `random=~x|f`;
>
> - for a normal or nonparametric random intercept model at the observation level no specification of `random` is needed.

`mixture`   The type of random effect:

> `gq` for normal random intercept models using Gaussian quadrature,
>
> `np` for nonparametric random effect models.

`K`   the number of points of the nonparametric mixture if `mixture="np"`, or the number of Gaussian quadrature points if `mixture="gq"`.

Next we will explain how to use `gamlssNP()`, firstly to fit normal random intercept models in the predictor for μ using Gaussian quadrature, and secondly to fit nonparametric random effect models for one or more of the parameters of the distribution.

10.4.1 Fitting a normal random intercept for μ

In order to define a normal random intercept in the predictor for μ, the user has to set the arguments `mixture="gq"` for the Gaussian quadrature and K for the number of Gaussian quadrature points. We recommend at least `K=10` for reasonable accuracy. Typically the minimal code for fitting a normal random intercept at the observational level is:

```
### random effect at the observation level
m1 <- gamlssNP(y~1, K=20, mixture="gq")
```

and at the factor level (with factor `f`):

```
### random effect at the factor level
m2 <- gamlssNP(y~1, K=10, random=~1|f, mixture="gq")
```

Note that a normal random intercept and slope model cannot be fitted using `gamlssNP()`.

10.4.2 Fitting nonparametric random effects

10.4.2.1 Fitting a nonparametric random intercept in the predictor for μ

This is the model defined in equations (10.2) and (10.5). Note that mass points u_k and their probabilities π_k are assumed to be fixed but unknown and need to be estimated from the data.

For random effects at the observation level (in the predictor for μ), the resulting (marginal) model for $\mathbf{Y}$ is just a finite mixture of GAMLSS models with parameters $(\beta_1, \beta_2, \beta_3, \beta_4)$ in common, but different intercept parameters $(u_1, u_2, \ldots, u_{\mathcal{K}})$ in the predictor for μ. Example code for fitting this model is:

```
#### random effects at the observation level
m3 <- gamlssNP(y~1, K=10, mixture="np")
```

Example code for fitting a nonparametric random effects model at the factor level (for factor `f`) in the predictor for μ is

```
#### random effects at the factor level
m4 <- gamlssNP(y~1,  K=10, mixture="np", random=~1|f)
```

10.4.2.2 Fitting nonparametric random intercept and slopes in the predictor for μ

This is the model defined in equation (10.2) with γ distributed as in (10.12). The following is example code for fitting this model using the factor `f` and explanatory variable `x`:

```
#### random effects (of slope and intercept) at the factor level
m5 <- gamlssNP(y~x, K=10, mixture="np", random=~x|f)
```

The fitted mass points and their estimated probabilities can be found in
`m5$mass.points` and `m5$prob`, respectively. Note that `m5$mass.points` in this case
is an array with two columns of length $\mathcal{K}$ with values u_{0k} and u_{1k} for $k = 1, 2, \ldots, \mathcal{K}$.
The function `plotMP()` produces a three-dimensional plot of the mass points and
their probabilities:

```
plotMP(m5$mass[,1], m5$mass[,2], m5$prob)
```

10.4.2.3   Fitting nonparametric random coefficients in the predictor for other distribution parameters

Model (10.2) can be amended to include nonparametric random coefficients (e.g.
intercepts and slopes) in the predictors for any one or more of the distributional
parameters μ, σ, ν and τ of a GAMLSS model. The model is fitted using the EM
algorithm of Section 10.2.2. For example, model (10.11) could be amended to model

$$
\begin{aligned}
g_1(\mu_{ij}) &= \mathbf{x}_{1ij}^\top \boldsymbol{\beta}_1 + \gamma_{10j} + \gamma_{11j} z_{ij} \\
g_2(\sigma_{ij}) &= \mathbf{x}_{2ij}^\top \boldsymbol{\beta}_2 + \gamma_{20j} \\
g_3(\nu_{ij}) &= \mathbf{x}_{3ij}^\top \boldsymbol{\beta}_3 \\
g_4(\tau_{ij}) &= \mathbf{x}_{4ij}^\top \boldsymbol{\beta}_4 \ .
\end{aligned}
\tag{10.15}
$$

The model for the nonparametric random effects distribution is given by assuming
$(\gamma_{10j}, \gamma_{11j}, \gamma_{20j})$ for $j = 1, 2, \ldots, J$ is a random sample from the nonparametric
(discrete) distribution

$$
\text{Prob}(\gamma_{10j} = u_{10k}, \gamma_{11j} = u_{11k}, \gamma_{20j} = u_{20k}) = \pi_k , \qquad \text{for } k = 1, \ldots, \mathcal{K} .
$$

The above model for σ can be fitted in `gamlssNP()` by adding the factor `MASS` in
the formula for the predictor of σ, e.g. `sigma.fo=~x+MASS`. For example in model
(10.15) (without ν and τ models), for a single explanatory variable `x` for μ and σ
and a single random effect explanatory variable $z = x$ at a factor `f` level for μ:

```
m6<-gamlssNP(y~x, data=mydata, random=~x|f,sigma.fo=~x+MASS,
            family=NO, K=5, mixture="np")
```

Note that in this case the `m6$mass.points` will only extract the mass points for
μ. The mass points for σ can be extracted from the coefficients for sigma, using
`coef(..., "sigma")`.

To fit a random intercept and slope for σ, the formula for σ is:

```
sigma.fo=~MASS*x
```

10.5   Examples using `gamlssNP()`

10.5.1   Example: Binary response with normal random intercept

Sommer et al. [1983] describe a cohort study of 275 Indonesian preschool children examined on up to six consecutive quarters for the presence of respiratory infection. Zeger and Karim [1991], Diggle et al. [2002], Skrondal and Rabe-Hesketh [2004, Section 9.2] and others have analysed the data.

R data file: `respInf` in package **gamlss.data** of dimensions 1200×14

variables

    `id` : a factor with 275 levels identifying the individual children

    `time` : the binary response variable identifying the presence of respiratory infection

    `resp` : a vector of ones (not used further)

    `age` : the age in months (centred around 36)

    `xero` : a dummy variable for the presence of xerophthalmia

    `cosine` : a cosine term of the annual cycle

    `sine` : a sine term of the annual cycle

    `female` : a gender dummy variable, 1=female, 0=male.

    `height` : height for age as percent of the National Centre for Health Statistics standard, centred at 90%

    `stunted` : a dummy variable whether below 85% in height for age

    `time.1` : the times that the child has been examined, 1 to 6

    `age1` : the age of the child at the first time of examination

    `season` : a variable taking the values 1,2,3,4 indicating the season

    `time2` : the time in months

purpose: to demonstrate the fitting of a binary response random effect model using `gamlssNP()`.

As the data are longitudinal we specify a random intercept at the subject (`id`) level. We fit a binomial distribution with a normal random intercept model at the factor `id` level in the predictor for μ, as in (10.2) and (10.13):

$$Y_{ij} \mid \gamma_j \sim \text{BI}(1, \mu_{ij}) , \qquad i = 1, \ldots, 6; j = 1, \ldots, 275$$

$$\text{logit}(\mu_{ij}) = \text{age}_j + \text{female}_j + \text{cosine}_i + \text{height}_j + \gamma_j$$

where Y_{ij} is the `time` variable, i indicates `time.1` and $\gamma_j \sim \mathcal{N}(0, \sigma_b^2)$ independently for subject $j = 1, \ldots, 275$.

```
library(gamlss.mx)
data(respInf)
# fit the model
m1<-gamlssNP(time~age+female+cosine+height, random=~1|id,
             K=20, mixture="gq",data=respInf, family=BI)

## 1 ..2 ..3 ..4 ..5 ..6 ..7 ..8 ..9 ..10 ..11 ..12 ..
## 13 ..
## EM algorithm met convergence criteria at iteration   13
## Global deviance trend plotted.

# display the parameters
m1

##
## Mixing Family:  "BI Mixture with NO"
##
## Fitting method: EM algorithm
##
## Call:
## gamlssNP(formula = time ~ age + female + cosine + height,
##     random = ~1 | id, family = BI, data = respInf,
##     K = 20, mixture = "gq")
##
## Mu Coefficients :
## (Intercept)           age        female1          cosine
##    -2.56357      -0.03262       -0.46361        -0.55114
##       height             z
##    -0.05419       0.83566
##
## Estimated probabilities: 0.0000 0.0000 0.0000 0.0000 0.0001 0.0018
## 0.0140 0.0615 0.1617 0.2608 0.2608 0.1617 0.0615 0.0140 0.0018
## 0.0001 0.0000 0.0000 0.0000 0.0000
##
##
## Degrees of Freedom for the fit: 6 Residual Deg. of Freedom    1194
## Global Deviance:       671.843
##            AIC:        683.843
##            SBC:        714.384
```

The fitted model is given by

$$\text{logit}(\hat{\mu}_{ij}) = -2.564 - 0.033\,\texttt{age}_j - 0.464\,(\texttt{female=1})_j - 0.551\,\texttt{cosine}_i$$
$$- 0.054\,\texttt{height}_j + \gamma_j$$

where $\gamma_j \sim \mathcal{N}(0, 0.836^2)$. Note that the estimate of σ_b is given by the coefficient of z. The fitted (marginal) global deviance is 671.843. The (marginal) AIC is 683.843, where an additional penalty 2 was added for parameter σ_b.

10.5.2 Example: Binomial response with nonparametric random intercept and slope

Silagy et al. [2004] report a meta-analysis of 27 clinical trials of nicotine replacement therapy for smoking cessation. In each trial, the participant was randomized to a treatment or control group. The treatment group were given a nicotine gum. In the majority of studies, the control group received gum having the same appearance but without the active ingredients, but in some trials they were given no gum. The outcome, whether the participant had quit smoking or not, was observed after six months. The data set **meta** consists of the variables below. The data have been analysed by Aitkin [1999b] and Skrondal and Rabe-Hesketh [2004, Chapter 9].

R data file: meta in package **gamlss.data** of dimensions 54×6

variables

> **studyname** : location of the study (note that **studyname** is the same for studies at the same place in different years)
>
> **year** : the year of the study
>
> **d** : the number of quitters (non-smokers) after six months
>
> **n** : the total number of participants
>
> **fac** : a factor with two levels indicating whether control (1) or treatment (2)
>
> **study** : numeric from 1 to 27 identifying the study (that is, the combination of **studyname** and **year**).

The first 27 cases record the variable values for the control groups, and the second 27 cases for the treatment groups.

First we input the data and display it graphically by plotting the sample log odds of quitting, i.e. $\log[d/(n-d)]$, against the treatment factor, **fac**, for each of the 27 studies.

```
data(meta)
library(lattice)
meta$logitd<-with(meta, log(d/(n-d)))
with(meta, xyplot(logitd~fac,OrchardSprays,groups=study,type="a",
  auto.key=list(space="no",points=FALSE,lines=TRUE),scales=
  list(cex=1.5,tick.number=5)),ylab=list(cex=1.5),xlab=list(cex=1.5))
```

Fig.
10.6

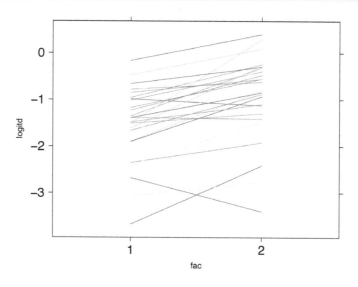

R
p. 34

FIGURE 10.6: Sample log odds ratios of quitting for the 27 smoking cessation studies.

This data set has a two-level structure, with **study** as the upper level and participants as the lower level. Following Aitkin [1999b], we fit a binomial model with nonparametric **study** random effects terms for each of the two levels of **fac**. We set the number of nonparametric mixture components to 10, as suggested by Skrondal and Rabe-Hesketh [2004, page 304].

```
me1<-gamlssNP(cbind(d,n-d)~fac,random=~fac|study,K=10,mixture="np",
            family=BI,data=meta, tol=.1)
```

```
## 1 ..2 ..3 ..4 ..5 ..6 ..7 ..8 ..9 ..10 ..11 ..12 ..
## 13 ..14 ..15 ..16 ..17 ..18 ..19 ..20 ..21 ..22 ..23 ..24 ..25 ..
## 26 ..27 ..28 ..29 ..30 ..31 ..32 ..33 ..34 ..35 ..36 ..37 ..
## EM algorithm met convergence criteria at iteration   37
## Global deviance trend plotted.
## EM Trajectories plotted.
```

```
summary(me1)
```

```
## ************************************************************
## Family:  c("BI Mixture with NP", "Binomial Mixture with NP")
##
```

```
## Call:
## gamlssNP(formula = cbind(d, n - d) ~ fac, random = ~fac | study,
##      family = BI, data = meta, K = 10, mixture = "np", tol = 0.1)
##
##
## Fitting method: RS()
##
## -----------------------------------------------------------------
## Mu link function:  logit
## Mu Coefficients:
##             Estimate Std. Error t value Pr(>|t|)
## (Intercept) -3.255e+00  6.245e-01  -5.213 2.69e-07 ***
## fac2         5.758e-01  8.296e-01   0.694 0.487987
## MASS2        3.062e-08  7.468e-01   0.000 1.000000
## MASS3        8.824e-01  6.511e-01   1.355 0.175941
## MASS4        1.808e+00  6.302e-01   2.869 0.004289 **
## MASS5        1.638e+00  6.478e-01   2.529 0.011733 *
## MASS6        2.125e+00  6.325e-01   3.360 0.000836 ***
## MASS7        1.776e+00  7.076e-01   2.510 0.012379 *
## MASS8        2.396e+00  6.365e-01   3.764 0.000186 ***
## MASS9        3.059e+00  7.051e-01   4.338 1.72e-05 ***
## MASS10       3.059e+00  7.790e-01   3.927 9.76e-05 ***
## fac2:MASS2  -5.178e-08  9.921e-01   0.000 1.000000
## fac2:MASS3   1.914e-01  8.570e-01   0.223 0.823342
## fac2:MASS4  -3.723e-01  8.377e-01  -0.444 0.656955
## fac2:MASS5   1.764e-01  8.556e-01   0.206 0.836752
## fac2:MASS6   5.776e-02  8.400e-01   0.069 0.945210
## fac2:MASS7   1.160e+00  9.348e-01   1.241 0.215250
## fac2:MASS8  -1.159e-01  8.441e-01  -0.137 0.890848
## fac2:MASS9  -5.297e-03  9.528e-01  -0.006 0.995567
## fac2:MASS10 -5.297e-03  1.064e+00  -0.005 0.996030
## ---
## Signif. codes:
## 0 '***' 0.001 '**' 0.01 '*' 0.05 '.' 0.1 ' ' 1
##
## -----------------------------------------------------------------
## No. of observations in the fit:  540
## Degrees of Freedom for the fit:  29
##         Residual Deg. of Freedom:  25
##                    at cycle:  1
##
## Global Deviance:     345.7512
##           AIC:       403.7512
##           SBC:       461.4318
```

**

The fitted model is

$$Y_{ij} \sim \mathtt{BI}(\mathbf{n}_{ij}, \mu_{ij}) \qquad i = 1, 2; \ j = 1, 2 \ldots, 27$$

where

$$\mathrm{logit}(\mu_{ij}) = \gamma_{0j} + \gamma_{1j}(\text{if } i = 2) \ ,$$

where i indicates the `fac` level and j indicates the study level. The mass points (u_{0k}, u_{1k}) and associated probabilities π_k for $k = 1, \ldots, 10$, for the nonparametric distribution of $(\gamma_{0j}, \gamma_{1j})$ are saved in `me1$mass.points` and `me1$prob`. These can be displayed or plotted. The fitted distribution of γ_0 is plotted as `mass.points[,1]` in Figure 10.7(b). Similarly the fitted distribution of γ_1 is plotted as `mass.points[,2]` in Figure 10.7(d).

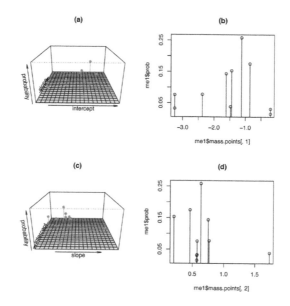

R
p. 34◀

FIGURE 10.7: Plots of the joint nonparametric distribution for (γ_0, γ_1) from different directions (a) and (c), and the marginal nonparametric distributions of γ_0 and γ_1, respectively, (b) and (d).

Figures 10.7(a) and (c) both plot the fitted bivariate distribution of (γ_0, γ_1), from different directions. The commands for plotting Figure 10.7 are

```
op<-par(mfrow=c(2,2))
plotMP(me1$mass.points[,1], me1$mass.points[,2], me1$prob,theta=0,
     phi=20, main="(a)")
plot(me1$mass.points[,1], me1$prob, type="h" , main="(b)")
points(me1$mass.points[,1], me1$prob)
plotMP(me1$mass.points[,1], me1$mass.points[,2], me1$prob, theta=90,
```

Fig.
10.7

```
        phi=20, main="(c)")
plot(me1$mass.points[,2], me1$prob, type="h", main="(d)" )
points(me1$mass.points[,2], me1$prob)
par(op)
```

The means, standard deviations and correlation of the fitted distribution of (γ_0, γ_1) are found using the commands:

```
vM <- cov.wt(me1$mass.points, me1$prob)
vM$cov
```

```
##             [,1]         [,2]
## [1,]   0.68894082 -0.03054541
## [2,]  -0.03054541  0.09848739
```

```
vM$center
```

```
## [1] -1.4978969  0.5978631
```

```
vM$n.obs
```

```
## [1] 10
```

```
vM$wt
```

```
##        MASS1        MASS2        MASS3        MASS4        MASS5
## 0.03262593 0.07586472 0.07704748 0.15312586 0.14349824
##        MASS6        MASS7        MASS8        MASS9       MASS10
## 0.25814324 0.03851473 0.17507287 0.03085242 0.01525452
```

```
# the correlation matrix
cov2cor(vM$cov)
```

```
##             [,1]        [,2]
## [1,]   1.000000 -0.117264
## [2,]  -0.117264  1.000000
```

```
# the standard errors
sqrt(diag(vM$cov))
```

```
## [1] 0.8300246 0.3138270
```

From the fitted nonparametric (discrete) distribution for (γ_0, γ_1), the fitted means of γ_0 and γ_1 (given by vM$center) are -1.4979 and 0.5979, and the fitted standard deviations (given by the square root of the diagonal of the covariance matrix) are 0.83 and 0.3138. The fitted correlation between γ_0 and γ_1 (given by cov2cor) is -0.1173. Hence the overall estimate of the treatment effect (on the logit scale) is 0.5979.

10.6   The function `random()`

The function `random()` is an additive smoothing function which allows random intercept terms to be fitted for any or all of the parameters of the distribution. For example the following model can be fitted using `random()`:

$$
\begin{aligned}
g_1(\boldsymbol{\mu}) &= \boldsymbol{\eta}_1 = \mathbf{X}_1\boldsymbol{\beta}_1 + \mathbf{Z}_1\boldsymbol{\gamma}_1 \\
g_2(\boldsymbol{\sigma}) &= \boldsymbol{\eta}_2 = \mathbf{X}_2\boldsymbol{\beta}_2 + \mathbf{Z}_2\boldsymbol{\gamma}_2 \\
g_3(\boldsymbol{\nu}) &= \boldsymbol{\eta}_3 = \mathbf{X}_3\boldsymbol{\beta}_3 + \mathbf{Z}_3\boldsymbol{\gamma}_3 \\
g_4(\boldsymbol{\tau}) &= \boldsymbol{\eta}_4 = \mathbf{X}_4\boldsymbol{\beta}_4 + \mathbf{Z}_4\boldsymbol{\gamma}_4
\end{aligned}
\tag{10.16}
$$

where the $\mathbf{Z}$'s are design matrices of a single factor, or the identity matrix for a random effect at the observational level; and $\boldsymbol{\gamma}_k \sim \mathcal{N}(\mathbf{0}, \sigma_k^2 \mathbf{I})$, for $k = 1, 2, 3, 4$. Note that for simplicity we have included only one grouping factor in (10.16), but if random intercept models are required for other factors they can be included too.

The function `random()` is based on the original function with the same name in the package **gam** [Hastie, 2015]. In the **gamlss** version, the function has been modified to allow a 'local' maximum likelihood estimation of the random effect parameter σ_k. This method is equivalent to the PQL method of Breslow and Clayton [1993], applied at the local iterations of the algorithm. In fact for a GLM and a simple random effect, it is equivalent to `glmmPQL()` in the **MASS** package [Venables and Ripley, 2002]. The function allows the fitted values for a factor term to be shrunk towards the overall mean, where the amount of shrinking depends either on the smoothing parameter lambda ($\lambda = \sigma_e/\sigma_b$, see Section 3.4) or the effective degrees of freedom or on the sigma parameter σ_b (default, where $\sigma_b = \sigma_k$ here). Similar in spirit to smoothing splines, this fitting method can be justified on Bayesian grounds or by a random effects model.

The fitted conditional likelihood (i.e. conditional on the fitted random effects $\hat{\boldsymbol{\gamma}} = (\hat{\boldsymbol{\gamma}}_1, \hat{\boldsymbol{\gamma}}_2, \hat{\boldsymbol{\gamma}}_3, \hat{\boldsymbol{\gamma}}_4)$) is given by

$$
L = \prod_{j=1}^{J} \left[\prod_{i=1}^{n_j} f(y_{ij} \mid \hat{\boldsymbol{\beta}}, \hat{\boldsymbol{\gamma}}) \right].
$$

It is important to note that this estimation procedure is approximate and can, for example, lead to negative bias in the estimation of a scale parameter σ, especially when the effective degrees of freedom for the random effects in the model for μ is not negligible or low relative to the sample size, as pointed out by Nelder [2006, p547]. This also applies to the function `re()` in Section 10.8.

The conditional AIC is given by $-2\log L + 2\sum_{k=1}^{4} \mathrm{df}_k$, where df_k is the effective degrees of freedom used in fitting $\hat{\boldsymbol{\gamma}}_k$ for $k = 1, 2, 3, 4$.

Note that the behaviour of `random()` is different from the original **gam** function. Here the function is implemented as follows:

1. If both `df` and `lambda` are NULL, then the PQL method is used to estimate λ (and hence σ_b).

2. If `lambda` is not NULL, then `lambda` is used for fitting.

3. If `lambda` is NULL and `df` is not NULL, then `df` is used for fitting.

`random()` has the following arguments:

x	the random effect factor.
df	the effective degrees of freedom, with default `df=NULL`.
lambda	the smoothing parameter, with default `lambda=NULL`.
start	The starting value for λ in the search for the smoothing parameter, with default `start=10`.

The function `random()` creates an object of type `random`, which can be accessed using `getSmo()`. The `random` object has `print()`, `coef()`, `fitted()` and `plot()` methods. Fitted $\hat{\gamma}$'s may be plotted using either `plot()` or `term.plot()`.

10.7 Examples using `random()`

10.7.1 The Hodges data

R data file: `hodges` in package **gamlss.data** of dimensions 341×6

variables

> `state` : a factor with 45 levels
>
> `plan` : a two-character code that identifies plans (within state) declared here as factor with 325 levels
>
> `prind` : the total premium for an individual
>
> `prfam` : the total premium for a family
>
> `enind` : the total enrolment of federal employees as individuals
>
> `enfam` : the total enrolment of federal employees as families.

purpose: to demonstrate the fitting of simple random effect model in **gamlss**

conclusion: A BCT distribution and random effect for μ is adequate

Here we consider a one-factor random effects model for the response variable health insurance premium (`prind`), with `state` as the random factor. The data were analysed in Hodges [1998], who used a conditional normal model for Y_{ij} given γ_j (the random effect in the mean for state j), and a normal distribution for γ_j, i.e.

$$Y_{ij} \mid \gamma_j \sim \mathcal{N}(\mu_{ij}, \sigma^2)$$
$$\mu_{ij} = \beta_1 + \gamma_j$$
$$\gamma_j \sim \mathcal{N}(0, \sigma_1^2)$$

independently for $i = 1, 2, \ldots, n_j$ and $j = 1, 2, \ldots, 45$.

Figure 10.8 provides box plots of `prind` against `state`, showing the variation in the location and scale of `prind` between states, and a positively skewed (and possible leptokurtic) distribution of `prind` within states. Although Hodges [1998] used an added variable diagnostic plot to identify the need for a Box–Cox transformation of `prind`, he did not model the data using a transformation of `prind`.

```
data(hodges)
with(hodges,plot(prind~state))
```

Fig. 10.8

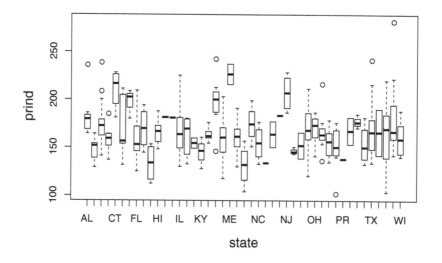

FIGURE 10.8: Hodges data: box plots of `prind` against `state`.

R

p. 348

In the discussion of Hodges [1998], Wakefield commented: "If it were believed that there were different within-state variances then one possibility would be to assume a hierarchy for these also", while in his reply Hodges also suggested treating the "within-state precisions or variances as draws from some distribution".

Hence we consider a Box–Cox t (BCT) distribution model [Rigby and Stasinopoulos, 2006], which allows for both skewness and kurtosis in the conditional distribution of Y and allows for differences between states, in the location, scale and shape of the conditional distribution of Y, by including a random effect term in each of the models for the parameters μ, σ, ν and τ:

$$Y_{ij} \,|\, \gamma_{1j}, \gamma_{2j}, \gamma_{3j}, \gamma_{4j} \sim \text{BCT}(\mu_{ij}, \sigma_{ij}, \nu_{ij}, \tau_{ij})$$

where

$$
\begin{aligned}
\mu_{ij} &= \beta_1 + \gamma_{1j} \\
\log \sigma_{ij} &= \beta_2 + \gamma_{2j} \\
\nu_{ij} &= \beta_3 + \gamma_{3j} \\
\log \tau_{ij} &= \beta_4 + \gamma_{4j} \\
\gamma_{kj} &\sim \mathcal{N}(0, \sigma_k^2)
\end{aligned}
$$

$$(10.17)$$

$$(10.18)$$

independently for $j = 1, 2, \ldots, 45$ and $k = 1, 2, 3, 4$.

Here we fit four models. The first, m0, fits μ with a random effect for state. Models m1, m2 and m3 fit additionally models for σ, ν and τ, respectively, with a random effect for state.

```
m0<-gamlss(prind~random(state), data=hodges,family=BCT, trace=FALSE)
m1<-gamlss(prind~random(state),sigma.fo=~random(state),
        data=hodges,family=BCT, trace=FALSE)
m2<-gamlss(prind~random(state),sigma.fo=~random(state),
            nu.fo=~random(state), data=hodges,family=BCT, trace=FALSE)
m3<-gamlss(prind~random(state),sigma.fo=~random(state),
            nu.fo=~random(state), tau.fo=~random(state), data=hodges,
        family=BCT, trace=FALSE)
AIC(m0,m1,m2,m3)

##          df        AIC
## m1 36.26933 3096.796
## m0 33.65337 3096.860
## m2 43.79012 3098.554
## m3 43.79012 3098.554
```

The AIC indicates that the random effects for ν and τ are not needed. Model m1 has the lowest AIC, although it is only very slightly better than m0. Indeed a slightly higher penalty κ than the default 2 in GAIC would result in the selection of model m0. In practice one would probably prefer m0 on the grounds of parsimony.

We show the fitted model m0 below, and we also use getSmo() to obtain the fitted degrees of freedom and the random effect parameter σ_1 (denoted as 'sigma_b', i.e. σ_b, in the output) for μ.

```
m0

##
## Family:  c("BCT", "Box-Cox t")
```

```
## Fitting method: RS()
##
## Call:
## gamlss(formula = prind ~ random(state), family = BCT,
##      data = hodges, trace = FALSE)
##
## Mu Coefficients:
##    (Intercept)  random(state)1
##           164.8                NA
## Sigma Coefficients:
## (Intercept)
##        -2.22
## Nu Coefficients:
## (Intercept)
##     -0.06086
## Tau Coefficients:
## (Intercept)
##        2.004
##
## Degrees of Freedom for the fit: 33.65 Resid. Deg. of Freedom 307.3
## Global Deviance:      3029.55
##              AIC:      3096.86
##              SBC:      3225.82
```

```
getSmo(m0)
```

```
## Random effects fit using the gamlss function random()
## Degrees of Freedom for the fit : 29.65337
## Random effect parameter sigma_b: 13.5603
## Smoothing parameter lambda    : 0.00594652
```

The fitted random effects parameter is $\hat{\sigma}_1 = 13.56$ for μ, corresponding to $\text{df}_1 = 29.6$. Since $\hat{\nu} = \hat{\beta}_3 = -0.061$ is close to zero, the fitted model for the conditional distribution of Y_{ij} given the random effect γ_{1j} is approximately $\hat{\sigma}^{-1} \log(Y_{ij} \,|\, \hat{\mu}_{ij}) \sim t_{\hat{\tau}}$ where $t_{\hat{\tau}}$ is a t distribution with $\hat{\tau} = \exp(\hat{\beta}_4) = \exp(2.004) = 7.42$ degrees of freedom; $\hat{\mu}_{ij} = \hat{\beta}_1 + \gamma_{ij}$; $\hat{\beta}_1 = 164.8$; and $\hat{\sigma} = \exp(\hat{\beta}_2) = \exp(-2.22) = 0.1086$. Hence the conditional distribution of $\log(Y_{ij})$ is approximately a scaled and shifted t distribution. Note that the model with effectively $\nu = 0$ can be fitted by

```
mOA<-gamlss(prind~random(state), data=hodges, family=BCT,
            nu.fix=TRUE, nu.start=0.000001)
```

The resulting parameters estimates change very slightly. (Note that we could fit model m0 using Gaussian quadrature; see Exercise 2. This is a useful check on the fitted model, especially to check if the fitted value of σ_1 is different.)

In Figure 10.9 the observed and fitted medians of `prind` (sorted by the observed median) are plotted against the state. The code is given below.

Fig.
10.9

```
# total no of obs. per state
 total <- with(hodges, tapply(prind,state,"length"))
# the median per state
 ho.median <- with(hodges, tapply(prind,state,"median"))
# quantities needed for plotting
      label <- attr(ho.median,"dimnames")
      lll <- label[[1]]
      index <- 1:45
# plot the median
 plot(ho.median[order(ho.median)],ylab="Median monthly premium ($)",
      xlab="State (sorted by sample median monthly premium)",
      ylim=c(90,260),xaxt="n",cex.lab=1.5,cex.axis=1,cex.main=1.2)
# getting fitted medians
   f.median <- qBCT(0.50,  mu=fitted(m0), sigma=fitted(m0, "sigma"),
                 nu=fitted(m0, "nu"), tau=fitted(m0, "tau"))
# fitted median by state
 fitted.m <-tapply(f.median, hodges$state,median)
points(fitted.m[order(ho.median)], pch=3, col="blue", cex=0.7)#
text(index,rep(95,45),lll[order(ho.median)],cex=0.65)
text(index,rep(100,45),total[order(ho.median)],cex=0.65)
legend("topleft", legend=c("observed","fitted"), pch=c(1,3),
      col=c("black","blue"))
```

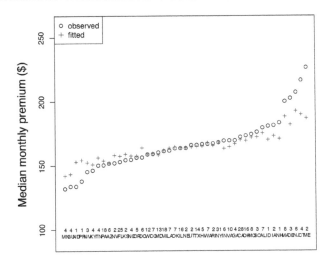

State (sorted by sample median monthly premium)

FIGURE 10.9: Hodges data: observed (o) and fitted (+) medians of `prind` against state.

10.7.2   Revisiting the respiratory infection in children

The fitted `m1` in Section 10.5.1 was obtained by maximizing the marginal likelihood of $\mathbf{Y}$ (integrating out the random intercepts using Gaussian quadrature) using `gamlssNP()`. The model can be fitted approximately as an ordinary GAMLSS
model by using `gamlss()` with an additive term `random()` for the factor `id`. In
this case $\hat{\sigma}_b$ is the local estimate of σ_b and the (joint) likelihood approximately
maximized is $f_{\mathbf{Y}}(\mathbf{y}\,|\,\boldsymbol{\beta},\boldsymbol{\gamma}) f_{\boldsymbol{\gamma}}(\boldsymbol{\gamma}\,|\,\hat{\sigma}_b)$. The empirical Bayes predictions (EBP) for μ
of the marginal model `m1` should be almost identical to those of the predictor for
μ of the ordinary GAMLSS model. Also the beta coefficients for the two models
should be very similar. We now fit the ordinary GAMLSS model for comparison.
The empirical Bayes predictions for `m1` are stored in `m1$ebp`.

```
m2 <- gamlss(time~age+female+cosine+height+random(id),
            data=respInf, family=BI)

## GAMLSS-RS iteration 1: Global Deviance = 468.6237
## GAMLSS-RS iteration 2: Global Deviance = 468.623

summary(m2)

## *********************************************************************
## Family:  c("BI", "Binomial")
##
## Call:
## gamlss(formula = time ~ age + female + cosine + height +
##     random(id), family = BI, data = respInf)
##
## Fitting method: RS()
##
## -------------------------------------------------------------------
## Mu link function:  logit
## Mu Coefficients:
##               Estimate Std. Error t value Pr(>|t|)
## (Intercept) -2.656604    0.151898 -17.489  < 2e-16 ***
## age         -0.028804    0.006828  -4.218 2.67e-05 ***
## female1     -0.456952    0.232332  -1.967  0.04946 *
## cosine      -0.561828    0.170875  -3.288  0.00104 **
## height      -0.062811    0.019503  -3.220  0.00132 **
## ---
## Signif. codes:
## 0 '***' 0.001 '**' 0.01 '*' 0.05 '.' 0.1 ' ' 1
##
## -------------------------------------------------------------------
## NOTE: Additive smoothing terms exist in the formulas:
##  i) Std. Error for smoothers are for the linear effect only.
## ii) Std. Error for the linear terms maybe are not accurate.
```

```
## -------------------------------------------------------------------
## No. of observations in the fit:  1200
## Degrees of Freedom for the fit:  123.1857
##        Residual Deg. of Freedom:  1076.814
##                      at cycle:  2
##
## Global Deviance:     468.623
##             AIC:     714.9944
##             SBC:     1342.019
## *********************************************************************
```

```
getSmo(m2)
```

```
## Random effects fit using the gamlss function random()
## Degrees of Freedom for the fit : 118.1857
## Random effect parameter sigma_b: 1.35294
## Smoothing parameter lambda     : 0.261632
```

We compare the fitted parameters of models m1 and m2, which are similar except for the fitted random effects parameter which is $\hat{\sigma}_\gamma = 0.836$ from m1, but $\hat{\sigma}_\gamma = 1.353$ from m2. Note also that the output from fitting m1 gives the marginal AIC, while m2 output gives the conditional AIC (given the fitted random effects parameters γ). The marginal and conditional AICs are not comparable.

```
# comparing the fitted coefficients
coef(m1)
```

```
## (Intercept)         age       female1      cosine       height
## -2.56356734  -0.03261940  -0.46361042  -0.55113738  -0.05419261
##           z
##   0.83565560
```

```
coef(m2)
```

```
## (Intercept)         age       female1      cosine       height
## -2.65660450  -0.02880369  -0.45695189  -0.56182782  -0.06281074
## random(id)1
##          NA
```

We now re-analyse using smoothers for one or both of the continuous variables age and height. We use the function pb():

```
m21 <- gamlss(time~pb(age)+female+cosine+pb(height)+random(id),
              data=respInf, family=BI, trace=FALSE)
m22 <- gamlss(time~age+female+cosine+pb(height)+random(id),
              data=respInf, family=BI, trace=FALSE)
m23 <- gamlss(time~pb(age)+female+cosine+height+random(id),
              data=respInf, family=BI, trace=FALSE)
```

```
AIC(m2,m21,m22,m23)

##              df       AIC
## m23 122.7400 709.4849
## m21 122.7400 709.4851
## m2  123.1857 714.9944
## m22 123.1858 714.9945
```

According to the AIC, model m23, which has a smoother for age only, is preferred.
The fitted terms are displayed in Figure 10.10, which indicates that the random
effects for id appear to be highly positively skewed. Perhaps one should try a
nonparametric random intercept model; see Exercise 4.

```
term.plot(m23, pages=1)
```

Fig.
10.1

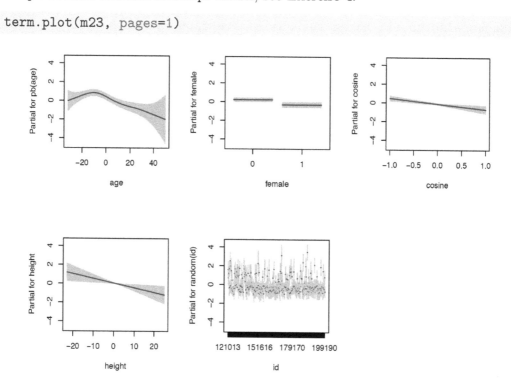

R
p. 35

FIGURE 10.10: Term plot of model m23.

10.8   The function re(), interfacing with lme()

The function re() is an interface for calling lme() of the **nlme** package [Pinheiro and
Bates, 2000]. This allows GAMLSS users to fit complicated random effect models
while the assumption of the normal distribution for the response variable is relaxed.
The theoretical justification comes from this method being a PQL approach [Breslow

and Clayton, 1993]. Users of this function should be familiar with both the random effect methodology and with the implementation of `lme()`. It is important to note that this estimation procedure is approximate; see the comment in Section 10.6.

The function `re()` can be used for any or all the parameters of the distribution and therefore the following GAMLSS model is appropriate:

$$
\begin{aligned}
g_1(\boldsymbol{\mu}) &= \boldsymbol{\eta}_1 = \mathbf{X}_1\boldsymbol{\beta}_1 + \mathbf{Z}_1\boldsymbol{\gamma}_1 + \boldsymbol{\varepsilon}_1 \\
g_2(\boldsymbol{\sigma}) &= \boldsymbol{\eta}_2 = \mathbf{X}_2\boldsymbol{\beta}_2 + \mathbf{Z}_2\boldsymbol{\gamma}_2 + \boldsymbol{\varepsilon}_2 \\
g_3(\boldsymbol{\nu}) &= \boldsymbol{\eta}_3 = \mathbf{X}_3\boldsymbol{\beta}_3 + \mathbf{Z}_3\boldsymbol{\gamma}_3 + \boldsymbol{\varepsilon}_3 \\
g_4(\boldsymbol{\tau}) &= \boldsymbol{\eta}_4 = \mathbf{X}_4\boldsymbol{\beta}_4 + \mathbf{Z}_4\boldsymbol{\gamma}_4 + \boldsymbol{\varepsilon}_4
\end{aligned}
\tag{10.19}
$$

where the $\mathbf{Z}$'s can now be complex design matrices depending on the number and type of random effects, and where $\boldsymbol{\gamma}_k \sim N(\mathbf{0}, \boldsymbol{\Psi}_k)$ and $\boldsymbol{\varepsilon}_k \sim \mathcal{N}(\mathbf{0}, \boldsymbol{\Lambda}_k)$ for $k = 1, 2, 3, 4$. The $\boldsymbol{\Psi}$'s are matrices associated with the variance-covariance structure of the random effects, while the $\boldsymbol{\Lambda}$'s are possible 'error' correlation structures which can occur at the observation level (as shown in (10.19)) or at the factor level.

The function `re()` has the following arguments:

`fixed` The fixed effect formula which passes to `lme`. In general, the fixed effect terms are declared within the `gamlss()` function using the relevant parameter formula, alternatively use `fixed` within `re()`. There are occasions, especially with complicated random effect models, where the use of the argument `fixed` helps convergence. Note that the formula in the `fixed` argument of the function `re()` does not require the response variable on the left side of the formula (as does the `fixed` argument of the `lme()` function).

`random` This is the random effect formula or a list specifying the structure of the random effects (i.e. for $\boldsymbol{\Psi}_k$). For more detail on how this argument can be used; see the help file of `lme()`. (In what follows we assume that `f1,...,fn` are factors and `x1,...,xn` are variables.) Here three alternative ways to specify the structure are mentioned:

  (i) A one-sided formula of the form

    `~x1 + ... + xn |f1/.../fm`

    with `x1 + ... + xn` specifying the model for the random effects and `f1/.../fm` the grouping (nested) structure. (Note that for a random effect at the observation level the latter can be omitted). The order of nesting will be assumed the same as the order of the elements in the list.

  (ii) A list of one-sided formulae of the form

    `~x1 + ... + xn |f`

    with possibly different random effects models for each grouping level.

  (iii) A named list of formulae, for example

```
list(f1 = pdDiag(~x1), f2=pdSymm(~x2))
```

This specification is useful when the random effect structure requires different variance-covariance specifications. For example, `f1=pdDiag(~x1)` indicates a random intercept and slope model in `x1` at the factor level `f1`, with the random intercept and slope parameters having covariance structure `pdDiag`, defined later, for Ψ_k. These specifications in the **nlme** package are given by the `pdMat` class objects.

The different forms of `pdMat` matrices for Ψ_k are described in Table 4.3, Pinheiro and Bates [2000, p158]. Below we display the different forms together with a small visual presentation of each matrix to help the user.

`pdSymm`: This is the default value, and it is a symmetric positive definite matrix structure, e.g.

$$\begin{bmatrix} \sigma_1^2 & \sigma_{21}^2 & \sigma_{31}^2 \\ \sigma_{21}^2 & \sigma_2^2 & \sigma_{32}^2 \\ \sigma_{31}^2 & \sigma_{32}^2 & \sigma_3^2 \end{bmatrix}$$

`pdBlocked`: For block diagonal structure of the type:

$$\begin{bmatrix} \sigma_1^2 & 0 & 0 & 0 \\ 0 & \sigma_1^2 & 0 & 0 \\ 0 & 0 & \sigma_2^2 & 0 \\ 0 & 0 & 0 & \sigma_2^2 \end{bmatrix} \quad \text{or} \quad \begin{bmatrix} \sigma_1^2 \mathbf{I} & 0 \\ 0 & \sigma_2^2 \mathbf{I} \end{bmatrix}$$

typically declared by something like:

```
random=list(f1=pdBlocked(list(pdIndent(~1),pdIdent(~f2-1))))
```

Note that when a factor is involved in a formula, the term `-1` ensures that all levels are considered.

`pdCompSymm`: Compound symmetry structure, e.g.

$$\begin{bmatrix} \sigma_1^2 + \sigma_2^2 & \sigma_1^2 & \sigma_1^2 \\ \sigma_1^2 & \sigma_1^2 + \sigma_2^2 & \sigma_2^2 \\ \sigma_1^2 & \sigma_1^2 & \sigma_1^2 + \sigma_2^2 \end{bmatrix}.$$

This is suitable for nested random effects models. For example, if factor `f2` is nested in factor `f1` a typical declaration is:

```
random=list(f1=pdCopmSymm(~f2-1))
```

`pdDiag`: Diagonal matrix structure, e.g.

$$\begin{bmatrix} \sigma_1^2 & 0 & 0 \\ 0 & \sigma_2^2 & 0 \\ 0 & 0 & \sigma_3^2 \end{bmatrix}.$$

This assumes independence between the random effects. Typically this will be of the form:

```
random=pdDiag(~x1+x2)
```

where the random slopes for `x1` and `x2` and the intercept are assumed to be independent.

`pdIdent`: Identity matrix structure, e.g.

$$\sigma_1^2 \begin{bmatrix} 1 & 0 & 0 \\ 0 & 1 & 0 \\ 0 & 0 & 1 \end{bmatrix}$$

This is useful when the levels of a factor `f1` have common variance:

```
random=pdIdent(~f1-1)
```

`correlation` The correlation structure of the within group or error variation (i.e. for Λ_k). The different correlation structures can be divided into time series and spatial structures and are described in detail in Pinheiro and Bates [2000, Section 5.3, p226]. Let ρ be the variance component of the random effect over the total variance (variance component from the random effect plus the error variance). The time series structures are:

   `corCompSymm`: compound symmetry structure corresponding to uniform correlation, e.g.

$$\begin{bmatrix} 1 & \rho & \rho \\ \rho & 1 & \rho \\ \rho & \rho & 1 \end{bmatrix}$$

   `corAR1`: autoregressive process of order 1, the correlation function is given by ρ^k, for $k = 0, 1, 2, \ldots$, e.g.

$$\begin{bmatrix} 1 & \rho & \rho^2 \\ \rho & 1 & \rho \\ \rho^2 & \rho & 1 \end{bmatrix}$$

   `corCAR1`: continuous autoregressive process (AR(1) process for a continuous time covariate). The correlation function is given by ρ^s, for $s \geq 0$.

   `corARMA`: autoregressive moving average process, with arbitrary orders for the autoregressive and moving average components, i.e. ARMA(p, q) corresponding to the combination of the p autoregressive parameters, $\phi = (\phi_1, \ldots, \phi_p)^\top$ and q moving average parameters $\lambda = (\lambda_1, \ldots, \lambda_q)^\top$. For instance, a correlation matrix for an ARMA(1,1) is given by

$$\begin{bmatrix} 1 & \lambda_1 & \lambda_1\phi_1 \\ \lambda_1 & 1 & \lambda_1 \\ \lambda_1\phi_1 & \lambda_1 & 1 \end{bmatrix}.$$

While for the spatial correlation structures we have the elements: (i) range (r), which must be greater than zero; (ii) nugget (λ_1), meaning that a nugget effect is present, which must be between zero and one. (The nugget effect is the variability at zero lag distance.) Let d denote the distance between the individuals. Some examples for the spatial correlation structures are:

corExp: exponential spatial correlation; $(1 - \lambda_1)\exp(-d/r)$;

corGaus: Gaussian spatial correlation; $(1 - \lambda_1)\exp[-(d/r)^2]$;

corLin: linear spatial correlation; $(1 - \lambda_1)[1 - (d/r)]$ and if $d >= r$ the correlation is zero;

corRatio: Rational quadratic spatial correlation; $(1 - \lambda_1)/[1 + (d/r)^2]$;

corSpher: spherical spatial correlation; $(1 - \lambda_1)[1 - 1.5(d/r) + 0.5(d/r)^3]$ and if $d >= r$ the correlation is zero.

method The method used by lme(), which is method="ML" or method="REML".

... This can be used to pass arguments to the function lmeControl(), for controlling lme().

The final fitted random effect model can be accessed using getSmo(). The object obtained by getSmo() is an lme, so all methods available for an lme object should work, in particular plot(), summary(), fitted(), fittef(), pairs(), qqnorm(), ranef() and resid(). Note that fitted() and resid() operate on different levels of grouping.

10.9 Examples using re()

10.9.1 Refitting Hodges data using re()

In Section 10.7.1 we fitted model m0 to the Hodges data which has a BCT distribution, using random(). Here we demonstrate how the same model is fitted using re(), and compare the results.

```
# using random()
m0<-gamlss(prind~random(state),family=BCT,data=hodges,trace=FALSE)
# using re()
m01<-gamlss(prind~re(random=~1|state),family=BCT,data=hodges,
          trace=FALSE)
AIC(m0, m01)

##              df      AIC
## m0    33.65337 3096.86
```

```
## m01 33.65373 3096.86
```

```
getSmo(m0)
```

```
## Random effects fit using the gamlss function random()
## Degrees of Freedom for the fit : 29.65337
## Random effect parameter sigma_b: 13.5603
## Smoothing parameter lambda      : 0.00594652
```

```
getSmo(m01)
```

```
## Linear mixed-effects model fit by maximum likelihood
##    Data: Data
##    Log-likelihood: -1546.286
##    Fixed: fix.formula
## (Intercept)
##      0.766318
##
## Random effects:
##   Formula: ~1 | state
##           (Intercept) Residual
## StdDev:     13.56032 1.045683
##
## Variance function:
##   Structure: fixed weights
##   Formula: ~W.var
## Number of Observations: 341
## Number of Groups: 45
```

The important point to note here is that the estimate of $\hat{\sigma}_b = 13.56$ is identical for the two models.

10.9.2   Fitting a P-spline smoother using re()

We have shown in equation (9.6) that a P-splines smoother can be written as a random effects model: $\mathbf{y} = \mathbf{Z}\gamma + \mathbf{e}$, where $\mathbf{e} \sim \mathcal{N}(\mathbf{0}; \sigma_e^2 \mathbf{W}^{-1})$; $\gamma \sim \mathcal{N}(\mathbf{0}, \sigma_b^2 \mathbf{G}^{-1})$; $\mathbf{G} = \mathbf{D}^\top \mathbf{D}$; and $\mathbf{D}$ is the difference matrix. Equation (9.6) can be transformed to a mixed effect model formulation:

$$\mathbf{y} = \mathbf{X}\beta + \tilde{\mathbf{Z}}\tilde{\gamma} + \mathbf{e}$$

for $\tilde{\gamma} \sim \mathcal{N}(\mathbf{0}, \sigma_b^2 \mathbf{I})$ and $\mathbf{e} \sim \mathcal{N}(\mathbf{0}, \sigma_e^2 \mathbf{W}^{-1})$. The mixed effect formulation allows fitting of a P-spline smoother using re(). To get the mixed model formulation we have to decompose the matrix $\mathbf{Z}$ into two components such that $\mathbf{Z}\gamma = \mathbf{X}\beta + \tilde{\mathbf{Z}}\tilde{\gamma}$. The first component, $\mathbf{X}$, contains the constant (a vector of ones) and the linear term of the explanatory variable x. The second component, $\tilde{\mathbf{Z}}$, is the linear subspace of $\mathbf{Z}$ orthogonal to $\mathbf{X}$. There are several ways to obtain $\tilde{\mathbf{Z}}$ but here we adopt the methodology described in Currie et al. [2006], which is based on the singular value decomposition of the difference matrix $\mathbf{D} = \mathbf{U}\boldsymbol{\Lambda}\mathbf{V}^{\top}$ resulting in $\tilde{\mathbf{Z}} = \mathbf{Z}\mathbf{U}\boldsymbol{\Lambda}^{-1}$.

For the purpose of creating the $\tilde{\mathbf{Z}}$ matrix the function `getZmatrix()` is provided. The function takes as the main argument the explanatory variable x. Other arguments have to do with the way that the original basis and difference matrices, $\mathbf{Z}$ and $\mathbf{D}$, respectively, are created.

The `abdom` data were introduced in Section 4.2, where a P-spline was fitted to `x`, using the function `pb()` in the models for both μ and σ. Here we fit a P-spline for `x` in the μ model, and show, in Figure 10.11, that identical fitted values are obtained using `re()`. Note that the random effect is declared at the observation level by creating and using the factor `Ida`; and that σ_b is estimated by declaring the identity matrix `pdIdent`.

```
a1 <- gamlss(y~pb(x), data=abdom, trace=FALSE)
Z <- getZmatrix(abdom$x)
Ida <- factor(rep(1,610))
a2 <- gamlss(y~1+re(fixed=~x, random=list(Ida=pdIdent(~Z-1)),
                 method="REML"), data=abdom, trace=FALSE)
# getting sigma_b
getSmo(a1)

## P-spline fit using the gamlss function pb()
## Degrees of Freedom for the fit : 5.508274
## Random effect parameter sigma_b: 1.12909
## Smoothing parameter lambda     : 0.791556

getSmo(a2)

## Linear mixed-effects model fit by REML
##    Data: Data
##    Log-restricted-likelihood: -2475.399
##    Fixed: fix.formula
## (Intercept)            x
##  -283.00450     10.29609
##
## Random effects:
##  Formula: ~Z - 1 | Ida
##  Structure: Multiple of an Identity
##               Z1        Z2        Z3        Z4        Z5        Z6
## StdDev: 1.12909   1.12909   1.12909   1.12909   1.12909   1.12909
##               Z7        Z8        Z9       Z10       Z11       Z12
## StdDev: 1.12909   1.12909   1.12909   1.12909   1.12909   1.12909
##              Z13       Z14       Z15       Z16       Z17       Z18
## StdDev: 1.12909   1.12909   1.12909   1.12909   1.12909   1.12909
##              Z19       Z20       Z21 Residual
```

```
## StdDev: 1.12909 1.12909 1.12909 1.004545
##
## Variance function:
##   Structure: fixed weights
##   Formula: ~W.var
## Number of Observations: 610
## Number of Groups: 1
```

```
plot(y~x,data=abdom,cex=.2)
lines(a1$mu.fv[order(abdom$x)]~sort(abdom$x),col="red",lty=2)
lines(a2$mu.fv[order(abdom$x)]~sort(abdom$x),col="blue",lty=3)
legend("topleft",legend=c("model a1 (using pb())","model a2
       (using re())"), lty=2:3,col=c("red","blue"),lwd=2,cex=1.5)
```

The function `getSmo()` shows that the estimates for σ_b in models **a1** and **a2** are identical.

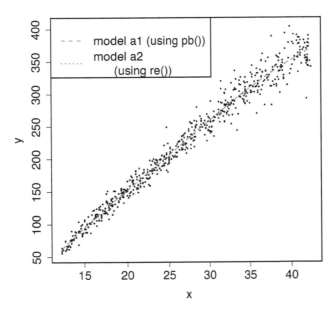

FIGURE 10.11: `abdom` data: equivalence of smoothing using `pb()` and `re()`.

10.9.3 Leukemia data

R data file: Leukemia in package **gamlss.data** of dimensions 1988×4

variables

    case : subject identifier

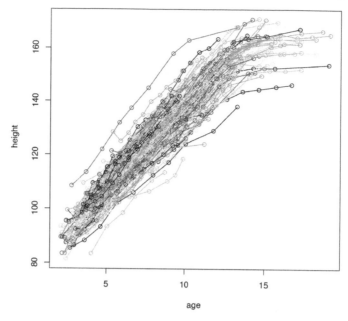

R
p. 3(

FIGURE 10.12: Leukemia data: `height` by `age`, for each of the 197 girls.

> `treatment` : 1, 2 or 3
>
> `height` : in centimetres
>
> `age` : in years

The data set is based on a clinical trial for treatments for acute lymphoblastic leukemia (ALL), conducted at Harvard University [Durbán et al., 2005]. Obesity and short stature are common side effects of leukemia treatments on children, and it is desirable that treatments minimize this type of side effect without effectiveness being compromised. Six hundred and eighteen (618) children were studied between 1987 and 1995 and three different treatments were applied: intracranial therapy without radiation, conventional intracranial radiation therapy and intracranial radiation therapy twice a day. The children's heights were measured approximately every six months.

In this example we are analysing simulated data of 1988 observations on 197 girls who were diagnosed with ALL between two and nine years old. The number of observations per child varies between 1 and 21. Figure 10.12 shows the growth of girls, which appears to flatten off around the age of 15 years.

```
library(gamlss)
data(Leukemia)
par.plot(height~age, subject=case, data=Leukemia, lty=1:10)
```

Fig.
10.12

To demonstrate how `re()` can be used within **gamlss**, we consider six different models for the μ parameter of response `height`. The six models can be described in a 'quasi' formula presentation as:

```
g1   height~treatment+age+random=~1|case
g2   height~treatment+s(age)+random=~1|case
g3   height~treatment+s(age)+random=~age|case
g4   height~treatment*s(age)+random=~age|case
g5   height~treatment+s(age)+random=~s(age)|case
g6   height~treatment*s(age)+random=~s(age)|case
```

The first model is a standard random effect model for μ, comprising a treatment factor effect, a linear term in age and a random intercept for subjects (i.e. case level). The second model introduces a smoothing term for age. This model fits a common smooth curve for all subjects, but it allows the curve to be shifted up and down (by the random intercept). The third model adds a linear random coefficient for age. The result of this is that the common smoothing curve for age can be lifted up and down (by the random intercept) or a steeper or flatter slope (by the random slope) between subjects. The fourth model adds a varying coefficient term which involves an interaction between the smoother for age and **treatment** (a factor). This term produces different smoothers for each level of **treatment**. The fifth model is the most interesting one in the sense that it involves random smooth curves between subjects. The model for μ, apart from the random up/down and steeper/flatter changes introduced through the random intercepts and slopes, respectively, also allows the shapes of the smoothing curves to change between subjects. Note however that this is not the same as a varying coefficient model where completely different smoothing curves are fitted for different levels of the factor involved. Here the smoothing curves are shrunk randomly towards a common smooth curve, the same way that a random intercept shrinks the coefficients towards a common mean. The sixth model contains both a varying coefficients term and a random smooth curve term.

```
# model 1 linear + random intercept
g1<-gamlss(height~treatment+age+re(random=~1|case), data=Leukemia,
         trace=FALSE)
# model 2 random intercepts + a smoother for age
g2<-gamlss(height~treatment+pb(age)+re(random=~1|case),
         data=Leukemia,trace=FALSE)
# model 3   random intercepts and slopes for age + smoother for age
g3<-gamlss(height~treatment+pb(age)+re(random=~age|case),
         data=Leukemia, n.cyc=200, trace=FALSE)
# model 4 smooth age interacts with treatment s(age)*treatment
g4<-gamlss(height~treatment+pvc(age, by=treatment)+
            re(random=~age|case), data=Leukemia, trace=FALSE)
# model 5 smooth age interacts with subjects
# it takes a long time to converge
Z.case <- with(Leukemia, getZmatrix(age,inter=10))
Z.block <- list(case=pdSymm(~age),case=pdIdent(~Z.case-1))
g5<-gamlss(height~treatment+re(fixed=~Z.case,
         random=Z.block), data=Leukemia,  bf.cyc=1, c.crit=0.1,
```

```
                d.tol=Inf,  method=mixed(3,10), trace=FALSE)
# model 6  smooth age interacts with both treatment and subjects
# it takes a long time to fit and convergence is difficult
g6<-gamlss(height~treatment+pvc(age, by=treatment)+re(fixed=~Z.case,
           random=Z.block), data=Leukemia,  bf.cyc=1, c.crit=0.1,
           gd.tol=Inf,  method=mixed(3,20), trace=FALSE)
```

Note that models **g1** and **g2** could alternatively have been fitted using `random` in either `gamlss()` or `gamlssNP()`. However models **g3** to **g6** can only be fitted using `re` in `gamlss()` as above. Note that for the last two models we had to reduce the number of backfitting cycles (`bf.cyc`) and the convergence criterion (`c.crit`) to speed up the fitting process.

We compare the models using both the AIC and SBC:

```
GAIC(g1, g2, g3, g4, g5, g6)

##          df       AIC
## g5 662.9459  6934.083
## g6 648.9152  7023.589
## g4 364.2249  8419.604
## g3 344.7642  8540.067
## g2 204.7149  9305.165
## g1 189.9907 10599.102

GAIC(g1, g2, g3, g4, g5, g6, k=log(nrow(Leukemia)))

##          df      AIC
## g2 204.7149 10450.52
## g4 364.2249 10457.40
## g3 344.7642 10468.98
## g5 662.9459 10643.19
## g6 648.9152 10654.19
## g1 189.9907 11662.08
```

The AIC selects model **g5**, which is rather complex, whereas the SBC selects the simpler model **g2**. Fitted values for the six models are plotted in Figure 10.13.

```
library(lattice)
with(Leukemia,xyplot(fitted(g1)+fitted(g2)+fitted(g3)+fitted(g4)+
    fitted(g5)+fitted(g6)~age,OrchardSprays,groups=case,
    type="a",auto.key=list(space="no",points=FALSE,lines=TRUE)))
```

Fig.
10.13

Unfortunately both **g2** and **g5** have rather undesirable residuals. Fitting a smooth curve for **age** in σ, i.e.

```
g21 <- gamlss(height~treatment+pb(age)+re(random=~1|case),
              sigma.fo=~pb(age), data=Leukemia, trace=FALSE)
```

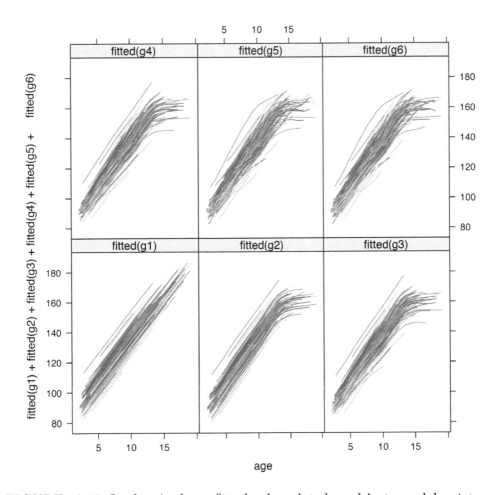

FIGURE 10.13: Leukemia data: fitted value plots by subject, models **g1** to **g6**.

improves the fit according to the GAIC but not the residuals. A different distribution is required here rather than the normal. After further investigation we found that the SHASH distribution with smooth curves for age for models for σ, ν and τ has adequate residuals (see Exercise 6).

10.10   Bibliographic notes

According to West et al. [2014], the first known formulation of a one-way random effects model is by Airy [1861], which was further clarified by Scheffé [1956]. Fisher [1925] summarized the general method for estimating variance components. According to Scheffé [1956], Jackson [1939] was the first to distinguish explicitly between fixed and random effects in writing down a model and assuming normality for random effects and residuals in a model for one random factor and one fixed factor. Eisenhart [1947] introduced the mixed model terminology and made explicit and formally the distinction between fixed and random effects models. According to Longford [1994], several estimation methods have been proposed for mixed effects models: maximum likelihood and restricted maximum likelihood [Harville, 1974] are generally adopted for linear mixed effects models. The algorithms used are usually the EM algorithm [Dempster et al., 1977] or Newton–Raphson methods [Thisted, 1988]. Jennrich and Schluchter [1986] consider maximum likelihood estimation in linear mixed effects models for repeated measures with structured variance-covariance matrices. Schall [1991] presents estimation in generalized linear models with random effects.

Sheiner and Beal [1980] were one of the first to develop nonlinear mixed effects models. Lindstrom and Bates [1990] also proposed a nonlinear mixed model. A nonparametric maximum likelihood method for nonlinear mixed effects models was proposed by Mallet et al. [1988]. Vonesh and Carter [1992] have developed a mixed effects model that is nonlinear in the fixed effects, but linear in the random effects. Davidian and Gallant [1992] introduced a smooth nonparametric maximum likelihood estimation method for nonlinear mixed effects.

A Bayesian approach for nonlinear mixed effects is described in Wakefield [1996]. Verbeke and Molenberghs [2000] present linear mixed models for longitudinal data. Pinheiro and Bates [2000] provide an overview of the theory and application of linear and nonlinear mixed-effects models in the analysis of grouped data. Gu and Ma [2005] present optimal smoothing in nonparametric mixed-effect models. Wu [2010] provides an overview of linear mixed models, generalized linear mixed models, nonlinear mixed effects models, frailty models and semiparametric and nonparametric mixed effects models, and also discusses appropriate approaches to address missing data, measurement errors, censoring and outliers in mixed effects models. Demidenko [2013] presents the theory of mixed models and its application in R, considering for example linear and nonlinear mixed models for statistical analysis of shape and statistical image analysis. Some open problems in mixed models are also mentioned. Fahrmeir and Tutz [2001] present linear random effects models for normal data, random effects in generalized linear models and Bayesian mixed models. West et al. [2014] present a brief history of linear mixed models.

Pinheiro et al. [2001] describe a robust version of the linear mixed-effects model in which the Gaussian distributions for the random effects and the within-subject errors are replaced by multivariate t-distributions. Guo [2002] presents functional mixed effects models. Angelini et al. [2003] present wavelet regression estimation in nonparametric mixed effect models. Eilers et al. [2015] established the connections between P-splines, mixed models and Bayesian analysis. Fabio et al. [2012] developed a Poisson mixed model with a nonnormal random effect distribution.

10.11   Exercises

1. **The Hodges data: a comparison of normal random effects in GAMLSS.** This is an exercise to demonstrate the different random effects methods in **gamlss**. The data used here are the **hodges** data introduced in Section 10.7. All models fitted in this question assume that the response variable is normally distributed. The purpose is to compare results from **gamlss()** with **lme()**.

   (a) Start by fitting the null model (only the constant) and the model using the main effect for the factor **state**.

   ```
   m0<-gamlss(prind~1, data=hodges)# null
   m1<-gamlss(prind~state, data=hodges)# main effect for state
   ```

   (b) Plot the observations and the fitted values for both models.

   ```
   plot(prind~unclass(state), data=hodges, pch=3)
   points(fitted(m1)~unclass(state),data=hodges,pch=5,col="red")
   lines(fitted(m0))
   ```

   The fitted values from model **m1** are the sample means for each of the levels of the factor **state**, while the fitted values from model **m0** are the overall sample mean.

   (c) Fit a random effect for **state** using both the **random()** and the **re()** functions, and compare the fitted values and fitted degrees of freedom for μ.

   ```
   m2<-gamlss(prind~random(state), data=hodges)
   m3<-gamlss(prind~re(random=~1|state), data=hodges)
   ```

```
m31<-gamlss(prind~re(random=~1|state, method="REML"),
            data=hodges)
```

(d) Get the estimates for the variance components. Note that getting σ_b from
the lme() output is rather awkward, so we have created a small function
to do that.

```
getSmo(m2)$sigb # sigma_b
getSigmab <- function(obj) #  new function
  {
  vc <- VarCorr(obj)
  suppressWarnings(storage.mode(vc) <- "numeric")
  vc[1:2,"StdDev"][1]
  }
getSigmab(getSmo(m3))
getSigmab(getSmo(m31))
# fitted sigma_e within the gamlss algorithm
getSmo(m2)$sige
getSmo(m3)$sigma
getSmo(m31)$sigma
```

How do the different estimates for each of σ_b and σ_e compare?

(e) Now use lme() to fit the random effects model, using ML and REML, and
compare the fitted σ_b and σ_e estimates.

```
l1 <- lme(prind~1, data=hodges, random=~1|state) # REML
l2 <- lme(prind~1, data=hodges, random=~1|state, method="ML")
# fitted sigma_b
getSigmab(l1)
getSigmab(l2)
# compare to gamlss
getSmo(m2)$sigb
getSigmab(getSmo(m3))
getSigmab(getSmo(m31))
# fitted sigma_e
l1$sigma
l2$sigma
#compare to gamlss
fitted(m2, "sigma")[1]
fitted(m2, "sigma")[1]*getSmo(m2)$sige #adjusted estimate of
                                       #sigma_e
```

Note that the last command adjusts the estimated gamlss σ by multiplying
it by the fitted σ_e from the lme model fitted within gamlss(). This is a

consequence of the MAP estimation of random effects within GAMLSS; see Nelder [2006].

(f) Compare the fitted values obtained from the two models. Note that `lme()` provides two fitted values: (i) `level=0`, the fitted values for the marginal model, and (ii) `level=1`, the fitted values from the conditional model (the default). The latter is referred to in the literature as BLUP (best linear unbiased predictors) or EBP (empirical Bayes predictors). The fitted values of the `gamlss` models `m3` and `m31` are the BLUP estimates, since they are conditional models.

(g) Random effects are shrinkage estimators which in our case, for the `hodges` data, means that they shrink the level means towards the overall mean. This can be seen in the following plot:

```
# shrinkage effect
plot(fitted(m1)~unclass(state),data=hodges,pch=5,col="red")
points(fitted(m3)~unclass(state),data=hodges,pch=3,col="blue")
lines(fitted(m0))
```

(h) Note that the overall mean and the fixed effect estimate we get from the `lme` fit are slightly different.

```
mean(hodges$prind)
fitted(m0)[1]
fitted(l1, level=0)[1]
fitted(l2, level=0)[1]
```

(i) The function `resid()` in **nlme** also has the option `level` referring to the marginal (`level=0`) or conditional (`level=1`, default) residuals, respectively. The two types of residuals are the difference of the data values from the fitted marginal and BLUP values, respectively. For example:

```
# conditional data-BLUP
res <- with(hodges, prind)-fitted(m2)
plot(resid(l2)~res)
# marginal data-fitted
res <- with(hodges, prind)-fitted(m0)
plot(resid(l2, level=0)~res)
```

The **gamlss** residuals are the normalized quantile residuals, which for a normal distribution with constant variance σ^2 are the standardized residuals $r = (y - \hat{\mu})/\hat{\sigma}$. The **lme** residuals also have the option to produce the Pearson residuals $(y - \hat{y})/\sqrt{V(\hat{\mu})}$. Since the **gamlss** models have a normal distribution with constant variance, $V(\hat{\mu}) = 1$, and so the Pearson's residuals are $(y - \hat{\mu})$.

```
r1 <- qNO(pNO(with(hodges, prind), mu=fitted(l2, level=1),
          sigma=l2$sigma))
```

```
head(r1)
head(resid(12,type="p")) #Pearson residuals
head(resid(m2))
plot(resid(m3)~r1)
plot(resid(12,type="p")~r1)
#level 0 residuals
r0 <- qNO(pNO(with(hodges, prind), mu=fitted(12, level=0),
            sigma=12$sigma))
head(r0)
head(resid(12, level=0, type="p"))
plot(resid(12, level=0, type="p")~r0)
```

Note that **gamlss** does not provide marginal residuals (but **gamlssNP()** does).

(j) The following are some of the **lme** functions (methods) which can be used if the **re()** function is used within **gamlss()**.

```
# summary
summary(getSmo(m3))
# residual level=1  plot
plot(getSmo(m3))
# random effect estimates
ranef(getSmo(m3))
# fitted coefficients
coef(getSmo(m3))
# Confidence Intervals
intervals(getSmo(m3))
# ANOVA
anova(getSmo(m3))
```

(k) Use the **gamlssNP()** function (**gamlss.mx** package) to fit the same model as above. Use $\mathcal{K} = 10$ and $\mathcal{K} = 20$ Gaussian quadrature points for comparison.

(l) Compare the parameter estimates for σ_b and σ_e from **gamlssNP()** and **lme()**.

(m) Compare the fitted values obtained from the **gamlssNP()** fits with those obtained from **lme()**.

2. **Hodges data revisited**. The Hodges data were analyzed in Section 10.7.1 and Exercise 1. Here we use **gamlssNP()** to fit a normal random intercept model for μ using Gaussian quadrature, with different conditional distributions given the random effects. Try different distributions like NO, LOGNO and BCT, e.g.

```
mBCT <- gamlssNP(prind~1, data=hodges, random=~1|state,
            mixture="gq", K=20, family=BCT)
```

3. **The epileptic data: count data random effect model.** The `epil` data from
the **MASS** package originate from Thall and Vail [1990]. They comprise four repeated measurements of seizure counts (each over a two-week period preceding a
clinical visit) for 59 epileptics, a total of 236 observations. Breslow and Clayton
[1993] and Lee and Nelder [1996] identified overdispersion in the counts, which
they modelled using a random effect for cases in the predictor for the mean in a
Poisson GLMM. Lee and Nelder [2000] additionally considered an overdispersed
Poisson GLMM, using extended quasi likelihood. The latter authors also identified random effects for subjects in the predictor for the mean. Within `gamlss`
we can directly model the casewise overdispersion in the counts by using a negative binomial (type I) model and consider random effects for subjects in the
predictors for both the mean, μ, and the dispersion, σ. Specifically we assume
that, conditional on μ_i and σ_i (i.e. conditional on the random effects) the seizure
counts Y_{ij} are independent over subjects $j = 1, 2, \ldots, 59$ and repeated measurements $i = 1, 2, 3, 4$ with a negative binomial (type I) distribution:

$$Y_{ij} \mid \mu_{ij}, \sigma_{ij} \sim \text{NBI}(\mu_{ij}, \sigma_{ij}) \, ,$$

where the log of the mean is modelled using explanatory terms and the log of
both the mean and dispersion include a random effects term for subjects. (Note
the conditional variance of Y_{ij} is given by $V(Y_{ij} \mid \mu_{ij}, \sigma_{ij}) = \mu_{ij} + \sigma_{ij}\mu_{ij}^2$.)

The data are described below.

R data file: `epil` in package MASS of dimensions 236×9

variables

> `Y` : the count for the 2-week period
>
> `trt` : treatment, `placebo` or `progabide`
>
> `base` : the counts in the baseline 8-week period
>
> `age` : subject's age, in years
>
> `V4` : 0/1 indicator variable of period 4
>
> `subject` : subject number, 1 to 59
>
> `period` : period, 1 to 4
>
> `lbase` : log counts for the baseline period, centred to have zero mean
>
> `lage` : log ages, centred to have zero mean

purpose: to demonstrate random effect models in GAMLSS.

The variable `visit` defined below is an extra variable needed to conform with
the notation used in Breslow and Clayton [1993] where the clinic visits were
coded (-0.3, -0.1, 0.1, 0.3).

(a) Input the data set epil from the **MASS** package, calculate the variable
visit and declare subject as a factor.

```
data(epil)
# tranformation of the data
epil<-transform(epil, visit=(2*epil$period-5)/10,
                Subject=factor(epil$subject))
```

(b) Following the approach of Breslow and Clayton [1993], fit a negative binomial type I ($NBI(\mu, \sigma)$) distribution for Y with the μ model
trt*lbase+visit+lage, with random intercept for Subject.

(c) Obtain the estimates of the random effect components.

(d) Use the worm plot to check the adequacy of the fitted model. Comment on
the plot.

(e) Fit a random intercept model for σ.

```
m4<-gamlss(y~trt*lbase+visit+lage+re(random=~1|Subject),
    sigma.fo=~re(random=~1|Subject), data=epil, family=NBI,
    gd.tol=Inf)
```

(f) Get the random effect estimates for σ_k for $k = 1, 2$ (i.e. for μ and σ,
respectively).

(g) Check the worm plot for model m4.

(h) It was suggested that a model for σ which discriminates between visit 4
and the rest could be appropriate.

```
m5<-gamlss(y~trt*lbase+visit+lage+re(random=~1|Subject),
    sigma.fo=~factor(V4), data=epil, family=NBI, gd.tol=Inf)
```

(i) Compare the worm plots of the model with random effects for σ (m4) and
the model with a fixed effect for σ (m5).

(j) Check whether we need a random effect for σ using the gamlssNP() function.

(k) Should we use smooth functions for the fixed effects?

```
m1s<-gamlss(y~pvc(lbase,by=trt)+pbz(lbase)+pbz(visit)+
    pbz(lage)+random(Subject),data=epil,family=NBI,gd.tol=Inf)
drop1(m1s)
term.plot(m1s, pages=1)
m3s<-gamlss(y~pvc(lbase, by=trt)+pbz(lbase)+pbz(visit)+
    pbz(lage)+random(Subject), sigma.fo=~random(Subject),
    data=epil,family=NBI, gd.tol=Inf)

m5s<-gamlss(y~pvc(lbase, by=trt)+pbz(lbase)+pbz(visit)+
```

```
        pbz(lage)+random(Subject), sigma.fo=~factor(V4), data=epil,
     family=NBI, gd.tol=Inf)
```

4. **Respiratory infection in children: nonparametric fit**. The data were analyzed in Section 10.5.1 using `gamlssNP()`, and in Section 10.7.2 using `random()`, i.e. a normal random effect. Here use a nonparametric (discrete) random intercept, and then a random intercept and slope.

```
m1<-gamlssNP(time~age+xero+female+cosine+sine+height+stunted,
   random=~1|id,K=10, mixture="np",data=respInf,family=BI)
plot(m1$mass,m1$prob, type="h")
m2<-gamlssNP(time~age+xero+female+cosine+sine+height+stunted,
   random=~age|id,K=10, mixture="np",data=respInf,family=BI)
plotMP(m2$mass[,1], m2$mass[,2], m2$prob)
```

Comment on the shape of the nonparametric random effect distribution. How will you choose between the models?

5. **The meta-analysis revisited**. Here we fit the model using standard GAMLSS.

The data were analyzed in Section 10.5.2 using `random()` with `mixture="np"`, i.e. a nonparametric (discrete) random effect. Here use normal random effect models.

```
me2<-gamlss(cbind(d,n-d)~fac+re(fixed=~fac,random=~fac|study),
   family=BI, data=meta)
me3<-gamlss(cbind(d,n-d)~fac+re(fixed=~fac,
   random=list(study=pdSymm(~fac))), family=BI, data=meta)
me4<-gamlss(cbind(d,n-d)~fac+re(fixed=~fac,
   random=list(study=pdIdent(~fac))), family=BI, data=meta)
me5<-gamlss(cbind(d,n-d)~fac+re(fixed= ~fac,
   random=list(study=pdDiag(~fac))), family=BI, data=meta)
me6<-gamlss(cbind(d,n-d)~fac+re(fixed=~fac,
   random=list(study=pdDiag(~fac))), family=BB, data=meta)
AIC(me2, me3,me4,me5,me6)
```

6. **Leukemia data revisited**. The leukemia data were analyzed in Section 10.9.3 using a normal conditional distribution given a random effect model for μ. Compare **g2**, which was the best model using SBC, with alternative conditional distributions. Do the following:

   (a) Add a smooth function for **age** in the σ model;

   (b) Fit different distributions to the data, i.e. LO, BCPE, ST3, SHASH;

   (c) For the **SHASH** fitted model, add a smooth function for ν, and sequentially for τ;

(d) Check the values of GAIC for the models and use diagnostics to check their adequacy.

Part V

Model selection and diagnostics

11

Model selection techniques

CONTENTS

11.1   Introduction: Statistical model selection

This chapter explains the model selection techniques in **gamlss**. In particular, it explains:

1. model selection techniques for a GAMLSS model;

2. stepwise selection functions for selecting explanatory terms;

3. techniques for selecting smoothing parameters.

This chapter is important for understanding the statistical modelling process within GAMLSS.

Model selection for regression models is primarily concerned with the problem of selecting relevant predictors from a larger set of potential predictors. Model selection for GAMLSS models involves the selection of (i) the distribution for the response variable and (ii) the predictors for the distribution parameters of the selected distribution.

In general, statistical models are built to:

- explore the data where no theory exists (*exploratory* models), and/or

- explain or verify a theory (*explanatory* models), and/or

- predict future values (*predictive* models).

The performance of a statistical model (*model assessment*) is usually related to its explanatory or predictive capability on an independent data set (often called the test data set). Thus, in general, it is recognized that *overfitted* (over-interpreting the current data) or *underfitted* (under-interpreting the current data) models are not very good for explanation or prediction. Different functions (see Table 11.1) have been implemented to support model selection in **gamlss**. The general aim of these functions is to avoid both overfitting and underfitting, in order to select a model with the best plausible performance. In statistical inference terms, this turns out to be a balance between *variance* (the variability of the parameter estimates) and *bias* (the amount by which the average of the estimates of the model parameters differs from their true values) of the estimators. Generally, variance and bias both affect the prediction error of the model. Overfitted model estimators (despite their lower bias) have large variance and so are often poor for prediction, while underfitted model estimators (despite their lower variance) are biased and also often poor for prediction. An objective is a model which neither overfits nor underfits the data. Typically, the more complex the model, the lower the bias but the higher the variance [Hastie et al., 2009].

Let $\mathcal{M}$ be a statistical model having parameter vector $\boldsymbol{\theta}$. For a parametric statistical model and within a likelihood-based inferential procedure, the fit of model $\mathcal{M}$ can be assessed by its fitted global deviance (GDEV), defined as GDEV $= -2\ell(\hat{\boldsymbol{\theta}})$.

Let $\mathcal{M}_0$ and $\mathcal{M}_1$ be statistical models with fitted global deviances GDEV$_0$ and GDEV$_1$ and degrees of freedom df$_0$ and df$_1$, respectively.

Definition: Model $\mathcal{M}_0$ is *nested* within $\mathcal{M}_1$ if $\mathcal{M}_0$ is a subclass of $\mathcal{M}_1$.

Two nested parametric statistical models, $\mathcal{M}_0$ and $\mathcal{M}_1$, may be compared using the generalized likelihood ratio test statistic

$$\Lambda = \text{GDEV}_0 - \text{GDEV}_1$$

which, under certain conditions, has asymptotically the χ_d^2 distribution under the null hypothesis that the correct model is $\mathcal{M}_0$, where $d = \text{df}_0 - \text{df}_1$. When $\mathcal{M}_0$ and $\mathcal{M}_1$ contain *nonparametric* additive terms, the same test can be used as a guide to model selection in the same way that Hastie and Tibshirani [1990, Section 3.9] compare 'nested' generalized additive model (GAM) fits. The *effective* degrees of freedom used here is the trace of the smoothing matrix $\mathbf{S}$ in the fitting algorithm. For details, see Chapter 9 and Hastie and Tibshirani [1990].

For comparing non-nested GAMLSS models, the generalized Akaike information criterion (GAIC) [Akaike, 1983] can be used to penalize overfitting. This is obtained by adding to the fitted deviance a penalty κ for each effective degree of freedom used in a model, i.e.

$$\text{GAIC}(\kappa) = \text{GDEV} + (\kappa \times \text{df}),\qquad (11.1)$$

where df denotes the total effective degrees of freedom of the model. The model with the smallest value of the criterion $\text{GAIC}(\kappa)$ is then selected. The Akaike information criterion (AIC) [Akaike, 1974] and the Bayesian information criterion (BIC) or Schwarz Bayesian criterion (SBC) [Schwarz, 1978] are special cases of $\text{GAIC}(\kappa)$ corresponding to $\kappa = 2$ and $\kappa = \log(n)$, respectively. The AIC and BIC/SBC are asymptotically justified as predicting the degree of fit in a new data set, i.e. as approximations to the average predictive error. Justification for the use of BIC/SBC comes also as a crude approximation to Bayes factors [Raftery, 1996, 1999]. In practice it is usually found that the AIC leads to overfitting in model selection, while the BIC/SBC leads to underfitting. Our experience suggests that a value of the penalty κ in the range $2.5 \leq \kappa \leq 4$ or $\kappa = \sqrt{\log(n)}$ (when $n \geq 1000$) often works well. We suggest that a selection of different values of κ, e.g. $\kappa = 2, 2.5, 3, 3.5, 4$, could be used in turn to investigate the sensitivity or robustness of the model selection to the choice of the value of κ.

Cross validation techniques can be employed as an alternative to the GAIC for the comparison of (nested or non-nested) GAMLSS models, particularly for predictive modelling. In particular, K-fold cross validation is a simple procedure for comparison of models. Firstly the data is randomly divided into K subsets $S_1, S_2, \ldots, S_K$. Then:

- For $j = 1, 2, \ldots, K$, omit S_j and fit a model to the rest of the data.
- Calculate a measure of goodness of predictive fit to S_j.
- Combine the K measures of goodness of fit.
- Choose between models using the combined measure of goodness of fit.

The measure of goodness of fit can be the global deviance or any other measure of disparity. The model with the smallest overall value of the chosen measure of disparity is the best for prediction purposes. A *simple cross validation* is defined when only one observation is omitted in each fit and hence $K = n$.

The GAIC and cross-validation techniques are used for relatively small or moderate size data sets where the full data sample is used for both model fitting (e.g. minimizing GDEV) and for model selection (e.g. minimizing a GAIC or a cross-validation criterion). For large data sets (e.g. $n \geq 5000$), the data could be randomly split into:

(i) a training data set;

(ii) a validation data set; and

(iii) a test data set.

The training data is used for model fitting, the validation data set is used for model selection and the test data set is used for model assessment. This split is routinely available in statistical packages that perform data mining. Some of these procedures are implemented in the **gamlss** packages and they are described later in this chapter.

It is not necessary for the model selection process to select only a 'best' model, giving the possibly false impression that this is the only model that predicts or explains the data. The practice of selecting only one model ignores uncertainty due to model choice, and can lead to over-confident inferences. In particular, inference about quantities of interest can be made either conditionally on a single selected 'final' or 'best' model, or by averaging between selected models. Conditioning on a single final model was criticized by Draper [1995] and Madigan and Raftery [1994] since it ignores model uncertainty and generally leads to the underestimation of the uncertainty about quantities of interest. Averaging between selected models can reduce this underestimation [Hjort and Claeskens, 2003]. Selection of the set of models to be used for averaging can be based upon a model selection criterion such as GAIC(κ) for different values of κ. Alternatively, for a specific value of κ, a weighted average of the models can be used in which the weights are a function of the GAIC(κ) (e.g. the model with the lowest GAIC is given the highest weight).

When using model selection strategies to build a statistical model, determination of the model adequacy should always be carried out with respect to the substantive questions of interest, and not in isolation. This means that different problems could possibly require different model selection strategies. For example, if the purpose of the study is to describe the data parsimoniously, then a single 'final' model is usually sufficient. After selecting a model, it is strongly advisable to use diagnostics to assess the selected model (see Chapter 12).

11.2 GAMLSS model selection

Let $\mathcal{M} = \{\mathcal{D}, \mathcal{G}, \mathcal{T}, \mathcal{L}\}$ represent a GAMLSS model as defined in Section 3.1. The components of $\mathcal{M}$ are defined as follows:

(i) $\mathcal{D}$ specifies the distribution of the response variable,

(ii) $\mathcal{G}$ specifies the set of link functions,

(iii) $\mathcal{T}$ specifies the terms appearing in all the predictors for μ, σ, ν and τ,

(iv) $\mathcal{L}$ specifies the smoothing hyperparameters which determine the amount of smoothing in the $s_{kj}(\cdot)$ functions of equation (3.1).

In the search for an appropriate GAMLSS model for any new data set, all of the above components have to be specified as objectively as possible. We next discuss how $\mathcal{D}$, $\mathcal{G}$, $\mathcal{T}$ and $\mathcal{L}$ are specified.

11.2.1 Component $\mathcal{D}$: Selection of the distribution

The selection of the appropriate distribution can be done in two stages, the *fitting* stage and the *diagnostic* stage. The fitting stage involves the comparison of fitted models from different distributions using the GAIC(κ) discussed above. The diagnostic stage involves the use of the worm plot, which allows detection of inadequacies in the model globally or within specific ranges of one (or two) explanatory variable. The worm plot is described in Section 12.4.

If no explanatory variables are involved, the `fitDist()` function may be used to fit all relevant parametric GAMLSS family distributions to the response variable. The quality of the fits is compared using the GAIC, and the distribution with the lowest GAIC score is selected. If there are explanatory variables, this procedure can be used at the exploratory stage of the analysis as long as the user is aware that `fitDist()` assesses the marginal distribution of the response, and that the conditional distribution given the explanatory variables can be different.

It is strongly advisable to use diagnostics to assess whether the selected model residuals are supportive of the assumed response distribution (see Chapter 12), both initially (the marginal distribution) and after selection of predictors for the distribution parameters (the conditional distribution).

11.2.2 Component $\mathcal{G}$: Selection of the link functions

The initial selection of the link function for each distribution parameter is usually determined by the range of the parameter in hand. For example for the normal distribution, the parameter ranges are $-\infty < \mu < \infty$ and $0 < \sigma < \infty$. The identity link for μ and the log link for σ ensure that μ and σ are always within their ranges. There are occasions in which the choice of the link function is important from the interpretation point of view. For example if we believe that the explanatory variables affect the distribution parameter multiplicatively, then a log link is appropriate. The choice of link function may affect the model fit considerably. Different link functions should be compared using the GAIC or worm plots.

11.2.3   Component $\mathcal{T}$: Selection of the additive terms in the model

The selection of explanatory variables is in practice one of the most important subjects in statistical modelling. Let $\mathcal{X}_k$ be a pool of explanatory terms available for consideration for modelling the parameter θ_k of a GAMLSS model, where $\theta = (\theta_1, \theta_2, \theta_3, \theta_4) = (\mu, \sigma, \nu, \tau)$. Typically $\mathcal{X}_k$ will contain factors and quantitative terms which can enter the model as linear or smoothing additive terms; see Section 8.1. How to select the terms and the form in which the terms enter in the model is the main task of this chapter.

Procedures in the current statistical literature for the selection of terms can be divided into three main categories.

criterion-based methods: A criterion such as the GAIC is used and the terms in $\mathcal{X}_k$ enter the model if the value of the criterion improves. Criterion-based methods are the main subject of this chapter.

regularization: Here all terms available in $\mathcal{X}_k$ enter the model (linearly) but their effect on the distribution parameter, that is their fitted coefficients, are damped by a shrinking method. Ridge and lasso regression are typical examples of regularization. Section 9.4.7 describes how these methods can be used within the GAMLSS framework.

dimension-reduction procedures: These procedures reduce the dimensionality of $\mathcal{X}_k$ by principal component analysis, and then a few chosen components enter the GAMLSS model. No functions are currently provided for this type of analysis.

Boosting, which is another very powerful method for variable selection, is discussed in Mayr et al. [2012] and Hofner et al. [2016].

Criterion-based methods

For a given distribution, the selection of the terms has to be done for all the distribution parameters, not only the location parameter μ. The usual *forward, backward* and *stepwise* procedures can be applied for each distribution parameter (explained below). The number of available terms in $\mathcal{X}_k$, relative to the number of observations in the sample, matters as far as selection of terms is concerned. For example, if the number of continuous explanatory variables is small, say five, all $2^5 = 32$ different combinations of those five variables can be assessed. On the other hand when we are dealing with say 50 continuous explanatory variables, there are $2^{50} = 1.13 \times 10^{15}$ different combinations. As it is not generally practical to fit such a large number of models, we have to implement a different strategy.

First consider procedures for choosing a model for a single distribution parameter θ (e.g. μ). In the usual forward procedure, each variable that is not already in the model for θ is tested for inclusion. The procedure stops when none of the remaining variables is significant, based on the model selection criterion, when added to the model. A key drawback of the forward procedure is the fact that each addition of a new variable may render one or more of the already included variables non-significant. On the other hand, the backward procedure starts with fitting a model with all the variables of interest included. Then the least significant variable is dropped, as long as it is not significant based upon the model selection criterion. One continues by successively re-fitting reduced models until all remaining variables in the model are significant. Finally, the stepwise procedure (used in GAMLSS function `stepGAIC()`) allows dropping or adding one variable at each step. All variables currently in the model are individually considered for dropping at each step, while all variables not currently in the model are considered for adding, The variable (to drop or add) that gives the minimum value of the model selection criterion is chosen at that step, provided it reduces the criterion.

There are many different strategies that could be applied for the selection of the terms used to model all the parameters μ, σ, ν and τ. In the current implementation we have two strategies for selecting terms for all the parameters, which we call strategy A and strategy B. These are implemented in `stepGAICAll.A()` and `stepGAICAll.B()`, respectively. (See Sections 11.5 and 11.6.)

There are several functions within **gamlss** to assist with selecting terms when all the data points are used. (See Section 11.2.5 for when this is not the case.) The basic functions are `addterm()` and `dropterm()`, which allow the addition or removal of a term for a distribution parameter, respectively. These functions are the building blocks for `stepGAIC()`, which is suitable for stepwise selection of terms for one of the distribution parameters. `stepGAIC()` is in turn the basic building block of `stepGAICAll.A()` and `stepGAICAll.B()`.

11.2.4 Component $\mathcal{L}$: Selection of the smoothing parameters

Each smoothing term selected for any of the distribution parameters has at least one smoothing parameter (or hyperparameter) λ associated with it. In Chapter 3 we refer to the hyperparameters as the 'lambdas' and here we use $\mathcal{L}$ to denote the set of all smoothing parameters in a model, e.g. $\mathcal{L} = \{\lambda_{\mu,1}, \lambda_{\mu,2}, \lambda_{\sigma,1}, \lambda_{\nu,1}\}$. The smoothing parameters can be fixed or estimated from the data. The standard way of fixing the smoothing parameters, as suggested in Hastie and Tibshirani [1990], is by fixing the effective degrees of freedom for smoothing. Many of the smoothing procedures within the **gamlss** packages allow the user to do that by using the argument `df`. More generally, it is desirable to estimate the smoothing parameters automatically. Section 3.4 describes the methods used within GAMLSS to estimate the smoothing parameters. These are called *local* methods when they are applied within the iterative GAMLSS algorithm, and *global* when applied outside the iterative GAMLSS algorithm. In our experience the local methods are much faster and often produce similar results to the global methods. However, the global methods can sometimes be more reliable.

11.2.5   Selection of all components using a validation data set

For large data sets, the statistical modeller can afford to split the data into different parts. For example:

i) the training data could be used for model fitting (minimizing its GDEV);

ii) the validation data could be used for model selection, in particular selection of the distribution, link functions, predictor terms and smoothing parameters, by minimizing the validation global deviance (VDEV);

iii) the test data could be used for the assessment of the predictive power of the model chosen by (ii) applied to the test data, by calculating the test global deviance (TDEV).

There are several functions within **gamlss** to assist model selection. For example, gamlssVGD() fits a model to the training data set and then calculates the VDEV for the validation data set. Different models fitted this way can be compared using VGD(), which behaves similarly to GAIC(). The key difference between GAIC() and VGD() is that GAIC() compares fitted models and VGD() compares validated models. If we already have GAMLSS fitted models and we want to see how well they perform on a new test data set, getTGD() is used to get their TDEV and TGD() is used to compare them.

The basic functions for selection of variables when we have training data sets are add1TGD() and drop1TGD(). These allow the inclusion or exclusion of a single term in the model based upon the TDEV. The function stepTGDAll.A() performs similarly to stepGAICAll.A() by selecting terms for all the parameters using TDEV as a criterion.

11.2.6   Summary of the GAMLSS functions for model selection

Table 11.1 provides a summary of the model selection functions described in this chapter, according to GAMLSS model component and data set-up. The functions with an asterisk are covered in other chapters or are standard **R** functions (in the case of optim()).

Sections 11.3 to 11.6 describe the **gamlss** functions which can be used for model selection if the whole sample is used for model selection. Section 11.3 describes addterm() and dropterm(), which are the building blocks for a full model selection strategy. Sections 11.4, 11.5 and 11.6 describe stepGAIC(), stepGAICAll.A() and stepGAICAll.B(), respectively. Section 11.7 describes model selection using K-fold cross validation, while Section 11.8 describes model selection using validation and test data sets.

TABLE 11.1: The model selection functions described in this chapter according to component of the GAMLSS model and data set-up

Component	All data	K-fold cross validation	Validation and test data
$\mathcal{D}$	GAIC()* wp() *	gamlssCV() CV()	gamlssVGD() VGD() getTGD() TGD()
$\mathcal{G}$	deviance() * GAIC() * wp() *	gamlssCV() CV()	as above
$\mathcal{T}$	drop1() add1() add1ALL() drop1ALL() stepGAIC() stepGAICAll.A() stepGAICAll.B()	gamlssCV() CV()	drop1TGD() add1TGD() stepTGD() stepTGDAll.A()
$\mathcal{L}$ (global)	findhyper() optim()*	optim()*	optim()*

Note: Functions with asterisk are not covered in this chapter.

11.3 The addterm() and dropterm() functions

The functions addterm() and dropterm() are generic **R** S3 object functions with their original definitions as in the package **MASS**. Note that addterm() and dropterm() have a **parallel** argument which can be used for parallel computations, assuming that the computer has multiple CPUs. This can be beneficial for large data sets, when the fitting of each individual model can take several minutes.

The dropterm() and addterm() functions in **gamlss** have the following arguments:

object   a **gamlss** object;

scope   a formula giving terms which might be dropped or added. For dropterm() the default is the model formula. For addterm(), scope is a formula specifying a maximal model which should include the current one. Only terms that can be dropped or added while maintaining marginality are actually tried;

what   the parameter of the distribution (equivalent to **parameter**);

parameter   equivalent to **what**;

scale   not used in `gamlss`;

test   `test="none"` for no test and `test="Chisq"` for a χ^2 test statistic relative to the original model. Note that the default is `test="none"` for `dropterm()` and `addterm()`, while it is `test="Chisq"` for the equivalent `drop1()` and `add1()` functions;

k   the penalty for each model degree of freedom in the GAIC. Note `k=2` gives AIC while `k=log(n)` gives SBC;

sorted   If `sorted=TRUE` (the default), the results are sorted in the order of the GAIC from the lowest (the best model) to the highest (the worst model);

trace   if `trace=TRUE` (the default), additional information may be given on the fits as they are tried;

parallel   The type of parallel operation to be used with alternatives `"no"` (the default), `"multicore"` and `"snow"`;

ncpus   the number of processes to be used in parallel operation. Typically one would choose this to be the number of available CPUs in the computer;

cl   This is the (optional) name of a parallel or snow cluster if `parallel="snow"` is used. If the argument is not supplied, a cluster on the local machine is created for the duration of the call;

...   arguments passed to or from other methods.

The `add1()` and `drop1()` functions are identical to `addterm()` and `dropterm()` except for having different default values for the argument `test`. By default `add1()` and `drop1()` report both GAIC and the likelihood ratio test with its χ^2 values, while `addterm()` and `dropterm()` report only the GAIC. `drop1()` has been suggested in Chapters 1, 2 and 5 as a way of testing terms. For large data sets with many predictors, `drop1()` and `dropterm()` might take a long time to complete.

In order to demonstrate `dropterm()` and `addterm()`, we use the US pollution data [Hand et al., 1994], first described in Chapter 9. Preliminary analysis has shown that it is better to model the distribution of the response variable y using the gamma rather than the normal distribution. We start by fitting the full linear model for μ including all six explanatory variables:

```
data(usair)
mod1<-gamlss(y~., data=usair, family=GA, trace=FALSE)
summary(mod1)

## *****************************************************************
## Family:  c("GA", "Gamma")
##
```

```
## Call:  gamlss(formula = y ~ ., family = GA, data = usair,
##     trace = FALSE)
##
## Fitting method: RS()
##
## -----------------------------------------------------------
## Mu link function:  log
## Mu Coefficients:
##               Estimate Std. Error t value Pr(>|t|)
## (Intercept)  7.3164944  1.3681744   5.348 6.61e-06 ***
## x1          -0.0622829  0.0168679  -3.692 0.000798 ***
## x2           0.0013416  0.0003506   3.826 0.000550 ***
## x3          -0.0008132  0.0003546  -2.294 0.028323 *
## x4          -0.1562766  0.0557698  -2.802 0.008429 **
## x5           0.0196006  0.0101839   1.925 0.062928 .
## x6           0.0002016  0.0047788   0.042 0.966601
## ---
## Signif. codes:
## 0 '***' 0.001 '**' 0.01 '*' 0.05 '.' 0.1 ' ' 1
##
## -----------------------------------------------------------
## Sigma link function:  log
## Sigma Coefficients:
##               Estimate Std. Error t value Pr(>|t|)
## (Intercept)  -0.9022      0.1713  -5.268 8.36e-06 ***
## ---
## Signif. codes:
## 0 '***' 0.001 '**' 0.01 '*' 0.05 '.' 0.1 ' ' 1
##
## -----------------------------------------------------------
## No. of observations in the fit:  41
## Degrees of Freedom for the fit:  8
##       Residual Deg. of Freedom:  33
##                       at cycle:  2
##
## Global Deviance:      303.1602
##             AIC:      319.1602
##             SBC:      332.8687
## ***********************************************************
```

The '.' in the model formula selects all variables other than y in the usair data frame as explanatory variables. Now we use drop1() to check whether any linear terms can be dropped:

```
dd<-drop1(mod1)
dd

## Single term deletions for
## mu
##
## Model:
## y ~ x1 + x2 + x3 + x4 + x5 + x6
##         Df    AIC     LRT  Pr(Chi)
## <none>       319.16
## x1       1 327.58 10.4245 0.001244 **
## x2       1 326.92  9.7557 0.001788 **
## x3       1 321.39  4.2299 0.039717 *
## x4       1 324.08  6.9247 0.008501 **
## x5       1 320.57  3.4141 0.064642 .
## x6       1 317.16  0.0017 0.966960
## ---
## Signif. codes:
## 0 '***' 0.001 '**' 0.01 '*' 0.05 '.' 0.1 ' ' 1
```

Note that the same output is obtained from

```
dd<-dropterm(mod1,test="Chisq")
```

The above output gives the generalized likelihood ratio test statistic (LRT) and its Chi-square p-values (Pr(Chi)) for removing each of the six variables from the full model. Given all other linear terms in the model, the variable x6 is the first to be dropped since it has the highest p-value.

For a full parametric model (such as the one above), the Chi-square test using drop1() and the t-values using summary() could produce different p-values and hence different conclusions. This is because drop1() provides the generalized likelihood test, which in general is preferable to the Wald test statistic provided by summary(). Note also that drop1() tests whether a factor term should be excluded from the model, given the rest of the terms in the model, while summary() only tests individual parameters of a factor. Also when smooth terms are in the model, summary() does not provide the correct information for testing if a term should be excluded given the rest of the terms in the model, as discussed later in this section. So drop1() is very useful in model selection.

Next the use of dropterm() when a smoother and a factor are in the model are demonstrated. The aids data set, first described in Chapter 4, is used.

```
data(aids)
aids1<-gamlss(y~qrt+pb(x), data=aids, family=NBI, trace=FALSE)
summary(aids1)

## ****************************************************************
```

```
## Family:  c("NBI", "Negative Binomial type I")
##
## Call:
## gamlss(formula = y ~ qrt + pb(x), family = NBI, data = aids,
##     trace = FALSE)
##
## Fitting method: RS()
##
## ------------------------------------------------------------
## Mu link function:  log
## Mu Coefficients:
##               Estimate Std. Error t value Pr(>|t|)
## (Intercept)  2.659646   0.057763  46.044  < 2e-16 ***
## qrt2        -0.162258   0.046546  -3.486  0.00139 **
## qrt3         0.024024   0.045395   0.529  0.60015
## qrt4        -0.121794   0.045385  -2.684  0.01124 *
## pb(x)        0.093858   0.001597  58.786  < 2e-16 ***
## ---
## Signif. codes:
## 0 '***' 0.001 '**' 0.01 '*' 0.05 '.' 0.1 ' ' 1
##
## ------------------------------------------------------------
## Sigma link function:  log
## Sigma Coefficients:
##               Estimate Std. Error t value Pr(>|t|)
## (Intercept)    -5.272      0.433  -12.18 7.82e-14 ***
## ---
## Signif. codes:
## 0 '***' 0.001 '**' 0.01 '*' 0.05 '.' 0.1 ' ' 1
##
## ------------------------------------------------------------
## NOTE: Additive smoothing terms exist in the formulas:
##  i) Std. Error for smoothers are for the linear effect only.
## ii) Std. Error for the linear terms maybe are not accurate.
## ------------------------------------------------------------
## No. of observations in the fit:  45
## Degrees of Freedom for the fit:  11.58886
##        Residual Deg. of Freedom:  33.41114
##                        at cycle:  5
##
## Global Deviance:     366.9258
##             AIC:     390.1036
##             SBC:     411.0407
## ************************************************************
```

```
dropterm(aids1,test="Chisq")

## Single term deletions for
## mu
##
## Model:
## y ~ qrt + pb(x)
##              Df     AIC      LRT    Pr(Chi)
## <none>           390.10
## qrt     4.0778 403.95   22.003 0.0002168 ***
## pb(x)   6.5889 576.91  199.983 < 2.2e-16 ***
## ---
## Signif. codes:
## 0 '***' 0.001 '**' 0.01 '*' 0.05 '.' 0.1 ' ' 1
```

The t-tests provided by `summary()` should be used with caution. For example the t-value of 0.529 of the third quarter `qrt3` is testing whether the third quarter is significantly different from the first quarter. This is not an overall test of significance of the factor `qrt`. The `dropterm()` Chi-square test for `qrt` approximates this, and with $p = 0.00022 < 0.001$ it suggests that `qrt` is highly significant.

A more serious problem arises for the interpretation of the value of the t-statistic provided for smoothing terms. In our case the smoothing term for `pb(x)` in the `summary()` table has a t-value. This is **not** a test of whether the overall smooth function for `x` is significant or not. Instead this test checks whether the linear part in `x` is significant, given the factor `qrt` and the nonlinear contribution of `x` are in the model. That is a rather peculiar test and arises due to the way that the `gamlss` backfitting algorithm works. The real question is whether the smoother for `x` is significant given `qrt`, and this is indicated by `dropterm()`. The $p \simeq 0$ for the smooth function for `x` is suggestive that it has a highly significant contribution to the model.

To demonstrate `add1()`, consider adding a two-way interaction term into the linear model `mod1` of the `usair` data. Note that when `add1()` is used, the `scope` argument has to be defined explicitly.

```
add1(mod1, scope=~(x1+x2+x3+x4+x5+x6)^2)

## Single term additions for
## mu
##
## Model:
## y ~ x1 + x2 + x3 + x4 + x5 + x6
##           Df     AIC      LRT   Pr(Chi)
## <none>       319.16
## x1:x2     1 320.09   1.0689 0.3012045
## x1:x3     1 319.40   1.7626 0.1843028
```

```
## x1:x4   1 320.60   0.5623 0.4533271
## x1:x5   1 316.94   4.2226 0.0398901 *
## x1:x6   1 320.93   0.2351 0.6277906
## x2:x3   1 320.48   0.6786 0.4100846
## x2:x4   1 319.75   1.4138 0.2344256
## x2:x5   1 318.17   2.9873 0.0839194 .
## x2:x6   1 321.13   0.0310 0.8603147
## x3:x4   1 317.38   3.7783 0.0519200 .
## x3:x5   1 320.19   0.9672 0.3253680
## x3:x6   1 320.85   0.3061 0.5800599
## x4:x5   1 307.07 14.0870 0.0001745 ***
## x4:x6   1 320.33   0.8346 0.3609322
## x5:x6   1 318.74   2.4188 0.1198894
## ---
## Signif. codes:
## 0 '***' 0.001 '**' 0.01 '*' 0.05 '.' 0.1 ' ' 1
```

Among the two-way interactions, x4:x5 is highly significant with $p < 0.001$.

The following code demonstrates how smoothers can be used in model selection. Initially we fit the null model containing only the intercept, and then each smoother is added, one at a time. The resulting LRT indicates whether the smooth function of each explanatory variable is predictive of the response variable.

```
mod0 <- gamlss(y~1, data=usair,family=GA, trace=FALSE )
addterm(mod0, scope=~pb(x1)+pb(x2)+pb(x3)+pb(x4)+pb(x5)+pb(x6),
        test="Chisq")

## Single term additions for
## mu
##
## Model:
## y ~ 1
##                Df    AIC     LRT   Pr(Chi)
## <none>             353.71
## pb(x1) 1.0000 338.04 17.6792 2.615e-05 ***
## pb(x2) 1.0000 343.05 12.6660 0.0003724 ***
## pb(x3) 1.6664 348.02  9.0249 0.0073307 **
## pb(x4) 3.3633 345.59 14.8494 0.0028119 **
## pb(x5) 3.0057 341.75 17.9793 0.0004471 ***
## pb(x6) 1.1863 343.82 12.2658 0.0006515 ***
## ---
## Signif. codes:
## 0 '***' 0.001 '**' 0.01 '*' 0.05 '.' 0.1 ' ' 1
```

All p-values being less than 0.05 indicates that each smoother of an explanatory variable is a significant predictor of the response (when entering in the model on its own).

11.4   The `stepGAIC()` function

The function `stepGAIC()` can be used in order to build a model for any of the distribution parameters using a forward, backward or stepwise procedure as described in Section 11.2.3. `stepGAIC()` is based on `stepAIC()` of the **MASS** library [Venables and Ripley, 2002] (where more details and examples of the function can be found). The additional argument `parameter` of `stepGAIC()` is designed to allow the selection of terms for a specific parameter of the distribution. The function has also been changed to allow parallel computation. The main arguments are:

object
: a `gamlss` object which is used as the initial model in the search;

scope
: The set of models searched is determined by the `scope` argument and its `lower` and `upper` components. The `scope` should be either a single formula, or a list containing components `upper` and `lower`, both formulae. The terms defined by the formula in `lower` are always included in the model. The formula in `upper` is the most complicated model that the procedure would consider. The `lower` model must be a sub-model of the `upper` model. The initial fitted model specified in the `object` option must be the `lower` or `upper` model, or a model lying between them. If `scope` is missing then backward elimination starts from the model defined by `object`;

direction
: the mode of stepwise search, which can be one of `"both"`, `"backward"` or `"forward"`, with a default of `direction="both"` which performs a stepwise model selection. If the `scope` argument is missing, the default is `direction="backward"`;

trace
: if positive, information is printed during the running of `stepGAIC`. Larger values may give more information on the fitting process;

keep
: a filter function whose input is a fitted model object and the associated **AIC** statistic, and whose output is arbitrary. Typically `keep` will select a subset of the components of the object and return them. The default is not to keep anything;

steps
: the maximum number of steps to be considered. The default is 1000 (essentially as many as required). It is typically used to stop the process early.

The extra arguments `what`, `parameter`, `k`, `parallel`, `ncpus` and `cl` operate similarly to those described in Section 11.3 for `dropterm()` and `addterm()`.

11.4.1 Selecting a model for μ

In the following example a backward elimination is performed on the model `mod1`. Note that `mod2` has a new component called `anova` showing the steps taken in the search of the model.

```
mod2<-stepGAIC(mod1)
```

```
## Distribution parameter:  mu
## Start:  AIC= 319.16
##  y ~ x1 + x2 + x3 + x4 + x5 + x6
##
## . . .
## Step:  AIC= 317.16
##  y ~ x1 + x2 + x3 + x4 + x5
##
##          Df    AIC
## <none>      317.16
## - x3     1 319.39
## - x4     1 322.48
## - x5     1 324.14
## - x2     1 324.92
## - x1     1 336.11
```

```
mod2$anova
```

```
## Stepwise Model Path
## Analysis of Deviance Table
##
## Initial
## mu
##   Model:
## y ~ x1 + x2 + x3 + x4 + x5 + x6
##
## Final
## mu
##   Model:
## y ~ x1 + x2 + x3 + x4 + x5
##
##
##    Step Df    Deviance Resid. Df Resid. Dev      AIC
## 1                            33   303.1602 319.1602
## 2 - x6  1 0.001715758        34   303.1619 317.1619
```

The above backward search procedure confirms that, if we want to include only linear terms in the model, the variable x6 is not needed. The default penalty of

$\kappa = 2$ has been used, implying use of the AIC. Note that changing to SBC the same conclusion is reached:

```
mod21<-stepGAIC(mod1, k=log(length(usair$y)))
```

As an example of using the **scope** argument explicitly, we consider whether two-way interactions are needed in the model. The initial model is **mod1**, containing linear terms in all six variables. The simplest model we consider is the intercept, i.e. **lower=~1**, and the most complicated is the one with all two-way interactions. The final model will be between those two. We use the SBC criterion.

```
mod3 <- stepGAIC(mod1, scope=list(lower=~1,
                  upper=~(x1+x2+x3+x4+x5+x6)^2), k=log(41))
```

```
## Distribution parameter:  mu
## Start:  AIC= 332.87
##  y ~ x1 + x2 + x3 + x4 + x5 + x6
##
## . . .
## - x3      1 321.51
## - x2      1 330.08
## - x4:x5   1 339.44
```

```
mod3$anov
```

```
## Stepwise Model Path
## Analysis of Deviance Table
##
## Initial
## mu
##   Model:
## y ~ x1 + x2 + x3 + x4 + x5 + x6
##
## Final
## mu
##   Model:
## y ~ x1 + x2 + x3 + x4 + x5 + x6 + x4:x5 + x1:x6 + x4:x6
##
##
##       Step Df  Deviance Resid. Df Resid. Dev      AIC
## 1                             33    303.1602 332.8687
## 2 + x4:x5  1 14.086994         32    289.0732 322.4953
## 3 + x1:x6  1  8.133304         31    280.9399 318.0756
## 4 + x4:x6  1  4.786482         30    276.1534 317.0027
```

Note that a more complicated model than **mod3** may be selected if the default value $\kappa = 2$ is used. Interactions of higher order than two-way are not permitted for continuous variables. A plot of the residuals of mod3 indicates possible heterogeneity in the variation of y. We shall deal with this problem later. For now, we show how to select smoothing terms. In order to do that we first create a formula containing all the linear main effects and second-order interactions plus smooth functions of the explanatory variables:

```
FORM <- as.formula("~(x1 + x2 + x3 + x4 + x5 + x6)^2 + pb(x1) +
        pb(x2) + pb(x3) + pb(x4) + pb(x5) + pb(x6)")
```

We use FORM as an **upper** argument for **scope**. It is important to start from mod1 (all 6 linear terms) so that all interactions are considered at the first step.

```
mod10<- stepGAIC(mod1, scope=list(lower=~1, upper=FORM), k=log(41))
```

```
## Distribution parameter:   mu
## Start:  AIC= 332.87
##  y ~ x1 + x2 + x3 + x4 + x5 + x6
##
## . . .
## - x4:x6  1.0000e+00 318.08
## + x3:x4  1.0000e+00 318.63
## - x1:x6  1.0000e+00 318.79
## + x2:x3  1.0000e+00 318.92
## + x3:x6  1.0000e+00 319.24
## + x1:x5  1.0000e+00 319.54
## + x2:x6  1.0000e+00 319.86
## + x1:x3  1.0000e+00 320.36
## + x1:x4  1.0000e+00 320.50
## + x1:x2  1.0000e+00 320.57
## + x2:x5  1.0000e+00 320.59
## + x2:x4  1.0000e+00 320.60
## + x3:x5  1.0000e+00 320.63
## + x5:x6  1.0000e+00 320.71
## - x3     1.0000e+00 321.51
## - x2     1.0000e+00 330.08
## - x4:x5  1.0000e+00 339.44
```

```
mod10$anova
```

```
## Stepwise Model Path
## Analysis of Deviance Table
##
## Initial
## mu
##   Model:
## y ~ x1 + x2 + x3 + x4 + x5 + x6
##
```

```
## Final
## mu
##   Model:
## y ~ x1 + x2 + x3 + x4 + x5 + x6 + x4:x5 + x1:x6 + x4:x6
##
##
##          Step Df  Deviance Resid. Df Resid. Dev       AIC
## 1                                33    303.1602 332.8687
## 2 + x4:x5   1 14.086994         32    289.0732 322.4953
## 3 + x1:x6   1  8.133304         31    280.9399 318.0756
## 4 + x4:x6   1  4.786482         30    276.1534 317.0027
```

The resulting model x1 + x2 + x3 + x4 + x5 + x6 + x4:x5 + x1:x6 + x4:x6
contains no smooth functions, linear terms for x1 to x6 and three interaction terms
x4:x5, x1:x6 and x4:x6.

Note that, when using pb(), the linear part of the explanatory variable is automatically fitted separately. Hence in stepGAIC() interactions involving the linear
component of smoothing terms never enter into consideration. This is a limitation
of stepGAIC(), of which the user has to be aware, and is the reason for starting
from mod1 (all six linear terms) above.

11.4.2   Selecting a model for σ

We now investigate the inclusion of linear terms in the σ model. Note that with only
41 observations and with a reasonably complicated model for μ, it is not advisable
to attempt to fit smoothing terms for σ. Here we check whether including linear
terms in the model for σ will improve mod1:

```
mod4 <- stepGAIC(mod1, parameter="sigma", scope=~x1+x2+x3+x4+x5+x6,
              k=log(41))
```

```
## Distribution parameter:  sigma
## Start:  AIC= 332.87
##   ~1
##
## . . .
## Step:  AIC= 331.34
##   ~x5 + x1
##
##          Df      AIC
## <none>       331.34
## - x1      1 332.74
## + x3      1 333.42
## + x2      1 334.43
```

```
## + x6      1 334.44
## - x5      1 334.71
## + x4      1 334.89
```

```
mod4$anova
```

```
## Stepwise Model Path
## Analysis of Deviance Table
##
## Initial
## sigma
##  Model:
## ~1
##
##
## Final
## sigma
##  Model:
## ~x5 + x1
##
##
##    Step Df Deviance Resid.  Df Resid. Dev      AIC
## 1                            33  303.1602 332.8687
## 2 + x5  1 3.838199          32  299.3220 332.7441
## 3 + x1  1 5.113662          31  294.2083 331.3440
```

According to the AIC, the model needs x1+x5 in the formula for σ. A method which selects terms for all the distribution parameters is described below.

11.5 Strategy A: The stepGAICAll.A() function

Strategies A and B are strategies for selecting additive terms using the GAIC for all distribution parameters, assuming a particular response distribution. Strategy A is described as follows:

(1) Use a forward GAIC selection procedure to select an appropriate model for μ, with σ, ν and τ fitted as constants.

(2) Given the model for μ obtained in (1) and for ν and τ fitted as constants, use a forward selection procedure to select an appropriate model for σ.

(3) Given the models for μ and σ obtained in (1) and (2), respectively, and with τ fitted as constant, use a forward selection procedure to select an appropriate model for ν.

(4) Given the models for μ, σ and ν obtained in (1), (2) and (3), respectively, use a forward selection procedure to select an appropriate model for τ.

(5) Given the models for μ, σ and τ obtained in (1), (2) and (4), respectively, use a backward selection procedure, from the model for ν given by (3), to select an appropriate model for ν,

(6) Given the models for μ, ν and τ obtained in (1), (5) and (4), respectively, use a backward selection procedure, from the model for σ given by (2), to select an appropriate model for σ.

(7) Given the models for σ, ν and τ obtained in (6), (5) and (4), respectively, use a backward selection procedure, from the model for μ given by (1), to select an appropriate model for μ and then stop.

The final model will contain possibly different subsets of terms for μ, σ, ν and τ. This is illustrated in Table 11.2 showing, for example, that among all the available variables, x_1 was chosen only for μ and ν but not for σ or τ.

TABLE 11.2: A possible result from a selection of variables using strategy A

	x_1	x_2	x_3	x_4	x_5	x_6
μ	✓		✓	✓		✓
σ			✓	✓		
ν	✓		✓			
τ				✓		

Note: Among all available variables $x_1, x_2, \ldots, x_6$, some were chosen for μ, some for σ, some for ν and some for τ.

The function which implements strategy A is `stepGAICAll.A()` and has the following arguments:

`object`	a gamlss object which is used as the initial model in the search;
`scope`	a list with elements `lower` and `upper` containing formulae;
`sigma.scope`	the scope of σ if different from `scope`;
`nu.scope`	the scope of ν if different from `scope`;
`tau.scope`	the scope of τ if different from `scope`;
`mu.try`	the default is `mu.try=TRUE`, can be set to `FALSE` if no model selection for μ is needed;
`sigma.try`	the default is `sigma.try=TRUE`, can be set to `FALSE` if no model selection for σ is needed;
`nu.try`	the default is `nu.try=TRUE`, can be set to `FALSE` if no model selection for ν is needed;

tau.try   the default is `tau.try=TRUE`, can be set to `FALSE` if no model se-
                 lection for τ is needed.

We demonstrate `stepGAICAll.A()` to select linear terms for both μ and σ. As the
`stepGAICAll.A()` output is lengthy we do not reproduce it here, but rather display
the resulting object:

```
m1<-gamlss(y~1, data=usair, family=GA, trace=FALSE, n.cyc = 50)
m2<-stepGAICAll.A(m1, scope=list(lower=~1, upper=~x1+x2+x3+x4+x5+x6),
                  k=log(41))
```

```
m2

##
## Family:  c("GA", "Gamma")
## Fitting method: RS()
##
## Call:
## gamlss(formula = y ~ x1 + x2 + x5 + x4 + x3, sigma.formula = ~x5 +
##     x1, family = GA, data = usair, trace = FALSE, n.cyc = 50)
##
##
## Mu Coefficients:
## (Intercept)          x1           x2           x5
##    7.5677204  -0.0557872    0.0013030    0.0194673
##           x4           x3
##   -0.2244998  -0.0007487
## Sigma Coefficients:
## (Intercept)          x5           x1
##      0.28178     0.04064   -0.04962
##
##   Degrees of Freedom for the fit: 9 Residual Deg. of Freedom    32
## Global Deviance:      296.084
##             AIC:      314.084
##             SBC:      329.507
```

The terms x1, x2, x3, x4 and x5 are selected for the model for μ while x1 and x5
are selected for σ.

11.6   Strategy B: The `stepGAICAll.B()` function

This strategy forces all the distribution parameters to have the same terms. That
is, if a term from $\mathcal{X}$ is selected it is included in the predictor of all parameters. The

inclusion using GAIC can be done using the forward, backward or stepwise selection procedures. Table 11.3 shows a possible result from strategy B.

TABLE 11.3: A possible result from a selection of variables using strategy B

	x_1	x_2	x_3	x_4	x_5	x_6
μ	✓		✓			✓
σ	✓		✓			✓
ν	✓		✓			✓
τ	✓		✓			✓

Note: Among all available variables $x_1, x_2, \ldots, x_6$, the selected terms are selected for all the parameters of the distribution.

The function which performs selection strategy B is stepGAICAll.B(). Its arguments object, scope, direction, trace, keep, steps, k, parallel, ncpus and cl are identical to those of stepGAICAll.A().

The following is an example of the use of stepGAICAll.B():

```
m4<-stepGAICAll.B(m1, scope=list(lower=~1, upper=~x1+x2+x3+x4+x5+x6),
                  k=log(41))
```

For brevity the stepGAICAll.B() output is not shown here. Instead we display m4:

```
m4

##
## Family:  c("GA", "Gamma")
## Fitting method: RS()
##
## Call:
## gamlss(formula = y ~ x1 + x2 + x4, sigma.formula = ~x1 +
##     x2 + x4, family = GA, data = usair, trace = FALSE,
##     n.cyc = 50)
##
## Mu Coefficients:
## (Intercept)          x1          x2          x4
##   6.5625855  -0.0608738   0.0003991  -0.0032512
## Sigma Coefficients:
## (Intercept)          x1          x2          x4
##  -2.0900413  -0.0200979  -0.0003764   0.2644994
##
##  Degrees of Freedom for the fit: 8 Residual Deg. of Freedom    33
## Global Deviance:     305.543
##             AIC:     321.543
##             SBC:     335.252
```

The variables x1, x2 and x4 were selected (for both μ and σ).

11.7   K-fold cross validation

Cross-validation modelling is achieved using `gamlssCV()`, and models fitted with `gamlssCV()` may be compared using `CV()`. The main arguments of `gamlssCV()` are the usual `gamlss()` function arguments, i.e. `formula`, `sigma.formula`, `family` etc. Also the parallel computations arguments `parallel`, `ncpus` and `cl` explained in Section 11.3 are available for speeding the procedure up. Random allocation of observations to the K cross validation subsets is achieved either manually, using the argument `rand`, or automatically using the `K.fold` argument. The arguments specific to this function are:

rand   a factor of length equal to the number of observations, and values in $\{1, 2, \ldots, K\}$, indicating to which cross-validation subset each observation should be assigned;

K.fold   the number of cross-validation subsets of the data. The cross-validation subsets will be created randomly;

set.seed   optional, if `K.fold` is used and the user wishes to reproduce results.

In the following code we use the **abdom** data to demonstrate `gamlssCV()` using 10-fold cross validation, to decide whether we should use the normal (`NO`), logistic (`LO`) or t family (`TF`) distribution for modelling the response variable.

```
set.seed(123)
# allocate the n=610 observations to K=10 cross-validation subsets
rand1 <- sample (10 , 610, replace=TRUE)

g1 <- gamlssCV(y~pb(x), sigma.formula=~pb(x), data=abdom,
            family=NO, rand=rand1)
g2 <- gamlssCV(y~pb(x), sigma.formula=~pb(x), data=abdom,
            family=LO, rand=rand1)
g3 <- gamlssCV(y~pb(x), sigma.formula=~pb(x), data=abdom,
            family=TF, rand=rand1)

CV(g1,g2,g3)

##     val[o.val]
## g3   4798.857
## g2   4799.018
## g1   4807.992
```

The t distribution (**g3**) is selected. (This choice may reverse under another random allocation of observations to subsets.)

Computation time is compared for `parallel` options `"no"`, `"multicore"` and `"snow"`:

```
## detect number of cores
nC <- detectCores()
system.time(capture.output(gamlssCV(y~pb(x),
          sigma.formula=~pb(x), data=abdom, family=LO,
          rand=rand1, parallel="no", ncpus = nC)))
```

```
##     user  system elapsed
##    7.114   0.698   7.815
```

```
system.time(capture.output(gamlssCV(y~pb(x),
          sigma.formula=~pb(x), data=abdom, family=LO,
          rand=rand1, parallel="multicore",ncpus = nC)))
```

```
##     user  system elapsed
##    5.350   1.682   4.644
```

```
system.time(capture.output(gamlssCV(y~pb(x),
          sigma.formula=~pb(x), data=abdom, family=LO,
          rand=rand1, parallel="snow", ncpus = nC)))
```

```
##     user  system elapsed
##    4.267   1.461   4.476
```

Of course these results are machine dependent. Note that, at time of writing, `"snow"` does not work in Windows.

11.8   Validation and test data

11.8.1   The gamlssVGD() and VGD() functions

Fitting a model to a training data set and validating it using a validation data set is achieved within GAMLSS using `gamlssVGD()`. Models fitted with `gamlssVGD()` may be compared using `VGD()`. The function `gamlssVGD()` works similarly to `gamlssCV()` (Section 11.7). The main arguments are the usual `gamlss` arguments; additional arguments are **rand** and **newdata**, which determine how the split into the two data sets (training and validation) is done. If the data are already split into two data frames, then the **data** and **newdata** arguments are used to specify the training and validation data sets, respectively. If, on the other hand, there is a single data set then the argument **rand** is used to define which part of the data will be used for training and which for validation/test.

rand   a factor of length equal to the number of observations, and values in
                $\{1, 2\}$, indicating whether an observation is assigned to the training

(1) or validation (2) data set, (assuming argument `newdata` is not specified).

We demonstrate these two approaches below.

First we generate a factor which splits the $n = 610$ observations of the data `abdom` into two groups containing approximately 60% and 40% of the observations. Then we fit three different models using the normal, logistic and t distributions and compare the validation global deviance using `VGD()`.

```
# generate the random split of the data
set.seed(4646)
rand2<-sample(2, 610, replace=TRUE, prob=c(0.6,0.4))
# the proportions in the sample
table(rand2)/610
```

```
## rand2
##             1             2
## 0.6163934 0.3836066
```

```
# using the argument rand
v1<-gamlssVGD(y~pb(x),sigma.formula=~pb(x), data=abdom,
              family=NO, rand=rand2)
v2<-gamlssVGD(y~pb(x),sigma.formula=~pb(x), data=abdom,
              family=LO, rand=rand2)
v3<-gamlssVGD(y~pb(x),sigma.formula=~pb(x), data=abdom,
              family=TF, rand=rand2)
```

```
VGD(v1,v2,v3)
```

```
##     val[o.val]
## v3    1835.254
## v2    1836.775
## v1    1840.346
```

Next, we repeat the same analysis but this time the data are split into two sets in advance.

```
# create training and validation data sets
training <- abdom[rand2==1,] # training data
validation <- abdom[rand2==2,] # validation data
v11<-gamlssVGD(y~pb(x),sigma.formula=~pb(x),
               data=training, family=NO, newdata=validation)
v12<-gamlssVGD(y~pb(x),sigma.formula=~pb(x),
               data=training, family=LO, newdata=validation)
v13<-gamlssVGD(y~pb(x),sigma.formula=~pb(x),
               data=training, family=TF, newdata=validation)
```

```
VGD(v11,v12,v13)

##      val[o.val]
## v13    1835.254
## v12    1836.775
## v11    1840.346
```

The t distribution model is supported by the data, although the difference between the t and logistic models is not large and the choice may in fact reverse under another randomization, as with K-fold cross validation.

11.8.2　The getTGD() and TGD() functions

The functions getTGD() and TGD() perform similar functions to gamlssVGD() and VGD(), respectively, with the difference that they assume that the models involved have been already fitted using the training data set. Given the fitted GAMLSS models, we compare them using the global deviance evaluated at the test data set, which is defined by the argument newdata:

```
# fit gamlss models first
g1<-gamlss(y~pb(x),sigma.formula=~pb(x), data=training,
           family=NO, trace=FALSE)
g2<-gamlss(y~pb(x),sigma.formula=~pb(x), data=training,
           family=LO, trace=FALSE)
g3<-gamlss(y~pb(x),sigma.formula=~pb(x), data=training,
           family=TF, trace=FALSE)
# and then use
gg1<-getTGD(g1, newdata=validation)
gg2<-getTGD(g2, newdata=validation)
gg3<-getTGD(g3, newdata=validation)
```

```
TGD(gg1,gg2,gg3)

##      val[o.val]
## gg3    1835.254
## gg2    1836.775
## gg1    1840.346
```

11.8.3　The stepTGD() function

The function stepTGD() behaves similarly to stepGAIC(), but it uses the test global deviance instead of GAIC as selection criterion. The arguments of stepTGD() are similar to those of stepGAIC(), with the addition of newdata which is the test data set.

To demonstrate the use of `stepTGD()`, we use the Munich rent data first used in Chapter 1. We split the data into training and test data sets and use the training data to fit a null model v0 and use it as initial model for `stepTGD()`.

```
# the data
set.seed(123)
rand <- sample(2, dim(rent)[1], replace=TRUE, prob=c(0.6,0.4))
# the proportions in the sample
table(rand)/dim(rent)[1]

## rand
##          1          2
## 0.6094464 0.3905536

oldrent <- rent[rand==1,] # training set
newrent <- rent[rand==2,] # validation set
# null model
v0 <- gamlss(R~1, data=oldrent, family=GA, trace=FALSE)
nC <- detectCores() # getting the number of CPU cores
```

We start from the null model:

```
v1 <- stepTGD(v0, scope=~pb(Fl)+pb(A)+H+loc,newdata=newrent,
              parallel="multicore", ncpus=nC)

## Distribution parameter:  mu
## Start:  TGD= 11241.88
##  R ~ 1
##
## . . .
## Step:  TGD= 10874.09
##  R ~ pb(Fl) + H + loc + pb(A)
##
## new prediction
## new prediction
##               Df    TGD
## <none>            10874
## - pb(A)   3.6739  10894
## - loc     1.8784  10906
## - H       1.6700  10937
## - pb(Fl)  1.6353  11113
```

All four terms are selected for the μ model. The following code demonstrates `drop1TGD()` and `add1TGD()`.

```
# drop1TGD starts from v1
(v2<- drop1TGD(v1, newdata=newrent,  parallel="multicore", ncpus=nC))
```

```
## new prediction
## Single term deletions for
## mu
##
## Model:
## R ~ pb(Fl) + H + loc + pb(A)
##           Df    TGD
## <none>          10874
## pb(Fl) 1.6353 11113
## H      1.6700 10937
## loc    1.8784 10906
## pb(A)  3.6739 10894
```

The test deviance from the full model is 1.0874×10^4 and there is no improvement when we drop each of four terms one at a time. Therefore we conclude that all terms are needed in the model.

```
# add1TGDP starts from null model v0
(v3<- add1TGD(v0, scope=~pb(Fl)+pb(A)+H+loc,newdata=newrent,
                        parallel="multicore", ncpus=nC))
```

```
## Single term additions for
## mu
##
## Model:
## R ~ 1
##           Df    TGD
## <none>          11242
## pb(Fl) 1.9378 11000
## pb(A)  3.9365 11225
## H      1.0000 11164
## loc    2.0000 11213
```

The null model test deviance is 1.1242×10^4 and adding any one of the four terms improves the test global deviance.

11.8.4   The stepTGDAll.A() function

The stepTGDAll.A() function performs a similar selection strategy to the stepGAICAll.A() function of Section 11.5 but it uses the test global deviance instead of GAIC as selection criterion. The following is an example of its use:

```
v5 <- stepTGDAll.A(v0, scope=~pb(Fl)+pb(A)+H+loc,newdata=newrent,
                        parallel="multicore", ncpus=nC)
v5
```

```
## -----------------------------------------------------
## Distribution parameter:  mu
## Start:   TGD= 11241.88
##  R ~ 1
##  . . .
## new prediction
## new prediction
## new prediction
## new prediction
##                Df    TGD
## <none>               10846
## - pb(A)  5.37093 10871
## - loc    2.25607 10874
## - H      0.98883 10888
## - pb(Fl) 1.80711 11101
## -----------------------------------------------------
##
## Family:  c("GA", "Gamma")
## Fitting method: RS()
##
## Call:
## gamlss(formula = R ~ pb(Fl) + H + loc + pb(A), sigma.formula =
##     ~pb(A) + loc + H, family = GA, data = oldrent, trace = FALSE)
##
## Mu Coefficients:
## (Intercept)        pb(Fl)            H1           loc2
##    1.659829      0.010087     -0.295978       0.196256
##        loc3         pb(A)
##    0.258968      0.002143
## Sigma Coefficients:
## (Intercept)        pb(A)          loc2           loc3
##    6.813790     -0.004017     -0.021067      -0.099187
##          H1
##    0.055614
##
## Degrees of Freedom for the fit: 16.44 Resid. Deg. of Freedom 1184
## Global Deviance:      16770.5
##             AIC:      16803.4
##             SBC:      16887.1
```

All four terms are selected for the μ model. pb(Fl) is excluded from the σ model.

11.9   The find.hyper() function

Estimation of the smoothing parameters has been discussed in Section 3.4 and also at the beginning of this chapter. We classify the methods used to two main categories: *global*, in which the methods are applied outside the GAMLSS algorithm, and *local* in which the methods are applied within. Local methods have been discussed in Chapters 3 and 9. Here we focus on the function find.hyper(), which is a global method for estimating smoothing parameters and appears to work well in searching for the optimum degrees of freedom for smoothing and/or nonlinear parameters (e.g. a power parameter ξ used to transform x to x^ξ). The function repetitively fits GAMLSS models and uses optim() to minimize the GAIC. For large data sets (e.g. $n > 5000$), the function is slow compared to local estimation methods and for this reason is not preferred. Here the results of find.hyper() are compared with local methods of estimation of the smoothing parameters.

The arguments of find.hyper() are:

model	this is a quoted (quote()) GAMLSS model in which the required hyperparameters are denoted by p[number], e.g. quote(gamlss(y~cs(x,df=p[1]), sigma.fo=~cs(x,df=p[2]), data=abdom))
parameters	the starting parameter values in the search for the optimum hyperparameters and/or nonlinear parameters, e.g. parameters=c(3,3);
other	this is used to optimize nonlinear parameter(s), for example a transformation of the explanatory variable of the kind $x^{p[3]}$, e.g. others=quote(nx<-x^p[3]) where nx is now in the model formula;
k	GAIC penalty, with default k=2;
steps	the steps in the parameter(s) taken during the optimization procedure (see for example the ndeps option in the control function for optim()), by default set to 0.1 for all hyperparameters and nonlinear parameters;
lower	the lower bounds on the permissible values of the parameters, e.g. for two parameters lower=c(1,1). This does not apply if a method other than the default "L-BFGS-B" is used;
upper	the upper bounds on the permissible values of the parameters, e.g. for two parameters upper=c(30,10). This does not apply if a method other than the default "L-BFGS-B" is used;
method	the method used in optim() to numerically minimize the GAIC over the hyperparameters and/or nonlinear parameters. By default

> this is `method="L-BFGS-B"` to allow box-restriction on the parameters;
>
> ... extra arguments in the `control` argument of `optim()`.

`find.hyper()` returns the same output as `optim()`.

In the following example we compare the local and global estimation of the smoothing parameters using the `abdom` data. Both models for μ and σ are fitted using P-splines. We first use the three local methods for estimating the smoothing parameters, i.e. `"ML"`, `"GCV"` and `"GAIC"`:

```
# fitting the model with pb()
a1<-gamlss(y~pb(x), sigma.fo=~pb(x), data=abdom, family=LO,
           trace=FALSE)
a2<- gamlss(y~pb(x, method="GCV"), sigma.fo=~pb(x, method="GCV"),
            data=abdom, family=LO, trace=FALSE)
a3<-gamlss(y~pb(x, method="GAIC", k=2), sigma.fo=~pb(x,
           method="GAIC",k=2), data=abdom, family=LO, trace=FALSE)
# the effective degrees of freedom used
edfAll(a1)

## $mu
##    pb(x)
## 5.796263
##
## $sigma
##    pb(x)
## 2.001641

edfAll(a2)

## $mu
## pb(x, method = "GCV")
##              4.842172
##
## $sigma
## pb(x, method = "GCV")
##              2.600804

edfAll(a3)

## $mu
## pb(x, method = "GAIC", k = 2)
##                      2.000007
##
## $sigma
## pb(x, method = "GAIC", k = 2)
##                      2.001502
```

We next demonstrate use of the global GAIC method to find the degrees of freedom. Initially the model is declared using `quote()`. For each hyperparameter to be estimated, `p[.]` is used. In this example we are optimizing over two degrees of freedom parameters, one in the model for μ and one in the model for σ, so the parameters are `p[1]` and `p[2]`.

The function `find.hyper()` minimizes GAIC with k=2 by default. The initial values for the parameters p are both set to 3 (`parameters=c(3,3)`), the minimum value for p for the search is set to 0 (`lower=c(0,0)`) and the steps in each parameter p used within the `optim()` search to 0.1 (`steps=c(0.1, 0.1)`). The default method used by `optim()` within `find.hyper()` is "L-BFGS-B", which starts with the initial parameter value(s), changes each parameter in turn by ± step for that parameter and then jumps to new value(s) for the set of parameter(s). This is repeated until convergence. See the help on `optim()` for details. Note that `df` as defined in `pb()` are the effective degrees of freedom on top of the constant and linear term(s), so `df=0` corresponds to a linear fit.

```
mod1 <- quote(gamlss(y ~ pb(x, df = p[1]), sigma.fo=~pb(x, df=p[2]),
          family = LO, data = abdom, trace=FALSE))
```

```
op <- find.hyper(model = mod1, par = c(3,3), lower = c(0,0),
          steps = c(0.1, 0.1), trace=FALSE)
```

```
## par 3 3 crit= 4798.775 with pen= 2
## par 3.1 3 crit= 4798.599 with pen= 2
## . . .
## par 3.712447 0.1 crit= 4795.033 with pen= 2
## par 3.712447 0 crit= 4794.866 with pen= 2
```

```
op
```

```
## $par
## [1] 3.712447 0.000000
##
## $value
## [1] 4794.866
##
## $counts
## function gradient
##       10       10
##
## $convergence
## [1] 0
##
## $message
## [1] "CONVERGENCE: REL_REDUCTION_OF_F <= FACTR*EPSMCH"
```

The resulting extra degrees of freedom are 3.71 for μ and 0 for σ, corresponding to total effective degrees of freedom 5.71 for μ and 2 for σ. These are close to the results obtained using the default local "ML" method in pb() in model a1.

11.10 Bibliographic notes

Voudouris et al. [2012] (and references therein) provide an overview of model selection decisions for a GAMLSS model. The selection of the appropriate theoretical distribution $\mathcal{D}$ is not trivial. Rigby et al. [2015] provide an overview of the ordering and comparison of continuous distributions as a guide to developing statistical models with appropriate theoretical continuous distributions. For a comprehensive encyclopaedia of continuous and discrete distributions; see Johnson et al. [1994, 1995, 2005]. Many approaches have been proposed over the years for dealing with the selection of additive terms. A standard introduction for statistical model selection (and assessment) is given by Hastie et al. [2009, Chapter 7], and references therein. Other references include Akaike [1983], Madigan and Raftery [1994], Nott and Jialiang [2010] and Wood et al. [2017]. In problems with a large number of observations (say $n \geq 10000$) and a large number of terms (over 500) the model selection functions presented here will be slow, unless the functions run on computers with many CPUs. In this instance, the use of machine learning techniques (e.g. neural networks, random forests, boosting methods) to pre-select plausible terms might be used. In fact, if a machine learning technique accepts prior weights, it can be incorporated within GAMLSS. For example, Section 9.6.1 demonstrates how to interface with neural networks using nnet(). For a general discussion of machine learning techniques; see Hastie et al. [2009, Chapters 10, 11, 12 and 15], and references therein. Boosting is one of the most popular techniques in machine learning. Mayr et al. [2012] introduced boosting to the GAMLSS model, and Hofner et al. [2016] provide a tutorial on how to use the **R** package **gamboostLSS**. Further applications of boosting in GAMLSS, for example analyzing measurement errors and permutation tests and using beta distribution models, are given in Mayr et al. [2015] and Schmid et al. [2013], respectively. General reviews on boosting can be found in Mayr et al. [2014a] and Mayr et al. [2014b].

11.11 Exercises

1. **The LGA claims data:** The LGAclaims data set [de Jong and Heller, 2008] contains the number of third party claims in a twelve-month period from 1984–1986 in each of 176 geographical areas (local government areas) in New

South Wales, Australia. Areas are grouped into thirteen statistical divisions (SD). Other recorded variables are the number of accidents, the number of people killed or injured and population in each area. The number of claims (Claims) is analyzed as the response variable.

> **R data file:** LGAclaims in package **gamlss.data** of dimensions 176×11
>
> **var** LGA : local government area name
>
>   SD : statistical division (1,2,...,13)
>
>   Claims : number of third party claims
>
>   Pop_density : population density
>
>   KI : number of people killed or injured
>
>   Accidents : number of accidents
>
>   Population : population size
>
>   L_Population : log population
>
>   L_Accidents : log number of accidents
>
>   L_KI : log KI
>
>   L_Popdensity : log population density
>
> **purpose:** to demonstrate selection of variables.

This exercise explores the use of stepGAIC() for the selection of terms for particular distribution parameters. Exercise 2 explores the automated function stepGAICAll.A() for selecting terms for all the distribution parameters.

(a) Input the data and plot them.

(b) Check whether a Poisson or negative binomial model for Claims (using the explanatory variables) is appropriate for the data.

(c) The function dropterm() provides a single term deletion facility in gamlss. Check whether any of the linear terms can be deleted from the model.

(d) The function addterm() provides the facility of adding single terms in the model. Use the function to check whether a two-way interaction is needed (from the model with the linear terms).

(e) The function stepGAIC() provides a mechanism for stepwise selection of appropriate linear terms for any of the parameters of the distribution. Use it here to select a model for μ. Note that the default argument gd.tol=Inf is crucial for some of the fitting at later stages since it prevents the algorithm from stopping if the deviance increases in any of the iterations.

(f) Use the function `stepGAIC()` to select an appropriate model for σ, given the model for μ.

(g) Conditional on the selected model for σ, explore if the model for μ can be simplified.

(h) Plot the fitted terms for μ and σ, respectively, using `term.plot()`.

(i) Use diagnostics to assess whether the model residuals are supportive of the assumed model (see Chapter 12 for details).

2. **The `oil` data: using model selection to discover what affects the price of oil.** The `oil` data set contains the daily prices of front month WTI (West Texas Intermediate) oil price traded by NYMEX (New York Mercantile Exchange). The front month WTI oil price is a futures contract with the shortest duration that could be purchased in the NYMEX market. The idea is to use other financially traded products (e.g. gold price) to discover what might affect the daily dynamics of the price of oil. The data set was downloaded from `https://www.quandl.com/` and `https://finance.yahoo.com/`.

R data file: `oil` in package **gamlss.data** of dimensions 1000×25

variables

    `OILPRICE` : the log price of front month WTI oil contract traded by NYMEX – in financial terms, this is the CL1. This is the response variable

    `CL2...CL15` : CL2 ... CL15 are the log prices of the 2 to 15 months ahead of WTI oil contracts traded by NYMEX. For example, for the trading day of 2nd June 2016, the CL2 is the WTI oil contract for delivery in August 2016

    `BDIY` : the Baltic Dry Index, which is an assessment of the price of moving the major raw materials by sea

    `SPX` : the S&P 500 Index

    `DX1` : the US Dollar Index

    `GC1` : the log price of front month gold price contract traded by NYMEX

    `HO1` : the log price of front month heating oil contract traded by NYMEX.

    `USCI` : the United States Commodity Index

    `GNR` : the S&P Global Natural Resources Index

    `SHCOMP` : the Shanghai Stock Exchange Composite Index

FTSE : the FTSE 100 Index

respLAG : the lag 1 of OILPRICE – lagged version of the response variable

purpose: to demonstrate selection of variables of moderate complexity.

(a) Use `fitDist()` to fit more than 50 different distributions to the response variable (`OILPRICE`) to select a plausible distribution for the GAMLSS model.

(b) For the `SHASHo` distribution, use the function `stepGAICAll.B()` to select appropriate linear and smooth terms for all the parameters of the distribution. Because of the size of the data, `stepGAICAll.B()` might be slow. Furthermore, the `stepGAICAll.B()` uses all the core available. First select linear terms.

```
m0<-gamlss(OILPRICE~1,family=SHASHo,data=oil,gd.tol=Inf)
nC<-detectCores()
FORM<-as.formula(paste("~",paste(paste(paste("(",
        names(oil)[-1], sep=""),")",sep=""), collapse="+")))
mStratB.SHASHo<-stepGAICAll.B(m0, scope=list(lower=~1,
            upper=FORM), ncpus=nC,k=sqrt(log(dim(oil)[1])))
summary(mStratB.SHASHo)
```

(c) Re-fit your model by including a smoother for the selected variables.

```
mStratB.SHASHo.1=gamlss(OILPRICE ~ pb(respLAG) + pb(HO1_log),
        sigma.formula = ~pb(respLAG) + pb(HO1_log),
        nu.formula = ~pb(respLAG) + pb(HO1_log),
        tau.formula = ~pb(respLAG) + pb(HO1_log),
        family = SHASHo, data = oil, gd.tol = Inf)
```

(d) Check, using differrent `k`, which model is the best (so far).

```
GAIC(mStratB.SHASHo,mStratB.SHASHo.1,k=2)
GAIC(mStratB.SHASHo,mStratB.SHASHo.1,k=2.5)
GAIC(mStratB.SHASHo,mStratB.SHASHo.1,k=3)
GAIC(mStratB.SHASHo,mStratB.SHASHo.1,k=3.5)
GAIC(mStratB.SHASHo,mStratB.SHASHo.1,k=4)
```

(e) For the `SHASHo` distribution, use the function `stepGAICAll.A()` to select appropriate linear terms for all the parameters for the distribution. Because of the size of the data, `stepGAICAll.A()` might be slow. Furthermore, the `stepGAICAll.A()` uses all the core available.

```
mStratA.SHASHo<-stepGAICAll.A(m0, scope=list(lower=~1,
            upper=FORM),ncpus=nC,k=sqrt(log(dim(oil)[1])))
summary(mStratA.SHASHo)
```

(f) Re-fit your model by including a smoother for the selected variables. (See Chapter 9 for the smoothing techniques supported by GAMLSS.)

```
mStratA.SHASHo.1<-gamlss(OILPRICE~pb(respLAG)+pb(CL2_log)+
    pb(SHCOMP_log), sigma.fo=~pb(CL2_log) + pb(CL15_log)+
    pb(CL8_log) + pb(SHCOMP_log)+pb(CL9_log)+pb(DX1_log),
    nu.fo=~pb(CL10_log) + pb(CL8_log),
    tau.fo=~pb(CL10_log) + pb(CL14_log),
    family = SHASHo, data = oil, gd.tol = Inf)
```

(g) Use diagnostics to assess whether the model residuals are supportive of the assumed response distribution. (See Chapter 12 for details.)

(h) If you assess that the model residuals are not supportive of the assumed response distribution, return to step (b) and repeat the process using the next best distribution (**SEP4**). Alternatively, use the **stepGAICAll.A()** function with an extended **scope** that includes smoothers. (This will be slow.)

```
FORM<-as.formula( paste("~",paste(paste(paste("pbz(",
    names(oil)[-1],sep=""),")", sep=""), collapse="+")))
mStratA.SHASHo.2<-stepGAICAll.A(m0, scope=list(lower=~1,
    upper=FORM), ncpus=nC,k=sqrt(log(dim(oil)[1])))
```

12

Diagnostics

CONTENTS

12.1   Introduction

This chapter provides:

1. the definition of normalized (randomized) quantile residuals; and

2. other diagnostic tools based on residuals, such as worm plots and Q statistics.

The proper use of diagnostics is important because it reveals the strengths and weaknesses of a model. Thus, this chapter is important for understanding the tools for checking the adequacy of a GAMLSS model. The use of diagnostics is an important step in model checking and hence model selection.

In the simple linear regression model

$$y_i = \beta_0 + \beta_1 x_i + \varepsilon_i \ ,$$

simple residuals are defined as the difference between observed and fitted values:

$$\hat{\varepsilon}_i = y_i - \hat{y}_i \ ,$$

where $\hat{y}_i = \hat{\beta}_0 + \hat{\beta}_1 x_i$ for $i = 1, 2, \ldots, n$. Sometimes the $\hat{\varepsilon}_i$'s are called the *raw* residuals to distinguish them from *standardized residuals* which are defined as

$$r_i = \frac{y_i - \hat{y}_i}{\hat{\sigma}\sqrt{(1 - h_{ii})}} \ ,$$

where h_{ii} are the diagonal values of the hat matrix. The problem with raw residuals is that they are difficult to generalize to distributions other than the normal. For example, within the GLM literature the *deviance residuals* $r_i^d = \text{sign}\left(y_i - \hat{\mu}_i\right)\sqrt{d_i}$ where $d_i = -2\log(L_i^c/L_i^s)$[1] and the *Pearson residuals* $r_i^P = (y_i - \hat{\mu}_i)/\sqrt{V\left(\hat{\mu}_i\right)}$ (where $V(\cdot)$ is the variance function, see Section 1.4) are often used. Unfortunately the deviance residuals are not in general well defined (e.g. when modelling multiple parameters for the response distribution), while the Pearson residuals can be far from normally distributed for a highly skewed or kurtotic response variable. Therefore for GAMLSS models we use the *normalized (randomized) quantile residuals* [Dunn and Smyth, 1996], which we refer to as 'residuals' throughout this book.

Section 12.2 introduces normalized quantile residuals for a continuous response variable and normalized *randomized* quantile residuals for a discrete response variable. Section 12.3 describes the `plot()` function of **gamlss**. Section 12.4 describes the worm plot function, `wp()`. Section 12.5 describes the detrended transformed Owen's plot (`dtop()`), which provides a method for visually checking the adequacy of a fitted model. The Q statistics function `Q.stats()` is considered in Section 12.6 and the `rqres.plot()` function in Section 12.7.

12.2   Normalized (randomized) quantile residuals

This section introduces normalized (randomized) quantile residuals, and explains how they are used within the **gamlss** package.

The main advantage of normalized (randomized) quantile residuals is that, whatever the distribution of the response variable, the true residuals always have a standard normal distribution when the assumed model is correct. Since the checking of model

---

[1]L_i^c represents the fitted likelihood for observation i from the current model and L_i^s from the saturated model, that is, when $\hat{\mu}_i = y_i$ [McCullagh and Nelder, 1989].

assumptions via normality of residuals is well established within the statistical literature, the normalized (randomized) quantile residuals provide us with a familiar way to check the adequacy of a GAMLSS fitted model.

Given that the distribution $f(y \mid \boldsymbol{\theta})$ is fitted to observations y_i for $i = 1, \ldots, n$, the fitted normalized (randomized) quantile residuals [Dunn and Smyth, 1996] are given by

$$\hat{r}_i = \Phi^{-1}(\hat{u}_i) \, ,$$

where $\Phi^{-1}(\cdot)$ is the inverse cdf of the standard normal distribution. The $\hat{u}_i$'s are *quantile residuals*, defined differently for continuous and discrete response variables as described below.

If y is an observation from a continuous response variable then let $u = F(y \mid \boldsymbol{\theta})$ and $\hat{u} = F(y \mid \hat{\boldsymbol{\theta}})$ be the model and fitted cdf's, respectively. If the model is correctly specified, u has the uniform distribution between zero and one. This is called the *probability integral transform*.[2] The process is described diagrammatically in Figure 12.1. The top panel shows the pdf for a specific observation y. The middle panel shows how, using the cdf, the observation y is mapped onto u. In the bottom panel, u is transformed to true residual r (called a *z-score*) by:

$$r = \Phi^{-1}(u)$$

which has a standard normal distribution if the model is correct. Similarly $\hat{u}$ is transformed to the fitted residual $\hat{r}$ by

$$\hat{r} = \Phi^{-1}(\hat{u}) = \Phi^{-1} \left[F(y \mid \hat{\boldsymbol{\theta}}) \right]$$

and $\hat{r}$ has an approximate standard normal distribution.

If y is an observation from a discrete integer response variable, then $F(y \mid \boldsymbol{\theta})$ is a step function with jumps at the integers $y \in R_Y$. The distribution of $u = F(y \mid \boldsymbol{\theta})$ has the range zero to one, but is discrete with positive probability at the points $F(y \mid \boldsymbol{\theta}), y \in R_Y$. In order to deal with this discreteness, u is defined as a random value from the uniform distribution on the interval $[u_1, u_2] = [F(y - 1 \mid \boldsymbol{\theta}), F(y \mid \boldsymbol{\theta})]$ and similarly $\hat{u}$ is a random value from a uniform distribution on $[\hat{u}_1, \hat{u}_2] = \left[F(y - 1 \mid \hat{\boldsymbol{\theta}}), F(y \mid \hat{\boldsymbol{\theta}}) \right]$. The process is depicted in Figure 12.2. For a given discrete probability function (top panel), the observed y value is transformed to an interval (u_1, u_2) (the shaded strip in the middle panel). Then u is selected randomly from a uniform distribution on the interval (u_1, u_2) and then u is transformed into the true residual (or z-score), $r = \Phi^{-1}(u)$ (bottom panel). The distribution of r is exactly standard normal if the model is correct [Dunn and Smyth, 1996]. Similarly, using the fitted cdf, y is transformed to $\hat{u}$, randomly chosen from $(\hat{u}_1, \hat{u}_2)$, and then transformed to the fitted residual $\hat{r} = \Phi^{-1}(\hat{u})$ and $\hat{r}$ has an approximate standard normal distribution.

---

[2]Proof of the probability integral transform (PIT) is given in Appendix 12.8.1. The u's are referred to in the econometric literature as PIT residuals.

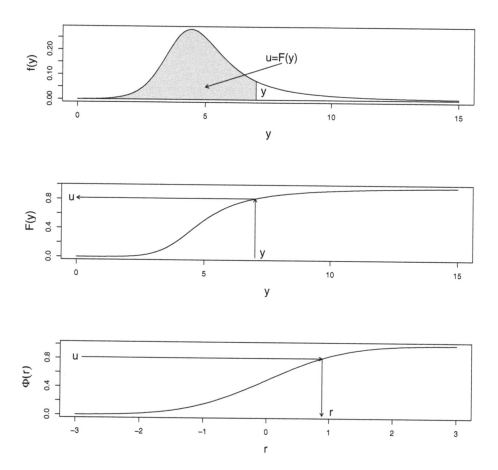

FIGURE 12.1: A depiction of how a normalized quantile residual r is obtained for a continuous distribution. The functions plotted are the model pdf $f(y)$, the cdf $F(y)$ and the cdf of a standard normal random variable $\Phi(r)$, showing an observed y transformed to u and then from u to r. The residual r is the z-score for the specific observation and has a standard normal distribution if the model is correct.

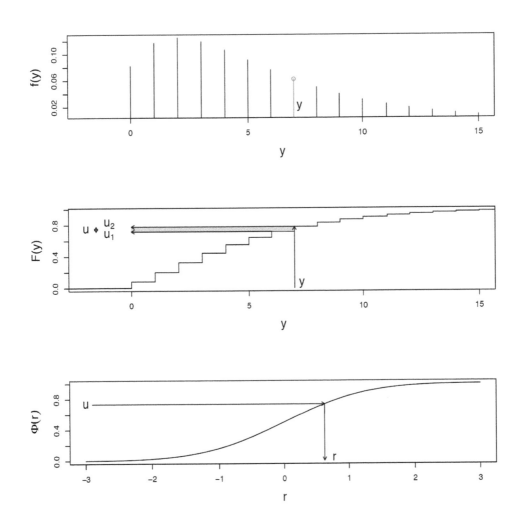

FIGURE 12.2: A depiction of how a normalized randomized quantile residual r is obtained for a discrete distribution. The functions plotted are probability function $f(y)$, the cdf $F(y)$ and the cdf of a standard normal variable $\Phi(r)$. The observed y is transformed to u, a random value between u_1 and u_2, then u is transformed to r. The residual r is a z-score for the specific observation and has a standard normal distribution if the model is correct.

Randomized quantile residuals are also appropriate for an interval or censored response variables [for a brief discussion of censored response variables see Chapter 6]. For example, for a continuous response right censored at y, $\hat{u}$ is defined as a random value from the uniform distribution on the interval $\left[F(y \mid \hat{\boldsymbol{\theta}}), 1\right]$.

When randomization is used, several randomized sets of residuals (or their average) should be studied before a decision about the adequacy of a fitted model is taken.

Note that the true residuals always have a standard normal distribution if the model is correct, and a standard normal distribution has mean 0, variance 1, (moment based) skewness and kurtosis 0 and 3, respectively.

The fitted normalized (randomized) quantile residuals can be obtained in the **gamlss** package using the function `resid()`. There are several other functions in the package using the normalized (randomized) quantile residuals:

- `plot()` is for general residual checking;

- the worm plot function `wp()` can be used to identify whether the fitted distribution is adequate, either overall or within non-overlapping ranges of either one or two explanatory variables;

- the Q statistics function `Q.stats()` is for detecting whether the residuals are 'significantly' different from a normal distribution in their mean, variance, skewness and kurtosis (and potentially which distribution parameters of the model failed to fit adequately), in ranges of an explanatory variable;

- `rqres.plot()` is designed for repeated randomization of the residuals (when the response variable is not continuous).

The above functions are explained below.

12.3   The `plot()` function

The full name of this function is `plot.gamlss()` but since it is a method function in **R** it can be called using `plot()`, provided its first argument is a fitted **gamlss** object. The function `plot()` produces four plots for checking the normalized (randomized) quantile residuals of a fitted **gamlss** object. Randomization is performed for discrete and mixed response variables and also for interval or censored data. The four plots are

- residuals against the fitted values of the μ parameter;

- residuals against an index or a specified covariate;

- a kernel density estimate of the residuals;

- a QQ-normal plot of the residuals.

When randomization is performed it is advisable to use `rqres.plot()`, described in Section 12.7.

The arguments of the `plot.gamlss()` function are

x	a `gamlss` fitted object;
xvar	an explanatory variable to plot the residuals against. By default the index $1 : n$ is plotted, where n is the number of observations;

parameters	this option can be used to change the default parameters in the plotting. The current default parameters are
	`par(mfrow=c(2,2), mar=par("mar")+c(0,1,0,0), col.axis=`
	`"blue4", col.main="blue4", col.lab="blue4", col=`
	`"darkgreen", bg="beige").`
	These parameters are not appropriate for inclusion of the plot into a document. We have found that the option
	`parameter=par(mfrow=c(2,2), mar=par("mar")+c(0,1,0,0),`
	`col.axis = "blue4", col ="blue4", col.main="blue4",`
	`col.lab ="blue4", pch = "+", cex =.45,`
	`cex.lab = 1.2, cex.axis=1, cex.main=1.2)`
	gives reasonable plots for printed documents;

ts	set this to **TRUE** if ACF and PACF plots of the residuals are required. This option is appropriate if the response variable is a time series. The ACF and PACF then replace the first two of the four plots listed above;
summaries	set this to **FALSE** if no summary statistics of the residuals are required. By default `plot.gamlss()` produces summary statistics for the residuals.

The following is an example of the use of `plot()` using the abdominal circumference data (see page 90):

```
data(abdom)
abd10<-gamlss(y~pb(x),sigma.fo=~pb(x),data=abdom,family=BCT,
              trace=FALSE)
```

```
plot(abd10)
```

Fig. 12.3
```
## ****************************************************************
##            Summary of the Quantile Residuals
##                             mean    =   0.001271073
##                         variance    =   1.002348
##                coef. of skewness    =   -0.003628787
##                coef. of kurtosis    =   2.992959
## Filliben correlation coefficient    =   0.9993249
```

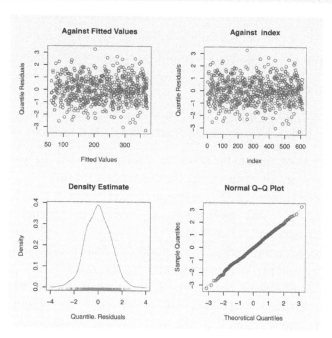

FIGURE 12.3: Residual plots from the BCT model `abd10`.

R
p. 42

The resulting plot is shown in Figure 12.3 Note that the residuals behave well, since the top two plots of the residuals against the fitted values of μ and against the index show a random scatter around the horizontal line at 0, while the kernel density estimate of the residuals is approximately normal and the normal Q-Q plot is approximately linear (with intercept 0 and gradient 1).

Looking at the summary statistics of the quantile residuals, their mean is nearly zero, their variance nearly one, their (moment-based) coefficient of skewness is near zero and their (moment-based) coefficient of kurtosis is near 3. The statistics suggest that the residuals are approximately normally distributed ($r \sim \mathcal{N}(0,1)$), as they should be for an adequate model. Furthermore, the Filliben correlation coefficient (or the normal probability plot correlation coefficient) is near 1.

Let us now use some of the options. Here we use the option `xvar` to change the top right-hand plot so the plot shows the residuals against age (`abdom$x`) instead of the index. Note though that this makes very little difference to the plot, since age is already ordered. Also we change the values of the plotting parameters. The plot is shown in Figure 12.4.

```
newpar<-par(mfrow=c(2,2),mar=par("mar")+c(0,1,0,0),col.axis="blue4",
            col="blue4", col.main="blue4",col.lab="blue4",pch="+",
            cex=.45, cex.lab=1.2, cex.axis=1, cex.main=1.2)
plot(abd10,xvar=abdom$x,par=newpar)
```

Fig.
12.4

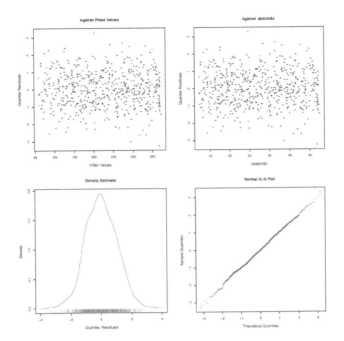

FIGURE 12.4: Residual plots from the BCT model `abd10`, where the `xvar` and `par` options have been modified.

In order to see an application of the option `ts=TRUE`, consider the `aids` data (see Section 4.3.2). Here we model the counts `y` using a negative binomial distribution with a (smooth) regression model (a cubic spline) in time `x` and a quarterly effect for the mean of `y`:

```
data(aids)
aids.1<-gamlss(y~cs(x,df=7)+qrt,family=NBI, data=aids, trace=FALSE)
```

The residual plot is shown in Figure 12.5. It appears from the ACF and PACF functions (top plots) that the residuals show only mild systematic autocorrelation (since most of the values lie within the confidence intervals).

```
plot(aids.1,ts=TRUE)
## ******************************************************************
##    Summary of the Randomised Quantile Residuals
##                            mean    =  -0.007775317
##                        variance    =  0.9522827
##              coef. of skewness    =  -0.6162532
##              coef. of kurtosis    =  3.244916
## Filliben correlation coefficient  =  0.9832549
## ******************************************************************
```

Note that since we are using a discrete response distribution, the residuals are randomized and the function `rqres.plot` should be used in addition to `plot`.

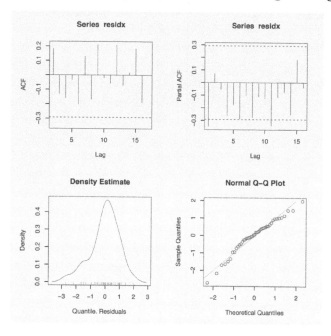

FIGURE 12.5: Residual plots from the NBI model fitted to the **aids** data.

R

p. 4?

12.4   The `wp()` function

Worm plots of the residuals were introduced by van Buuren and Fredriks [2001] in order to identify regions (intervals) of an explanatory variable within which the model does not adequately fit the data (which they called 'model violation'). The **R** function `wp()` (which is based on the original **S-PLUS** function given in van Buuren and Fredriks [2001]) provides *single* or *multiple* worm plots for **gamlss** fitted objects. This is a diagnostic tool for checking the residuals for different ranges (by default not overlapping) of one or two explanatory variables. The worm plot is a detrended QQ-plot and the name comes from the worm-like appearance of the plotted points.

12.4.1   Single worm plot

If the `xvar` argument of the `wp()` function is not specified then a single worm plot is produced. The following is an example of a single worm plot from model `abd10` fitted in Section 12.3.

```
wp(abd10)
```

Fig.
12.6

The resulting plot in Figure 12.6 is equivalent to the normal Q-Q plot in Figure 12.4, detrended by subtracting the line (intercept 0 and slope 1). This allows the worm

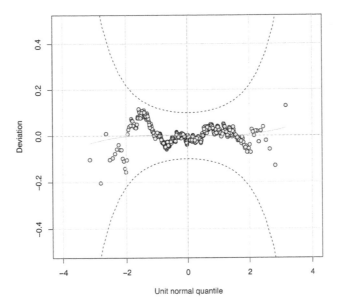

FIGURE 12.6: Worm plot from the BCT model `abd10` at default values.

plot to focus on deviations from a standard normal distribution in the distribution of the residuals. There are several important features:

- The points of the plot (the worm): These points show how far the ordered residuals are from their (approximate) expected values represented in the figure by the horizontal dotted line. The closer the points are to the horizontal line, the closer the distribution of the residuals is to a standard normal distribution.

- The approximate point-wise 95% confidence intervals given by the two elliptic curves in the figure. If the model is correct we would expect approximately 95% of the points to lie between the two elliptic curves and 5% outside. A higher percentage of the points outside the two elliptic curves (or a clear systematic departure from the horizontal line) indicates that the fitted distribution (or the fitted terms) of the model are inadequate to explain the response variable. The plot may also identify outliers lying well outside the two elliptic curves.

- The fitted curve to the points of the worm: This curve is a cubic fit to the worm plot points. The shape of this cubic fit reflects different inadequacies in the model, described in Table 12.1 and illustrated in Figure 12.7. For example if the level of plotting points in the worm plot is above a horizontal line at the origin, as illustrated in Figure 12.7(a), this indicates that the residual mean is too high, which implies that the location of the fitted distribution is too low, as shown in Table 12.1. (This may be corrected by increasing the parameter μ, if it is a location parameter, or improving the model for μ (e.g. making it more flexible), or changing the model distribution.) Similar concerns apply if the level of the worm plot is below a horizontal line at the origin. Furthermore a linear trend (positive or negative), quadratic shape (U or inverse U) or cubic shape (S shape,

TABLE 12.1: The different shapes for the worm plot of the residuals (first column) and the corresponding deficiency in the residuals (second column) and deficiency in the fitted distribution (third column)

Shape of worm plot (or its fitted curve)	Residuals	Fitted distribution
level: above the origin	mean too high	fitted location too low
level: below the origin	mean too low	fitted location too high
line: positive slope	variance too high	fitted scale too low
line: negative slope	variance too low	fitted scale too high
U-shape	positive skewness	fitted skewness too low
inverted U-shape	negative skewness	fitted skewness too high
S-shape with left bent down	leptokurtosis	tails of fitted distribution too light
S-shape with left bent up	platykurtosis	tails of fitted distribution too heavy

as illustrated in Figure 12.7) indicates a problem with the variance, skewness or kurtosis of the residuals, respectively. This in turn highlights a problem with the fitted distribution (and/or models for the parameters), as shown in Table 12.1.

As far as Figure 12.6 is concerned, since all the observations fall in the "acceptance" region inside the two elliptic curves and no specific shape is detected in the points, the model appears to fit well overall.

12.4.2   Multiple worm plot

If the **xvar** argument of **wp()** is specified then we have as many worm plots as argument **n.inter** indicates. A quantitative explanatory variable is cut into **n.inter** non-overlapping intervals with equal numbers of observations in each, and the worm plots of the residuals for each interval are plotted. A categorical explanatory variable (factor) splits the data into its different levels and plots one worm plot for each level. This is a way of highlighting failures of the model within different ranges of the explanatory variable. This should be applied to each explanatory variable in turn and is especially important when one of the explanatory variables is dominant in the analysis (as can occur for example in centile estimation or in time series data). The parameters of the fitted cubic polynomials to the residuals in the worm plot are obtained by, e.g.

```
Res <- wp(model1, xvar=x, n.inter=9)
```

and can be used as a way of checking the region of the explanatory variable in which the model does not fit adequately.

In the abdominal circumference example we are interested in whether the model fits

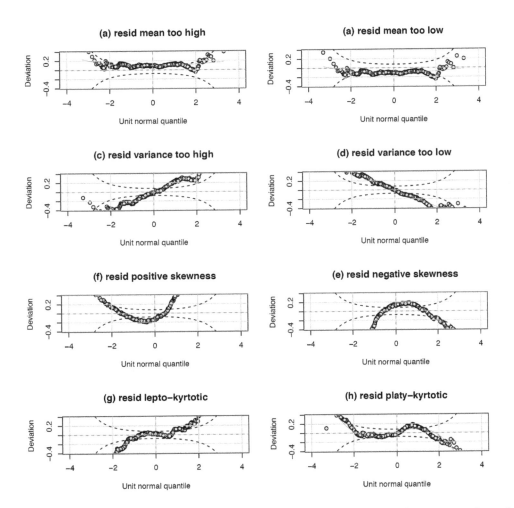

FIGURE 12.7: Different type of model failures indicated by the worm plot: (a) and (b) indicate failure to correctly fit the location; (c) and (d) indicate failure to correctly fit the scale; (e) and (f) indicate failure to correctly fit the skewness; and (g) and (h) indicate failure to correctly fit the kurtosis.

well at the different regions of age. Here we use the option `xvar` to specify age, and
`n.inter` to specify nine intervals of age with equal numbers of observations, for the
worm plot. We also save the coefficient parameters of the fitted cubic polynomials
for further diagnostics.

```
coef.1 <- wp(abd10,xvar=abdom$x,n.inter=9)
```

```
coef.1
```

```
## $classes
##          [,1]   [,2]
##   [1,]  12.22  16.36
##   [2,]  16.36  19.50
##   [3,]  19.50  22.50
##   [4,]  22.50  25.21
##   [5,]  25.21  28.36
##   [6,]  28.36  32.07
##   [7,]  32.07  35.21
##   [8,]  35.21  38.78
##   [9,]  38.78  42.50
##
## $coef
##                 [,1]            [,2]            [,3]            [,4]
##   [1,]   0.043594305   0.043925177  -0.005328564  -0.010167197
##   [2,]   0.020768938   0.042150024  -0.013714482  -0.009325333
##   [3,]  -0.065344250   0.156093848   0.028205518  -0.067684581
##   [4,]  -0.039820449  -0.059832136   0.028895496   0.029799686
##   [5,]  -0.015744439  -0.013758616   0.011531208  -0.040762801
##   [6,]   0.009523141   0.027326782   0.045198638   0.006814800
##   [7,]   0.001338649  -0.077452581  -0.022366387   0.034188200
##   [8,]   0.014868877   0.005635431  -0.038020639   0.015777066
##   [9,]   0.024495114  -0.068089549  -0.020443334   0.022093903
```

```
wp(abd10,xvar=abdom$x,n.inter=9)
```

```
## number of missing points from plot= 0  out of  68
## number of missing points from plot= 0  out of  71
## number of missing points from plot= 0  out of  67
## number of missing points from plot= 0  out of  67
## number of missing points from plot= 0  out of  66
## number of missing points from plot= 0  out of  71
## number of missing points from plot= 0  out of  65
## number of missing points from plot= 0  out of  69
## number of missing points from plot= 0  out of  66
```

Fig.
12.8

The resulting plot is shown in Figure 12.8. The table of intervals (`$classes`) gives
the nine non-overlapping x (i.e. age) ranges in weeks, which are also plotted in steps

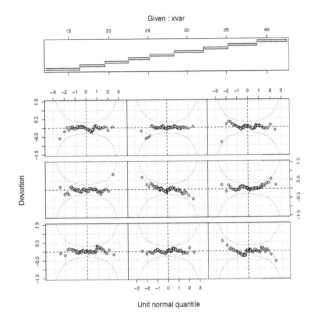

FIGURE 12.8: Worm plot from the BCT model `abd10`.

above the worm plot in Figure 12.8. The individual worm plots in Figure 12.8 are read along rows from bottom left to top right, corresponding to the nine age intervals from left to right in the steps above the worm plot and listed by `coef.1$classes`. Since the points in each of the nine individual worm plots in Figure 12.8 lie within the two elliptic curves, this suggests an adequate model.

The table of coefficients (`$coef`) gives the fitted constant, linear, quadratic and cubic coefficients $\hat{\beta}_0$, $\hat{\beta}_1$, $\hat{\beta}_2$ and $\hat{\beta}_3$, respectively, for each of the nine cubic polynomials fitted to thenine worm plots. van Buuren and Fredriks [2001] categorize absolute values of $\hat{\beta}_0$, $\hat{\beta}_1$, $\hat{\beta}_2$ and $\hat{\beta}_3$ in excess of threshold values 0.10, 0.10, 0.05 and 0.03, respectively, as misfits or model violations, indicating differences between the mean, variance, skewness and kurtosis of the theoretical model residuals and the fitted residuals, respectively, within the particular age range. Following these criteria in the above table of coefficients, there are no misfits in $\hat{\beta}_1$, one misfit 0.15609 in $\hat{\beta}_2$ (age group 3), no misfits in $\hat{\beta}_3$, and three misfits in $\hat{\beta}_4$ at 3rd, 5th and 7th ranges of age. With respect to the misfits at the 3rd and 5th range of age, the model seems to suggest that the tails of the BCT are too heavy with respect to the data in these age ranges. However, the 7th range of the data suggests otherwise. Plausible solutions are to provide a model for the τ parameter, e.g. `pb(x)`, or increase the degrees of freedom in the model for τ, e.g. `pb(x, df=4)`, or assume a different distribution. However, occasional misfits may occur (due to random variation) even when the model is correct.

In general, it may not always be possible to build a model without areas of misfits. One solution is to use calibration in order to minimize misfits (see Appendix 12.8.2 for a derivation). Exercise 2 in this chapter shows one way to calibrate the distribution of the response variable. (See also the function `calibration()` in centile estimation Section 13.3.2.)

In any case, extra care is needed when a model with many areas of misfits is used to support conclusions.

12.4.3   Arguments of the `wp` function

For completeness we provide here all the arguments of the `wp()` function:

object
: a `gamlss` fitted object or any other fitted model where the `resid()` method works (it should produce normalized quantile residuals).

xvar
: the explanatory variable(s) against which the worm plots will be plotted. If only one quantitative variable is involved use `xvar=x1`. Factors can be used in the formula, `xvar=~f1` is allowed, but not `xvar=f1`. If two variables are involved, quantitative variables and factors may be combined by using `xvar=~x1+x2`, `xvar=~x1+f1` or `xvar=~f1+f2`;

resid
: if `object` is missing this argument can be used to specify the residual vector (it should be normalized quantile (or z-score) residuals, e.g. residuals from a `gamlss` fit).

n.inter
: the number of intervals into which the explanatory variable `xvar` is cut, with default `n.inter=4`;

xcut.points
: the explanatory variable cut-off points, e.g. `c(20,30)`. If `xcut.points=NULL` then the `n.inter` argument is activated;

overlap
: how much overlapping in the `xvar` intervals. Default value is `overlap=0` for non-overlapping intervals;

xlim.all
: for a single worm plot this value is the x-axis limit, default is `xlim.all=4`.

xlim.worm
: for multiple worm plots this value is the x-axis limit, default is `xlim.worm=3.5`;

show.given
: whether to show the x-variable intervals in the top of the graph, default is `show.given=TRUE`;

line
: whether to plot the fitted cubic polynomial curve in each individual worm plot, default value is `line=TRUE`;

ylim.all
: for a single plot this value is the y-axis limit, default value is `ylim.all=12*sqrt(1/length(fitted(object)))`;

ylim.worm
: for multiple plots this value is the y-axis limit, default value is `ylim.worm=12*sqrt(n.inter/length(fitted(object)))`;

cex   plotting parameter with default `cex=1`;

pch   plotting parameter with default `pch=21`.

Exercise 1 shows how to use multiple worm plots with two explanatory variables (xvar=~x1+x2).

12.5   The `dtop()` function

The function `dtop()` provides a method for visually checking the adequacy of a fitted model. The function is based on the Owen [1995] construction of a nonparametric confidence interval for a true distribution function, given the empirical distribution function of the sample. Aitkin et al. [2009] used the empirical confidence intervals to check the adequacy of a fitted distribution. Here we use a detrended transformed Owen's plot (DTOP) applied to the fitted normalized (randomized) quantile residuals of the fitted model for the same purpose. Jager and Wellner [2004] corrected the approximation formula given by Owen [1995] (also in Owen [2001, Chapter 7]) for obtaining the 95% and 99% empirical confidence intervals. Our function `dtop()` has both formulae as options.

Using the `abd10` model, a multiple DTOP plot is given at Figure 12.9 by:

Fig.
2.9

```
dtop(abd10,xvar=abdom$x,n.inter=9)
```

R

433

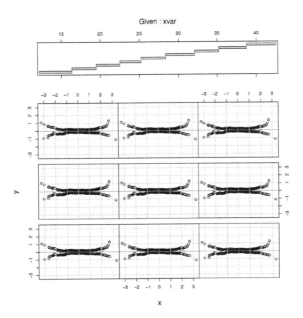

FIGURE 12.9: DTOP plot from the BCT model `abd10`.

TABLE 12.2: The different shapes of DTOP plots interpreted with respect to the residuals and the fitted distribution

Shape of DTOP	Residuals	Fitted distribution
level: above the origin	mean too low	fitted location too high
level: below the origin	mean too high	fitted location too low
line: positive slope	variance too low	fitted scale too high
line: negative slope	variance too high	fitted scale too low
U-shape	negative skewness	fitted skewness too high
inverted U-shape	positive skewness	fitted skewness too low
S-shape with left bent down	platykurtosis	tails of fitted distribution too heavy
S-shape with left bent up	leptokurtosis	tails of fitted distribution too light
left bent down (up)	short (long) left tail	left tail of the fitted distribution is too heavy (light)
right bent down (up)	long (short) right tail	right tail of the fitted distribution is too light (heavy)

Since the horizontal line of each DTOP plot lies within the 95% confidence bands, we conclude that the normalized residuals could have come from a normal distribution and consequently the assumed response variable distribution (BCT) is reasonable. See Table 12.2 for the interpretation of different shapes of DTOP plots.

12.5.1   Arguments of the dtop function

For completeness we provide here all the arguments of the dtop() function:

object   a gamlss fitted object or any other fitted model where the resid() method works (it should produce normalized quantile residuals);

xvar   the explanatory variable against which the detrended Owen's plots will be plotted;

resid   if object is missing this argument can be used to specify the residual vector. (It should be normalized quantile (or z-score) residuals, e.g. residuals from a gamlss fit);

type   whether to use Owen [1995] ("Owen") or Jager and Wellner (2004) ("JW") approximate formula;

conf.level 95 (default) or 99 percent confidence interval for the plots;

n.inter   the number of intervals into which the explanatory variable xvar is cut;

xcut.points	the explanatory variable cut-off points, e.g. `c(20,30)`. If `xcut.points=NULL` then the `n.inter` argument is activated;
overlap	how much overlapping in the `xvar` intervals. Default value is `overlap=0` for non-overlapping intervals;
show.given	whether to show the x-variable intervals in the top of the graph, default is `show.given=TRUE`;
line	whether to plot the fitted cubic polynomial curve in each individual plot, default value is `line=TRUE`;
cex	plotting parameter with default `cex=1`;
pch	plotting parameter with default `pch=21`.

12.6 The `Q.stats()` function

This function calculates and prints the Z and Q statistics which are useful for testing normality of the residuals within ranges of an independent x-variable, for example age in centile estimation, see Royston and Wright [2000].

In order to explain what a Z statistic is, let us consider the situation where `age` is our main explanatory variable. Let G be the number of age groups and let $\{r_{gi}, i = 1, 2, .., n_i\}$ be the residuals in age group g, with mean $\bar{r}_g$ and standard deviation s_g, for $g = 1, 2, .., G$. The statistics $Z_{g1}, Z_{g2}, Z_{g3}, Z_{g4}$ are calculated from the residuals in group g to test whether the residuals in group g have population mean 0, variance 1, moment-based skewness 0 and moment-based kurtosis 3 (the values of a standard normal distribution for the true residuals assuming the model is correct), where from Royston and Wright [2000]

$$Z_{g1} = n_g^{1/2}\bar{r}_g,$$

$$Z_{g2} = \left\{ s_g^{2/3} - [1 - 2/(9n_g - 9)] \right\} / \{2/(9n_g - 9)\}^{1/2}$$

and Z_{g3} and Z_{g4} are test statistics for skewness and kurtosis given by D'Agostino et al. [1990] in their equations (13) and (19), respectively. The D'Agostino K^2 statistic, given by

$$K^2 = Z_{g3}^2 + Z_{g4}^2,$$

is for jointly testing whether the skewness of the residuals is different from 0 and the kurtosis is different from 3. The Q statistics of Royston and Wright [2000] are then calculated by

$$Q_j = \sum_{g=1}^{G} Z_{gj}^2, \quad j = 1, 2, 3, 4.$$

Royston and Wright [2000] discuss approximate distributions for the Q statistics under the null hypothesis that the true residuals are normally distributed and suggest Chi-squared distributions with adjusted degrees of freedom $G - \mathrm{df}_\mu$ (where df_μ is the effective degrees of freedom used for modelling the μ parameter), $G - [\mathrm{df}_\sigma + 1]/2$ and $G - \mathrm{df}_\nu$ for Q_1, Q_2 and Q_3, respectively. By analogy we suggest degrees of freedom $G - \mathrm{df}_\tau$ for Q_4. The resulting significance levels should be regarded as providing a rough guide to model inadequacy, rather than exact formal test results.

Significant Q_1, Q_2, Q_3 or Q_4 statistics indicate possible inadequacies in the models for parameters μ, σ, ν and τ, which may be overcome in the model for the particular parameter, e.g. by increasing the degrees of freedom used in a smoothing function.

The Z_{gj}^2 statistic provides the contribution from age group g to the statistic Q_j, and hence helps identify which age group(s) are causing the Q_j statistic to be significant and therefore in which age group(s) the model is unacceptable.

Provided the number of groups G is sufficiently large (relative to the degrees of freedom adjustment for the parameter, e.g. df_μ), the Z_{gj} values should have approximately standard normal distributions under the null hypothesis that the true residuals are standard normally distributed. We suggest as a rough guide values of $|Z_{gj}|$ greater than 2 be considered as indicative of significant inadequacies in the model. Note that significant positive (or negative) values $Z_{gj} > 2$ (or $Z_{gj} < 2$) for $g = 1, 2, 3$ or 4 indicate, respectively, that the residuals have a higher (or lower) mean, variance, skewness or kurtosis than the standard normal distribution. The model for parameter μ, σ, ν or τ may need more degrees of freedom to overcome this. For example if the residual mean in an age group is too high, the model for μ (assuming μ is a location parameter) may need more smoothing degrees of freedom in order for the fitted μ from the model to increase within the age group. Note the D'Agostino K^2 statistic should be compared to the upper 5% point of a χ_2^2 distribution, i.e. 6.0.

12.6.1   Examples

The following output is produced using `Q.stats()` on the `abd10` model.

```
qstats<-Q.stats(abd10,xvar=abdom$x,n.inter=9)
print(qstats, digits=3)
```

Fig.
12.10

```
##                     Z1       Z2       Z3       Z4
## 12.22 to 16.36    0.3164   0.1858  -0.0621  -0.2573
## 16.36 to 19.50    0.0615   0.1837  -0.2403  -0.3440
## 19.50 to 22.50   -0.3083  -0.2341   0.4754  -2.4603
## 22.50 to 25.21   -0.0939   0.2972   0.7637   1.2636
## 25.21 to 28.36   -0.0360  -1.4162   0.2148  -1.6872
## 28.36 to 32.07    0.4543   0.5490   0.8652   0.3267
## 32.07 to 35.21   -0.1660   0.1988  -0.5205   0.9768
```

```
## 35.21 to 38.78      -0.1865  0.5799 -0.8696  0.6295
## 38.78 to 42.50       0.0361 -0.0294 -0.7508  1.2268
## TOTAL Q stats        0.4791  2.8952  3.2564 13.6428
## df for Q stats       3.1434  6.5475  8.0000  8.0000
## p-val for Q stats    0.9346  0.8654  0.9173  0.0916
##                     AgostinoK2    N
## 12.22 to 16.36        0.0701     68
## 16.36 to 19.50        0.1761     71
## 19.50 to 22.50        6.2789     67
## 22.50 to 25.21        2.1798     67
## 25.21 to 28.36        2.8928     66
## 28.36 to 32.07        0.8553     71
## 32.07 to 35.21        1.2250     65
## 35.21 to 38.78        1.1525     69
## 38.78 to 42.50        2.0687     66
## TOTAL Q stats        16.8992    610
## df for Q stats       16.0000      0
## p-val for Q stats     0.3922      0
```

FIGURE 12.10: A visual presentation of the Z statistics for the `abdom` model for easy identification of misfits in the data.

The intervals on the left of the above output are the 9 age intervals., i.e. `abdom$a`. The resulting plot of the Z statistics is shown in Figure 12.10. A light colour indicates a negative value and a dark colour indicates a positive Z value, while the size of the circle is proportional to the Z value. The square in the middle of the circle indicates

$|Z| > 2$, i.e. a possible misfit. In particular, a misfit in the kurtosis statistic **Z4** in the third age range 19.50 to 22.50 is easily identified, as in the **Q.stats()** tabular output $Z4 = -2.46 \, (< -2)$. However if the model was correct, then in a table of 36 Z statistics we might expect about two to be significant at the 5% level, by chance.

The original **Q.stats()** function was designed for checking centile curve fitting, where a large number of data points are expected. The current version of **Q.stats()** is more flexible, allowing the direct input of normalized (randomized) quantile (or z-score) residuals from GAMLSS or non-GAMLSS models, and also for smaller data sets (where tests for Q statistics maybe be less reliable). This happens with the use of the argument **resid** rather than **obj**. The resulting output includes the Z statistics, but not the Q statistics. The following is an example of the use of **Q.stats()** with the small data set **aids**.

```
a1<-gamlss(y~pb(x)+qrt, family=PO, data=aids, trace=FALSE)
Q.stats(resid=resid(a1), xvar=aids$x, n.inter=5)
```

Fig.
12.1

```
##                    Z1          Z2         Z3         Z4
##  0.5 to  9.5  0.40173816  0.1031412 -0.3810178 -0.6585527
##  9.5 to 18.5 -0.75077632  0.3316442 -0.3348597  0.5057579
## 18.5 to 27.5 -0.01249615  0.7873977  0.3915875 -0.3543823
## 27.5 to 36.5  0.21195602  4.5025437 -1.1197619 -0.1324640
## 36.5 to 45.5 -0.25487889  0.9608719 -1.3467842  0.5000232
```

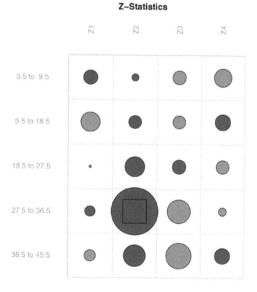

FIGURE 12.11: A visual presentation of the Z statistics for the **aids** model.

R
p. 43

The graphical presentation of the Z statistics is shown in Figure 12.11, where a misfit in the variance of the residuals in the fourth interval 27.5 to 36.5 is identified.

12.6.2   Arguments of the Q.stats function

The Q.stats function has the following arguments:

obj
: a gamlss object. The obj argument produces the Z and Q statistics table;

xvar
: the explanatory variable against which the Z and/or Q statistics will be calculated;

resid
: normalized (randomized) quantile (or z-score) residuals can be given here instead of a gamlss object in obj. Note that the function Q.stats behaves differently depending whether the obj or the resid argument is set. The argument resid allows any model residuals (not necessary GAMLSS), suitably normalized, and is appropriate for any size of data. The resulting table contains only the individual Z statistics;

xcut.points
: the explanatory variable cut off points, e.g. c(20,30). If xcut.points=NULL then the n.inter argument is activated;

n.inter
: the number of intervals into which the explanatory variable xvar is cut;

zvals
: if TRUE the output matrix contains the individual Z statistics, not the Q statistics;

save
: whether to save the Z and/or Q statistics or not, with default save=TRUE. In this case the function produces a matrix giving individual Z and/or Q statistics;

plot
: whether to plot a visual version of the Z statistics (default is plot=TRUE).

12.7   The rqres.plot() function

The function rqres.plot() is used to create multiple realizations of the normalized randomized quantile residuals when the distribution of the response variable is discrete, and plot them using worm plots or QQ-plots. These plots help visually to decide whether the chosen model is an adequate representation of the data.

12.7.1   Example

As an example we use the function rqres.plot() to plot residuals from a fitted model using the aids data:

```
m1 <- gamlss(y~pb(x)+qrt, data=aids, family=NBI, trace=FALSE)
rqres.plot(m1)
```

Fig.
12.12

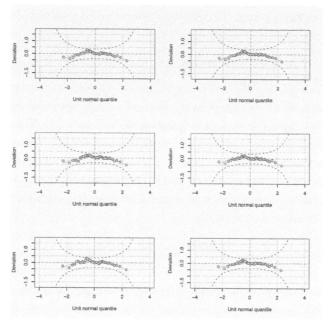

R
p. 440

FIGURE 12.12: Worm plots from the NBI model fitted to the aids data.

The resulting plot is shown in Figure 12.12, which shows six realizations of the worm plots of the normalized randomized quantile residuals from the fitted model m1. All six realizations show reasonable behaviour. For creating QQ-plots instead of worm plots, use

```
rqres.plot(m1, type="QQ")
```

12.7.2 Arguments of the rqres.plot() function

The following are the arguments of rqres.plot():

obj   a gamlss fitted model object, e.g. from a discrete family;

howmany   the number of worm or QQ-plots required, up to ten, with default howmany=6;

plot   whether to plot all plots, i.e. all the residual realizations "all" or just the mean "average";

type   whether to plot worm plots "wp"or QQ plots "QQ", with default type="wp".

12.8 Appendix

12.8.1 Proof of probability integral transform: Continuous case

Let Y be a continuous random variable with cdf $F_Y(y)$. By definition, $F_Y(y) = \text{Prob}(Y \leq y)$. Let U be a random variable defined by the transformation $U = F_Y(Y)$. Let U have cdf $F_U(u)$ then:

$$
\begin{aligned}
F_U(u) &= \text{Prob}(U \leq u) \\
&= \text{Prob}(F_Y(Y) \leq u) \\
&= \text{Prob}(Y \leq F_Y^{-1}(u)) \\
&= F_Y\left(F_Y^{-1}(u)\right) \\
&= u \; .
\end{aligned}
$$

So we have

$$
F_U(u) = \begin{cases} 0 & u < 0 \\ u & 0 \leq u \leq 1 \\ 1 & u > 1 \end{cases}
$$

which is the cdf of a uniform (0,1) random variable. See also Angus [1994].

12.8.2 Proof of calibration: Calibrating the pdf

Suppose a model with cdf $F_{Y_{oi}}(y)$ for Y_i for $i = 1, 2, \ldots, n$ is used to model the continuous response variable Y giving normalized quantile residuals $r_1, r_2, \ldots, r_n$. Furthermore, we assume that the model misfits, resulting in the residuals of the distribution being different from the assumed standard normal distribution.

In order to calibrate the fitted model, we assume that the residuals r_i come from the same distribution with pdf $f_R(r)$. From the definition of residual R, we have that $R = \Phi^{-1}[F_{Y_{oi}}(Y_i)]$, where $\Phi(\cdot)$ is the cdf of a standard normal distribution. Then Y_i is related to R by transformation:

$$
Y_i = F_{Y_{oi}}^{-1}[\Phi(R)] \; .
$$

For the univariate transformation, we have

$$
f_{Y_i}(y) = f_R(r)\frac{dr}{dy} \; . \tag{12.1}
$$

Since

$$
y = F_{Y_{oi}}^{-1}[\Phi(r)]
$$

we have

$$F_{Y_{oi}}(y) = \Phi(r)$$

$$\frac{d}{dy}F_{Y_{oi}}(y) = \frac{d}{dr}\Phi(r)\frac{dr}{dy}$$

$$f_{Y_{oi}}(y) = \phi(r)\frac{dr}{dy}$$

where $\phi(\cdot)$ is the pdf of a standard normal distribution. Hence

$$\frac{dr}{dy} = \frac{f_{Y_{oi}}(y)}{\phi(r)} \ . \tag{12.2}$$

Substituting (12.2) to (12.1) gives:

$$f_{Y_i}(y) = \frac{f_R(r)f_{Y_{oi}}(y)}{\phi(r)}$$

where $r = \Phi^{-1}[F_{Y_{oi}}(y)]$. In practice $f_R(r)$ can be estimated by fitting a flexible distribution to the fitted residuals $\hat{r}_1, \hat{r}_2, \ldots, \hat{r}_n$. See Exercise 2 for an example.

12.9   Bibliographic notes

There are a variety of ways to develop a model and different modelling choices can lead to different chosen models. It is therefore sensible to use diagnostics to guide the choice of model in fitting a particular set of data.

GAMLSS uses the normalized (randomized) quantile residuals for residual based diagnostics. Dunn and Smyth [1996] provide a discussion of normalized (randomized) quantile residuals, while Angus [1994] provides a simple proof of the probability integral transform theorem.

Within GAMLSS, there are a number of visual tools to check the normalized (randomized) quantile residuals. These tools include the worm plots, DTOP plots and Q statistics. For worm plots, see the work of van Buuren and Fredriks [2001]. For Q statistics, see Royston and Wright [2000]. The DTOP plot is a detrended transformed Owen's plot as is used in Aitkin et al. [2009]. Owen [1995] provided confidence intervals from the empirical cdf of the data. Those intervals were corrected by Jager and Wellner [2004]. The book by Owen [2001] looks at the more general problem of empirical likelihood. Jager and Wellner [2007] consider a unified family of goodness-of-fit tests.

Once we have the normalized (randomized) quantile residuals, we can also use a range of other tests to check for normality such as Shapiro–Wilk, Anderson–Darling or Cramér–von Mises tests for normality. Note however the tests assume the model parameters are known, while in general the model parameters are estimated. Thode [2002] provides a detailed discussion of the selection, design, theory and application of tests for normality.

12.10 Exercises

1. **The 1990s film data:** The `film90` data set from **gamlss.data**, was first introduced in Section 2.3. We use the data to demonstrate the use of diagnostics in GAMLSS.

   (a) Fit an additive smoothing model for the μ parameter.

   ```
   m1<-gamlss(lborev1~pb(lboopen)+pb(lnosc)+dist, data=film90,
               family=NO)
   ```

   (b) Check the residuals of the model. The model is particularly poor in terms of capturing the skewness and kurtosis in the data.

   ```
   plot(m1,xvar=film90$lboopen)
   ```

   (c) Re-confirm the misfits of the model using worm plots. From bottom left to top right in rows, the first and last worm plots show significant misfits of the kurtosis while the third plot shows a misfit with respect to skewness.

   ```
   wp(m1,xvar=~lboopen+lnosc,n.inter = 2,ylim.worm=4)
   ```

   (d) Next, we model the response variable using the **BCPE** distribution. First we start with a relatively simple model for the σ, ν and τ parameters.

   ```
   m2<-gamlss(lborev1~pb(lboopen) +pb(lnosc)+dist,
               sigma.fo=~pb(lboopen), nu.fo=~pb(lboopen),
               tau.fo=~pb(lboopen), data=film90, family=BCPE)
   ```

   (e) Check the residuals of **m2**. What do you think?

   ```
   plot(m2,xvar=film90$lboopen)
   wp(m2,xvar =~lboopen+lnosc,n.inter = 2,ylim.worm=2)
   ```

   (f) Although the residuals show a significantly improved model, it seems that there is a problem with the skewness, particularly for the second and third worm plots. Thus, we fit a new model for the ν parameter:

   ```
   m3<-gamlss(lborev1~pb(lboopen)+pb(lnosc)+dist,
               sigma.fo=~pb(lboopen),nu.fo=~pb(lboopen)+pb(lnosc),
               tau.fo=~pb(lboopen),data=film90,family=BCPE,
   ```

```
        start.from=m2)
```

(g) Check the residuals of `m3`. What do you think?

```
plot(m3,xvar=film90$lboopen)
wp(m3,xvar =~lboopen+lnosc,n.inter = 2, ylim.worm=1)
```

(h) Although the model shows minor misfits, it suggests the second and third worm plots are problematic in terms of the mean (second) and kurtosis (third). We fit a new model:

```
m4<-gamlss(lborev1~pvc(lboopen,by=dist)+pvc(lnosc,by=dist)+
          dist, sigma.fo=~pb(lboopen),nu.fo=~pb(lboopen)+
          pb(lnosc),tau.fo=~pb(lboopen)+pb(lnosc),
          data=film90, family=BCPE, start.from=m3)
```

(i) Use a normality test for the residuals of `m4`.

```
shapiro.test(resid(m4)) #Shapiro-Wilk normality test
#if you have the package "nortest", you can also use:
library(nortest)
ad.test(resid(m4)) #Anderson-Darling normality test
cvm.test(resid(m4)) #Cramer-von Mises normality test
```

(j) Using worm plots, check the residuals of `m4` and compare them with the residuals of the first model `m1`. *Note:* It seems that we still have a problem with the μ parameter because of the second worm plot. (However occasional misfits may occur due to random variation, even when the model is correct.) In practice, it is not always possible to build a model without areas of misfit. See Section 14.4 for alternative models for the film data.

(k) Using the 1930s film data, `film30` from package **gamlss.data**, develop and test a new GAMLSS model using the `BCPE` distribution.

2. **Calibration to address model misfits**. Using `film90`, we demonstrate the use of calibration to address model misfits. (See Appendix 12.8.2.) For GAMLSS, we are interested in calibrating the fitted (or forecasted) distribution of the response. Here, we want to calibrate the pdf corresponding to a particular film. (In time series, the calibration will correspond to the pdf of a point in time.)

(a) Check model `m2` from above.

```
dtop(m2,xvar=film90$lboopen,n.inter = 9)
```

(b) For demonstration purposes, we will assume that this is the best fitted model. How can we improve the best fitted (or forecasted) model distribution for Y for a particular observation (e.g. the last observation in the data set)? Use calibration.

(c) Create a sequence of $m = 100$ values **y** in a suitable range for the response variable Y at the last observation, i.e. $i = n$ and get the corresponding normalized quantile residuals $\mathbf{r} = (r_1, r_2, \ldots, r_m)$ using the fitted cdf $\hat{F}_{Y_{oi}}(y) = F_{Y_{oi}}(y|\hat{\mu}_i, \hat{\sigma}_i, \hat{\nu}_i, \hat{\tau}_i)$, i.e. using the following equation $r = \Phi^{-1}[\hat{F}_{Y_{oi}}(y)]$. Note that `residuals.normalized` below correspond to `r` in Appendix 12.8.2.

```
y <- seq(from=5,to =17,length.out = 100)
#Use the BCPE distribution and take the last observation
#(since we want to calibrate the distribution of the last
#film) The interval 5 to 17 is a suitable range for the
#pdf of the last observation
mu<-tail(fitted(m2,parameter='mu'),1)
sigma<-tail(fitted(m2,parameter='sigma'),1)
nu<-tail(fitted(m2,parameter='nu'),1)
tau<-tail(fitted(m2,parameter='tau'),1)
# calculate r values corresponding to y values
residuals.quantile<-pBCPE(y,mu=mu,sigma=sigma,nu=nu,tau=tau)
# calculate phi(r) values
residuals.normalized<-qNO(residuals.quantile)
```

(d) Get the pdf of the last observation $i = n$, evaluated at the sequence **y**. Note that this is the $f_{Y_{oi}}(y)$ in Appendix 12.8.2.

```
density.y<-dBCPE(y,mu=mu,sigma=sigma,nu=nu,tau=tau)
```

(e) Fit a univariate distribution to all the fitted model **m2** residuals (i.e. `resid(m2)` below) and get its fitted distribution parameters. Here, we are using the **SHASHo** distribution.

```
residuals.m2<-resid(m2)
model.residuals.m2<-gamlssML(residuals.m2,family=SHASHo)
#get the fitted parameters
mu0<-tail(fitted(model.residuals.m2,parameter='mu'),1)
sigma0<-tail(fitted(model.residuals.m2,parameter='sigma'),1)
nu0<-tail(fitted(model.residuals.m2,parameter='nu'),1)
tau0<-tail(fitted(model.residuals.m2,parameter='tau'),1)
```

(f) Get the pdf $f_R(r)$ evaluated at **r**.

```
residual.density<-dSHASHo(residuals.normalized,mu=mu0,
                    sigma=sigma0,nu=nu0,tau=tau0)
```

(g) Calibrate the original pdf $f_{Y_{oi}}(y)$ to get the calibrated $f_{Y_i}(y) = \frac{f_R(r)}{\phi(r)} f_{Y_{oi}}(y)$ (see Appendix 12.8.2).

```
calibrated.density<-(residual.density/dNO(residuals.normalized,
                    mu=0,sigma=1))*density.y
```

(h) Plot the calibrated $f_{Y_i}(y)$ and original $f_{Y_{oi}}(y)$.

```
plot(density.y~y,type='l')
lines(y=calibrated.density,x=y,col='red')
```

(i) Calibrate the density using the residuals of the second worm plot in `wp()` below. (We call this the local calibration.)

```
wp(m2,xvar=~lboopen+lnosc,n.inter = 2,ylim.worm=2)
```

Part VI

Applications

13

Centile estimation

CONTENTS

13.1   Introduction

This chapter explains how to create centile curves using **gamlss**. In particular it explains:

1. the LMS (lambda, mu, sigma) method of centile estimation;

2. the different functions for centile estimation within **gamlss**; and

3. how to use the functions effectively.

This chapter is important for practitioners involved in centile estimation since GAMLSS has become one of the standard tools for creating centile curves.

Centile estimation includes methods for estimating the age-related distribution of human growth. The standard estimation of centile curves usually involves two continuous variables:

1. the *response* variable, that is, the variable we are interested in and for which we are trying to find the centile curves, e.g. weight, BMI, head circumference.

2. the *explanatory* variable, e.g. age.

The $100p$ centile of a continuous random variable Y is the value y_p such that $\text{Prob}(Y \leq y_p) = p$. Then $y_p = F_Y^{-1}(p)$, and y_p is the inverse cdf of Y applied to p. In this chapter we consider the conditional centile of Y given explanatory variable $X = x$, i.e. $y_p(x) = F_{Y|x}^{-1}(p)$. By varying x, a $100p$ centile curve of $y_p(x)$ against x is obtained. Centile curves can be obtained for different values of p. The World Health Organization uses the values $100p=(3,15,50,85,97)$ in its charts and $100p=(1,3,5,15,25,50,75,85,95,97,99)$ in its tables; see WHO [2006, 2007, 2009]. Centile estimation can be extended to more than one explanatory continuous variable, e.g. age and height; see for example Cole et al. [2009] and Quanjer et al. [2012a] and Exercise 4 of this chapter. For explanatory categorical variables such as gender, the usual practice is to produce separate charts for each level of the categorical variable.

Note that a z-score given the values of y and x is defined by $z = \Phi^{-1}\left[F_{Y|x}(y)\right]$, where Φ^{-1} is the inverse cdf of the standard normal distribution. For the values of y and x used in the estimation of a model, the z-scores are the residuals of a fitted GAMLSS model. (See the definition of normalized quantile residuals in Section 12.2.)

The creation of sensible centile curves for Y against x usually relies on nonparametric smoothing methods, since parametric methods, e.g. polynomials, or even fractional polynomials [Royston and Altman, 1994], are not in general flexible enough to capture the features of the data. In smoothing methods the amount of smoothing depends on smoothing parameters and varies from data set to data set. The determination of the smoothing parameters is a crucial component of centile estimation. In the past several methods have been suggested, and these can be classified as:

• Subjective (but structured) methods: The statistician (or practitioner) uses prior knowledge and experience in conjunction with some broad guidelines to choose the smoothing parameters and create the centile curves. For example, one possible structured method is to first obtain a good smooth model for the location parameter, then for the scale parameter and finally for the shape parameter(s). Erratic centile curves may indicate the need to increase the values of the smoothing parameters.

- Automatic methods: In an automatic procedure a criterion such as, for example, the AIC, or generalizations of it, can be used to select the smoothing parameters.

- Methods based on diagnostics: Diagnostic tools such as the worm plots of van Buuren and Fredriks [2001] or Q statistics of Royston and Wright [2000] can be used to determine the amount of smoothing. Poor worm plots or Q statistics may indicate the need to decrease the values of the smoothing parameters; see for example Rigby and Stasinopoulos [2006].

In reality a combination of all of the above procedures is good practice. The popular methodology for creating centile references for individuals from a population comprises two different methods:

1. the nonparametric method of quantile regression [Koenker, 2005, Koenker and Bassett, 1978, Koenker and Ng, 2005, He and Ng, 1999, Np and Maechler, 2007]; and

2. the parametric LMS (lambda, mu, sigma) method of Cole [1988], Cole and Green [1992] and its extensions. For example, see Wright and Royston [1997] and Rigby and Stasinopoulos [2004, 2006].

In the next two sections we describe the two approaches.

13.1.1 Quantile regression

Standard quantile regression methods estimate each quantile (i.e. centile) separately. Koenker et al. [1994], He and Ng [1999] and Np and Maechler [2007] model smooth quantile curves, e.g. using B-splines with a smoothness penalty. **R** packages **quantreg** and **cobs** are available for quantile regression smoothing; see Koenker [2016] and Ng and Maechler [2015], respectively. The following are features associated with quantile regression modelling:

- The quantile regression model does not assume a distribution for the response variable, therefore it is flexible and also in general reduces the bias caused by assuming a (possibly wrong) distribution. This of course comes with a possible increase in the variability of the quantile curves (the usual bias against variance balance).

- The quantile curves near the extremes (i.e. for p close to 0 or 1) vary more (i.e. are more erratic) than those in the centre of the distribution of Y, due to the fact that those curves are supported by fewer observations. van Buuren [2007] commented that 'curves produced by the quantile model are irregular near the extremes, and are generally less aesthetically pleasing' than the ones produced by parametric methods. This is more obvious using the **cobs** and **quantreg** packages since each quantile curve is fitted separately. Quantile sheets [Schnabel and Eilers, 2013a,b] do not suffer from this problem, since the estimation of the quantiles is done simultaneously. A function to fit quantile sheets, `quantSheets()`, is available in **gamlss** and is discussed and demonstrated in Section 13.5.

- A possible problem with quantile regression is that different quantile curves $y_p(x)$ for different values of p may cross (implying negative probability). There are several papers using quantile regression as a method to fit centile curves jointly, in order to overcome the problem; see for example Gannoun et al. [2002]; He [1997]; Heagerty and Pepe [1999]; Kapitula and Bedrick [2005]; Wei et al. [2006]. However they result in restrictions on the quantile curves, thereby reducing their flexibility and possibly increasing their bias.

- The quantile regression model does not allow for interpolation between quantile curves (for different p's), nor for extrapolations beyond the outer centile curves which is desirable for example for tracking children with extreme growth.

- The fitted quantile regression model does not have an overall (likelihood-based) measure of fit, such as GAIC, and this creates difficulties in comparing competitive models.

- It is difficult to define the residuals of a fitted quantile regression model. Within **gamlss** and for a fitted `quantSheets` object, this is achieved using an approximation. This approximation involves the function `flexDist()` which allows the user to reconstruct a distribution given the quantiles (and/or the expectiles). See Section 13.5.

- The fitted quantile regression model lacks an explicit formula allowing the calculation of quantile $y_p(x)$ given p and x, or the z-score given y and x. This was one of the requirements set by a World Health Organisation expert committee [Borghi et al., 2006] for the adoption of a method for the construction of the world standard curves. This problem is related to the previous one and is solved approximately within **gamlss** using `flexDist()` for a fitted `quantSheets` object.

- The link function relating the quantile to explanatory variables is usually the identity link, leading to an additive model, while a log link function leading to a multiplicative model may be more appropriate.

13.1.2   The LMS method and extensions

The LMS method was developed by Cole [1988] and Cole and Green [1992] to create centile curves for a response variable Y against a single explanatory variable x (e.g. age). Because the LMS method assumes that the Y variable has a specific distribution, centile (quantile) curves for all p can be obtained simultaneously and do not cross. Calculation of the quantile $y_p(x)$, given p and x, or the z-score, given y and x, is available for the LMS models.

The LMS method can be fitted within **gamlss** by assuming that the response variable has a Box–Cox Cole and Green (BCCG) distribution, which is suitable for positively or negatively skewed data with $Y > 0$ and is defined as follows:

Let the positive random variable $Y > 0$ be defined through the transformed random variable $\mathcal{Z}$ given by

$$
\mathcal{Z} = \begin{cases} \frac{1}{\sigma\nu}\left[\left(\frac{Y}{\mu}\right)^{\nu} - 1\right] & \text{if } \nu \neq 0 \\[2mm] \frac{1}{\sigma}\log\left(\frac{Y}{\mu}\right) & \text{if } \nu = 0 \end{cases}
\tag{13.1}
$$

where $\mu > 0$, $\sigma > 0$ and $-\infty < \nu < \infty$, and where $\mathcal{Z}$ is assumed to follow a truncated standard normal distribution. The condition $0 < Y < \infty$ (required for Y^ν to be real for all ν) leads to the condition $-1/(\sigma\nu) < \mathcal{Z} < \infty$ if $\nu > 0$ and $-\infty < \mathcal{Z} < -1/(\sigma\nu)$ if $\nu < 0$, which necessitates the truncated standard normal distribution for $\mathcal{Z}$.

Rigby and Stasinopoulos [2004, 2006] extended the LMS method (which models skewness but not kurtosis in the data), by introducing the Box–Cox power exponential (BCPE) and Box–Cox t (BCT) distributions for Y and called the resulting centile estimation methods LMSP and LMST, respectively. The BCPE assumes that the transformed random variable $\mathcal{Z}$ has a truncated power exponential distribution, while BCT assumes that $\mathcal{Z}$ has a truncated t distribution. The BCCG, BCPE and BCT distributions are available in **gamlss**. In the case of centile estimation for Y given x, the GAMLSS model is

$$
\begin{aligned}
Y &\sim \mathcal{D}(\mu, \sigma, \nu, \tau) \\
g_1(\mu) &= s_1(u) \\
g_2(\sigma) &= s_2(u) \\
g_3(\nu) &= s_3(u) \\
g_4(\tau) &= s_4(u) \\
u &= x^\xi
\end{aligned}
\tag{13.2}
$$

where the distribution $\mathcal{D}$ typically represents the BCCG, BCPE or BCT distribution, for which μ, σ, ν and τ are the approximate median, approximate coefficient of variation, skewness and kurtosis parameters of the distribution, respectively. (Note that BCCG does not have the parameter τ.) The $g(\cdot)$ functions represent appropriate link functions, the $s(\cdot)$ are nonparametric smoothing functions and ξ is the power exponent of x.

The power transformation for x, i.e. x^ξ, is usually needed when the response variable has an early or late spell of fast growth. In those cases the transformation of x can stretch the x scale, improving the fit of the smooth curve.

Each link function $g(\cdot)$ is usually chosen to ensure that the parameters are defined appropriately. For example a log link ensures that the parameter in question remains positive. Note however, that the original formulation of the LMS method introduced by Cole and Green [1992] uses identity links for all the parameters of BCCG. For historical reasons the first formulation of the BCCG, BCPE and BCT distributions has an identity link function for μ as a default, even though μ should always be positive. The default link functions for σ, ν and τ are `log`, `identity` and `log`, respectively. In **gamlss**, the distributions BCCGo, BCPEo and BCTo all have a log link as the default for μ (but are otherwise identical to BCCG, BCPE and BCT, respectively).

The nonparametric smoothing functions $s(\cdot)$ usually require the specification of a smoothing parameter λ or the equivalent effective degrees of freedom to be used; see for example Hastie and Tibshirani [1990] and Wood [2006a], and Chapter 9. Next we describe the methods used within **gamlss** to estimate these.

Model selection procedures for the LMS method

The selection of the link functions $g_k(\cdot)$, for $k = 1, 2, 3, 4$ usually does not create a problem. Log link functions are preferable for σ and τ (to ensure $\sigma > 0$ and $\tau > 0$). The identity link function is appropriate for ν since $-\infty < \nu < \infty$. For μ the safe option is to use the log link by using the BCCGo, BCPEo and BCTo distributions, but for most cases the identity link also works (BCCG, BCPE and BCT). The preferred link function is the one for which the fitted model has the smallest value of GAIC(κ) for a particular penalty κ.

Given a chosen distribution in (13.2) and its chosen link functions, the model specification comprises finding the effective degrees of freedom for modelling μ, σ, ν and τ, and ξ in the transformation for x. That is, we have to select the five hyperparameters (df$_\mu$, df$_\sigma$, df$_\nu$, df$_\tau$ and ξ).

Over the years different procedures have been considered by the authors. Here we explain three procedures used for choosing the hyperparameters.

Method 1: This method minimizes GAIC(κ), for a fixed κ, over the five hyperparameters. Rigby and Stasinopoulos [2006] used an automatic procedure `find.hyper()`, which is based on the numerical optimization function `optim()`, to minimize GAIC(κ). They used the BCT distribution model (13.2) and different fixed values of the penalty κ including $\kappa = 2$ (AIC) and $\kappa = \log n$ (SBC). They found that using $\kappa = 2$ overfitted the data leading to erratic centile curves, while $\kappa = \log n$ underfitted the data leading to oversmooth (and hence biased) centile curves and unsatisfactory residual diagnostics. They found that $\kappa = 3$ gave a good compromise between these and produced smooth growth curves which appeared to neither overfit nor underfit the data.

Method 2: This method minimizes the validation global deviance (VDEV) over the five hyperparameters [Stasinopoulos and Rigby, 2007]. In this procedure the data are split randomly into, for example, 60% training and 40% validation data sets. For each specific set of hyperparameters, model (13.2) is fitted to the training data and the resulting validation global deviance VDEV $= -2\tilde{\ell}$ (see equation (3.20)) is calculated, where $\tilde{\ell}$ is the log-likelihood of the validation data. VDEV is then minimized over the five hyperparameters using `optim()`. This method was found to moderately overfit the data set used.

Method 3: This method has two steps. In the first step, if a transformation on the x-axis is needed, then, for a simple normal distribution model with $g(\mu) = s_1(x^\xi)$ and constant σ, the GAIC(κ) is minimized over ξ, for a fixed value of κ. Given the estimated ξ, the second step involves fitting model (13.2) for particular distributions $\mathcal{D}$ (e.g. BCCGo, BCPEo and BCTo) and estimating (at most) the four degrees of freedom hyperparameters (df$_\mu$, df$_\sigma$, df$_\nu$, df$_\tau$) using a local ML procedure see; Sections 3.4.2.1 and 9.4 and also [Rigby and Stasinopoulos, 2013]. The chosen distribution is the one that has the smallest value of GAIC(κ). This is the fastest method and resulted in a model with similar centiles to the two previous methods in the data set used.

13.1.3 Example: The Dutch boys BMI data

For the remainder of this chapter we will use data from the Fourth Dutch Growth Study, [Fredriks et al., 2000a,b], described in Chapter 9. Here we use the BMI (y) and age (x) of the boys, originally plotted in Figure 9.5 and reproduced here in Figure 13.1(a). We are interested in a transformation of age to $u = age^\xi$ in (13.2). To demonstrate the use of some of the functions of this chapter, we have taken a sample of 1,000 observations from the original 7,040 observations; see Figure 13.1(b):

Fig.
13.1

```
data(dbbmi)
plot(bmi~age, data=dbbmi, pch = 15, cex = 0.5, col = gray(0.5))
title("(a)")
set.seed(2803)
IND<-sample.int(7040, 1000, replace=FALSE)
dbbmi1 <- dbbmi[IND,]
plot(bmi~age, data=dbbmi1, pch = 15, cex = 0.5, col = gray(0.5))
title("(b)")
```

13.2 Fitting centile curves

Centile curves for y against a single (quantitative) explanatory variable x can be obtained by fitting model (13.2) using either the lms(), gamlss() or find.hyper() functions as discussed in this section. However centile curves for y against multiple explanatory variables (i.e. centile surfaces) are obtained using gamlss() or find.hyper(), but not lms(); see Exercise 4.

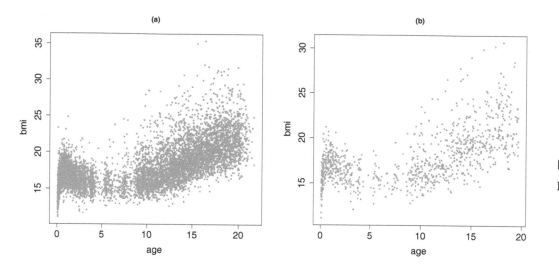

FIGURE 13.1: BMI against age, Dutch boys data (a) the full data set of 7,040 observations, (b) random sample of 1,000 observations.

R

p. 45

13.2.1 The `lms()` function

The `lms()` function is designed to facilitate the automatic selection of an appropriate LMS type centile estimation given by model (13.2) by constructing growth curves using Method 3 above. In particular, the following are determined: (i) the distribution of the response variable, (ii) the appropriate degrees of freedom for all the parameters of the distribution and (iii) the power parameter ξ. This avoids a global GAMLSS model selection as described in Methods 1 and 2, with the result that `lms()` is much faster (but possibly less reliable). Note, however, that `lms()` is applicable with one explanatory variable only. The function has the following arguments:

y	the response variable;
x	the unique explanatory variable, usually age;
families	a list of `gamlss.families` with default `families=c("BCCGo",` `"BCPEo"`, `"BCTo")`. Note that this default list is appropriate for positive response variables;
data	the data frame;
k	the penalty to be used in the GAIC, with default `k=2`;
cent	a vector with elements the centile values for which the centile curves are evaluated, with default `cent = c(0.4, 2, 10, 25, 50, 75, 90, 98, 99.6)`;

`calibration`	whether calibration is required, with default `calibration=TRUE` (see Section 13.3.2);
`trans.x`	whether to transform x to x^ξ, with default `trans.x=FALSE`;
`lim.trans`	the limits for the search of the power parameter ξ for transforming x to x^ξ;
`legend`	whether a legend is required in the plot, with default `legend=FALSE`;
`mu.df`	effective degrees of freedom for smoothing (i.e. on top of the 2 degrees of freedom for the constant and linear parameters) used for μ, in the smooth function $s_1(x^\xi)$, if required, otherwise it is estimated;
`sigma.df`	effective degrees of freedom for smoothing used for σ, in the smooth function $s_2(x^\xi)$, if required, otherwise it is estimated;
`nu.df`	effective degrees of freedom for smoothing used for ν, in the smooth function $s_3(x^\xi)$, if required, otherwise it is estimated;
`tau.df`	effective degrees of freedom for smoothing used for τ, in the smooth function $s_4(x^\xi)$, if required, otherwise it is estimated;
`method.pb`	the method used in the penalized B-splines function `pb()` for local estimation of the smoothing parameters. The default is local maximum likelihood `"ML"`. `"GAIC"` is also permitted where `k` is specified (below);
`k`	specified when `method.pb="GAIC"`;
`...`	extra arguments which can be passed to `gamlss`.

Note that the total effective degrees of freedom for μ is $\mathrm{df}_\mu = \texttt{df.mu} + 2$. Similarly for df_σ, df_ν and df_τ.

We illustrate the use of `lms()` on the sample of Dutch boys data. Figure 13.1 shows a fast growth in BMI for children during the first year of life, indicating that a power transformation for age could be appropriate for these data. Therefore the argument `trans.x=TRUE` is used.

```
m0 <- lms(bmi,age, data=dbbmi1, trans.x=TRUE, k=2)

m0$family
## [1] "BCCGo"                    "Box-Cox-Cole-Green-orig."

m0$power
## [1] 0.3294879
```

The chosen power transformation for age is given by m0$power, i.e. $u = \text{age}^{0.329}$ and the best distribution, according to $\text{GAIC}(\kappa = 2)$, is given by m0$family, and is BCCGo. The chosen model m0 can be fitted directly by

```
dbbmi1$Tage<-(dbbmi1$age)^(m0$power)
m0A<-gamlss(bmi~pb(Tage), sigma.formula=~pb(Tage),
            nu.formula=~pb(Tage), family=BCCGo,
            trace=FALSE, data=dbbmi1)
```

Alternatively m0 can be fitted by

```
m0B<-gamlss(bmi~pb(age^m0$power), sigma.formula=~pb(age^m0$power),
            nu.formula=~pb(age^m0$power), family=BCCGo,data=dbbmi1,
            trace=FALSE)
```

The effective degrees of freedom used for the smooth functions in model m0 (including 2 for the constant and linear terms) are given by

```
edfAll(m0)

## $mu
## pb(x, df = mu.df)
##          9.169865
##
## $sigma
## pb(x, df = sigma.df)
##             4.221547
##
## $nu
## pb(x, df = nu.df)
##          5.797055
```

Hence $\text{df}_\mu = 9.2$, $\text{df}_\sigma = 4.2$, $\text{df}_\nu = 5.8$. Note the total effective degrees for each parameter can be also obtained using:

```
m0$mu.df; m0$sigma.df; m0$nu.df
```

13.2.2   Estimating the smoothing degrees of freedom using a local GAIC

An alternative method of estimation of the smoothing degrees of freedom is using a local GAIC criterion with penalty κ. For example, if $\kappa = 4$ the command is

```
dbbmi1$Tage<-(dbbmi1$age)^(m0$power)
m0C<-gamlss(bmi~pb(Tage, method="GAIC", k=4),
            sigma.formula=~pb(Tage, method="GAIC", k=4),
            nu.formula=~pb(Tage, method="GAIC", k=4),
            family=BCCGo,data=dbbmi1, trace=FALSE)
edfAll(m0C)
```

```
## $mu
## pb(Tage, method = "GAIC", k = 4)
##                         6.836944
##
## $sigma
## pb(Tage, method = "GAIC", k = 4)
##                         3.515268
##
## $nu
## pb(Tage, method = "GAIC", k = 4)
##                         5.143173
```

A model with smoother centile curves may be obtained by increasing the value of κ in the above command, but this may fit the data less well, resulting in poorer residual diagnostics. Conversely a model which fits the data better, with better residual diagnostics, may be obtained by decreasing the value of κ, but this may result in less smooth centile curves.

13.2.3 The find.hyper() function

The function find.hyper() can be used to estimate the smoothing parameters (and hence their effective degrees of freedom) and the power parameter ξ using a global GAIC criterion with penalty κ, as described in Method 1 of Section 13.1.2. It is preferable to use cubic splines (cs()) rather than P-splines (pb()) for the smoothing functions. For example if $\kappa = 4$ the commands are

```
mod<-quote(gamlss(bmi~cs(Tage,df=p[1]),sigma.formula=~cs(Tage,
    df=p[2]),nu.formula=~cs(Tage,df=p[3]),c.spar=c(-1.5,2.5),
    family=BCCGo,data=dbbmi1,control=gamlss.control(trace=FALSE,
    n.cyc=1000,gd.tol=Inf)))
op<-find.hyper(model=mod, other=quote(Tage<-age^p[4]),
    par=c(6,2,2,0.1),lower=c(0.1,0.1,0.1,0.001),
    steps=c(0.1,0.1,0.1,0.005),factr=2e9,parscale=c(1,1,1), k=4)
```

The find.hyper() function is explained in Section 11.9.

```
op$par
```

```
## [1] 4.6278596 1.3926286 2.8454589 0.2134324
```

```
op$value
```

```
## [1] 4007.622
```

The effective degrees of freedom for μ, σ and ν (on top of 2 for the constant and linear terms) that minimize GAIC(4) are stored in op$par[1], op$par[2] and op$par[3], respectively. op$par[4] stores the chosen power parameter ξ. op$value stores the minimized value of GAIC(4). The effect of increasing or decreasing κ in the above command is the same as when using a local GAIC criterion, as explained in Section 13.2.2.

The chosen model can now be fitted by

```
Tage<-(dbbmi1$age)^(op$par[4])
mOD<-gamlss(bmi~cs(Tage,df=op$par[1]), sigma.formula=~cs(Tage,
        df=op$par[2]), nu.formula=~cs(Tage,df=op$par[3],
        c.spar=c(-1.5,2.5)), family=BCCGo, data=dbbmi1)
```

The effective degrees of freedom (*including* 2 for the constant and linear terms) for μ, σ and ν are

```
mOD$mu.df
mOD$sigma.df
mOD$nu.df
```

Hence, for example, df_μ is given by `mOD$mu.df=op$par[1]+2`. Model `mOD` is best according to GAIC($\kappa = 4$).

```
GAIC(mOA, mOB, mOC, mOD, k=4)
```

Note that in the resulting table the extra one degree of freedom for estimating the power parameter is not included in df or GAIC.

13.2.4   Residual diagnostics

Checking the fitted model `m0` using residual diagnostics is very important for the creation of centile curves. Q statistics and worm plots, described in Chapter 12, are two methods:

```
round(Q.stats(m0, xvar=dbbmi1$age),3)
```

		Z1	Z2	Z3	Z4	AgostinoK2
## 0.055 to	0.265	0.209	0.958	-0.158	0.009	0.025
## 0.265 to	0.875	-0.358	-0.552	1.255	0.539	1.867
## 0.875 to	1.625	0.508	-0.613	-0.749	0.047	0.563
## 1.625 to	3.035	0.186	-0.998	0.731	-0.001	0.534
## 3.035 to	7.435	-1.139	0.599	-0.428	0.087	0.191
## 7.435 to	9.965	0.891	-0.154	-1.033	0.826	1.750
## 9.965 to	11.415	-1.381	0.673	1.694	-0.277	2.945
## 11.415 to	13.095	-0.217	0.490	0.022	0.472	0.223
## 13.095 to	14.425	1.205	-0.341	-0.096	0.413	0.179
## 14.425 to	15.865	-0.398	0.407	-1.318	1.250	3.298
## 15.865 to	17.715	0.415	-0.848	0.912	0.217	0.878
## 17.715 to	19.645	0.041	0.182	0.216	-1.086	1.227
## TOTAL Q stats		6.294	4.703	9.438	4.243	13.681

Fig. 13.2

```
## df for Q stats        2.830  9.389  6.203 12.000     18.203
## p-val for Q stats     0.087  0.881  0.164  0.979      0.761
##                           N
##   0.055 to  0.265       92
##   0.265 to  0.875       75
##   0.875 to  1.625       83
##   1.625 to  3.035       83
##   3.035 to  7.435       85
##   7.435 to  9.965       82
##   9.965 to 11.415       83
## 11.415 to 13.095        84
## 13.095 to 14.425        83
## 14.425 to 15.865        83
## 15.865 to 17.715        84
## 17.715 to 19.645        83
## TOTAL Q stats         1000
## df for Q stats           0
## p-val for Q stats        0
```

The table above indicates that the Q statistics are reasonable (since all their p-values are greater than 0.05). The plot of the Z statistics in Figure 13.2 indicates an adequate model, since all the values of $|Z|$ are less than 2, shown by no squares within the circles. Next we check the worm plots.

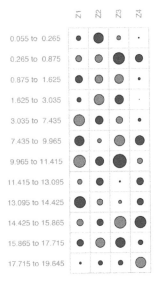

FIGURE 13.2: Plot of Q statistics for the fitted `lms` object `m0`.

```
wp(m0, xvar=dbbmi1$age, n.inter=9)
## number of missing points from plot= 0   out of   112
## number of missing points from plot= 0   out of   111
## number of missing points from plot= 0   out of   110
## number of missing points from plot= 0   out of   112
## number of missing points from plot= 0   out of   111
## number of missing points from plot= 0   out of   111
## number of missing points from plot= 0   out of   112
## number of missing points from plot= 0   out of   110
## number of missing points from plot= 0   out of   111
```

Fig.
13.3

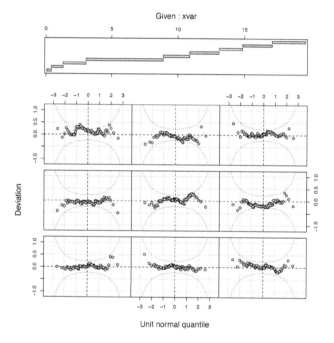

FIGURE 13.3: Worm plot for the fitted lms object m0.

R
p. 46

Figure 13.3 shows that the residuals look good for all nine intervals of age (since most points lie between the approximate 95% pointwise elliptical confidence bands), indicating that the model is adequate.

13.2.5   The fittedPlot() function

The function fittedPlot() provides a convenient way of plotting the fitted distribution parameters (in general μ, σ, ν and τ) if the model involves only one explanatory variable. Therefore it can be used after the lms() function.

Fig.
3.4 `fittedPlot(m0, x=dbbmi1$age)`

The plot is given in Figure 13.4. These is no fitted τ because the chosen distribution BCCGo(μ, σ, ν) only has three parameters.

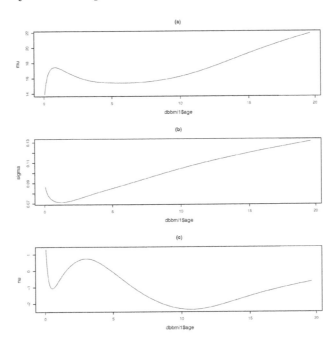

FIGURE 13.4: Fitted values for (a) μ (b) σ and (c) ν, against age, from a Box–Cox Cole and Green (BCCGo) distribution, Dutch boys BMI data.

`fittedPlot()` has the following arguments:

`object`	a fitted **gamlss** model object (with one explanatory variable);
`...`	optionally more fitted **gamlss** model objects;
`x`	the unique explanatory variable;
`color`	whether the fitted lines in the plot are shown in colour with default `color=TRUE`;
`line.type`	whether the line types should be different or not with default `line.type=FALSE`;
`xlab`	the label for the x-axis.

The fitted values of more than one model can be plotted together using `fittedPlot()`. For example, here we compare `m0` with `m1`, which is fitted using the BCPEo distribution and `pb()` for each distribution parameter.

```
dbbmi1$Tage<-(dbbmi1$age)^(m0$power)
m1<-gamlss(bmi~pb(Tage), sigma.formula=~pb(Tage),
```

464   **Flexible Regression and Smoothing: Using GAMLSS in** R

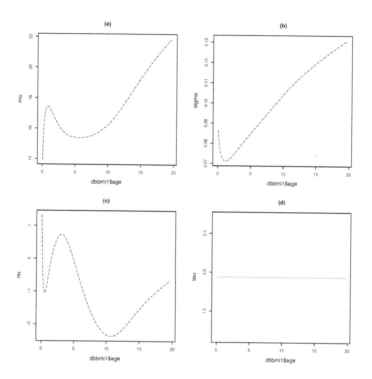

R
p. 46

FIGURE 13.5: Comparing the fitted values for all parameters against age, for the BCCGo model m0 (dashed line), and the BCPEo model m1 (solid line): (a) μ (b) σ (c) ν (d) τ.

```
            nu.formula=~pb(Tage), tau.fomula=~pb(Tage),
            family=BCPEo, data=dbbmi1)
```

```
## GAMLSS-RS iteration 1: Global Deviance = 3967.614
## . . .
## GAMLSS-RS iteration 7: Global Deviance = 3939.411
```

ig.
3.5 `fittedPlot(m1,m0, x=dbbmi1$age, line.type=c(1,2))`

The plot is given in Figure 13.5. The fitted values for μ, σ and ν are very similar for the two models. Note that the fitted values for τ for BCPEo are flat, indicating that the fitted τ is (approximately) constant. BCCGo has no τ parameter.

13.3 Plotting centile curves

Centile plots are currently provided for all the continuous distributions in Table 6.1. There are five functions for plotting centiles: `centiles()`, `calibration()`, `centiles.fan()`, `centiles.split()` and `centiles.com()`. These are described in Sections 13.3.1 to 13.3.5, respectively.

13.3.1 centiles()

A simple use of `centiles()` is demonstrated below. This plots fitted centile curves for y against x for different centile percentages. Note that the *sample* percentage of observations below each of the fitted centile curves from the fitted model are printed (at the end of each line), so comparisons with *nominal model* centiles (printed after **below** on each line) can be made. In this example the sample percentages are reasonably close to the nominal model centiles. Figure 13.6 shows the plot.

Fig.
3.6 `centiles(m0,dbbmi1$age)`

```
## % of cases below  0.4 centile is  0.4
## % of cases below  2 centile is  2.4
## % of cases below  10 centile is  10.6
## % of cases below  25 centile is  24.5
## % of cases below  50 centile is  50.1
## % of cases below  75 centile is  75.2
## % of cases below  90 centile is  89.9
## % of cases below  98 centile is  97.7
## % of cases below  99.6 centile is  99.8
```

The following are the arguments of `centiles()`:

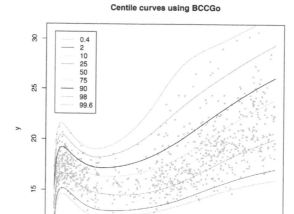

R
p. 4●

FIGURE 13.6: Centile curves using the Box–Cox Cole and Green (BCCGo) distribution for the Dutch boys BMI data.

obj	a fitted **gamlss** object;
xvar	the unique explanatory variable for which the fitted model centiles are calculated;
cent	a vector with elements the centile values for which the fitted model centile curves are evaluated. For example, if you wish centiles at 5% and 95% only, use **cent=c(5,95)**;
legend	whether a legend is required within the plot or not. The default is **legend=TRUE**. This legend identifies the different centile curves;
ylab	the y-variable label;
xlab	the x-variable label;
main	the main title. If **main=NULL** the default title "Centile curves using NO" (or the relevant distribution) is shown;
main.gsub	if the character "@" appears in **main** then the default title is added;
xleg	position of the legend in the x-axis;
yleg	position of the legend in the y-axis;
xlim	limits of the x-axis;
ylim	limits of the y-axis;
save	whether to save the sample percentages or not, with default

save=FALSE. In this case the sample percentages are printed but
are not saved;

plot   whether to plot the centile curves;

pch   the character to be used as the default in plotting points; see the
options for par();

cex   size of character, see par();

col   the colour of points, see par();

col.centiles   the colours for the centile curves;

lty.centiles   the line types for the centile curves;

lwd.centiles   the line width for the centile curves;

points   whether the data points should be plotted;

...   for extra arguments.

As an example, a modified version of the centiles in Figure 13.7 is given below.

Fig.
13.7
```
centiles(m0,dbbmi1$age,cent=c(5,25,50,75,95), ylab="bmi", xlab="age",
        col.centiles = c(2,3,1,3,2), lty.centiles = c(2,4,1,4,2),
        lwd.centiles =c(2,2,2.5,2,2))
```

```
## % of cases below  5 centile is  4.7
## % of cases below  25 centile is  24.5
## % of cases below  50 centile is  50.1
## % of cases below  75 centile is  75.2
## % of cases below  95 centile is  95
```

Note that the output obtained from the centiles() function can be useful to get
information on how well a distribution fits at a particular age. We demonstrate this
by selecting the cases from the Dutch boys data set at the rounded age of 10 (i.e.
between 9.5 and 10.5 years), and fitting a BCCGo distribution to the sample.

Fig.
13.8
```
sub1<-subset(dbbmi, (age > 9.5 & age < 10.5))
h1 <- gamlssML(bmi, data=sub1, family=BCCGo)
centiles(h1,sub1$age,cent=c(1,2.5, 10, 25, 50, 75, 90, 97.5, 99))
```

```
## % of cases below  1 centile is  0.2777778
## % of cases below  2.5 centile is  3.611111
## % of cases below  10 centile is  8.888889
## % of cases below  25 centile is  26.11111
## % of cases below  50 centile is  50.83333
## % of cases below  75 centile is  77.5
## % of cases below  90 centile is  89.72222
## % of cases below  97.5 centile is  97.22222
## % of cases below  99 centile is  99.16667
```

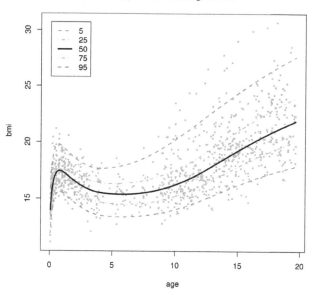

FIGURE 13.7: Centile curves using Box–Cox Cole and Green (`BCCGo`) distribution for the Dutch boys BMI data, modified plot.

R

p. 46

The plot is shown in Figure 13.8. The table above shows that the fit did not capture the lower tail of the BMI distribution well, but captured the upper tail far better.

If no variable is available the user can create an index variable and use this in the centiles command:

```
index<-1:n
centiles(h1,index)
```

where `n` is the number of observations. An alternative way to check the fit of model `m0` between ages 9.5 to 10.5 is to use `centiles.split()` with argument `xcut.points=c(9.5,10.5)`; see Section 13.3.4.

13.3.2 calibration()

This function can be used when the fitted model centiles differ from the sample centiles and it is assumed that this failure is the same for all values of the explanatory variable. The `calibration` function automatically adjusts the values selected for argument `cent` so that the sample percentage of cases below each centile curve is 'correct', i.e. equal to the selected `cent` values. Consider the following example:

```
calibration(m0,xvar=dbbmi1$age, cent=c(5,25,50,75,95))
## % of cases below  5.249092 centile is  5
## % of cases below  25.47616 centile is  25
```

Fig.
13.9

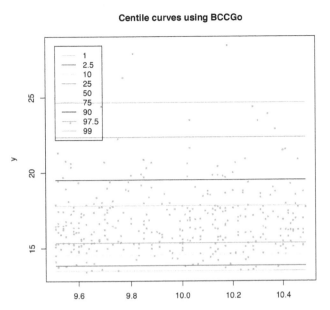

FIGURE 13.8: Centile curves using the Box–Cox Cole and Green distribution to fit BMI at rounded age 10, Dutch boys BMI data.

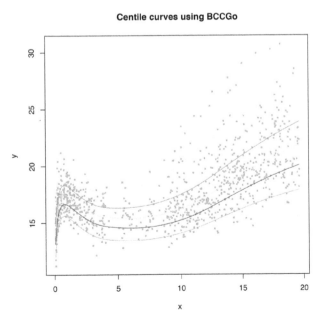

FIGURE 13.9: Calibration curves using the Box–Cox Cole and Green distribution for the Dutch boys BMI data.

```
## % of cases below  49.94412 centile is  50
## % of cases below  74.69884 centile is  75
## % of cases below  95.0125 centile is  95
```

The plot is shown in Figure 13.9. In this case `calibration()` produces similar results to `centiles()`. The `calibration()` function has as arguments, apart from `object`, `xvar` and `cent`:

legend   whether legend is required (default is `legend=FALSE`);

fan   for fan plots (default is `fan=FALSE`).

13.3.3   centiles.fan()

The function `centiles.fan()` plots a fan-chart of the centile curves.

```
centiles.fan(m0,dbbmi1$age,cent=c(5,25,50,75,95), ylab="bmi",
             xlab="age")
```

Fig. 13.1●

The different colour schemes available for the fan-chart are `"cm"`, `"gray"`, `"rainbow"`, `"heat"`, `"terrain"` and `"topo"`, using the argument `colours`.

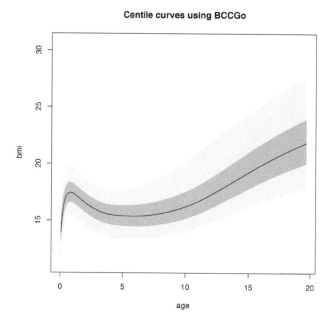

R
p. 47

FIGURE 13.10: A fan-chart (centile) curves using the Box–Cox Cole and Green distribution for the Dutch boys BMI data.

13.3.4 `centiles.split()`

The function `centiles.split()` splits the fitted centile curves according to different cut points in the x-variable (age). Here we split the centiles plot at age 2 (i.e. `xcut.points=c(2)`):

Fig.

5.11

```
centiles.split(m0,xvar=dbbmi1$age,xcut.points=c(2))
```

```
##           0.06 to 2   2 to 19.64
## 0.4        0.00000    0.5464481
## 2          2.61194    2.3224044
## 10        11.19403   10.3825137
## 25        23.88060   24.7267760
## 50        49.25373   50.4098361
## 75        74.62687   75.4098361
## 90        91.04478   89.4808743
## 98        97.76119   97.6775956
## 99.6      99.25373  100.0000000
```

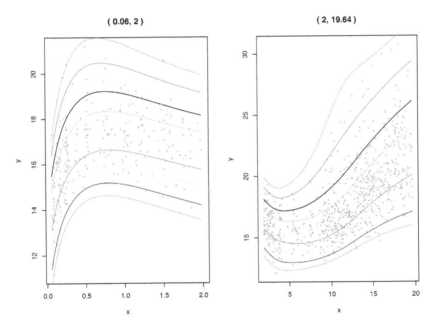

FIGURE 13.11: Centile curves for two age ranges using Box–Cox Cole and Green distribution fitted to the sample of 1000 observations from the BMI data.

The table above gives the sample percentage of cases below the $0.4, 2, 10, \ldots, 99.6$ centile curves for each of the two age ranges in the split, i.e. age range (0.06 to 2) and age range (2 to 19.64), where 0.06 and 19.64 are the minimum and maximum ages in the data set.

The arguments for `centiles.split()` are

obj   a fitted `gamlss` object;

xvar	the unique explanatory variable;
xcut.points	the x-axis cut point(s), e.g. xcut.points=c(20,30). If xcut.points=NULL then the n.inter argument is activated;
n.inter	if xcut.points=NULL this argument gives the number of intervals (with equal numbers of cases) in which the x-variable will be split, with default n.inter=4;
cent	a vector with elements the centile values for which the centile curves are evaluated;
legend	whether a legend is required in the plots or not, with default legend=FALSE;
ylab	the y-variable label;
xlab	the x-variable label;
overlap	how much overlapping in the xvar intervals. Default value is overlap=0 for non-overlapping intervals;
save	whether to save the sample percentages or not, with default save=TRUE. In this case the function produces a matrix giving the sample percentages for each interval;
plot	whether to plot the centiles. This option is useful if the sample statistics only are required. Default is plot=TRUE;
...	for extra arguments in the par() plotting function.

For example a split of the observations into four equally-sized groups based on age is achieved using:

```
centiles.split(m0,dbbmi1$age)
```

Fig. 13.12

```
##         0.05500 to 1.625 1.625 to 9.965 9.965 to 14.425
## 0.4                  0.0            0.8             0.4
## 2                    2.4            2.8             1.6
## 10                  10.8           10.4            11.2
## 25                  24.0           26.0            25.2
## 50                  48.8           50.0            51.6
## 75                  74.4           74.8            77.6
## 90                  90.4           89.6            88.4
## 98                  97.6           98.8            96.4
## 99.6                99.2          100.0           100.0
##         14.425 to 19.645
## 0.4                  0.4
## 2                    2.8
## 10                  10.0
## 25                  22.8
```

```
## 50               50.0
## 75               74.0
## 90               91.2
## 98               98.0
## 99.6            100.0
```

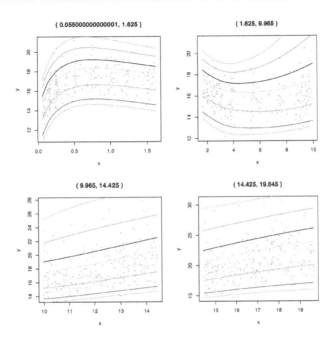

FIGURE 13.12: Centile curves for four age ranges using Box–Cox Cole and Green distribution for the BMI data.

13.3.5 Comparing centile curves: `centiles.com()`

This function is useful for comparison of centile curves produced by different fitted models. Here we fit a new `lms` object using the `SHASH` distribution and compare the result with the original `lms` model m0.

```
m2<-lms(bmi, age, data=dbbmi1, trans.x=TRUE , families=c("SHASH"),
        n.cyc=100)
```

```
## *** Checking for transformation for x ***
## *** power parameters   0.3294879 ***
## *** Initial  fit***
## GAMLSS-RS iteration 1: Global Deviance = 4260.768
## . . .
## % of cases below  98.08233 centile is  97.7
## % of cases below  99.51668 centile is  99.6
```

```
centiles.com(m0, m2, xvar=dbbmi1$age, legend=TRUE, color=TRUE)

## ********  Model 1 ********
## % of cases below  0.4 centile is  0.4
## % of cases below  10 centile is  10.6
## % of cases below  50 centile is  50.1
## % of cases below  90 centile is  89.9
## % of cases below  99.6 centile is  99.8
## ********  Model 2 ********
## % of cases below  0.4 centile is  0.2
## % of cases below  10 centile is  11.2
## % of cases below  50 centile is  49.8
## % of cases below  90 centile is  89.9
## % of cases below  99.6 centile is  99.7
```

Fig.
13.1

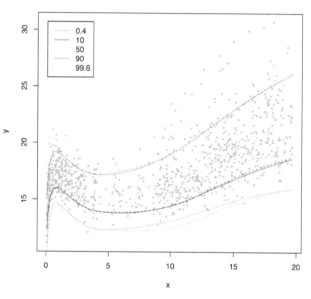

R
p. 47

FIGURE 13.13: Comparison of centile curves using the BCCGo (solid lines) and SHASH (Sinh-Arcsinh, dashed lines) distributions.

Most of the arguments of the function are similar to the ones in `centiles()`. We draw the reader's attention to the argument `no.data`, which is useful for excluding data points from the plot.

13.3.6   Plot of distribution of y for specific values of x

Another informative plot is given by `plotSimpleGamlss()`, introduced in Section 2.3.7. While this plot does not involve centiles, it is useful to contrast it with the centile plots shown in this chapter. We plot the fitted distribution of `bmi` for the values of `age` specified by `x.val=seq(5,20,5)`, i.e. $age = 5, 10, 15, 20$. Argument `val` controls the height of the probability density functions in the plot. The resulting plot is given in Figure 13.14.

Fig.
3.14

```
library(gamlss.util)
plotSimpleGamlss(bmi,age,m0,data=dbbmi1,x.val=seq(5,20,5),
                 xlim=c(-3,23), val=5)
```

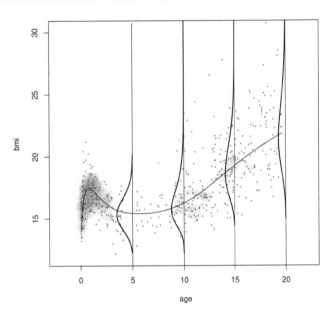

FIGURE 13.14: Plot of fitted distribution of `bmi` for specific values of `age`.

13.4   Predictive centile curves: `centiles.pred()`, `z.scores()`

The function `centiles.pred()` is designed to create predictive centile curves for new x-values, given a `gamlss` fitted model. It has three different functionalities which are described below:

Case 1: For given new x-values and given centile percentages, calculates a matrix containing the centile values for y.

Case 2: For given new x-values and standard normalized centile values (i.e. z-scores), calculates a matrix containing the centile values for y.

Case 3: For given new x-values and new y-values calculates the z-scores (one z-score for each (x, y) pair).

The first two options are useful for creating (growth curve) centile tables and plots useful for publication purposes. The third option is useful for checking where new observations are lying within the standard growth charts. Because of the importance of this task, the function z.scores() has been created to provide the same functionality. As with all the rest of the functions in this chapter, centiles.pred() and z.scores() apply to models with only one explanatory variable.

13.4.1   Case 1: Centile for y given x and centile percentage

To demonstrate the first case, we start by creating new values for the x variable, i.e. age (stored in newage) and use them to find the corresponding centiles for the y variable, i.e. bmi, which are stored in a matrix. The default centile percentage values are c(0.4,2,10,25,50,75,90,98,99.6). The centiles can then be plotted using the argument plot=TRUE. Note that we can use the model fitted by lms().

```
newage<-seq(0.05,2,.05)
mat1<-centiles.pred(m0, xname="age", xvalues=newage,
                cent=c(5,25,50,75,95))
head(mat1)
```

```
##     age      C5      C25      C50      C75      C95
## 1 0.05 11.70660 12.93057 13.74896 14.54431 15.65289
## 2 0.10 12.70627 13.85829 14.67872 15.51494 16.74513
## 3 0.15 13.40397 14.51738 15.33935 16.20276 17.52130
## 4 0.20 13.92418 15.01355 15.83630 16.71824 18.10166
## 5 0.25 14.32197 15.39543 16.21821 17.11237 18.54181
## 6 0.30 14.63125 15.69387 16.51617 17.41801 18.87869
```

```
mat1<-centiles.pred(m0, xname="age", xvalues=newage, plot=TRUE,
                ylab="BMI", xlab="age",cent=c(5,25,50,75,95),
                legend=FALSE)
```

Fig. 13.15

The resulting plot is given in Figure 13.15. The centiles can also be computed from model m0A by

```
mat1A<-centiles.pred(m0A, xname="Tage", xvalues=newage,
                power=m0$power,cent=c(5,25,50,75,95),
                data=dbbmi1)
```

The results are slightly different for age close to zero, because prediction for a gamlss model (e.g. m0A) involves some refitting (see Section 5.4), while for an lms model (e.g. m0) prediction involves interpolating or extrapolating the fitted spline functions, which may be preferable. Note also that including newage=0 in the prediction range may lead to an unreliable prediction at age=0 because its power transformed age is well outside the power transformed age range of the original data. For model m0A this also leads to slightly changed predictions within the original age range, because

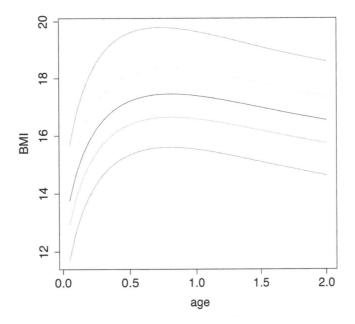

FIGURE 13.15: A plot of centile curves in the age range 0 to 2 using selected centiles.

the default 20 P-spline knots are spread over an expanded transformed age range, resulting in fewer knots within the transformed original age range. This change within the original age range may be reduced by using more knots, e.g. inter=50, in fitting the gamlss model mOA.

Model mOB can be used for prediction instead of mOA, leading to very similar predictions, by

```
mat1B<-centiles.pred(mOB, xname="age", xvalues=newage,
                 cent=c(5,25,50,75,95), data=dbbmi1)
```

13.4.2   Case 2: Centile for y given x and centile z-score

In the second case the objective is to create centiles based not on percentages but on standard normalized values or Z values. These are using the centiles.pred argument dev with default z-score (Z) values

```
dev=c(-4,-3,-2,-1,0,1,2,3,4)
```

Note that the corresponding centile percentages for the standard normalized values can be obtained by applying the standard normal cdf $\Phi(Z)$ (pNO(dev)). The resulting percentages are

```
round(100*pNO(dev),3)
```

```
## [1]  0.003  0.135  2.275 15.866 50.000 84.134 97.725 99.865
## [9] 99.997
```

We use the same **newage** values as above, but this time we use the argument
`type="standard-centiles"`:

```
mat2<-centiles.pred(m0, xname="age",xvalues=newage,
      type="standard-centiles",dev=c(-3,-2,-1,0,1,2,3))
head(mat2)
```

```
##     age         -3        -2        -1         0         1
## 1 0.05   9.883591 11.24295 12.52651 13.74896 14.92063
## 2 0.10  11.155005 12.29316 13.46809 14.67872 15.92407
## 3 0.15  11.976989 13.01599 14.13498 15.33935 16.63481
## 4 0.20  12.567656 13.55108 14.63623 15.83630 17.16658
## 5 0.25  13.008780 13.95829 15.02162 16.21821 17.57196
## 6 0.30  13.345671 14.27371 15.32256 16.51617 17.88504
##           2        3
## 1 16.04912 17.14025
## 2 17.20324 18.51538
## 3 18.02738 19.52343
## 4 18.64483 20.29173
## 5 19.11242 20.87661
## 6 19.46864 21.31886
```

```
mat2 <- centiles.pred(m0, xname="age", xvalues=newage, type="s",
            dev=c(-3,-2,-1,0,1,2,3), plot=TRUE)
```

Fig.
13.1(

The resulting plot is given by Figure 13.16.

13.4.3   Case 3: z-score given y and x

Case 3 is used in the situation in which one or more new individuals are available
for whom we know the value of the y-variable (e.g. BMI), and x-variable (e.g. age),
and we want to classify whether the individual is at risk. This is done by obtaining
the individual's z-score. For the true model, the z-scores have a standard normal
distribution and hence z-scores below -2 and above 2 are potentially of concern.
Here we show how to obtain the z-scores using `centiles.pred()` and `z.scores()`,
respectively. Consider four new cases with the following values of `bmi` and `age`:

Case	age	bmi
1	2	20
2	5	18
3	10	25
4	15	14

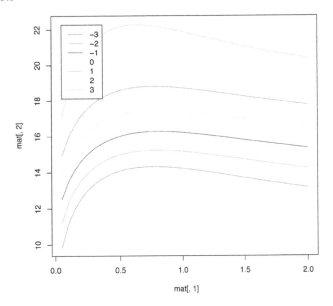

FIGURE 13.16: A plot of prediction centile curves using selected standard normalized deviates (i.e. Z values).

```
centiles.pred(m0,xname="age",xval=c(2,5,10,15),yval=c(20,18,25,14),
         type="z-scores")
## [1]   2.707422  1.813789  2.618234 -3.435872
z.scores(m0, x=c(2,5,10,15), y=c(20,18,25,14))
## [1]   2.707422  1.813789  2.618234 -3.435872
```

The outputs of the two functions are identical. They indicate that cases 1 and 3 are potentially of concern because their BMIs are too high for their ages (since z-score> 2), while case 4 is potentially of concern because his BMI is too low for his age (since z-score< -2).

The arguments for `centiles.pred()` are

obj   a fitted `gamlss` object;

type   the default `type="centiles"` gets the centile values given in the option `cent`.
           `type="standard-centiles"` gets the standard centiles given in `dev`.
           `type="z-scores"` gets the z-scores for given `xval` and `yval` new values;

xname   the name of the unique explanatory variable (it has to be the same as in the original fitted model);

xval   the new values for the explanatory variable where the prediction will take place;

power   if a power transformation is needed;

yval   the response variable values (corresponding to explanatory variable values given by xval), required for the calculation of z-scores;

cent   a vector with elements the centile values for which the centile curves have to be evaluated;

dev   a vector with elements the standard normalized deviate values (or z-score or Z values) for which the centile curves have to be evaluated using the option type="standard-centiles";

plot   whether to produce a plot (of the "centiles" or "standard-centiles"), with default plot=FALSE;

legend   whether a legend is required in the plot or not, with default legend=TRUE;

...   for extra arguments.

The z.scores() function has only three arguments: object for a fitted lms model; y for new y values; and x for new x values.

13.5   Quantile sheets: quantSheets()

In this section we describe the use of quantile sheets regression for constructing (growth) centile curves. Quantile sheets were developed by Schnabel and Eilers [2013a,b] in order to overcome some of the problems associated with quantile regression. In particular the main advance of the quantile sheets is the simultaneous estimation of the centiles, and the introduction of a smoothing parameter in the response variable direction. This reduces the wide variability (i.e. erratic behaviour) of the quantile curves at the extremes (i.e. for probability p close to 0 or 1) and makes them look more realistic. It also usually avoids (but does not guarantee to eliminate) the problem of crossing quantiles.

Quantile sheets can be fitted within **gamlss** using the function quantSheets(). This is a modified version of an earlier **R** function given to the authors by Paul Eilers. In its current form it can take only one explanatory variable. The function is fast (compared to lms()), even for large data sets.

13.5.1 Smoothing parameters

The fit of quantiles (i.e. centiles) depends on two smoothing parameters:

1. `x.lambda`, smoothing parameter in the direction of the explanatory variable; and

2. `p.lambda`, smoothing parameter in the direction of the response variable.

The smoothing parameters in `quantSheets()` are *not* estimated automatically and should be chosen by the user either by inspection or by some other criterion. Unfortunately since quantile sheets estimation does not easily provide an overall measurement of fit, e.g. GAIC, the selection of the two smoothing parameters has to be done by other means. Here we propose a heuristic method of choosing the smoothing parameters based on residuals, where the adequacy of the model is checked via residual diagnostics.

13.5.2 Residuals

The calculation of the normalized quantile residuals[1] or z-scores within a quantile regression is not straightforward. The following approach is used by the generic function `resid()` applied to a `quantSheets` object here to approximate the z-scores.

The fitted `quantSheets` model provides a fitted quantile value for y for each distinct x-value in the data and each chosen p-value in the `quantSheets` command. Given the fitted quantile values, a nonparametric distribution is constructed for each distinct x-value, using the **gamlss** function `flexDist()`. This constructs a (nonparametric) distribution using known quantile (or expectile) values of the distribution, and provides approximate numerically calculated pdf and cdf values for the distribution. Given the estimated cdf at points x, the probability integral transform (PIT) residuals can be found, and therefore the normalized quantile residuals. (The function `z.scoresQS()` performs those steps.) Note that because the cdf is different for each distinct value of the explanatory variable, for large data sets the calculation of the approximate quantile residuals can take several minutes. To avoid this problem the function `residuals.quantSheets()` provides, as a default, a quicker way of calculating the residuals. It starts by binning the observations in the x-direction, to say 100 intervals (option `inter=100`), with equal numbers of observations in each bin. It then calculates (using `flexDist()`) the cdf at the midpoints of the intervals and evaluates the PITs and quantile residuals of all the observations that fall in the bin. This reduces the time for calculating the quantile residuals considerably. The full quantile residuals can be obtained using the option `all=TRUE`.

While the approach described above seems to work well, the user should be aware that `flexDist()` uses penalties which themselves depend on smoothing parameters.

---

[1]Familiarity with material in Section 12.2 is assumed.

While the default smoothing parameters appear to work well, a visual inspection of
at least some distinct values of x is recommended.

13.5.3   Fitting the model

Here we fit a quantile sheets model to the 1,000 sampled observations from the
dbbmi data. We first use findPower() to find a suitable power transformation for
age, and use it in the option power in the quantile sheets fitting. The smoothing
parameter values x.lambda=1 and p.lambda=10 are chosen arbitrarily. Figure 13.17
gives the plot of the quantile sheet curves.

```
ppp<-findPower(dbbmi1$bmi,dbbmi1$age)

## *** Checking for transformation for x ***
## *** power parameters   0.3294879 ***
```

Fig.
13.1

```
qs1<-quantSheets(bmi, age, data = dbbmi1,
               cent=c(0.4, 2, 10, 25, 50, 75, 90, 98, 99.6),
               x.lambda=1, p.lambda=10, logit=TRUE, power=ppp)

## % of cases below  0.4 centile is  0
## % of cases below  2 centile is  0.4
## % of cases below  10 centile is  8.4
## % of cases below  25 centile is  27.4
## % of cases below  50 centile is  53.8
## % of cases below  75 centile is  75.9
## % of cases below  90 centile is  88.7
## % of cases below  98 centile is  96.8
## % of cases below  99.6 centile is  98.6
```

Note that there are considerable differences between the sample percentage of cases
below a centile curve (at end of each line) and the nominal model centile (after
below on each line).

To check the model we first calculate the residuals and then use the worm plot and
Q statistics as diagnostics.

```
res1 <- resid(qs1)
wp(resid=res1, xvar=dbbmi1$age, n.inter = 9 )

## number of missing points from plot= 0   out of   112
## number of missing points from plot= 0   out of   111
## number of missing points from plot= 0   out of   110
## number of missing points from plot= 0   out of   112
## number of missing points from plot= 0   out of   111
## number of missing points from plot= 0   out of   111
## number of missing points from plot= 0   out of   112
```

Fig.
13.18

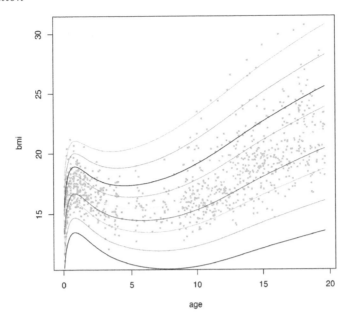

FIGURE 13.17: Quantile sheet curves fitted to the sample of the Dutch boys BMI data, using smoothing parameters x.lambda=1 and p.lambda=10.

```
## number of missing points from plot= 0   out of   110
## number of missing points from plot= 0   out of   111
```

round(Q.stats(resid=res1, xvar=dbbmi1$age), 3)

Fig.
3.19

```
##                        Z1       Z2       Z3       Z4
##   0.055 to   0.285 -0.019    1.198    0.433  -0.136
##   0.285 to   1.155  0.585    0.567    1.885  -0.433
##   1.155 to   2.525 -0.215   -0.845    0.508  -0.867
##   2.525 to   5.915  0.146   -0.681   -0.600  -1.210
##   5.915 to   9.965 -0.141   -1.966    2.434   0.535
##   9.965 to  11.705 -1.010    0.515    4.005   1.554
##  11.705 to  13.605 -0.230   -0.036    3.006   0.876
##  13.605 to  15.135  0.840    0.611    3.281   1.846
##  15.135 to  17.385 -1.214   -0.005    2.985   1.531
##  17.385 to  19.645  0.892    0.509    1.291  -1.279
```

Both diagnostic plots, Figure 13.18 for the worm plots and the left panel of Figure 13.19 for the Z statistics, show evidence that the skewness of the response distribution is not modelled properly. For example, the worm plots show quadratic shapes, while the skewness column of the Z statistics, Z3, has five values larger than 2. From the two smoothing parameters, p.lambda is the one that is most likely to affect the shape of the distribution of the response variable, so the next step is to decrease the value of p.lambda while simultaneously checking the Z3 column of the Q statistics. The following combination of smoothing parameters works well.

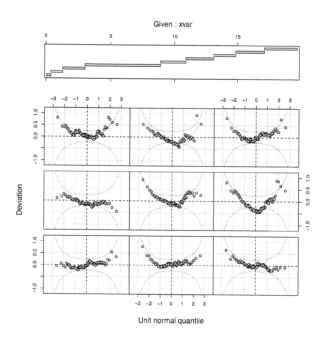

FIGURE 13.18: Worm plots from quantile sheet curves fitted to the sample of Dutch boys BMI data, using smoothing parameters `x.lambda=1` and `p.lambda=10`.

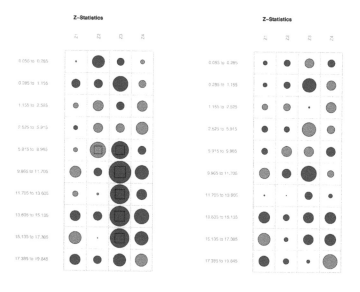

FIGURE 13.19: Q statistics plots from the two quantile sheets models fitted to the sample of Dutch boys BMI data. Left panel: `x.lambda=1` and `p.lambda=10`; right panel: `x.lambda=1` and `p.lambda=0.05`.

R
p. 48

R
p. 48

```
qs2<-quantSheets(bmi, age, data = dbbmi1,
                 cent=c(0.4, 2, 10, 25, 50, 75, 90, 98, 99.6),
                 x.lambda=1, p.lambda=.05, logit=TRUE, power=ppp)

## % of cases below  0.4 centile is  0.4
## % of cases below  2 centile is  2
## % of cases below  10 centile is  10.7
## % of cases below  25 centile is  25.7
## % of cases below  50 centile is  50
## % of cases below  75 centile is  74.4
## % of cases below  90 centile is  89.7
## % of cases below  98 centile is  97.9
## % of cases below  99.6 centile is  99.6
```

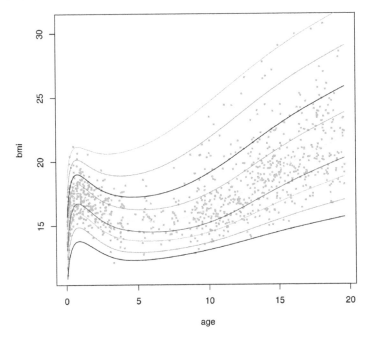

FIGURE 13.20: Quantile sheet curves fitted to the sample of Dutch boys BMI data, using smoothing parameters x.lambda=1 and p.lambda=.05.

It seems that decreasing p.lambda to 0.05 provides a model with good residual diagnostics, as shown in Figure 13.21 and the right panel of Figure 13.19, and no Z statistics greater than 2 in absolute value.

```
res2 <- resid(qs2)
wp(resid=res2, xvar=dbbmi1$age, n.inter = 9 )

## number of missing points from plot= 0  out of  112
## number of missing points from plot= 0  out of  111
## number of missing points from plot= 0  out of  110
```

```
## number of missing points from plot= 0  out of  112
## number of missing points from plot= 0  out of  111
## number of missing points from plot= 0  out of  111
## number of missing points from plot= 0  out of  112
## number of missing points from plot= 0  out of  110
## number of missing points from plot= 0  out of  111
```

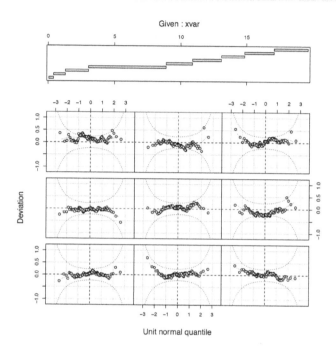

R
p. 48

FIGURE 13.21: Worm plots from the fitted quantile sheet to the sample of Dutch boys BMI data, using smoothing parameters x.lambda=1 and p.lambda=0.05.

```
round(Q.stats(resid=res2, xvar=dbbmi1$age),3)
```

```
##                     Z1      Z2      Z3      Z4
##  0.055 to  0.285  0.143   0.332  -0.669   0.439
##  0.285 to  1.155  0.142   0.349   1.547  -0.726
##  1.155 to  2.525 -0.274  -0.403   0.019  -0.859
##  2.525 to  5.915  0.305   0.247  -1.524  -0.499
##  5.915 to  9.965  0.310  -0.834  -0.750   0.690
##  9.965 to 11.705 -0.900   0.662   1.925  -0.357
## 11.705 to 13.605  0.013  -0.003   0.463   0.124
## 13.605 to 15.135  1.021   0.311   0.876   1.029
## 15.135 to 17.385 -1.230   0.139   0.808   1.058
## 17.385 to 19.645  0.908   0.257   0.090  -1.714
```

13.6 Bibliographic notes

The LMS method of centile estimation for a continuous response variable, based on the Box–Cox Cole and Green (BCCG) distribution, together with smooth spline functions for modelling each parameter of the distribution as a function of a single continuous explanatory variable, was developed by Cole [1988] and Cole and Green [1992]. The method was extended to the LMSP and LMST methods of centile estimation, based on the Box–Cox power exponential (BCPE) and Box–Cox t (BCT) distributions by Rigby and Stasinopoulos [2004, 2006], respectively, also allowing multiple explanatory variables in the model.

The World Health Organisation (WHO) compared over forty methods of centile estimation for a response variable based on a single continuous explanatory variable, for their WHO Multicentre Growth Reference Study (MGRS), and selected the LMSP method. This comparison was described by Borghi et al. [2006]. The LMSP method was used in the WHO investigations, although the final growth charts were produced using the LMS method, with GAMLSS software [Stasinopoulos and Rigby, 2007].

Many other authors have used the LMS and/or LMSP and/or LMST methods of centile estimation (using GAMLSS software) for a response variable based on a single explanatory variable, e.g. Visser et al. [2009], Kempny et al. [2011] and Bonafide et al. [2013]. Villar et al. [2014] compared the three methods with their preferred alternative based on the skew t type 3 distribution together with fractional polynomial functions of the explanatory variable for modelling the parameters of the distribution, fitted using GAMLSS software.

Centiles for a response variable based on two continuous explanatory variables have been obtained by several authors. Stanojevic et al. [2008] used the BCCG distribution and GAMLSS software to obtain centiles for a spirometry (lung function) response variable based on age, height and centre (a factor indicating the source of the data). Their modelling approach was described by Cole et al. [2009] and used by Quanjer et al. [2012a] in a major study to develop the Global Lung Function Initiative 2012 equations. Quanjer et al. [2012b] also provided annotated examples of their use of GAMLSS. Neuhauser et al. [2011] and Yan et al. [2013] obtained centiles for blood pressure based on age and height.

Centiles for a response variable based on multiple explanatory variables were investigated by Beyerlein et al. [2008], Kneib [2013a,b], Rigby and Stasinopoulos [2013] and Koenker [2016]. All these authors compared the GAMLSS and quantile regression models for centile estimation. The quantile regression model for centile estimation has a long history originating with quantile linear models [Koenker and Bassett, 1978] and including quantile smoothing models [Koenker et al., 1994]. **R** packages for quantile regression are **quantreg** [Koenker, 2016] and **cobs** [Np and Maechler, 2007].

The **gamlss** package can be used to obtain centiles for any of the distributions available in the package (continuous, discrete or mixed). The continuous distributions can have range $(-\infty, \infty)$, $(0, \infty)$, or $(0, 1)$. The mixed distributions can have range $[0, \infty)$, $[0, 1)$, $(0, 1]$ or $[0, 1]$. Hossain et al. [2016] obtained centiles for a proportion response variable with range $(0, 1]$.

13.7   Exercises

1. **The Dutch boys head circumference data** The Fourth Dutch Growth Study [Fredriks et al., 2000a,b] also recorded head circumference. In the data file **db** from **gamlss.data**, we have head circumference and age of the Dutch boys. Cases with missing values have been removed. There are 7,040 observations.

> **R data file:** db in package **gamlss.data** of dimensions 7040×2
>
> **variables**
>
> > head : head circumference in *cm*
> >
> > age : age in years
>
> **purpose:** to demonstrate centile estimation.

Familiarize yourself with centile estimation by using the **R** commands given in this chapter, using the Dutch boys head circumference against age as follows.

(a) Input and plot the data.

```
data(db)
names(db)
plot(head~age,data=db)
```

(b) To obtain centile curves for head circumference (**head**) against **age**, use the automated function **lms()**. This function performs the following:

    i. first chooses an appropriate power transformation of **age**, $u = age^\xi$, by default **trans.x=TRUE**;

    ii. fits each of a list of families of distributions to the response variable **head**. Each parameter of a distribution is fitted locally using the P-spline smoothing function **pb()** in the transformed explanatory variable. **pb()** automatically chooses the smoothing parameter;

    iii. chooses the best distribution from the list of families according to criterion $\text{GAIC}(\kappa)$.

```
m0<-lms(head,age,families=c("BCCGo","BCPEo","BCTo"),data=db,
        k=4,calibration=F, trans.x=T)
m0$family
m0$power
```

This function takes a few minutes, so read on while you wait.

Note that the best distribution family, according to GAIC(4), is stored in m0$family, i.e. BCTo. The power transformation chosen for age is stored in m0$power, i.e. $u = \text{age}^{\text{m0\$power}}$.

(c) You can refit the chosen model using the gamlss function:

```
db$Tage<-(db$age)^(m0$power)
m1<-gamlss(head~pb(Tage),sigma.fo=~pb(Tage),nu.fo=~pb(Tage),
           tau.fo=~pb(Tage), family=BCTo,data=db)
```

Alternatively use

```
m2<-gamlss(head~pb(age^m0$power),sigma.fo=~pb(age^m0$power),
           nu.fo=~pb(age^m0$power), tau.fo=~pb(age^m0$power),
           family=BCTo,data=db)
GAIC(m0,m1,m2,k=4)
```

The alternative method is not recommended for large data since the calculation $\text{age}^{\text{m0\$power}}$ is performed at each iteration of the gamlss fitting algorithm.

(d) The centile curves are given by

```
centiles(m0,xvar=db$age)
centiles.fan(m0,xvar=db$age)
```

To split the centile plot at age $= 3$ (in order to see the centiles for age < 3 more clearly):

```
centiles.split(m0,xvar=db$age,xcut.points=c(3))
```

A plot showing the distribution of head circumference (vertically) for specific values of age is given by

```
library(gamlss.util)
plotSimpleGamlss(head,age,m0,data=db,x.val=seq(5,20,5),
                 xlim=c(-3,23))
plotSimpleGamlss(head,age,m0,data=db,x.val=seq(1,22,7),
                 xlim=c(-8,23))
```

(e) Look at the fitted parameters μ (the approximate median), σ (the approximate coefficient of variation), ν (the skewness parameter) and τ (the kurtosis parameter) of the BCTo distribution plotted against age.

```
fittedPlot(m0,x=db$age)
```

(f) Check the residuals of the model to see if the model is adequate. The function plot() gives a QQ plot of the residuals, and wp() gives a worm plot.
Approximately 95% of the residuals should be between the 95% pointwise
interval bands in the worm plot.

```
plot(m0)
wp(m0,ylim.all=1)
```

Looking at the single worm plot, we find seven outliers in the upper tail.
Now split the range of age into 16 intervals and obtain a QQ plot of the
residuals within each age range. This allows the identification of any regions
of age where the model is inadequate.

```
wp(m0,xvar=db$age,ylim.worm=1.5,n.inter=16)
```

Now obtain Q statistics for the 16 regions of age. This also allows the
identification of any regions of age where the model is inadequate.

```
Q.stats(m0,xvar=db$age,n.inter=20)
```

(g) The centiles values can be calculated and plotted for new values of age:

```
nage<-seq(0,20,0.1)
centiles.pred(m0,xname="age",xvalues=nage,plot=T,ylab="head",
              xlab="age",legend=F)
```

Also the z-scores for three new people with (head, age)=(45,5), (50,10)
and (60,15), respectively, is obtained by:

```
newhead<- c(45,50,60)
newage<-c(5,10,15)
centiles.pred(m0,xname="age",xvalues=newage,yval=newhead,
              type="z-scores",plot=T, ylab="head",xlab="age")
```

A individual with a z-score < -2 indicates the person has an unusually low
head circumference for his age, while a z-score > 2 indicates the person has
an unusually high head circumference for his age.

(h) Remove (or weight out) extreme outliers in head circumference (given age)
as follows below.

From the worm plot of the residuals from model m0 there are seven extreme
outliers in the upper tail, with residuals greater than 3.5. There are also
two cases with residuals less than -3.5. They are causing a distortion in
the worm plot, resulting in a distortion in the fitted model as seen by the
Q statistics and hence a distortion in the centile curves.

One solution to the distorted centile percentages (i.e. a difference between the nominal model percentages and the sample percentages below the centile curves) is to use calibration.

```
calibration(m0,xvar=db$age)
```

An alternative, possibly better, approach is to remove these outliers, i.e. seven extreme outliers in the upper tail and two in the lower tail.

```
which(resid(m0)>3.5)
which(resid(m0)< -3.5)
dbsub <- subset(db, (resid(m0)> -3.5)&(resid(m0)< 3.5))
```

Refit the model.

```
m3<-gamlss(head~pb(age^m0$power),sigma.fo=~pb(age^m0$power),
          nu.fo=~pb(age^m0$power),tau.fo=~pb(age^m0$power),
          family=BCTo,data=dbsub)
wp(m3,ylim.all=1)
```

The resulting fit to the data, worm plot and Q statistics are substantially improved. If the nine outliers are believed to be errors in the data set, then centile curves can be obtained directly from m3. However if the outliers are believed to be genuine observations, then centile curves should be obtained for the full data set db. The centile curve percentages from m3 need to be adjusted for the cases removed from each tail. To obtain centile curves for the full data set db at centiles given by cent use the following:

```
cent<- c(0.4,2,10,25,50,75,90,98,99.6)
a<- (2/7040)*100 # lower percentage removed
b<- (7/7040)*100 # upper percentage removed
newcent<-(cent-a)/(1-(a+b)/100)
centiles(m3,xvar=dbsub$age,cent=newcent, legend=FALSE)
```

2. **BMI Dutch boys (full) data**: Find centiles for bmi given age for the full Dutch boys BMI data, dbbmi:

(a) Use lms() to find a transformation of age and a distribution for bmi (from BCCGo, BCPEo and BCTo) and use it to find centiles for bmi given age.

(b) Using the transformation for age and distribution for bmi chosen in (a), fit a model for bmi using pb() smoothing functions for each of μ, σ, ν and τ, each with a local GAIC(4) to estimate the smoothing degrees of freedom. Obtain the resulting centiles for bmi.

(c) Use find.hyper() to select the transformation for age and the smoothing degrees of freedom using cubic spline cs() smoothing functions for each of μ, σ, ν and τ with a global GAIC(4). This can take some time (about 1 or 2 hours!). Obtain the resulting centiles for bmi.

(d) Check the adequacy of the model for `bmi` from (a) by checking the residuals using worm plots and Q statistics.

3. **The hand grip strength of English boys data:** The `grip` data are analysed by Cohen et al. [2010]. The response variable is hand grip (HG) strength, with explanatory variables gender and age, in English schoolchildren. Here we analyse the $n = 3,766$ observations.

> **R data file:** `grip` in package **gamlss.data** of dimensions 3766×2
>
> **variables**
>
> > `age` : age of the participant
> >
> > `grip` : hand grip strength
>
> **purpose:** for centile estimation.

(a) Input the data.

(b) Plot `grip` against `age`. State which statistical model you need to fit here. There is no need to power transform `age` in these data. Explain why.

(c) Use the LMS method to fit the data using the `BCCGo` distribution for `grip`.

(d) How many degrees of freedom were used for smoothing (including 2 for the constant and linear terms) for each distribution parameter in the model? Use the function `edf()` or `edfAll()`.

(e) Use the fitted values from the LMS model in (d) as starting values for fitting the `BCTo` and the `BCPEo` distributions to the data.

(f) Use GAIC to compare the three models.

(g) Plot the fitted parameters for the fitted models in (c) and (e) using the function `fitted.plot()`.

(h) Obtain a centile plot for the fitted models in (c) and (e) using `centiles()` or `centiles.split()`, and compare them.

(i) Investigate the residuals from the fitted models in (c) and (e) using, e.g. `plot()`, `wp()` and `Q.stats()`.

(j) Choose between the models and give a reason for your choice.

4. **The Global Lung Function Initiative data, males:** This analysis finds centiles of a response variable dependent on two quantitative explanatory variables. The data are provided by the Global Lung Function Initiative, and are accessed at

`www.ers-education.org/guidelines/global-lung-function-initiative/`
`statistics.aspx`

The response variable is the forced expired volume (**fev**) and the explanatory variables are **height** and **age**.

(a) i. Input the data into data frame **lung** and select the males into data frame **dm**.

```
dm<-subset(lung, sex==1)
dim(dm)
```

The number of male cases is $n = 5,723$.

ii. Obtain a scatterplot of **fev** against **height** and **age** and both.

```
plot(fev~height,data=dm)
plot(fev~age,data=dm)
height <-dm$height
age <- dm$age
fev <- dm$fev
library(lattice)
cloud(fev~height*age)
# or more detailed
library(rgl)
plot3d(height,age,fev)
library(car)
scatter3d(dm$height,dm$age,dm$fev,xlab="height",ylab="age",
          zlab="fev",ticktype="detailed")
```

iii. Following Cole et al. [2009] and Quanjer et al. [2012a], apply a log transformation to height and age.

```
dm <- transform(da, la= log(age),lh=height)
```

(b) Use **stepGAICAll.A()** to search for a suitable model for **fev** using the BCTo distribution (starting from a model **m1** with constant parameters). Use a *local* SBC to choose the effective degrees of freedom for smoothing in the smoothing functions **pb**. Also use a *global* SBC criterion to select terms in the **stepGAICAll.A** procedure. The reason for using SBC (i.e. $\kappa = \log(5723)$) is to achieve smooth centiles. A lower value of κ (e.g. $\kappa = 4$) would result in less smooth centiles but a better fit to the data, while a higher value of κ would result in even smoother centiles, but a worse fit to the data.

```
m1<-gamlss(fev~1,sigma.fo=~1,nu.fo=~1,tau.fo=~1, family=BCTo,
           data=dm,n.cyc=100)
k1<-log(5723)
m2<-stepGAICAll.A(m1,scope=list(lower=~1,upper=~pb(lh,
    method="GAIC",k=k1) + pb(la,method="GAIC",k=k1)), k=k1)
```

This will take about five minutes to complete. See the chosen model by

```
summary(m2)
```

(c) i. Refit the chosen model, but replacing lh and la by log(height) and
log(age) in order to use predictAll() in (f) below.

```
m3<-gamlss(fev~pb((log(height)),method="GAIC",k=k1)+
              pb((log(age)),method="GAIC",k=k1),
         sigma.fo=~pb((log(height)),method="GAIC",k=k1)+
              pb((log(age)),method="GAIC",k=k1),
         nu.fo=~1,tau.fo=~1, family=BCTo,data=dm, n.cyc=100)
```

ii. Amend model m3 to fit distribution BCCGo and then BCPEo and show
that m3 has the lowest SBC.

iii. Check the adequacy of model m3 using residual diagnostics.

```
plot(m3)
wp(m3,ylim.all=0.6)
wp(m3, xvar=~age, n.inter=9, ylim.worm=0.8)
wp(m3, xvar=~height, n.inter=9, ylim.worm=0.8)
wp(m3, xvar=~age+height, n.inter=4, ylim.worm=1)
Q.stats(m3,xvar=dm$height,n.inter=25)
```

iv. Output the effective degrees of freedom (including 2 for the constant
and linear terms) used for each smoothing function in model m3.

```
edfAll(m3)
```

v. Look at the fitted smooth functions in model m3.

```
term.plot(m3,what="mu", pages=1)
term.plot(m3,what="sigma", pages=1)
```

(d) An alternative method of choosing the effective degrees of freedom for the
smoothing functions is by minimizing a global SBC, instead of a local SBC
in (c), using the find.hyper() function. This should use cubic splines
instead of penalized splines. This takes about 60 minutes.

```
mod<-quote(gamlss(fev~cs((log(height)),df=p[1])+
   cs((log(age)),df=p[2]),sigma.fo=~cs((log(height)),
   df=p[3])+cs((log(age)),df=p[4]),nu.fo=~1,tau.fo=~1,
   family=BCTo,data=dm, control=gamlss.control(trace=FALSE,
                                      n.cyc=100)))

best<-find.hyper(model=mod,par=c(6,6,3,3),
        lower=c(0.01,0.01,0.01,0.01),
        steps=c(0.1,0.1,0.1,0.1), k=k1)
best
```

The resulting effective degrees of freedom are very similar to model m3.

(e) i. Now fit a model for **height** against **age**. The purpose of this is to find lower and upper centile limits (0.1% and 99.9%) of **height** for each **age** (to be used for the contour plot of the 5% centile of **fev** against **height** and **age**. in (f) below.

```
mh<-gamlss(height~pb(log(age),method="GAIC",k=k1),
           sigma.fo=~pb(log(age),method="GAIC",k=k1),
           nu.fo=~pb(log(age),method="GAIC",k=k1),
           tau.fo=~pb(log(age),method="GAIC",k=k1),
           family=BCTo, data=dm)
```

Plot the centiles for **height** against **age** for model **mh**.

```
centiles(mh,xvar=dm$age,cent=c(0.1,0.4,2,10,25,50,75,90,
         98,99.6,99.9),ylab="height",xlab="age",legend=FALSE)
```

ii. Now find lower (0.1%) and upper (99.9%) limits for **height** given **age**, stored in **maty[,2]** and **maty[,4]**.

```
newage<- seq(5,90,0.1)
newcent<- c(0.1,50,99.9)
maty<-centiles.pred(mh,xname="age",xvalues=newage,
                    cent=newcent,plot=TRUE)
maty[1:10,]
```

(f) Construct a contour plot of the 5th centile of **fev** against **height** and **age**:

i. Expand a grid of values of **age** from 5 to 90 years and **height** from 100 to 210 cm to cover the limits of **height** in (e)(ii) above.

```
newdata<-expand.grid(age=seq(5,90,0.1),
                     height=seq(100,210,1))
```

ii. Use the chosen model **m3** for **fev** to predict all the parameters μ, σ, ν and τ of the distribution BCTo for the values of **age** and **height** in **newdata**.

```
m3p<-predictAll(m3, newdata=newdata, type="response")
```

iii. Calculate the 5th centile of **fev** for all cases in **newdata**.

```
fev5<-qBCPE(0.05,m3p$mu,m3p$sigma,m3p$nu,m3p$tau)
```

iv. For all cases of **newdata** with values of **height** outside the lower (0.1%) and upper (99.9%) bounds for **height**, replace the value of **fev5** with a missing value (NaN).

```
lower<-rep(maty[,2],111)
upper<-rep(maty[,4],111)
```

```
fev5a<-ifelse(((newdata$height<lower)|
            (newdata$height>upper)),NaN,fev5)
```

v. Obtain a contour plot of the 5th centile of `fev` against `height` and `age`.

```
newheight<-seq(100,210,1)
newage<-seq(5,90,0.1)
mfev5<-matrix(data=fev5a,nrow=851,ncol=111)
contour(newage,newheight,mfev5,nlevels=40,
        xlab="age(years)",ylab="height(cm)")
```

5. The Global Lung Function Initiative data, females:

(a) Input the Global Lung Function Initiative data from question 4 into data frame `lung` and select the females into data frame `df`.

```
df<-subset(lung,sex==2)
dim(df)
```

The number of female cases is $n = 7,106$.

(b) Follow the analysis in the previous question to obtain a contour plot for the 5th centile of `fev` against `height` and `age`.

6. The Global Lung Function Initiative data, males modelled with centre:

Use the data frame `dm` from Question 4. Model `fev` using `height`, `age` and the `centre` factor which has four levels. Obtain a contour plot for the 5th centile of fev against `height` and `age` for `centre` = 2, which has the most cases.

14

Further applications

CONTENTS

14.1 Introduction

This chapter provides further applications of GAMLSS modelling. In particular:

- the fish species data as an example of fitting different discrete count distributions to data;

- the hospital stay data as an example of fitting binomial type distributions; and

- the film data as an example of fitting smooth two-dimensional surfaces to a continuous response variable.

This chapter presents some interesting applications of GAMLSS models which could be useful to the reader. The authors have used these data sets over the years in the development of GAMLSS, showing the flexibility of the models and the **R** software.

The first example, the fish species data, is a very simple regression example where the response variable is an overdispersed count. It is used to demonstrate the different overdispersed distributions which can be fitted within GAMLSS.

The second example is the hospital stay data, which has an overdispersed binomial type response variable. Historically it was very important for the GAMLSS development, since it was the first data set in which the originators of GAMLSS realised that the extended quasi likelihood, used at the time for overdispersed response data, was producing different results to a full likelihood approach.

The third example, the film data, illustrates the fitting of a smooth bivariate surface
(i.e. an interaction between two explanatory variables) to a continuous response
variable.

14.2   Count data: The fish species data

The number of different fish species (`fish`) was recorded for 70 lakes of the world
together with explanatory variable x = log lake area. The data were first introduced
in Section 3.2.3 and are plotted in Figure 14.1.

```
library(gamlss)
data(species)
# creating the log(lake)
species <- transform(species, x=log(lake))
plot(fish~x,data=species)
```

Fig.
14.1

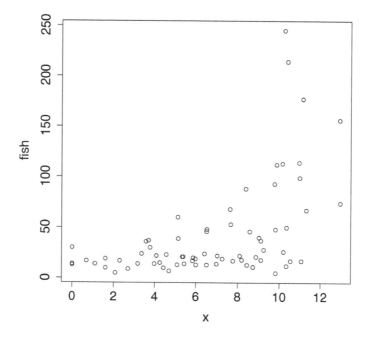

FIGURE 14.1: The fish species data.

R

p. 49

The data are given and analysed by Stein and Juritz [1988] using a Poisson inverse
Gaussian ($PIG(\mu, \sigma)$) distribution for `fish`, with a linear model in `log(lake)` for
$\log \mu$, and a constant for σ. Rigby et al. [2008] analysed this data set and identified
the following questions that need to be answered. Note that the same questions

could apply to any regression situation where the response variable is a count and x represents a set of explanatory variables.

- How does the mean of the response variable depend on x?

- Is the response variable overdispersed Poisson?

- How does the variance of the response variable depend on its mean?

- What is the conditional distribution of the response variable given x?

- Do the scale and shape parameters of the response variable distribution depend on x?

Here we will model the data using different discrete distributions and consider flexible models for the distribution parameters, where any or all of them may depend on the explanatory variable log(lake). We start by fitting seven different count distributions to the data:

- Poisson (PO),

- double Poisson (DPO),

- negative binomial types I and II (NBI, NBII),

- Poisson inverse Gaussian (PIG),

- Delaporte (DEL) and

- Sichel (SICHEL).

We first use a linear and then a quadratic polynomial in x=log(lake). The AIC of each model is printed for comparison.

```
# the count distributions
fam<-c("PO","DPO", "NBI", "NBII", "PIG", "DEL", "SICHEL")
#creating lists to keep the results
m.l<-m.q<-list()
# fitting the linear in x models
for (i in 1:7) {
m.l[[fam[i]]]<-GAIC(gamlss(fish~x,data=species, family=fam[i],
               n.cyc=60, trace=FALSE),k=2)}
# fitting the quadratic in x models
for (i in 1:7) {
m.q[[fam[i]]]<-GAIC(gamlss(fish~poly(x,2),data=species,
               family=fam[i], n.cyc=60, trace=FALSE), k=2)}
# print the AICs
unlist(m.l)
```

```
##         PO        DPO        NBI       NBII        PIG        DEL
## 1900.1562   654.1616   625.8443   647.5359   623.4632   626.2330
##     SICHEL
```

```
##   625.3923
```

```
unlist(m.q)
```

```
##        PO        DPO       NBI       NBII       PIG        DEL
## 1855.2965  655.2520  622.3173  645.0129  621.3459  623.5816
##     SICHEL
##   623.0995
```

The Poisson model has a very large AIC compared to the rest of the distributions so we can conclude that the data are overdispersed. The quadratic polynomial in x seems to fit better than the linear term across the different count distributions (except for DPO), as judged by AIC. The best model at this stage is the Poisson inverse Gaussian (PIG) model with a quadratic polynomial in x. We now compare the AIC of a PIG model with a P-spline smoother, instead of a quadratic polynomial, in x. The total effective degrees of freedom for x is calculated automatically using pb(x) by the local ML method described in Section 3.4.

```
GAIC(m.pb<-gamlss(fish~pb(x), data=species, family=PIG, trace=FALSE))
```

```
## [1] 623.4637
```

```
m.pb$mu.df
```

```
## [1] 2.000016
```

The P-spline smoothing does not seem to improve the model, so we keep the quadratic polynomial in x. We now model $\log(\sigma)$ as a linear function of x in the six remaining count distributions (after excluding the Poisson distribution which does not have a σ parameter).

```
# redefine the list of distributions
fam<-c("DPO","NBI", "NBII", "PIG", "DEL", "SICHEL")
m.ql<-list()
for (i in 1:6) {
m.ql[[fam[i]]]<-GAIC(gamlss(fish~poly(x,2),data=species,
        sigma.fo=~x, family=fam[i], n.cyc=60, trace=FALSE))}
unlist(m.ql)
```

```
##       DPO       NBI      NBII       PIG       DEL    SICHEL
## 626.4056  614.9565  615.1250  612.3667  614.6059  613.7327
```

Modelling $\log(\sigma)$ as a linear function of x improves the AIC for all models. The PIG model is still the 'best'. Since the Sichel and the Delaporte distributions have three parameters we will try to model the predictor of the third parameter ν as a linear function of x. The Sichel uses the identity as the default link for ν while the Delaporte uses the logit.

```
fam<-c("DEL", "SICHEL")
m.qll<-list()
```

```
for (i in 1:2) {
m.q11[[fam[i]]]<-GAIC(gamlss(fish~poly(x,2),data=species,
                  sigma.fo=~x, nu.fo=~x, family=fam[i], n.cyc=60,
                  trace=FALSE))}
unlist(m.q11)

##        DEL    SICHEL
## 614.7376 611.6346
```

Modelling the predictor of ν as a linear function of x improves the Sichel model (which now has a lower AIC than the PIG model) but not the Delaporte model. A further simplification of the Sichel model can be achieved by dropping the linear term in x for the $\log(\sigma)$ model which does not contribute anything to the fit (at least according to the AIC):

```
mSI<-gamlss(fish~poly(x,2),data=species, sigma.fo=~1, nu.fo=~x,
        family=SICHEL, n.cyc=60, trace=FALSE)
GAIC(mSI)

## [1] 609.7268
```

Fig.
14.2
```
plot(fish~log(lake), data=species)
lines(species$x[order(species$lake)],    fitted(mSI)[order(
        species$lake)], col="red")
```

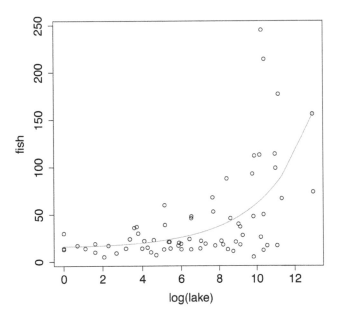

FIGURE 14.2: Fitted mean number of fish species against log lake area.

The fitted μ model together with the data are shown in Figure 14.2. Figures 14.3(a) and 14.3(b) give the fitted distribution of the number of fish species for observation 7, with lake area of $44\,\text{km}^2$, i.e. $x = \log(44) = 3.74$, and $(\hat{\mu}, \hat{\sigma}, \hat{\nu}) = (19.37, 1.44, -7.18)$, and observation 68, with lake area $9{,}065\,\text{km}^2$, i.e. $x = \log(9065) = 9.11$ and $(\hat{\mu}, \hat{\sigma}, \hat{\nu}) = (48.86, 1.44, -1.10)$, respectively. Note that the vertical scale is different for the two plots in Figure 14.3.

```
pdf.plot(mSI,c(7,68), min=0, max=120, step=1)
```

Fig.
14.3

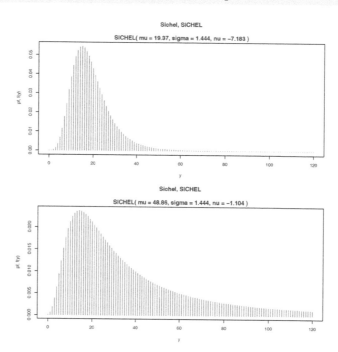

R
p. 5(

FIGURE 14.3: Fitted Sichel distributions for observations (a) 7 and (b) 68.

Table 14.1 (effectively Table 2 from Rigby et al. [2008]), gives GDEV, AIC and SBC for specific models fitted to the fish species data, and is used to answer the questions at the start of this section. The terms 1, x and x<2> indicate constant, linear and quadratic terms, respectively, while the term cs(x,3) indicates a cubic smoothing spline with three degrees of freedom on top of the linear term x. Table 14.1 includes additional distributions to those previously fitted.

The following four paragraphs are taken from Rigby et al. [2008]. "Comparing models 2, 3 and 4 indicates that a quadratic model for $\log \mu$ is found to be adequate (while the linear and the cubic spline models were found to be inappropriate here). Comparing model 1 and 3 indicates that Y has a highly overdispersed Poisson distribution. Comparing model 3 with models 5 and 6 shows that either a linear model in x for $\log(\sigma)$ or a different variance-mean relationship from that of the negative binomial (NBI) [i.e. $V[Y] = \mu + \sigma\mu^2$] is required. In particular the estimated ν parameter in the negative binomial family (NBF) of model 6 is $\hat{\nu} = 2.9$ suggesting a possible variance-mean relationship $V[Y] = \mu + \sigma\mu^3$. Modelling σ in the NBF did not improve the fit greatly, as shown by model 7."

"A search of alternative mixed Poisson distributions included the Poisson-inverse Gaussian (`PIG`), the Sichel (`SI`) and the Delaporte (`DEL`). The models with the best AIC for each distribution were recorded" in Table 14.1 models 8 to 11. "A normal random effect mixture distribution was fitted" (using 20 Gaussian quadrature points) "to the Poisson and NBI conditional distributions giving models 12 and 13, i.e. Poisson-Normal and NBI-Normal, respectively. 'Non-parametric' random effects (effectively finite mixtures) (NPFM) were also fitted to Poisson and `NBI` conditional distributions giving models 14 and 15", i.e. `PO-NPFM(6)` and `NB-NPFM(2)` with 6 and 2 components, respectively. "Efron's double exponential (Poisson) distribution was fitted giving model 16" (DPO). "The best discretized continuous distribution fitted was a discrete inverse Gaussian distribution giving model 17 (`IGdisc`), again suggesting a possible cubic variance-mean relationship." Note that the Table 14.1 model 14 gives results for model `PO-NPFM(6)` instead of `PO-NPFM(5)` in Table 2 of Rigby et al. [2008].

"Overall the best model according to Akaike information criterion (AIC) is model 9, the Sichel model, followed closely by model 11, a Delaporte model. According to the Schwarz Bayesian criterion (SBC) the best model is model 17, the discretized inverse Gaussian distribution, again followed closely by model 11." In model 11, σ was fixed to 1.

The following code reproduces the results of Table 14.1.

```
library(gamlss.mx)
m1 <- gamlss(fish~poly(x,2), data=species, family=PO, trace=FALSE)
m2 <- gamlss(fish~x, data=species, family=NBI, trace=FALSE)
m3 <- gamlss(fish~poly(x,2), data=species, family=NBI, trace=FALSE)
m4 <- gamlss(fish~cs(x,3), data=species, family=NBI, trace=FALSE)
m5 <- gamlss(fish~poly(x,2), sigma.fo=~x, data=species, family=NBI,
             trace=FALSE)
m6 <- gamlss(fish~poly(x,2), sigma.fo=~1, data=species, family=NBF,
             n.cyc=200, trace=FALSE)
m7 <- gamlss(fish~poly(x,2), sigma.fo=~x, data=species, family=NBF,
             n.cyc=100,  trace=FALSE)
m8 <- gamlss(fish~poly(x,2),  data=species, family=PIG, trace=FALSE)
m9 <- gamlss(fish~poly(x,2), nu.fo=~x, data=species, family=SICHEL,
             trace=FALSE)
m10 <- gamlss(fish~poly(x,2), nu.fo=~x, data=species, family=DEL,
              n.cyc=50, trace=FALSE)
m11 <- gamlss(fish~poly(x,2), nu.fo=~x, data=species, family=DEL,
              sigma.fix=TRUE, sigma.start=1, n.cyc=50, trace=FALSE)
m12 <- gamlssNP(fish~poly(x,2), data=species, mixture = "gq", K=20,
                family=PO, control=NP.control(trace=FALSE))
m13 <- gamlssNP(fish~poly(x,2), sigma.fo=~x, data=species,
                mixture = "gq", K=20, family=NBI,
                control=NP.control(trace=FALSE))
```

```
m14 <- gamlssNP(fish~poly(x,2), data=species, mixture = "np", K=6,
        tol=0.1,family=PO, control=NP.control(trace=FALSE))
m15 <- gamlssNP(fish~poly(x,2), data=species, mixture = "np", K=2,
        family=NBI, control=NP.control(trace=FALSE))
m16 <- gamlss(fish~poly(x,2), nu.fo=~x, data=species, family=DPO,
        trace=FALSE)
library(gamlss.cens)
m17 <- gamlss(Surv(fish,fish+1,type= "interval2")~x+I(x^2),
        sigma.fo=~1, data=species,
        family=cens(IG, type="interval"), trace=FALSE)
GAIC(m1, m2, m3, m4, m5, m6, m7, m8, m9, m10, m11, m12, m13, m14,
        m15, m16, m17)

##            df        AIC
## m9    6.00000   609.7268
## m11   5.00000   610.6493
## m17   4.00000   611.2793
## m10   6.00000   612.6593
## m5    5.00000   614.9565
## m13   6.00000   615.7281
## m6    5.00000   616.0828
## m7    6.00000   616.9229
## m8    4.00000   621.3459
## m3    4.00000   622.3173
## m14  13.00000   622.8926
## m12   4.00000   623.2455
## m15   6.00000   623.8794
## m4    5.99924   623.9083
## m2    3.00000   625.8443
## m16   4.00000   655.2520
## m1    3.00000  1855.2965

GAIC(m1, m2, m3, m4, m5, m6, m7, m8, m9, m10, m11, m12, m13, m14,
        m15, m16, m17, k=log(70))

##            df        AIC
## m17   4.00000   620.2733
## m11   5.00000   621.8918
## m9    6.00000   623.2178
## m10   6.00000   626.1503
## m5    5.00000   626.1990
## m6    5.00000   627.3253
## m13   6.00000   629.2191
## m8    4.00000   630.3399
## m7    6.00000   630.4138
## m3    4.00000   631.3113
```

TABLE 14.1: Comparison of models for the fish species data

Model	Response distribution	μ	σ	ν	GDEV	df	AIC	SBC
1	PO	x<2>	-	-	1849.3	3	1855.3	1862.0
2	NBI	x	1	-	619.8	3	625.8	632.6
3	NBI	x<2>	1	-	614.3	4	622.3	631.3
4	NBI	cs(x,3)	1	-	611.9	6	623.9	637.4
5	NBI	x<2>	x	-	605.0	5	615.0	626.2
6	NB family	x<2>	1	1	606.1	5	616.1	627.3
7	NB family	x<2>	x	1	604.9	6	616.9	630.4
8	PIG	x<2>	1	-	613.3	4	621.3	630.3
9	SICHEL	x<2>	1	x	597.7	6	609.7	623.2
10	DEL	x<2>	1	x	600.7	6	612.7	626.2
11	DEL	x<2>	-	x	600.6	5	610.6	621.9
12	PO-Normal	x<2>	1	-	615.2	4	623.2	632.2
13	NBI-Normal	x<2>	x	1	603.7	6	615.7	629.2
14	PO-NPFM(6)	x<2>	-	—	596.9	13	622.9	652.1
15	NB-NPFM(2)	x<2>	1	—	611.9	6	623.9	637.4
16	DPO	x<2>	x	-	647.3	5	655.3	664.2
17	IGdisc	x<2>	1	-	603.3	4	611.3	620.3

```
## m12  4.00000  632.2395
## m2   3.00000  632.5898
## m15  6.00000  637.3704
## m4   5.99924  637.3975
## m14 13.00000  652.1230
## m16  4.00000  664.2460
## m1   3.00000 1862.0420
```

Fig. 4.4

```
wp(m9) ; title("(a)")
wp(m11); title("(b)")
```

The 'best' fitted models are `m9` and `m17`, as suggested by AIC and SBC, respectively, Their worm plots are shown in Figure 14.4, indicating that both models have adequate fits. The fitted parameters of the Sichel model `m9` are shown below. They are obtained by refitting the model using an ordinary quadratic polynomial in x for $\log(\mu)$, rather than the orthogonal quadratic polynomial produced by `poly(x,2)`:

```
mSI<- gamlss(fish~x+I(x^2), sigma.fo=~1, nu.fo=~x, data=species,
            family=SICHEL, trace=FALSE)
summary(mSI)

## ************************************************************
## Family:  c("SICHEL", "Sichel")
```

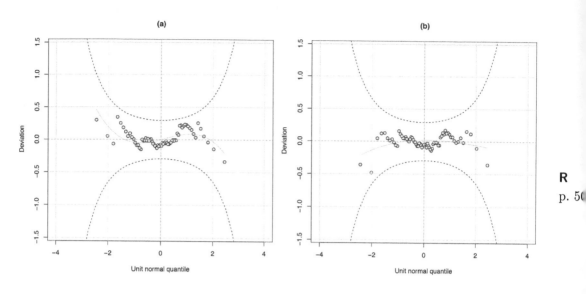

R

p. 5(

FIGURE 14.4: Worm plots for the chosen model (a) m9 using AIC and (b) m17 using SBC.

```
##
## Call:
## gamlss(formula = fish ~ x + I(x^2), sigma.formula = ~1,
##     nu.formula = ~x, family = SICHEL, data = species,
##     trace = FALSE)
##
## Fitting method: RS()
##
## --------------------------------------------------------------
## Mu link function:  log
## Mu Coefficients:
##              Estimate Std. Error t value Pr(>|t|)
## (Intercept)  2.788203   0.171613  16.247  <2e-16 ***
## x           -0.006376   0.066870  -0.095  0.9243
## I(x^2)       0.013957   0.005503   2.536  0.0137 *
## ---
## Signif. codes:
## 0 '***' 0.001 '**' 0.01 '*' 0.05 '.' 0.1 ' ' 1
##
## --------------------------------------------------------------
## Sigma link function:  log
## Sigma Coefficients:
##              Estimate Std. Error t value Pr(>|t|)
```

```
## (Intercept)     0.3674      0.4632    0.793     0.431
##
## -------------------------------------------------------------------
## Nu link function:  identity
## Nu Coefficients:
##                Estimate Std. Error t value Pr(>|t|)
## (Intercept) -11.5009       3.1110  -3.697 0.000455 ***
## x              1.1410       0.3249   3.512 0.000822 ***
## ---
## Signif. codes:
## 0 '***' 0.001 '**' 0.01 '*' 0.05 '.' 0.1 ' ' 1
##
## -------------------------------------------------------------------
## No. of observations in the fit:  70
## Degrees of Freedom for the fit:  6
##        Residual Deg. of Freedom:  64
##                        at cycle:  7
##
## Global Deviance:     597.7268
##            AIC:     609.7268
##            SBC:     623.2178
## *********************************************************************
```

The above output gives the fitted Sichel model m9 (with lowest AIC), i.e.

$$\texttt{fish} \sim \text{SICHEL}(\hat{\mu}, \hat{\sigma}, \hat{\nu})$$
$$\log \hat{\mu} = 2.788 - 0.00638\text{x} + 0.0140\text{x}^2$$
$$\log \hat{\sigma} = 0.367$$
$$\hat{\nu} = -11.501 + 1.141\text{x} .$$

For comparison model 11 gives the Delaporte model. Note that $\sigma = 1$ is fixed in the Delaporte distribution, corresponding to a Poisson-shifted exponential distribution, which when refitted using an ordinary quadratic polynomial in x for the $\log(\mu)$ model gives the fitted model

$$\texttt{fish} \sim \text{DEL}(\hat{\mu}, \sigma = 1, \hat{\nu})$$
$$\log \hat{\mu} = 2.787 - 0.004208\text{x} + 0.013959\text{x}^2$$
$$\text{logit } \hat{\nu} = 1.066 - 0.2853\text{x} .$$

14.3   Binomial data: The hospital stay data

R data file: aep in package **gamlss.data** of dimensions 1383×8

source: Gange et al. [1996]

var los : total number of days

   noinap : number of inappropriate days patient stay in hospital

   loglos : the log of **los/10**

   sex : the gender of patient

   ward : type of ward in the hospital (medical, surgical or other)

   year : 1988 or 1990

   age : age of the patient subtracted from 55

   y : the response variable, a matrix with columns **noinap**, **los-noinap**

purpose: to demonstrate the fitting of a beta binomial distribution

The data, 1,383 observations, are from a study at the Hospital del Mar, Barcelona, during the years 1988 and 1990; see Gange et al. [1996]. The response variable is the number of inappropriate days (**noinap**) out of the total number of days (**los**) patients spent in hospital. Each patient was assessed for inappropriate stay on each day by two physicians who used the appropriateness evaluation protocol (AEP); see Gange et al. [1996] and their references for more details. The following variables were used as explanatory variables: **age**, **sex**, **ward**, **year** and **loglos**.

A plot of the inappropriateness rates **ninap/los** against **age**, **sex**, **ward** and **year** are shown in Figure 14.5 obtained by:

```
data(aep)
prop<-with(aep, noinap/los)
par(mfrow = c(2, 2))
plot(prop~age, data=aep, cex=los/30)
plot(prop~sex,data=aep)
plot(prop~ward,data=aep)
plot(prop~year,data=aep)
```

Fig. 14.5

The data were analyzed by Gange et al. [1996] and later by Rigby and Stasinopoulos [2005]. Gange et al. [1996] used a logistic regression model for the number of inappropriate days, with binomial and beta binomial response distributions, and found that the latter provided a better fit to the data. They modelled both the mean μ and the dispersion σ of the beta binomial distribution (BB) as functions of explanatory variables using a logit link for μ and an identity link for σ. Their final

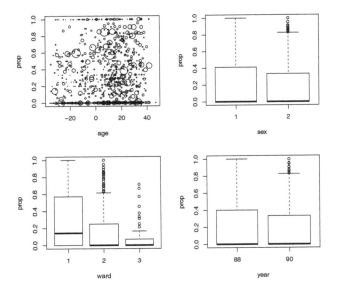

FIGURE 14.5: The rate of inappropriateness against `age`, `sex`, `ward` and `year`.

model was beta binomial $BB(\mu, \sigma)$, with terms `ward`, `year` and `loglos` in the model for $\text{logit}(\mu)$ and `year` for the σ model. Here to show the flexibility of GAMLSS we follow the analysis of Rigby and Stasinopoulos [2005] and fit the following four models:

```
m1 <- gamlss(y~ward+year+loglos, sigma.fo=~year, family=BB,
             data=aep, trace=FALSE)
m2 <- gamlss(y~ward+year+loglos, sigma.fo=~year+ward, family=BB,
             data=aep, trace=FALSE)
m3 <- gamlss(y~ward+year+cs(loglos,df=1), sigma.fo=~year+ward,
             family=BB, data=aep, trace=FALSE)
m4 <- gamlss(y~ward+year+cs(loglos, df=1)+cs(age, df=1),
             sigma.fo=~year+ward, family=BB, data=aep, trace=FALSE)
GAIC(m1,m2,m3,m4, k=0) # the global deviance

##          df      AIC
## m4 12.00010 4454.362
## m3 10.00045 4459.427
## m2  9.00000 4483.020
## m1  7.00000 4519.441

GAIC(m1,m2,m3,m4) # AIC

##          df      AIC
## m4 12.00010 4478.362
## m3 10.00045 4479.427
## m2  9.00000 4501.020
## m1  7.00000 4533.441
```

TABLE 14.2: Models for the AEP data

Model	Link	Terms	GDEV	AIC	SBC
m1	$\text{logit}(\mu)$	ward+loglos+year	4519.4	4533.4	4570.1
	$\log(\sigma)$	year			
m2	$\text{logit}(\mu)$	ward+loglos+year	4483.0	4501.0	4548.1
	$\log(\sigma)$	year+ward			
m3	$\text{logit}(\mu)$	ward+cs(loglos,1)+year	4459.4	4479.4	4531.7
	$\log(\sigma)$	year+ward			
m4	$\text{logit}(\mu)$	ward+cs(loglos,1)+year+	4454.4	4478.4	4541.1
		cs(age,1)			
	$\log(\sigma)$	year+ward			

```
GAIC(m1,m2,m3,m4, k=log(length(aep$age)))

##           df       AIC
## m3 10.00045 4531.750
## m4 12.00010 4541.147
## m2  9.00000 4548.108
## m1  7.00000 4570.065
```

Table 14.2 (which is effectively the same as Table 2 of Rigby and Stasinopoulos [2005]) shows the results for the four models fitted. Model m1 is the final model of Gange et al. [1996]. Model m2 adds ward to the model for σ. Nonlinearities in the μ model for loglos and age are investigated in models m3 and m4. A cubic smoothing spline (cs()) with one effective degree of freedom for smoothing on top of the linear term is used for loglos and age. There is strong support for the inclusion of a smoothing term for loglos as indicated by the reduction in the AIC and SBC for m3 compared to m2. The inclusion of a smoothing term for age is not so clear since there is some marginal support from the AIC, but rejected strongly by the SBC.

Note also that m4 can also be improved marginally by changing the logistic link for the mean to a probit link, giving GDEV = 4452.4, AIC = 4476.4 and SBC = 4539.1 as shown below:

```
(m41 <- gamlss(y~ward+year+cs(loglos,1)+cs(age,1),
          sigma.fo=~year+ward, family=BB(mu.link="probit"),
          data=aep, trace=FALSE))

##
## Family:  c("BB", "Beta Binomial")
## Fitting method: RS()
##
## Call:  gamlss(formula = y ~ ward + year + cs(loglos, 1) +
##     cs(age, 1), sigma.formula = ~year + ward, family =
##     BB(mu.link = "probit"), data = aep, trace = FALSE)
```

```
##
## Mu Coefficients:
##    (Intercept)           ward2              ward3            year90
##      -0.667316        -0.244238          -0.473429          0.151170
## cs(loglos, 1)        cs(age, 1)
##      0.240327          0.002647
## Sigma Coefficients:
## (Intercept)           year90             ward2             ward3
##      0.2953           -0.3729           -0.7172           -1.1713
##
##   Degrees of Freedom for the fit: 12 Residual Deg. of Freedom    1371
## Global Deviance:       4452.36
##            AIC:        4476.36
##            SBC:        4539.14
```

The fitted functions for all the terms for μ in model **m4** are shown in Figure 14.6, and the fitted terms for σ in Figure 14.7. They have been obtained using **term.plot()**:

Fig.
14.6
14.7

```
term.plot(m4,  pages=1)
term.plot(m4, "sigma", pages=1)
```

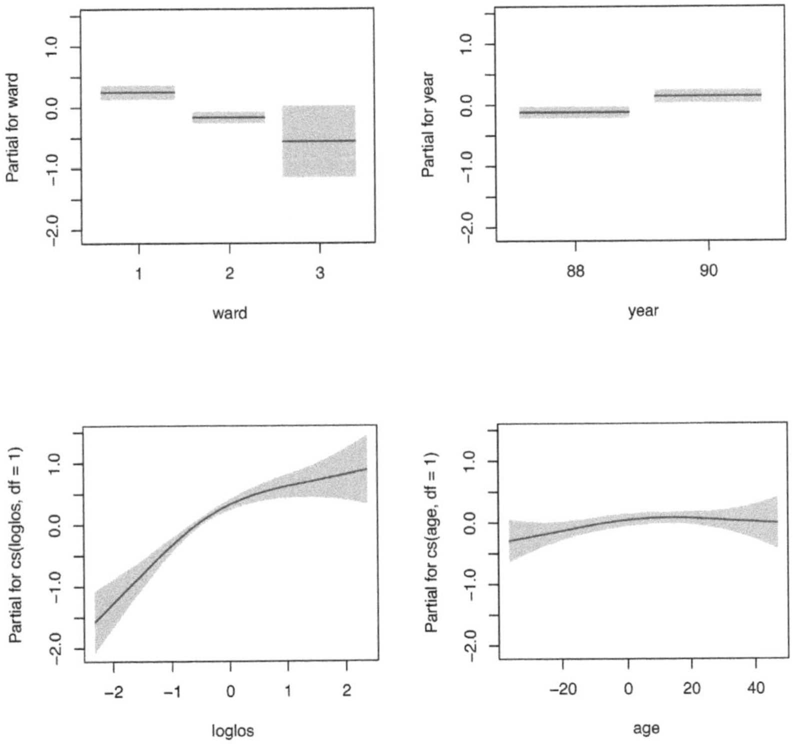

R
511

FIGURE 14.6: The fitted terms for μ in model **m4**.

Fig.
14.8

```
rqres.plot(m4)
```

Figure 14.8 displays six instances of the normalized randomized quantile residuals

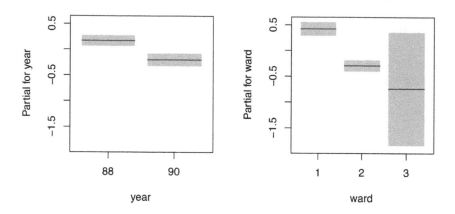

R

p. 51

FIGURE 14.7: The fitted terms for σ in model m4.

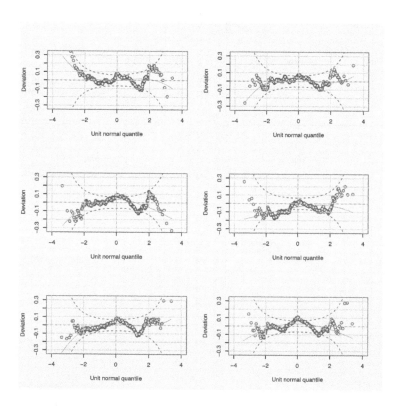

R

p. 51

FIGURE 14.8: Six instances of the normalized randomized quantile residuals for model m4.

(see Section 12.2) from model **m4**. The residuals seem to be reasonably (though not entirely) satisfactory.

14.4 Continuous data: Revisiting the 1990s film data

The film revenue data from the 1990s were analysed in Chapter 2. The data are an anonymized and randomized version of the data used by Voudouris et al. [2012] and are used here for demonstrating some of the features of GAMLSS, and in particular for exploring smooth interactions of explanatory variables. Information about the data can be found in Section 2.3.

14.4.1 Preliminary analysis

Here we demonstrate how the data can be plotted in two- and three-dimensional plots. In Figure 14.9 we plot the response variable against (a) the log of the number of screens and (b) the log of box office opening revenues. The major and independent distributors are represented with different symbols.

```
data(film90)
names(film90)

## [1] "lnosc"   "lboopen" "lborev1" "dist"
```

Fig. 4.9

```
with(film90, plot(lnosc,lborev1,pch=c(21,24)[unclass(dist)],
   bg=c("red","lightgray")[unclass(dist)],
   xlab="log no of screens", ylab="log extra revenue", main="(a)"))
legend("bottomright",legend=c("Independent","Major"),pch=c(21,24),
   pt.bg=c("red","lightgray"),cex=1.5)
with(film90, plot(lboopen,lborev1,pch=c(21,24)[unclass(dist)],
   bg=c("red","lightgray")[unclass(dist)],
   xlab="log opening revenue", ylab="log extra revenue", main="(b)"))
legend("bottomright",legend=c("Independent","Major"),pch=c(21,24),
   pt.bg=c("red","lightgray"),cex=1.5)
```

A good way of inspecting the data in three dimensions is with the package **rgl**. The following commands show how this can be done. The user may increase the size (by clicking and expanding the border), and rotate the figure:

```
library(rgl)
with(film90, plot3d(lboopen, lnosc, lborev1,
      col=c("red","green3")[unclass(dist)]))
```

To show a linear least squares fit to the data, the **rpanel** package may be used:

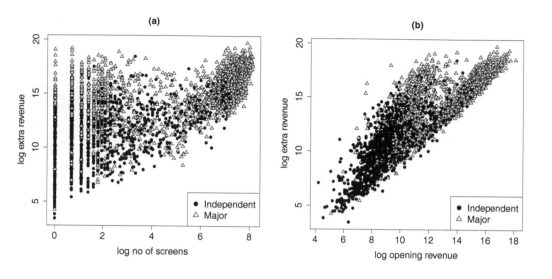

FIGURE 14.9: Showing (a) `lborev1` against `lnosc` (b) `lborev1` against `lboopen`.

R

p. 5

```
library(rpanel)
with(film90, rp.regression(cbind(lboopen, lnosc), lborev1))
```

14.4.2 Modelling the data using the normal distribution

To start the analysis we assume a normal distribution for the response variable and
check whether the mean model needs:

- a simple linear interaction model for the two explanatory variables `lboopen` and
  `lnosc`,

- an additive smoothing model for each of `lboopen` and `lnosc` or

- a fitted smooth surface model (using a tensor product spline) for `lboopen` and
  `lnosc`.

We also check whether we should include or exclude the factor `dist` in the mean
model. Note that in order to fit a smooth surface to the data, we use the function
`ga()` which is an interface to `gam()` from the **mgcv** package [Wood, 2001]. Note
that `te()` gives a tensor product spline with five knots for each variable (which may
need to be increased). For more details about the interface see Section 9.5.2.

```
library(gamlss.add)
# linear interaction model
 m1 <- gamlss(lborev1~lboopen*lnosc, data=film90, trace=FALSE)
 m2 <- gamlss(lborev1~lboopen*lnosc+dist, data=film90, trace=FALSE)
# additive model using the pb() function
 m3 <- gamlss(lborev1~pb(lboopen) +pb(lnosc), data=film90,
```

```
                trace=FALSE)
  m4 <- gamlss(lborev1~pb(lboopen) +pb(lnosc)+dist, data=film90,
                trace=FALSE)
# fitting a surface using ga()
  m5 <- gamlss(lborev1~ga(~te(lboopen,lnosc)), data=film90,
                trace=FALSE)
  m6 <- gamlss(lborev1~ga(~te(lboopen,lnosc))+dist, data=film90,
                trace=FALSE)
```

```
GAIC(m1, m2, m3, m4, m5, m6)
```

```
##            df       AIC
## m6 16.01650 11779.76
## m4 18.53520 11828.59
## m5 15.91276 11843.78
## m3 18.12674 11908.73
## m2  6.00000 12080.99
## m1  5.00000 12226.84
```

```
GAIC(m1, m2, m3, m4, m5, m6, k=log(4031))
```

```
##            df       AIC
## m6 16.01650 11880.69
## m5 15.91276 11944.06
## m4 18.53520 11945.39
## m3 18.12674 12022.96
## m2  6.00000 12118.80
## m1  5.00000 12258.35
```

The best model appears to be m6, which fits a surface for lboopen and lnosc and an additive term for dist. Unfortunately a look at its residuals reveals that the normal distribution model fits very badly. The following worm plot shows this clearly, since most of the points lie outside the approximate pointwise 95% confidence interval bands (shown as dashed elliptical curves).

Fig.
4.10
```
wp(m6, ylim.all=1.1)
```

Note that in order to visualize the fitted surface, plot() or vis.gam() of **mgcv** may be used. The **gam** object fitted within the backfitting algorithm is saved under the name **g4$mu.coefSmo** and is retrieved using the function getSmo():

Fig.
4.11
```
library(mgcv)
plot(getSmo(m6))
vis.gam(getSmo(m6),theta = 0, phi = 30)
```

To check whether we need to model σ as a function of the explanatory variables:

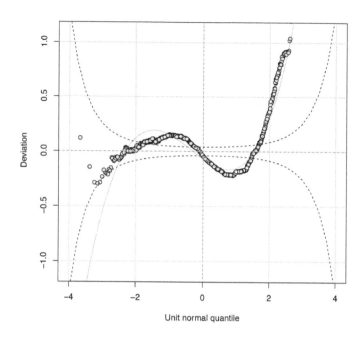

R
p. 51

FIGURE 14.10: The worm plot from the normal distribution model m6, in which a fitted surface was used for μ.

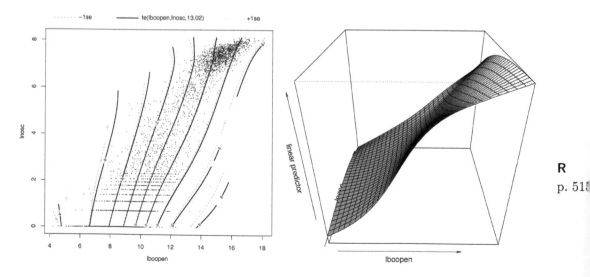

R
p. 51

FIGURE 14.11: The fitted contour and surface plot from model m6.

```
m7<- gamlss(lborev1~ga(~te(lboopen,lnosc))+dist,
            sigma.fo=~ga(~te(lboopen,lnosc))+dist,
            data=film90, trace=FALSE)

AIC(m6, m7)

##              df       AIC
## m7 27.64729 10043.89
## m6 16.01650 11779.76

AIC(m6, m7, k=log(4031))

##              df       AIC
## m7 27.64729 10218.12
## m6 16.01650 11880.69
```

We find that model m7 is superior to m6, using either AIC or SBC. A worm plot of the residuals (Figure 14.12) is used to check the adequacy of the model. This indicates that model m7, while an improvement compared to m6, still does not adequately explain the response variable.

```
wp(m7, ylim.all=1.1)
```

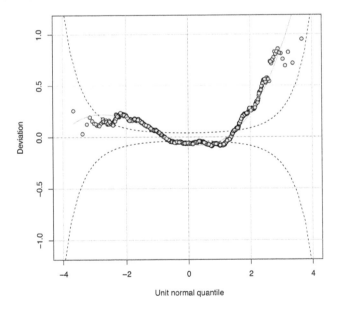

FIGURE 14.12: The worm plot from the normal distribution model m7, in which a fitted surface is used for both μ and σ.

518   Flexible Regression and Smoothing: Using GAMLSS in R

14.4.3   Modelling the data using the BCPE distribution

Next we model the response variable using the BCPE distribution [Rigby and
Stasinopoulos, 2004], which is a four-parameter distribution defined on the positive real line. Model m8 fits additive terms using pb(), while model m9 fits smooth
surfaces using ga() for all four distribution parameters.

```
m8 <- gamlss(lborev1 ~ pb(lboopen)+pb(lnosc) + dist,
          sigma.fo = ~ pb(lboopen)+pb(lnosc) + dist,
             nu.fo = ~ pb(lboopen)+pb(lnosc) + dist,
            tau.fo = ~ pb(lboopen)+pb(lnosc) + dist,
          family = BCPE, data = film90, trace=FALSE)
m9 <- gamlss(lborev1 ~ ga(~te(lboopen,lnosc)) + dist,
          sigma.fo = ~ ga(~te(lboopen,lnosc)) + dist,
             nu.fo = ~ ga(~te(lboopen,lnosc)) + dist,
            tau.fo = ~ ga(~te(lboopen,lnosc)) + dist,
          family = BCPE, data = film90, n.cyc=20, trace=FALSE)
```

```
AIC(m6, m7, m8, m9)

##          df        AIC
## m9 41.83029   9836.412
## m8 44.95828   9980.948
## m7 27.64729  10043.889
## m6 16.01650  11779.759

AIC(m6, m7, m8, m9, k=log(4031))

##          df        AIC
## m9 41.83029  10100.02
## m7 27.64729  10218.12
## m8 44.95828  10264.26
## m6 16.01650  11880.69
```

The model m9 seems superior according to AIC and SBC, but it is more complicated
(using far more degrees of freedom) and may be overfitting the data. Next we plot
the worm plots for m8 and m9.

```
wp(m8, ylim.all=0.5)
wp(m9, ylim.all=0.5)
```

Fig.
14.13

The worm plot of m8 (left panel of Figure 14.13) looks slightly better than that of
m9 (on the right), but it is hard to decide. We can get a better idea of how the
model fits in the joint ranges of the two explanatory varibles lboopen and lnosc
by using a worm plot with two explanatory variables:

```
wp(m9, xvar=~lboopen+lnosc, ylim.worm=1)
```

Fig.
14.14

In the resulting worm plot given in Figure 14.14, the four columns correspond to

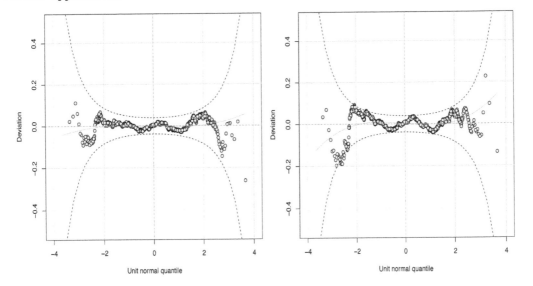

FIGURE 14.13: Worm plots from the BCPE distribution models. Left: m8, right: m9.

the four ranges of `lboopen` displayed above the plot, and the four rows correspond
to the four ranges of `lnosc` displayed to the right of the plot. Within the plot there
are 16 individual worm plots of the residuals corresponding to the 16 joint ranges
of `lboopen` and `lnosc`. Some joint ranges have no observations within them. The
worm plots generally indicate an adequate fit within the joint ranges.

The fitted smooth surfaces for μ, σ, ν and τ for model `m9` are plotted in Figure
14.15 by using the following commands:

Fig.
14.15

```
vis.gam(getSmo(m9,what="mu"), theta=30, phi=10)
title("mu")
vis.gam(getSmo(m9,what="sigma"), theta=30, phi=15)
title("sigma")
vis.gam(getSmo(m9,what="nu"), theta=30, phi=15)
title("nu")
vis.gam(getSmo(m9,what="tau"), theta=30, phi=15)
title("tau")
```

We leave further simpification of the model to the reader.

14.5 Epilogue

GAMLSS is now widely used for statistical modelling and learning. Its great advantages are that it enables flexible regression and smoothing models to be fitted to the

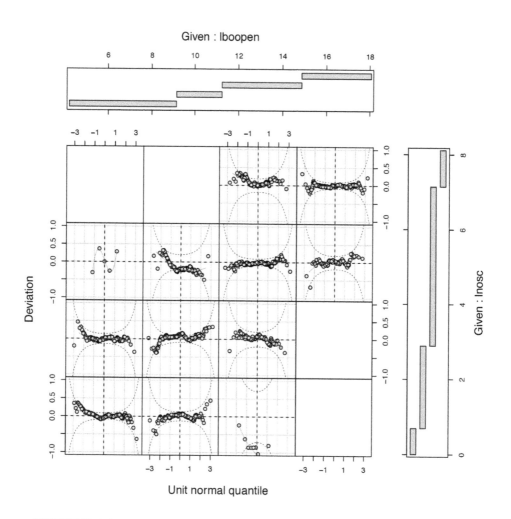

FIGURE 14.14: The worm plot for model m9, by lboopen and lnosc.

R

p. 51⬛

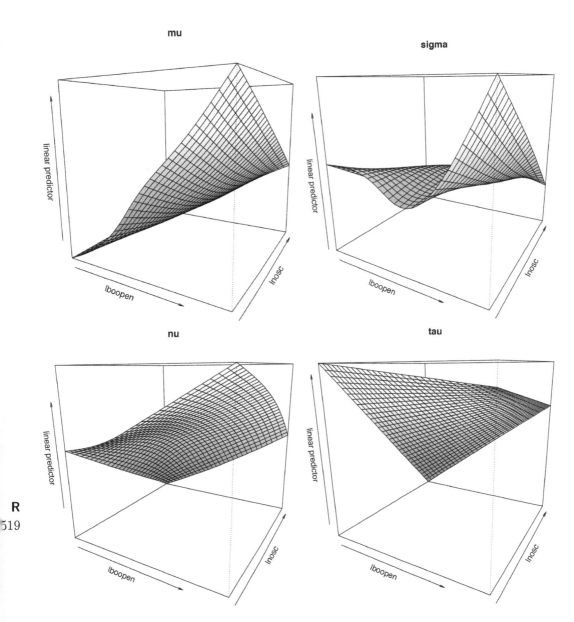

FIGURE 14.15: The fitted smooth surfaces for μ, σ, ν and τ of model m9.

data, in which the distribution of response variable has any parametric distribution (heavy or light-tailed, and positively or negatively skewed) and all parameters of the distribution can be modelled.

The applications given in this book are a small sample of the potential of GAMLSS models for learning from data.

GAMLSS has been used in a variety of applied fields including: social science, environmental science, finance, actuarial science, biology, biosciences, energy, genomics, fisheries, food consumption, growth curve estimation, marine research, medicine, meteorology, rainfalls, vaccines. By way of specific examples:

- World Health Organisation (WHO): The Department of Nutrition for Health and Development of the WHO adopted GAMLSS to construct worldwide standard growth (centile) curves charts for children; see WHO [2006, 2007, 2009]. Those charts are used in more than 140 countries.

- The International Monetary Fund (IMF): IMF used GAMLSS for *stress testing* the U.S. financial system [International Monetary Fund, 2015].

- European Parliament: GAMLSS was used for the European Parliament's "Making the European Banking Union Macro-Economically Resilient: Cost of Non-Europe Report" led by Gael Giraud (Chief Economist, Agence Française de Développement).

- ABM Analytics Ltd: ABM (a boutique research company) is using GAMLSS for predictive analytics and risk quantification (e.g. exposure at default or loss given default). For example, ABM is using GAMLSS for its oil trading.

With rapid changes in computing technology and the big data age, the field of statistical learning is constantly challenged. This book and the associated GAMLSS software is our attempt to support practitioners and researchers to learn from data. As we move forward, more about latest developments, further examples and exercises using GAMLSS can be found at `www.gamlss.org`.

Bibliography

L. S. Aiken and S. G. West. *Multiple regression: Testing and interpreting interactions*. Newbury Park: Sage Publications, 1991.

G. B. Airy. *On the algebraical and numerical theory of errors of observations and the combination of observations*. Macmillan & Company, 1861.

M. Aitkin. Modelling variance heterogeneity in normal regression using GLIM. *Applied Statistics*, 36:332–339, 1987.

M. Aitkin. A general maximum likelihood analysis of overdispersion in generalized linear models. *Statistics and Computing*, 6:251–262, 1996.

M. Aitkin. A general maximum likelihood analysis of variance components in generalized linear models. *Biometrics*, 55:117–128, 1999a.

M. Aitkin. Meta-analysis by random effect modelling in generalized linear models. *Statistics in Medicine*, 18:2343–2351, 1999b.

M. A. Aitkin, B. Francis, J. Hinde, and R. Darnell. *Statistical Modelling in R*. Oxford University Press Oxford, 2009.

H. Akaike. A new look at the statistical model identification. *IEEE Transactions on Automatic Control*, 19(6):716–723, 1974.

H. Akaike. Information measures and model selection. *Bulletin of the International Statistical Institute*, 50:277–290, 1983.

G. Ambler. *fracpoly(): Fractional Polynomial Model*, 1999. URL `http://lib.stat.cmu.edu/S/fracpoly`. S-PLUS.

C. Angelini, D. De Canditiis, and F. Leblanc. Wavelet regression estimation in nonparametric mixed effect models. *Journal of Multivariate Analysis*, 85(2):267–291, 2003.

J. E. Angus. The probability integral transform and related results. *SIAM Review*, 36:652–654, 1994.

K. Attebo, P. Mitchell, and W. Smith. Visual acuity and the causes of visual loss in Australia: the Blue Mountains Eye Study. *Ophthalmology*, 103(3):357–364, 1996.

R. J. Baker and J. A. Nelder. *The GLIM System, Release 3*. Oxford: Numerical Algorithms Group, 1978.

F. H. Barajas, M. Torres, L. Arteaga, and C. Castro. GAMLSS models applied in the treatment of agro-industrial waste. *Comunicaciones en Estadística*, 8(2): 245–254, 2015.

R. A. Becker and J. M. Chambers. *S: an interactive environment for data analysis and graphics*. CRC Press, 1984.

T. Benaglia, D. Chauveau, D. Hunter, and D. Young. mixtools: An R package for analyzing finite mixture models. *Journal of Statistical Software*, 32(6):1–29, 2009.

A. Beyerlein, L. Fahrmeir, U. Mansmann, and A.M. Toschke. Alternative regression models to assess increase in childhood BMI. *BMC Medical Research Methodology*, 8(1):59, 2008.

C. M Bishop. *Neural networks for pattern recognition*. Clarendon Press Oxford, 1995.

D. Böhning. Computer-assisted analysis of mixtures and applications. *Technometrics*, 42(4):442–442, 2000.

C. P. Bonafide, P. W. Brady, R. Keren, P. H. Conway, K. Marsolo, and C. Daymont. Development of heart and respiratory rate percentile curves for hospitalized children. *Pediatrics*, 131(4):1150–1157, 2013.

E. Borghi, M. de Onis, C. Garza, J.E. Van den Broeck, E. A. Frongillo, L. Grummer-Strawn, S. Van Buuren, H. Pan, L. Molinari, R. Martorell, A. W. Onyango, and J. C. Martines. Construction of the World Health Organization child growth standards: selection of methods for attained growth curves. *Statistics in Medicine*, 25:247–265, 2006.

G. E. P. Box. Robustness in the strategy of scientific model building. *Robustness in Statistics*, 1:201–236, 1979.

L. Breiman and J. H. Friedman. Estimating optimal transformations for multiple regression and correlations (with discussion). *Journal of the American Statistical Association*, 80(391):580–619, 1985.

L. Breiman, J. Friedman, R. Olshen, and C. Stone. *Classification and regression trees*. Wadsworth, New York, 1984.

N. E. Breslow. Extra-Poisson variation in log-linear models. *Applied Statistics*, pages 38–44, 1984.

N. E. Breslow and D. G Clayton. Approximate inference in generalized linear mixed models. *Journal of the American Statistical Association*, 88:9–25, 1993.

J. M. Chambers. *Computational methods for data analysis*. New York: Wiley, 1977.

J. M. Chambers and T. J. Hastie. *Statistical Models in S*. Chapman & Hall, London, 1992.

G. Claeskens and N. L. Hjort. The focused information criterion. *Journal of the American Statistical Association*, 98:900–916, 2003.

Gerda Claeskens, Nils Lid Hjort, et al. *Model selection and model averaging*, volume 330. Cambridge University Press Cambridge, 2008.

W. S. Cleveland. Robust locally-weighted regression and smoothing scatter plots. *Journal of the American Statistical Association*, 74:829–836, 1979.

W. S. Cleveland and S. J. Devlin. Robust locally-weighted regression: an approach to regression analysis by local fitting. *Journal of the American Statistical Association*, 83:597–610, 1988.

D. D. Cohen, C. Voss, M. J. D. Taylor, D. M. Stasinopoulos, A. Delextrat, and G. R. H. Sandercock. Handgrip strength in English schoolchildren. *Acta Paediatrica*, 99:1065–1072, 2010.

T. J. Cole. Fitting smoothed centile curves to reference data (with discussion). *Journal of the Royal Statistical Society, Series A*, 151:385–418, 1988.

T. J. Cole and P. J. Green. Smoothing reference centile curves: The LMS method and penalized likelihood. *Statistics in Medicine.*, 11:1305–1319, 1992.

T.J. Cole, S. Stanojevic, J. Stocks, A.L. Coates, J.L. Hankinson, and A.M. Wade. Age-and size-related reference ranges: A case study of spirometry through childhood and adulthood. *Statistics in Medicine*, 28(5):880–898, 2009.

D. D. Cox and F. O'Sullivan. Asymptotic analysis of penalized likelihood and related estimators. *Annals of Statistics*, 18:1676–1695, 1990.

D. R. Cox. Tests of separate families of hypotheses. In *Proceedings of the fourth Berkeley symposium on mathematical statistics and probability*, volume 1, pages 105–123, 1961.

D. R. Cox and D. V. Hinkley. *Theoretical Statistics*. Chapman & Hall/CRC, 1979.

D. R. Cox and N. Reid. Parameter orthogonality and approximate conditional inference (with discussion). *Journal of the Royal Statistical Society, Series B*, 49: 1–39, 1987.

S. L. Crawford. An application of the Laplace method to finite mixture distributions. *Annals of Statistics*, 89(425):259–267, 1994.

S. L. Crawford, M. H. DeGroot, J. B. Kadane, and M. J. Small. Modeling lake-chemistry distributions: Approximate Bayesian methods for estimating a finite-mixture model. *Technometrics*, 34(4):441–453, 1992.

A. Crisp and J. Burridge. A note on nonregular likelihood functions in heteroscedastic regression models. *Biometrika*, 81:585–587, 1994.

I. D. Currie, M. Durbán, and P. H. C. Eilers. Generalized linear array models with applications to multidimensional smoothing. *Journal of the Royal Statistical Society, Series B (Statistical Methodology)*, 68(2):259–280, 2006.

R. B. D'Agostino, A. Balanger, and R. B. D'Agostino Jr. A suggestion for using powerful and informative tests of normality. *American Statistician*, 44:316–321, 1990.

M. Davidian and A. R. Gallant. Smooth nonparametric maximum likelihood for population pharmacokinetics, with application to quinidine. *Journal of Pharmacokinetics and Biopharmaceutics*, 20:529–556, 1992.

A. C. Davison and D. V. Hinkley. *Bootstrap methods and their application*, volume 1. Cambridge University Press, 1997.

F. De Bastiani, D. M. Stasinopoulos, R. A. Rigby, A. H. M. A. Cysneiros, and M. A. Uribe-Opazo. Gaussian markov random field spatial models in gamlss. *Journal of Applied Statistics*, online:1–19, 2016.

C. de Boor. *A Practical Guide to Splines*. Springer-Verlag, New York, revised edition, 2001.

M. de Castro, V.G. Cancho, and J. Rodrigues. A hands-on approach for fitting long-term survival models under the GAMLSS framework. *Computer Methods and Programs in Biomedicine*, 97(2):168–177, 2010.

P. de Jong and G. Z. Heller. *Generalized Linear Models for Insurance Data*. Cambridge University Press, 2008.

C. Dean, J. F. Lawless, and G. E. Willmot. A mixed Poisson-inverse-Gaussian regression model. *Canadian Journal of Statistics*, 17(2):171–181, 1989.

E. Demidenko. *Mixed models: Theory and applications with R*. John Wiley & Sons, 2013.

A. Dempster, N. Laird, and D. Rubin. Maximum likelihood from incomplete data via EM algorithm (with discussion). *Journal of the Royal Statistical Society*, 39: 1–38, 1977.

J. Diebolt and C. P. Robert. Estimation of finite mixture distributions through Bayesian sampling. *Journal of the Royal Statistitical Society, Series B (Methodological)*, pages 363–375, 1994.

P. Dierckx. *Curve and surface fitting with splines*. Oxford University Press, 1995.

P. J. Diggle, P. Heagerty, K. Y. Liang, and S. L. Zeger. *Analysis of longitudinal data*. Oxford University Press, Oxford, 2nd edition, 2002.

A. Dobson and A. Barnett. *An Introduction to Generalized Linear Models*. CRC Press, 3rd edition, 2008.

D. Draper. Assessment and propagation of model uncertainty (with discussion). *Journal of the Royal Statistical Society, Series B*, 57:45–97, 1995.

N. R. Draper, H. Smith, and E. Pownell. *Applied regression analysis*, volume 3. Wiley New York, 1966.

P. K. Dunn and G. K. Smyth. Randomized quantile residuals. *Journal of Computational and Graphical Statistics*, 5:236–244, 1996.

M. Durbán, J. Harezlak, M. Wand, and R. Carroll. Simple fitting of subject-specific curves for longitudinal data. *Statistics in Medicine*, 24:1153–1167, 2005.

B. Efron. The 1977 RIETZ lecture. *The Annals of Statistics*, 7(1):1–26, 1979.

B. Efron. *Bootstrap methods: another look at the jackknife*. Springer, 1992.

B. Efron and R.J. Tibshirani. *An introduction to the bootstrap*. Chapman & Hall, New York, 1993.

P. H. C. Eilers. A perfect smoother. *Analytical Chemistry*, 75(14):3631–3636, 2003.

P. H. C. Eilers and B. D. Marx. Flexible smoothing with B-splines and penalties (with comments and rejoinder). *Statistical Science*, 11:89–121, 1996.

P. H. C. Eilers and B. D. Marx. Multivariate calibration with temperature interaction using two dimensional penalized signal regression. *Chemometrics and Intelligent Laboratory Systems*, 66:159–174, 2003.

P. H. C. Eilers and B. D. Marx. Splines, knots, and penalties. *Wiley Interdisciplinary Reviews: Computational Statistics*, 2(6):637–653, 2010.

P. H. C. Eilers, B. D Marx, and M. Durbán. Twenty years of P-splines. *SORT-Statistics and Operations Research Transactions*, 39(2):149–186, 2015.

J. Einbeck, R. Darnell, and J. Hinde. *npmlreg: Nonparametric maximum likelihood estimation for random effect models*, 2014. URL http://CRAN.R-project.org/package=npmlreg. R package version 0.46-1.

C. Eisenhart. The assumptions underlying the analysis of variance. *Biometrics*, 3: 1–21, 1947.

R. F. Engle. Autoregressive conditional heteroscedasticity with estimates of the variance of United Kingdom inflation. *Econometrica: Journal of the Econometric Society*, pages 987–1007, 1982.

R. F. Engle. *ARCH: selected readings*. Oxford University Press, Oxford, 1995.

B. S. Everitt. *Finite mixture distributions*. Wiley Online Library, 1981.

L. C. Fabio, G. A. Paula, and M. de Castro. A Poisson mixed model with nonnormal random effect distribution. *Computational Statistics and Data Analysis*, 56(6): 1499–1510, 2012.

L. Fahrmeir and G. Tutz. *Multivariate statistical modelling based on generalized linear models.* Springer, New York, 2nd edition, 2001.

L. Fahrmeir and S. Wagenpfeil. Penalized likelihood estimation and iterative Kalman smoothing for non-Gaussian dynamic regression models. *Computational Statistics and Data Analysis*, 24:295–320, 1997.

L. Fahrmeir, T. Kneib, and S. Lang. Penalized structured additive regression for space-time data: a Bayesian perspective. *Statistica Sinica*, 14:731–761, 2004.

L. Fahrmeir, T. Kneib, S. Lang, and B. Marx. *Regression: models, methods and applications.* Springer, 2013.

J. J. Faraway. *Linear models with R.* Chapman & Hall/CRC, 2009.

C. Fernandez and M. F. J. Steel. On Bayesian modelling of fat tails and skewness. *Journal of the American Statistical Association*, 93:359–371, 1998.

C. Fernandez, J. Osiewalski, and M. J. F. Steel. Modeling and inference with v-spherical distributions. *Journal of the American Statistical Association*, 90: 1331–1340, 1995.

R. A. Fisher. *Statistical methods for research workers.* Oliver and Boyd, Edinburgh, 1925.

C. Fraley and A. E Raftery. Model-based clustering, discriminant analysis, and density estimation. *Journal of the American Statistical Association*, 97(458): 611–631, 2002.

B. J. Francis, M. Green, and C. Payne. *GLIM 4: The statistical system for generalized linear interactive modelling.* Clarendon Press, Oxford, 1993.

A. M. Fredriks, S. van Buuren, R. J. F. Burgmeijer, J. F. Meulmeester, R. J. Beuker, E. Brugman, M. J. Roede, S. P. Verloove-Vanhorick, and J. M. Wit. Continuing positive secular change in The Netherlands, 1955-1997. *Pediatric Research*, 47: 316–323, 2000a.

A. M. Fredriks, S. van Buuren, J. M. Wit, and S. P. Verloove-Vanhorick. Body index measurements in 1996-7 compared with 1980. *Archives of Childhood Diseases*, 82: 107–112, 2000b.

D. A. Freedman. On the so-called Huber sandwich estimator and robust standard errors. *The American Statistician*, 60(4), 2006.

J. H. Friedman. Multivariate adaptive regression splines. *The Annals of Statistics*, 19(1):1–67, 03 1991.

S. J. Gange, A. Munoz, M. Saez, and J. Alonso. Use of the beta-binomial distribution to model the effect of policy changes on appropriateness of hospital stays. *Applied Statistics*, 45:371–382, 1996.

A. Gannoun, S. Girard, C. Cuinot, and J. Saracco. Reference curves based on non-parametric quantile regression. *Statistics in Medicine*, 21:3119–3135, 2002.

C. F. Gauss. *Methodus nova integralium valores per approximationem inveniendi.* APVD Henricvm Dieterich, 1815.

J. D. Gergonne. Application de la méthode des moindres quarrés a l'interpolation des suites. *Annales de mathématiques pures et appliquées*, 6:242–252, 1815.

J. Gertheiss and G. Tutz. Regularized regression for categorical data. *Statistical Modelling*, 16:161–200, 2016a.

J. Gertheiss and G. Tutz. Rejoinder: Regularized regression for categorical data. *Statistical Modelling*, 16:249–260, 2016b.

R. Gilchrist, A. Kamara, and J. Rudge. An insurance type model for the health cost of cold housing: An application of GAMLSS. *REVSTAT-Statistical Journal*, 7(1):55–66, 2009.

P. E. Gill and W. Murray. Newton-type methods for unconstrained and linearly constrained optimization. *Mathematical Programming*, 7:311–350, 1974.

I. J. Good and R. A. Gaskins. Nonparametric roughness penalties for probability densities. *Biometrika*, 58(2):255–277, 1971.

C. Gourieroux, A. Monfort, and A. Trognon. Pseudo maximum likelihood methods: Theory. *Econometrica: Journal of the Econometric Society*, pages 681–700, 1984.

P. J. Green. Iterative reweighted least squares for maximum likelihood estimation, and some robust and resistant alternatives (with discussion). *Journal of the Royal Statistical Society, Series B*, 46:149–192, 1984.

P. J. Green. Penalised likelihood for general semi-parametric regression models. *International Statistical Review*, 55:245–260, 1987.

P. J. Green and B. W. Silverman. *Nonparametric regression and generalized linear models*. Chapman & Hall, London, 1994.

B. Grün and F. Leisch. Fitting finite mixtures of generalized linear regressions in R. *Computational Statistics & Data Analysis*, 51(11):5247–5252, 2007.

C. Gu and P. Ma. Optimal smoothing in nonparametric mixed-effect models. *Annals of Statistics*, 33(3):1357–1379, 06 2005.

W. Guo. Functional mixed effects models. *Biometrics*, 58:121–128, 2002.

D. J. Hand, F. Daly, A. D. Lunn, K. J. McConway, and E. Ostrowski. *A handbook of small data sets*. Chapman & Hall, London, 1994.

J. W. Hardin and J. Hilbe. *Generalized linear models and extensions, 3rd edition*. StataCorp LP, 2012.

W. Härdle. *Smoothing techniques, with implementation in S.* Springer-Verlag, New York, 1990.

A. C. Harvey. Estimating regression models with multiplicative heteroscedasticity. *Econometrica*, 41:461–465, 1976.

D. A. Harville. Bayesian inference for variance components using only error contrasts. *Biometrika*, 61(2):383–385, 1974. doi: 10.1093/biomet/61.2.383.

T. Hastie. *gam: Generalized additive models*, 2015. URL `http://CRAN.R-project.org/package=gam`. R package version 1.12.

T. Hastie and A. Herman. An analysis of gestational age, neonatal size and neonatal death using nonparametric logistic regression. *Journal of Clinical Epidemiology*, 1990.

T. J. Hastie and R. J. Tibshirani. *Generalized additive models.* Chapman & Hall, London, 1990.

T. J. Hastie and R. J. Tibshirani. Varying coefficient models (with discussion). *Journal of the Royal Statistical Society, Series B*, 55:757–796, 1993.

T. J. Hastie, R. J. Tibshirani, and J. Friedman. *The elements of statistical learning: Data mining, inference and prediction.* Springer, New York, 2nd edition, 2009.

X. He. Quantile curves without crossing. *The American Statistician*, 51:186–192, 1997.

X. He and P. Ng. Cobs: Qualitative constained smoothing via linear programming. *Computational Statistics*, 14:315–337, 1999.

P. J. Heagerty and M. S. Pepe. Semiparametric estimation of regression quantiles with applications. *Applied Statistics*, 48:533–551, 1999.

G. Z. Heller, D. L. Couturier, and S. Heritier. Parameter orthogonality for Poisson-inverse Gaussian regression. *Macquarie University Technical Report*, 2016.

C. Hennig. *fpc: Flexible procedures for clustering*, 2015. URL `https://CRAN.R-project.org/package=fpc`. R package version 2.1-10.

J. M. Hilbe. *Negative binomial regression.* Cambridge University Press, 2011.

J. Hinde. Compound Poisson regression models. In R. Gilchrist, editor, *GLIM 82, Proceedings of the International Conference on Generalised Linear Models*, pages 109–121. Springer, New York, 1982.

N. L. Hjort and G. Claeskens. Frequentist model average estimators. *Journal of the American Statistical Association*, 98:879–899, 2003.

J. S. Hodges. Some algebra and geometry for hierarchical models, applied to diagnostics. *Journal of the Royal Statistical Society, Series B*, 60:497–536, 1998.

A. E. Hoerl and R. Kennard. Ridge regression: biased estimation for nonorthogonal problems. *Technometrics*, 12:55–67, 1970.

B. Hofner, A. Mayr, and M. Schmid. gamboostLSS: An R package for model building and variable selection in the GAMLSS framework. *Journal of Statistical Software*, 74(1):1–31, 2016.

A. Hossain, R. A. Rigby, D. M. Stasinopoulos, and M. Enea. Centile estimation for a proportion response variable. *Statistics in Medicine*, 35(6):895–904, 2016. URL http://dx.doi.org/10.1002/sim.6748.

W. Hu, B. A. Swanson, and G. Z. Heller. A statistical method for the analysis of speech intelligibility tests. *PLoS ONE*, 10(7), 2015.

P. J. Huber. The behavior of maximum likelihood estimates under nonstandard conditions. In *Proceedings of the Fifth Berkeley Symposium on Mathematical Statistics and probability*, volume 1, pages 221–233, 1967.

I. L. Hudson, S. W. Kim, and M. R. Keatley. Climatic influences on the flowering phenology of four eucalypts: a GAMLSS approach. *Phenological Research*, pages 209–228, 2010.

International Monetary Fund. Stress Testing, Technical Note. *Country Report No. 15/173*, 2015.

R. W. B. Jackson. Reliability of mental tests. *British Journal of Psychology*, 29: 267–287, 1939.

L. Jager and J. A. Wellner. A new goodness of fit test: the reversed Berk Jones statistic, 2004. URL http://www.stat.washington.edu/www/research/reports/2004/tr443.pdf.

L. Jager and J. A. Wellner. Goodness-of-fit tests via phi-divergences. *The Annals of Statistics*, pages 2018–2053, 2007.

R. I. Jennrich and M. D. Schluchter. Unbalanced repeated-measures models with structured covariance matrices. *Biometrics*, 42(4):805–820, 1986.

N. L. Johnson, S. Kotz, and N. Balakrishnan. *Continuous univariate distributions*, volume 1. Wiley, New York, 2nd edition, 1994.

N. L. Johnson, S. Kotz, and N. Balakrishnan. *Continuous univariate distributions*, volume 2. Wiley, New York, 2nd edition, 1995.

N. L. Johnson, A. W. Kemp, and S. Kotz. *Univariate discrete distributions*. Wiley, New York, 3rd edition, 2005.

M. C. Jones and A. Pewsey. Sinh-arcsinh distributions. *Biometrika*, 96:761–780, 2009.

B. Jørgensen. *Statistical properties of the generalized inverse Gaussian distribution.* Lecture Notes in Statistics No. 9. Springer-Verlag, New York, 1982.

L.R. Kapitula and E.J. Bedrick. Diagnostics for the exponential normal growth curve model. *Statistics in Medicine,* 24:95–108, 2005.

A. Kempny, K. Dimopoulos, A. Uebing, P. Moceri, L. Swan, M. A. Gatzoulis, and G.-P. Diller. Reference values for exercise limitations among adults with congenital heart disease. relation to activities of daily life-single centre experience and review of published data. *European Heart Journal,* 461:1386–1396, 2011.

G. Kitagawa. Non-Gaussian state-space modeling of nonstationary time series. *Journal of the American Statistical Association,* 82(400):1032–1041, 1987.

N. Klein and T. Kneib. Scale-dependent priors for variance parameters in structured additive distributional regression. *Bayesian Analysis,* 11(4):1071–1116, 2016a.

N. Klein and T. Kneib. Simultaneous inference in structured additive conditional copula regression models: A unifying Bayesian approach. *Statistics and Computing,* 26(4):841–860, 2016b.

N. Klein, M. Denuit, S. Lang, and T. Kneib. Nonlife ratemaking and risk management with bayesian additive models for location, scale and shape. *Insurance: Mathematics and Economics,* 55:225–249, 2014.

N. Klein, T. Kneib, S. Klasen, and S. Lang. Bayesian structured additive distributional regression for multivariate responses. *Journal of the Royal Statistical Society, Series C (Applied Statistics),* 64:569–591, 2015a.

N. Klein, T. Kneib, and S. Lang. Bayesian generalized additive models for location, scale and shape for zero-inflated and overdispersed count data. *Journal of the American Statistical Association,* 110:405–419, 2015b.

N. Klein, T. Kneib, S. Lang, and A. Sohn. Bayesian structured additive distributional regression with an application to regional income inequality in Germany. *Annals of Applied Statistics,* 9:2014–1052, 2015c.

T. Kneib. Beyond mean regression (with discussion and rejoinder). *Statistical Modelling,* 13(4):275–303, 2013a.

T. Kneib. Rejoinder: Beyond mean regression. *Statistical Modelling,* 13(4):373–385, 2013b.

T. Kneib and L. Fahrmeir. Structured additive regression for categorical space-time data: A mixed model approach. *Biometrics,* 62:109–118, 2006.

R. Koenker. *Quantile regression.* Cambridge University Press, Cambridge, 2005.

R. Koenker. *quantreg: Quantile Regression,* 2016. URL https://CRAN.R-project.org/package=quantreg. R package version 5.21.

R. Koenker and G. Bassett. Regression quantiles. *Econometrica*, 46:33–50, 1978.

R. Koenker and P. Ng. Inequality constrained quantile regreesion. *Sankhya, The Indian Journal of Statistics*, 67:418–440, 2005.

R. Koenker, P. Ng, and S. Portnoy. Quantile smoothing splines. *Biometrika*, 81(4): 673–680, 1994. doi: 10.1093/biomet/81.4.673.

S. Lang and A. Brezger. Bayesian P-splines. *Journal of Computational and Graphical Statistics*, 13:183–212, 2004.

K. L. Lange, R. J. A. Little, and J. M. G. Taylor. Robust statistical modelling using the *t* distribution. *Journal of the American Statistical Association*, 84:881–896, 1989.

J. F. Lawless. Negative binomial and mixed Poisson regression. *Canadian Journal of Statistics*, 15(3):209–225, 1987.

Y. Lee and J. A. Nelder. Hierarchical generalized linear models (with discussion). *Journal of the Royal Statistical Society, Series B*, 58:619–678, 1996.

Y. Lee and J. A. Nelder. Two ways of modelling overdispersion in non-normal data. *Applied Statistics*, 49:591–598, 2000.

Y. Lee, J.A. Nelder, and Y. Pawitan. *Generalized linear models with random effects: Unified analysis via H-likelihood*. CRC Press, 2006.

E. L. Lehmann. On the history and use of some standard statistical models. *Probability and Statistics*, 2:114–126, 2008.

F. Leisch. FlexMix: A general framework for finite mixture models and latent class regression in R. *Journal of Statistical Software*, 11(1):1–18, 2004. ISSN 1548-7660. doi: 10.18637/jss.v011.i08. URL https://www.jstatsoft.org/index.php/jss/article/view/v011i08.

B. Li and P. K. Goel. Additive regression trees and smoothing splines, predictive modeling and interpretation in data mining. *Contemporary Mathematics*, 443: 83–101, 2007.

B. G. Lindsay. Mixture models: Theory, geometry and applications. In *NSF-CBMS regional conference series in probability and statistics*, pages i–163. JSTOR, 1995.

M. J. Lindstrom and D. M. Bates. Nonlinear mixed effects models for repeated measures data. *Biometrics*, 46:673–687, 1990.

N. T. Longford. Logistic regression with random coefficients. *Computational Statistics & Data Analysis*, 17(1):1–15, 1994.

A. Lopatatzidis and P. J. Green. Nonparametric quantile regression using the gamma distribution. *Private Communication*, 2000.

D. Madigan and A. E. Raftery. Model selection and accounting for model uncertainly in graphical models using Occam's window. *Journal of the American Statistical Association*, 89:1535–1546, 1994.

A. Mallet, F. Mentre, J-L. Steimer, and F. Lokiec. Nonparametric maximum likelihood estimation for population pharmacokinetics, with application to cyclosporin. *Journal of Pharmacokinetics and Biopharmaceutics*, 16(3):311–327, 1988.

B. H. Margolin, N. Kaplan, and E. Zeiger. Statistical analysis of the Ames Salmonella/microsome test. *Proceedings of the National Academy of Sciences*, 78(6):3779–3783, 1981.

B. D. Marx and P. H. C. Eilers. Direct generalized additive modeling with penalized likelihood. *Computational Statistics and Data Analysis*, 28:193–209, 1998.

A. Mayr, H. Binder, O. Gefeller, and M. Schmid. The evolution of boosting algorithms, from machine learning to statistical modelling. *Methods of Information in Medicine*, 53(6):419–427, 2014a.

A. Mayr, H. Binder, O. Gefeller, and M. Schmid. Extending statistical boosting, an overview of recent methodological developments. *Methods of Information in Medicine*, 53(6):428–435, 2014b.

A. Mayr, M. Schmid, A. Pfahlberg, W. Uter, and O. Gefeller. A permutation test to analyse systematic bias and random measurement errors of medical devices via boosting location and scale models. *Statistical Methods in Medical Research*, 2015.

N. Mayr, A .and Fenske, B. Hofner, T. Kneib, and M. Schmid. Generalized additive models for location, scale and shape for high-dimensional data, a flexible approach based on boosting. *Journal of the Royal Statistical Society, Series C (Applied Statistics)*, 61(3):403–427, 2012.

P. McCullagh. Quasi-likelihood functions. *Annals of Statistics*, 11:59–67, 1983.

P. McCullagh and J. A. Nelder. *Generalized linear models*. Chapman & Hall, London, 2nd edition, 1989.

G. McLachlan and D. Peel. *Finite mixture models*. John Wiley & Sons, 2004.

K. E. Muller and P. W. Stewart. *Linear model theory: univariate, multivariate and models*. John Wiley & Sons, New Jersey, 2006.

J. A. Nelder. Contribution to the discussion of Rigby and Stasinopoulos, Generalized additive models for location, scale and shape. *Applied Statistics*, 54:547, 2006.

J. A. Nelder and Y. Lee. Likelihood, quasi-likelihood and psuedolikelihood: Some comparisons. *Journal of the Royal Statistical Society, Series B*, 54:273–284, 1992.

J. A. Nelder and D. Pregibon. An extended quasi-likelihood function. *Biometrika*, 74:221–232, 1987.

J. A. Nelder and R. W. M. Wedderburn. Generalized linear models. *Journal of the Royal Statistical Society, Series A*, 135:370–384, 1972.

D. B. Nelson. Conditional heteroskedasticity in asset returns: a new approach. *Econometrica*, 59:347–370, 1991.

H. K. Neuhauser, M. Thamm, U. Ellert, H. W. Hense, and A. S. Rosario. Blood pressure percentiles by age and height from nonoverweight children and adolescents in Germany. *Pediatrics*, 127:978–988, 2011.

P. T. Ng and M. Maechler. *cobs: COBS, Constrained B-splines (Sparse matrix based)*, 2015. URL http://CRAN.R-project.org/package=cobs. R package version 1.3-1.

D.J. Nott and L. Jialiang. A sign based loss approach to model selection in non-parametric regression. *Statistics and Computing*, 20(4):485–498, 2010.

P. Np and M. Maechler. A fast and efficient implementation on qualitatively constrained quantile smoothing splines. *Statistical Modelling*, 7:315–328, 2007.

D. W. Nychka. *Smoothing and regression: Approaches, computation and application*, chapter Spatial-process estimates as smoothers, pages 393–424. Wiley, New York, 2000.

M. R. Oelker, W. Pößnecker, and G. Tutz. Selection and fusion of categorical predictors with-type penalties. *Statistical Modelling*, pages 389–410, 2015.

M. R. Osborne. Fisher's method of scoring. *International Statistical Institute*, 60 (1):99–117, 1992.

R. Ospina and S. L. P. Ferrari. Inflated beta distributions. *Statistical Papers*, 51: 111–126, 2010.

R. Ospina and S. L. P. Ferrari. A general class of zero-or-one inflated beta regression models. *Computational Statistics and Data Analysis*, 56:1609–1623, 2012.

A. B. Owen. Nonparametric likelihood confidence bands for a distribution function. *Journal of the American Statistical Association*, 90(430):516–521, 1995.

A. B. Owen. *Empirical likelihood*. CRC Press, 2001.

B. U. Park, E. Mammen, Y. K. Lee, and E. R. Lee. Varying coefficient regression models: A review and new developments. *International Statistical Review*, 83(1): 36–64, 2015.

R. Parker and J. Rice. Discussion of Silverman (1985). *Journal of the Royal Statistical Society, Series B*, 47(1):40–42, 1985.

Y. Pawitan. *In all likelihood: Statistical modelling and inference using likelihood.* Oxford University Press, USA, 2001.

J. C. Pinheiro and D. M. Bates. *Mixed-effects models in S and S-Plus.* Springer, 2000. URL http://nlme.stat.wisc.edu/MEMSS/. ISBN 0-387-98957-0.

J. C. Pinheiro, C. Liu, and Y. N. Wu. Efficient algorithms for robust estimation in linear mixed-effects models using the multivariate t distribution. *Journal of Computational and Graphical Statistics*, 10:249–276, 2001.

P. H. Quanjer, S. Stanojevic, T. J. Cole, X. Baur, G. L. Hall, B. H. Culver, P. L. Enright, J. L. Hankinson, M. S. Ip, J. Zheng, J. Stocks, and ERS Global Lung Function Initiative. Multi-ethnic reference values for spirometry for the 3-95-yr age range: The global lung function 2012 equations. *The European Respiratory Journal*, 40(6):1324–1343, 2012a. URL http://view.ncbi.nlm.nih.gov/pubmed/22743675.

P. H. Quanjer, S. Stanojevic, T. J. Cole, and J. Stocks. GLI-2012 - GAMLSS in action. *Global Lung Function Initiative*, 2012b.

R Core Team. *R: A language and environment for statistical computing.* R Foundation for Statistical Computing, Vienna, Austria, 2016. URL https://www.R-project.org/.

A. E. Raftery. Approximate Bayes factors and accounting for model uncertainty in generalised linear models. *Biometrika*, 83:251–266, 1996.

A. E. Raftery. Bayes Factors and BIC, comment on: A critique of the Bayesian Information Criterion for Model Selection. *Sociological Methods & Research*, 27: 411–427, 1999.

R. A. Rigby and D. M. Stasinopoulos. MADAM macros to fit Mean and Dispersion Additive Models. In T. Scallon and G. Morgan, editors, *GLIM4, Macro Library Manual, Release 2.0*, pages 68–84. NAG, Oxford, 1996.

R. A. Rigby and D. M. Stasinopoulos. A semi-parametric additive model for variance heterogeneity. *Statististics and Computing*, 6:57–65, 1996a.

R. A. Rigby and D. M. Stasinopoulos. Mean and dispersion additive models. In W. Hardle and M. G. Schimek, editors, *Statistical theory and computational aspects of smoothing*, pages 215–230. Physica, Heidelberg, 1996b.

R. A. Rigby and D. M. Stasinopoulos. Smooth centile curves for skew and kurtotic data modelled using the Box Cox power exponential distribution. *Statistics in Medicine*, 23:3053–3076, 2004.

R. A. Rigby and D. M. Stasinopoulos. Generalized additive models for location, scale and shape, (with discussion). *Applied Statistics*, 54:507–554, 2005.

R. A. Rigby and D. M. Stasinopoulos. Using the Box-Cox t distribution in GAMLSS to model skewness and kurtosis. *Statistical Modelling*, 6(3):209, 2006. ISSN 1471-082X.

R. A. Rigby and D. M. Stasinopoulos. Automatic smoothing parameter selection in GAMLSS with an application to centile estimation. *Statistical Methods in Medical Research*, 23(4):318–332, 2013. doi: 10.1177/0962280212473302.

R. A. Rigby, D. M. Stasinopoulos, and C. Akantziliotou. A framework for modelling overdispersed count data, including the Poisson-shifted generalized inverse Gaussian distribution. *Computational Statistics & Data Analysis*, 53(2):381–393, 2008. ISSN 0167-9473.

R. A. Rigby, D. M. Stasinopoulos, and V. Voudouris. Flexible statistical models: Methods for the ordering and comparison of theoretical distributions. *MPRA Paper*, (63620), 2015.

R. A. Rigby, D. M. Stasinopoulos, G. Z. Heller, V. Voudouris, and F. De Bastiani. Distributions for modelling location, scale, and shape: Using GAMLSS in R, 2017. URL www.gamlss.org.

B. D. Ripley. Statistical aspects of neural networks. *Networks and Chaos-Statistical and Probabilistic Aspects*, 50:40–123, 1993.

B. D. Ripley. *Pattern recognition and neural networks*. Cambridge University Press, Cambridge, 1996.

G. K. Robinson. That BLUP is a good thing: The estimation of random effects. *Statistical Science*, 6(1):15–32, 1991.

C. B. Roosen and T. J. Hastie. Automatic smoothing spline projection pursuit. *Journal of Computational and Graphical Statistics*, 3(3):235–248, 1994.

P. Royston and D. G. Altman. Regression using fractional polynomials of continuous covariates: parsimonious parametric modelling (with discussion). *Applied Statistics*, 43:429–467, 1994.

P. Royston and E. M. Wright. Goodness-of-fit statistics for age-specific reference intervals. *Statistics in Medicine*, 19:2943–2962, 2000.

H. Rue and L. Held. *Gaussian Markov random fields: Theory and applications*. CRC Press, 2005.

D. Ruppert, M. P. Wand, and R. J. Carroll. *Semiparametric regression*. Cambridge University Press, 2003.

D. Ruppert, M. P. Wand, and R. J. Carroll. Semiparametric regression during 2003-2007. *Electronic Journal of Statistics*, 3:1193–1256, 2009.

K. Saha and S. Paul. Bias-corrected maximum likelihood estimator of the negative binomial dispersion parameter. *Biometrics*, 61(1):179–185, 2005.

R. Schall. Estimation in generalized linear models with random effects. *Biometrika*, 78:719–727, 1991.

H. Scheffé. Alternative models for the analysis of variance. *Annals of Mathematical Statistics,*, 27:251–271, 1956.

M. Schmid, F. Wickler, K. O. Maloney, R. Mitchell, N. Fenske, and A. Mayr. Boosted beta regression. *PLoS ONE*, 8(4):e61623, 2013.

S. K. Schnabel and P. H. C. Eilers. A location-scale model for non-crossing expectile curves. *Stat*, 2(1):171–183, 2013a.

S. K. Schnabel and P. H. C. Eilers. Simultaneous estimation of quantile curves using quantile sheets. *AStA Advances in Statistical Analysis*, 97(1):77–87, 2013b.

I. J. Schoenberg. Contribution to the problem of approximation of equidistant data by analytic functions, part b: On the problems of osculatory interpolation, a second class of analytic approximation formulae. *Quarterly of Applied Mathematics*, 4:112–141, 1946a.

I. J. Schoenberg. Contribution to the problem of approximation of equidistant data by analytic functions, part a: On the problem of smoothing graduation, a first class of analytic approximation. *Quarterly of Applied Mathematics*, 4:45–99, 1946b.

L. Schumaker. *Spline functions: Basic theory*. Cambridge Mathematical Library, 2007.

G. E. Schwarz. Estimating the dimension of a model. *Annals of Statistics*, 6(2): 461–464, 1978.

L. B. Sheiner and S. L. Beal. Evaluation of methods for estimating population pharmacokinetics parameters. I. Michaelis Menten model: Routine clinical pharmacokinetic data. *Journal of Pharmacokinetics and Pharmacodynamics*, 8(6): 533–571, 1980.

C. Silagy, T. Lancaster, L. Stead, D. Mant, and G. Fowler. Nicotine replacement therapy for smoking cessation. *The Cochrane Library*, 2004.

A. Skrondal and S. Rabe-Hesketh. *Generalized latent variable modelling*. Chapman & Hall, 2004.

P. L. Smith. Splines as a useful and convenient statistical tool. *American Statistician*, 33:57–62, 1979.

G. K. Smyth. Generalized linear models with varying dispersion. *Journal of the Royal Statistical Society, Series B*, 51:47–60, 1989.

A. Sohn, N. Klein, and T. Kneib. A semiparametric analysis of conditional income distributions. *Schmollers Jahrbuch*, 135(1):13–22, 2016.

A. Sommer, G. Hussaini, I. Tarwotjo, and D. Susanto. Increased mortality in children with mild vitamin A deficiency. *The Lancet*, 322(8350):585–588, 1983.

J. G. Staniswalis. Local bandwidth selection for kernel estimates. *Journal of the American Statistical Association*, 84(405):284–288, 1989.

S. Stanojevic, A. Wade, J. Stocks, J. Hankinson, A. L. Coates, H. Pan, M. Rosenthal, M. Corey, P. Lebecque, and T. J. Cole. Reference ranges for spirometry across all ages: a new approach. *American Journal of Respiratory and Critical Care Medicine*, 177:253–260, 2008.

D. M. Stasinopoulos and R. A. Rigby. Detecting break points in generalised linear models. *Computational Statistics and Data Analysis*, 13:461–471, 1992.

D. M. Stasinopoulos and R. A. Rigby. Generalized additive models for location scale and shape (GAMLSS) in R. *Journal of Statistical Software*, 23(7):1–46, 2007.

D. M. Stasinopoulos, R. A. Rigby, and L. Fahrmeir. Modelling rental guide data using mean and dispersion additive models. *Statistician*, 49:479–493, 2000.

G. Z. Stein and J. M. Juritz. Linear models with an inverse Gaussian Poisson error distribution. *Communications in Statistics, Theory and Methods*, 17:557–571, 1988.

S. M. Stigler. Gergonne's 1815 paper on the design and analysis of polynomial regression experiments. *Historia Mathematica*, 1(4):431 – 439, 1974.

P. F. Thall and S. C. Vail. Some covariance models for longitudinal count data with overdispersion. *Biometrics*, 46:657–671, 1990.

R. A. Thisted. *Elements of statistical computing: Numerical computation*. Chapman & Hall, New York, 1988.

H. C. Thode. *Testing for normality*. CRC Press, 2002.

R. Tibshirani. Regression shrinkage and selection via the lasso. *Journal of the Royal Statistical Society, Series B*, 58:267–268, 1996.

R. J. Tibshirani and T. J. Hastie. Local likelihood estimation. *Journal of the American Statistical Association*, 82:559–568, 1987.

D. M. Titterington, A. F. M. Smith, and U. E. Makov. *Statistical analysis of finite mixture distributions*. Wiley, 1985.

M. C. K. Tweedie. An index which distinguishes between some important exponential families. In *Statistics: Applications and New Directions: Proceedings of the Indian Statistical Institute Golden Jubilee International Conference*, pages 579–604, 1984.

S. van Buuren. Worm plot to diagnose fit in quantile regression. *Statistical Modelling*, 7:363–376, 2007.

S. van Buuren and M. Fredriks. Worm plot: A simple diagnostic device for modelling growth reference curves. *Statistics in Medicine*, 20:1259–1277, 2001.

W. N. Venables and B. D. Ripley. *Modern Applied Statististics with S*. Springer, 4th edition, 2002.

G. Verbeke and G. Molenberghs. *Linear mixed models for longitudinal data.* Springer, 2000.

A. P. Verbyla. Modelling variance heterogeneity: Residual maximum likelihood and diagnostics. *Journal of the Royal Statistical Society, Series B*, 55:493–508, 1993.

J. Villar, I. L Cheikh, C. G. Victora, E. O. Ohuma, E. Bertino, D.G. Altman, A. Lambert, A. T. Papageorghiou, M. Carvalho, Y. A. Jaffer, M. G. Gravett, M. Purwar, I.O. Frederick, A. J. Noble, F. C. Pang, R. Barros, Z. A. Chumlea, C. Bhutta, and S. H. Kennedy. International standards for newborn weight, length, and head circumference by gestational age and sex: The Newborn Cross-Sectional Study of the INTERGROWTH-21st Project. *The Lancet*, 384(9946): 857–868, 2014.

R. L. Visser, J. E. M. Watson, C. R. Dickman, R. Southgate, D. Jenkins, and C. N. Johnson. A national framework for research on trophic regulation by the Dingo in Australia. *Pacific Conservation Biology*, 15:209–216, 2009.

E. F. Vonesh and R. L. Carter. Mixed-effects nonlinear regression for unbalanced repeated measures. *Biometrics*, 48(1):1–17, 1992.

V. Voudouris, R. Gilchrist, R. Rigby, J. Sedgwick, and D. Stasinopoulos. Modelling skewness and kurtosis with the BCPE density in GAMLSS. *Journal of Applied Statistics*, 39(6):1279–1293, 2012.

V. Voudouris, R. Ayres, A. C. Serrenho, and D. Kiose. The economic growth enigma revisited: The EU-15 since the 1970s. *Energy Policy*, 2015.

A. M. Wade and A. E. Ades. Age-related reference ranges : Significance tests for models and confidence intervals for centiles. *Statistics in Medicine*, 13:2359–2367, 1994.

G. Wahba. Improper priors, spline smoothing and the problem of guarding against model errors in regression. *Journal of the Royal Statistical Society, Series B*, 40: 364–372, 1978.

G. Wahba. A comparison of GCV and GML for choosing the smoothing parameter in the generalized spline smoothing problem. *Annals of Statistics*, 4:1378–1402, 1985.

G. Wahba. *Spline Models for Observational Data*. Society for Industrial and Applied Mathematics, Philadelphia, Pennsylvania, 1990.

J. Wakefield. The Bayesian analysis of population pharmacokinetic models. *Journal of the American Statistical Association*, 91(433):62–75, 1996.

M. Wand. Smoothing and mixed models. *Computational Statistics*, 18:223–249, 2003.

M. P. Wand and M. C. Jones. *Kernel smoothing*. Chapman & Hall, Essen, Germany, 1999.

M. P. Wand and J. T. Ormerod. On semiparametric regression with O'Sullivan penalised splines. *Australian and New Zealand Journal of Statistics*, 50:179–198, 2008.

Y. Wang. *Smoothing splines: methods and applications*. CRC Press, 2011.

R. W. M. Wedderburn. Quasi-likelihood functions, generalised linear models and the Gauss-Newton method. *Biometrika*, 61:439–447, 1974.

Y. Wei, A. Pere, R. Koenker, and X. He. Quantile regression methods for reference growth charts. *Statistics in Medicine*, 25(8):1369–1382, 2006.

S. Weisberg. *Applied linear regression*. Wiley, New York, 1980.

B. T. West, K. B. Welch, and A. T. Galecki. *Linear mixed models: A practical guide using statistical software*. Chapman & Hall/CRC, 2014.

H. White. A heteroskedasticity-consistent covariance matrix estimator and a direct test for heteroskedasticity. *Econometrica: Journal of the Econometric Society*, pages 817–838, 1980.

E. T. Whittaker. On a new method of graduation. *Proceedings of the Edinburgh Mathematical Society*, 41:63–75, 1922.

Multicentre Growth Reference Study Group WHO. *WHO Child Growth Standards: Length/height-for-age, weight-for-age, weight-for-length, weight-for-height and body mass index-for-age: Methods and development*. Geneva: World Health Organization, 2006.

Multicentre Growth Reference Study Group WHO. *WHO Child Growth Standards: Head circumference-for-age, arm circumference-for-age, triceps circumference-for-age and subscapular skinford-for-age: Methods and development*. Geneva: World Health Organization, 2007.

Multicentre Growth Reference Study Group WHO. *WHO Child Growth Standards: Growth velocity based on weight, length and head circumference: Methods and development*. Geneva: World Health Organization, 2009.

G. N. Wilkinson and C. E. Rogers. Symbolic description of factorial models for analysis of variance. *Applied Statistics*, 22:392–399, 1973.

G. Wimmer and G. Altmann. *Thesaurus of univariate discrete probability distributions*. Stamm Verlag, Essen, Germany, 1999.

S. N. Wood. mgcv: GAMs and generalized ridge regression for R. *R News*, 1:20–25, 2001.

S. N. Wood. *Generalized additive models. An introduction with R.* Chapman & Hall, 2006a.

S. N. Wood. On confidence intervals for generalized additive models based on penalized regression splines. *Australian and New Zealand Journal of Statistics*, 48:445–464, 2006b.

S. N. Wood, N. Pya, and B. Säfken. Smoothing parameter and model selection for general smooth models. *Journal of the American Statistical Association*, 111 (516):1548–1563, 2017.

E. M. Wright and P. Royston. A comparison of statistical methods for age-related reference intervals. *Journal of the Royal Statistical Society, Series A*, 160(2): 47–69, 1997.

L. Wu. *Mixed effects models for complex data.* Chapman & Hall, Boca Raton, 2010.

W. Yan, F. Liu, L. Li, X.and Wu, Y. Zhang, Y. Cheng, W. Zhou, and G. Huang. Blood pressure percentiles by age and height for non-overweight Chinese children and adolescents: Analysis of the China health and nutrition surveys 1991–2009. *BMC Pediatrics*, pages 13–195, 2013.

S. L. Zeger and M. R. Karim. Generalized linear models with random effects: A Gibbs sampling approach. *Journal of the American Statistical Association*, 86: 79–95, 1991.

Index

Milton Keynes UK
Ingram Content Group UK Ltd.
UKHW051904071024
449327UK00025B/2084